REPRODUCTIVE SUCCESS

Few studies of seabirds have been able to cover the lifetime of individuals. The difficulties are illustrated by George Dunnet's study of fulmar petrels (chap. 17), begun in 1951 (*upper left*) and still running today (*upper right*). The bottom photograph, taken in 1984, is of a fulmar first caught as a breeding adult in 1951.

REPRODUCTIVE SUCCESS

Studies of Individual Variation in Contrasting Breeding Systems

Edited by T. H. Clutton-Brock
With Fifty-three Contributors

The University of Chicago Press □ Chicago and London

T. H. Clutton-Brock is a Royal Society research fellow in biology and a lecturer in zoology at the University of Cambridge. He is the coauthor with F. E. Guinness and S. D. Albon of *Red Deer*, also published by the University of Chicago Press, and the editor of *Primate Ecology* and *Readings in Sociobiology*.

The University of Chicago Press, Chicago 60637
The University of Chicago Press, Ltd., London

Printed in the United States of America

97 96 95 94 93 92 91 90 89 88 54321

Library of Congress Cataloging in Publication Data

Reproductive success.

Bibliography: p.
Includes indexes.
1. Sexual selection in animals. 2. Reproduction.
I. Clutton-Brock, T. H.
QL761.R47 1988 575'.5 87-16257
ISBN 0-226-11058-3
ISBN 0-226-11059-1 (pbk.)

Contents

Editor's Acknowledgments

I would like to thank Steve Albon, Steve Arnold, Jon Endler, Michael Wade, Russ Lande, Paul Harvey, David Brown, Mark Elgar, Tim Caro, and Alan Grafen for encouragement, advice, or, criticism; Steve Albon, Anthony Arak, Mark Avery, Carol Berman, Neil Chalmers, Nick Davies, David Gibbons, Robert Gibson, Kathy Kennedy, Kate Lessells, Seamus McCann, R. H. K. Mann, Peter O'Donald, Myfryn Owen, Geoff Parker, Ian Patterson, Marion Petrie, Michael Simpson, Bob Wootton, and the contributors for refereeing manuscripts; and Jacqueline Hodkinson for assistance with the considerable secretarial load.

1 Introduction

T. H. Clutton-Brock

THE PURPOSE OF THIS book is to investigate the extent and causes of individual differences in breeding success among male and female animals in natural populations. The authors of the twenty-five empirical studies it contains were invited to answer four principal questions about reproductive success in the species they had studied. How widely does breeding success vary between individuals of each sex? How much of the variance in success is contributed by the different components of breeding success (survival to breeding age, reproductive life span, fecundity or mating success, and offspring survival)? To what extent does reproductive success change with age? And what environmental, phenotypic, developmental, or genetic factors affect breeding success in each sex? Answers to all four questions are important in studies of population demography (Begon 1984), selection (Endler 1986), and adaptation (Williams 1966a), yet until recently knowledge of the extent and causes of individual differences in reproductive success in natural populations has been rudimentary.

Though the aims of the twenty-five studies vary, the common factor in all but one of them is that their authors have measured the reproductive success of recognizable individuals in successive breeding attempts, in some cases over the normal life span of the species. Longitudinal studies of this kind are unusual. Despite a considerable body of research on individual variation in reproductive success in plants, from Carl Tamm's classic studies (Tamm 1948, 1972) to more recent work (see Harper 1983), and Tinkle's pioneering work on the side-blotched lizard, *Uta stansburiana* (Tinkle 1967), most studies of reproductive success in natural populations of animals have been based on successive cross-sectional samples of data collected at particular points in time from different sets of individuals. It is only comparatively recently that field studies (mostly of vertebrates) have tracked the reproductive success of individual animals through most or all of their natural life spans (see Howard 1979; McVey

1981; McGregor, Krebs, and Perrins 1981; Fincke 1982; Hoogland and Foltz 1982; Clutton-Brock, Guinness, and Albon 1982; McCauley 1983; Coulson and Thomas 1985; Newton 1985).

Despite their logistic drawbacks, longitudinal studies of individuals have four important advantages. First, they allow us to compare the total fitness of different categories of individuals. Most studies of selection and adaptation rely on estimates of fitness measured over a small fraction of an individual's life span (see Endler 1986). However, differences in morphology or behavior commonly affect several components of fitness, and important costs or benefits can easily be missed by studies confined to single episodes of selection. In particular, longitudinal studies allow us to measure components of reproductive success, such as the survival of offspring after the end of parental care, that cannot easily be investigated in the field using cross-sectional sampling techniques.

Second, longitudinal data can be used to guard against changes in the distribution of successive samples. For example, attempts to show that reproductive success changes with age using cross-sectional data can easily be confused by consistent associations between breeding success and longevity (see below).

Third, because the same individuals are followed through several breeding attempts, longitudinal studies can reduce the proportion of variation in breeding success that is caused by short-term environmental variation. This often makes it easier to measure the effects on breeding success of differences in phenotype. And fourth, the detailed observation of individuals that longitudinal studies involve can provide invaluable insights into why particular phenotypic traits affect components of reproductive success.

1.1 Variance in Reproductive Success

One of the aims of this book is to determine the extent to which reproductive success varies in males and females in species with contrasting breeding systems. Though it is commonly argued that variation in reproductive success is similar in the two sexes in monogamous species but is greater among males than females in species with promiscuous or polygynous mating systems, almost all attempts to compare variation in breeding success between the sexes have relied on short-term data, usually collected from individuals of unknown age (see Trivers 1976; Payne and Payne 1977; Payne 1979). Since intense competition between males typically compresses their effective breeding life spans into a small part of their total adult life span (see chaps. 18, 21, and 22), much of the observed variation may be due to age, and such measures may grossly overestimate variation in lifetime success (Clutton-Brock 1983). Conversely, the extent to which female success can vary within seasons in birds and mammals is constrained by the range of clutch or litter size. However, when breeding success is calculated over the life span, consistent differences in fecundity or offspring survival can generate substantial differences in breeding success that are not apparent within seasons.

Following Crow (1958) and Arnold and Wade (1984a,b), many of the studies in this book have used variance in mating or breeding success divided by the square of the mean ($\sigma^2/\bar{x}^2$) as an estimate of the variability of success. This ratio is the square of the coefficient of variation and is denoted by I. It is important to emphasize that I defined in this way is a measure of the *opportunity* or *potential* for selection, not of the *actual* intensity of selection on any character or group of characters, though it can set an upper limit on selection intensities (Arnold 1986). Even in a population of phenotypically identical individuals, variance in breeding success will arise as a consequence of environmental variation (Cundiff 1972; Clutton-Brock 1983) or chance differences in mating access (Sutherland 1985a,b; Hubbell and Johnson 1987) and will be incorporated in measures of variance in breeding success. As McVey points out (chap. 4), unless differences in reproductive success or survival are caused by phenotypic variation, no selection will occur, and the "opportunity for selection" is the opportunity for genetic drift. One important question, consequently, is what proportion of the opportunity for selection is caused by random factors (see below).

In arguments about selection it is also important to realize that several definitions of selection are in current use. Some biologists restrict selection to cases where fitness differences between phenotypes have a heritable basis (see Endler 1986), while others equate fitness variation between phenotypes with selection, distinguishing this from the genetic response to selection, which will depend on the heritability of the trait in question (see Fisher 1930; Haldane 1954; Lande and Arnold 1983). Most of the authors of this book use it in the latter sense.

1.2 Components of Variation in Reproductive Success

A second objective of the book is to determine the extent to which the different components of reproductive success (longevity, fecundity, mating success, and offspring survival) contribute to fitness variation in the two sexes and to examine relationships between different components. Comparison of the relative contributions of different components can be helpful in identifying the stages at which selection may be operating, providing a stimulus to improve estimates of the costs and benefits of particular traits or strategies. It is most useful where it prompts us to consider fitness components that might otherwise be ignored.

Several methods of estimating the relative contributions of different stages of the life history or components of breeding success are available (see Prout 1971a,b; Christiansen and Frydenberg 1973, 1976; Manly 1985). The two commonly used in this book are those suggested by Arnold and Wade (1984a,b) and by Brown (this volume, chap. 27). Arnold and Wade suggest that the opportunity for selection should be calculated for each stage of each successive breeding attempt during the life span. The contributions of each stage and the covariances between them can then be expressed as percentages of the total opportunity for selection. Where applied to long-lived iteroparous species the number of terms in this analysis can become unwieldy, and an alternative approach, suggested by Brown (chap. 27) and used by many of the authors of this book, is to concentrate on average values for each of the main components of breeding success, calculated across breeding attempts. One advantage of both methods is that they force us to appreciate that covariance between components of success can account for a substantial proportion of the observed variation. However, as chapters 27 and 29 stress, all current methods of partitioning components of success have their disadvantages, and their results can be difficult to interpret.

A common reason for measuring the components of breeding success is to compare the opportunity for natural and sexual selection. Wade and Arnold (1980) advocate the use of the ratio of standardized variance in mate number (I_s) to the total opportunity for selection in males (I_m) or females (I_f) to measure the intensity of sexual selection in each sex (Wade 1979; Wade and Arnold 1980; Arnold and Wade, 1984a,b). In chapter 29 I briefly review the use of this and other measures and argue that we need to reexamine the reasons for wishing to measure the opportunity for sexual selection.

1.3 The Effects of Age on Breeding Success

The third aim of this book is to document the effects of age on components of breeding success in the two sexes in different breeding systems and to investigate their causes. An understanding of the effects of age, the causes of these effects, and the extent to which they vary with environmental circumstances is of widespread importance in studies of population demography (Caughley 1977) as well as in understanding the evolution of life histories (Pianka 1978).

Longitudinal studies that can compare the breeding success of the same individuals at different ages have an important advantage over the usual practice of comparing the reproductive performance or survival of separate samples of animals observed (or collected) at different ages. Since individual differences in reproductive rate are often correlated with longevity (see Bryant 1979; Smith 1981b; Bell 1980; Clutton-Brock 1983), changes in the frequency of superior or inferior phenotypes in samples of animals at successive ages may generate or obscure relationships between breeding success and age. For example, an initial improvement in breeding success may be caused by the progressive elimination of inferior phenotypes from the population (see chap. 4). Similarly, if individuals with superior reproductive performance have shorter life spans, reproductive success may appear to decline with age in situations where age has no effect on reproduction. Conversely where individuals with superior phenotypes show high reproductive performance and live longer than average, cross-sectional samples may fail to reveal the effects of increasing age. Where longitudinal data are available, it is possible to avoid problems of this kind by comparing the reproductive success of the same individuals at different ages (see chap. 14).

1.4 Causes of Variation in Reproductive Success

The last aim of the book is to investigate and compare the determinants of breeding success in males and females in contrasting breeding systems.

A knowledge of the relationships between particular variables and components of breeding success or survival is basic to all attempts to measure selection in progress or to understand the adaptive significance of particular traits. In some cases, measuring the effects of particular traits on components of reproductive success is necessary to show what their adaptive significance may be (Arnold 1983a), while in others it is needed to test adaptive explanations of their form, frequency, or distribution (see Endler 1986). Many aspects of morphology, physiology, and behavior vary between breeding systems (Clutton-Brock and Harvey 1984), and studies of the determinants of fitness may provide important insights into the reasons underlying these associations as well as into the evolution of the breeding systems themselves.

In particular, studies of the factors affecting reproductive success are often necessary to determine what variables can be used as substitutes for fitness. Many of the most convincing tests of adaptive hypotheses are those that compare an animal's observed form or behavior with a priori predictions based on some type of optimal design (see, for example, Krebs et al. 1977; Pyke 1978; Wilson 1980 Arnold 1983a). Insectivorous birds feeding on small, cryptic items have trouble satisfying their daily food requirements (Gibb 1956), and it is reasonable to assume that behavior that maximizes food intake increases survival and breeding success. This assumption makes it possible to formulate quantitative predictions about the items the birds should select and the time they should allocate to particular feeding sites (Stephens and Krebs 1987). However, the intermediate variables that animals are likely to be maximizing are seldom as obvious as in this case. What principles should guide the feeding decisions of predator-limited herbivores that depend on fibrous foods? What might a bird be expected to maximize when choosing a nest site—access to food, or freedom from predation, or a mixture of the two? What criteria should determine its choice of mates? Its own speed of pairing? The potential partner's likely contributions to parental care? Or the partner's size, condition, parasite load, or genetic quality? In such cases, knowledge of the factors that affect breeding success and of their relative importance can help us formulate qualitative and quantitative predictions concerning the design of morphology or behavior (see Beatty 1980; Krebs and McCleery 1984; Pulliam and Caraco 1984).

The most important questions concerning the factors affecting reproductive success obviously vary between species and between studies. They include: To what extent is variation in reproductive success caused by phenotypic differences, by chance or by short-term environmental variation? Does fitness show low heritability as Fisher (1930) predicted, and if so, what environmental factors affect the individual's reproductive capacity? Does an individual's choice of mate have important consequences for its fitness or for that of its offspring? And to what extent do sex differences in morphology or behavior reflect differences in the selection pressures currently operating on males and females? None of the twenty-five empirical studies that follow have covered all these questions, but all have been examined in at least one study.

Since the type of data available and the questions that can be approached vary between major groupings of animals, I have divided the chapters on a taxonomic basis. The bias towards birds and mammals reflects the current distribution of studies and is likely to change in the future. Chapters 27 and 28 review methods for partitioning fitness and the uses and limitations of longitudinal studies of natural variation, and in the last chapter I briefly review the generalizations emerging from the empirical studies of reproductive success and some of their implications for future research.

1 Insects

THIS PART INCLUDES STUDIES of five insects with contrasting breeding systems: a fruitfly, *Drosophila melanogaster*, in which males compete to mate with receptive females at feeding sites; a damselfly, *Enallagma hageni*, in which males guard females after mating but do not defend territories; a dragonfly, *Erythemis simplicicollis*, in which males defend territories used by females for oviposition; an ant-tended lycaenid butterfly, *Jalmenus evagoras*, in which males gather in clusters around pupae that are ready to eclose, competing to mate with emerging females; and a social paper wasp, *Polistes annularis*, in which breeding groups consist of one or more queens assisted by nonbreeding workers. The research on *Drosophila* was carried out partly in the laboratory and partly in the field; the other four studies were done in the field.

As the studies of *Drosophila* and *Jalmenus* demonstrate, egg production commonly increases with body size among animals where reproductive output is not limited by parental care, and this probably explains why females are the larger sex in many of these species. Where males are larger than females, as in the giant damselfly studied by Fincke (see chap. 3), we can expect male size to exert an unusually strong influence on fitness.

The first four studies provide a chance to compare the total opportunity for selection in the two sexes, though close comparisons are complicated by differences in measurement of reproductive success, sampling methodology, and analysis (see chap. 3). In all four field studies, variation in male breeding success exceeded variation in female success, though the former varied widely with the local sex ratio. In contrast, in *Drosophila*, where males were kept alone, variation in female breeding success exceeded variation in male success (see chap. 2).

There was no consistent tendency for variation in male breeding success to be higher in species where males defended territories than in those where they did not. Estimates of the total opportunity for selection among breeding males were approximately similar in *Erythemis, Enallagma,* and *Coenagrion puella*, a damselfly in which males do not defend territories (Banks and Thompson 1985), but they were substantially higher in *Jalmenus* as well as in a tropical damselfly, *Megaloprepus coerulatus*, in which males defend territories around water-filled tree holes (see chap. 3). These results emphasize that it is the form and intensity of direct competition between males that is likely to determine the extent to which mating success differs between individuals rather than the gross nature of the breeding system.

The first four studies underline the dangers of estimating the total opportunity for selection using data collected over a part of the life span (see chap. 1). Both in *Enallagma* and in *Erythemis*, standardized variance in the daily breeding success of females underestimated variance in lifetime reproductive success, while, in males, standardized variance calculated within days exceeded lifetime variance by nearly 100%.

Both studies of odonates assess the intensity of sexual versus natural selection in breeding males. McVey (chap. 4) uses the methodology suggested by Wade and Arnold (1980), calculating the opportunity for sexual selection on males as I_s/I_m, where $I_s = I_m - R\ I_f$ and R is the ratio of breeding males to breeding females. Fincke (chap. 3) uses a modification of the same technique, calculating the opportunity for sexual selection from the ratio of standardized variance in fertilizations per mate plus variance in mating rate over I_m. In contrast, Elgar and Pierce use the methods suggested by Brown (chap. 27) to partition variation in lifetime mating success into its three main components: longevity, encounter rate, and mating efficiency. Though it is difficult to compare the results of the three studies directly, the relative opportunity for sexual selection within days was higher in *Erythemis* (65%–79%) than in *Enallagma* (47%). In *Jalmenus*, mating efficiency and longevity were closely correlated, and the covariance between them was the single most important source of variation in lifetime mating success, making it impossible to separate their contributions.

In *Enallagma*, as well as in *Coenagrion* (see Banks and Thompson 1985), observed variation in male mating success did not exceed that predicted by a random model, and in *Jalmenus* the difference was marginal. However, at least a part of the observed variation was evidently not of random origin, for male reproductive success was related to phenotypic differences in all three species.

In *Drosophila*, male body size was positively correlated with lifetime mating success and longevity in the laboratory as well as with mating success in the field, while in the other species the effects of phenotype were more subtle. In both *Enallagma* and *Jalmenus* there was no simple relationship between male size and mating efficiency: in *Enallagma*, intermediate-sized males had the highest number of mates per life span, while in *Jalmenus*, mating success increased with the body size of males relative to that of other individuals hatching at the same time. In *Erythemis*, McVey was unable to identify the phenotypic factors affecting male mating success, though experimental manipulation of male territories revealed that the same males consistently held the best territories. Qualitative comparisons among the different species of odonates that have been studied (see above) suggest that the effects of male size on mating success may increase with the amount of direct competition between males.

The final chapter on *Polistes* illustrates the problems of calculating selection opportunities and differentials in species where inclusive fitness has to be considered. The total opportunity for selection in females within colonies was higher than in any of the other four species, probably partly because reproductive success was calculated in terms of offspring surviving to reproductive age and partly because a successful foundress benefits from the efforts of other individuals. Even so, the opportunity for selection within colonies was only about half as large as the total opportunity for selection between colonies. Partitioning variance in success showed that differences in reproductive rate were responsible for most of the variation in fitness between colonies and were negatively related to the initial number of foundresses. Since subordinate females in larger colonies had lower fitness than solitary foundresses, this raised the question why subordinates did not nest alone. However, this choice may not have been open to them. As Queller and Strassman emphasize (this volume, chap. 6), questions of this kind are likely to be answered only by measures of the fitness consequences of the alternative options available to an individual.

2 Lifetime Reproductive Success in *Drosophila*

Linda Partridge

FRUITFLIES OF THE GENUS *Drosophila* have a holometabolous life history, the eggs producing larvae that in general burrow and feed in decaying plant tissue and give rise to a puparium. Metamorphosis then occurs to two-winged adults that feed on various microorganisms on the surface of vegetation and soil. The details of the feeding and breeding sites used differ between species, as do characteristics such as humidity and temperature requirements, circadian rhythms of mating, length of life cycle, and presence or absence of a diapause.

Drosophila has both outstanding advantages and maddening disadvantages as a subject for the study of lifetime reproductive success. The wealth of accumulated knowledge about its biology, the availability of sophisticated genetic techniques, and the ease with which the flies can be cultured and manipulated in the laboratory make it possible to disentangle processes that are at present confounded in larger and less well-studied animals. On the other hand, flies of most *Drosophila* species are too small and vagile for field studies on individually marked adults to be convenient, and fieldwork on individually marked eggs and larvae is currently out of the question.

Most studies have therefore been done in the laboratory under conditions that may have at best limited applicability to the field. Fortunately, workers in this area have started to appreciate the need for basic ecological information on the commonly used laboratory species, and as this starts to accumulate it should be possible to ensure that laboratory conditions reflect those the flies are likely to encounter in the field.

The data to be described in this chapter have all been gathered in the laboratory, mostly on a single species, *D. melanogaster*. The advantage these observations have over those from many field studies is the experimental approach, which allows us to disentangle the effects of variables such as age, reproductive effort, and body size. I shall first describe the effects of age on breeding success, costs of reproduction, and constraints on longevity. I shall then discuss proximate causes of variation in reproductive success and, last, take up the effects of size and courtship feeding on reproductive success.

2.1 Age and Breeding Success

Natural selection seems often to have caused age-related changes in reproductive activity and probability of survival. In *Drosophila* two

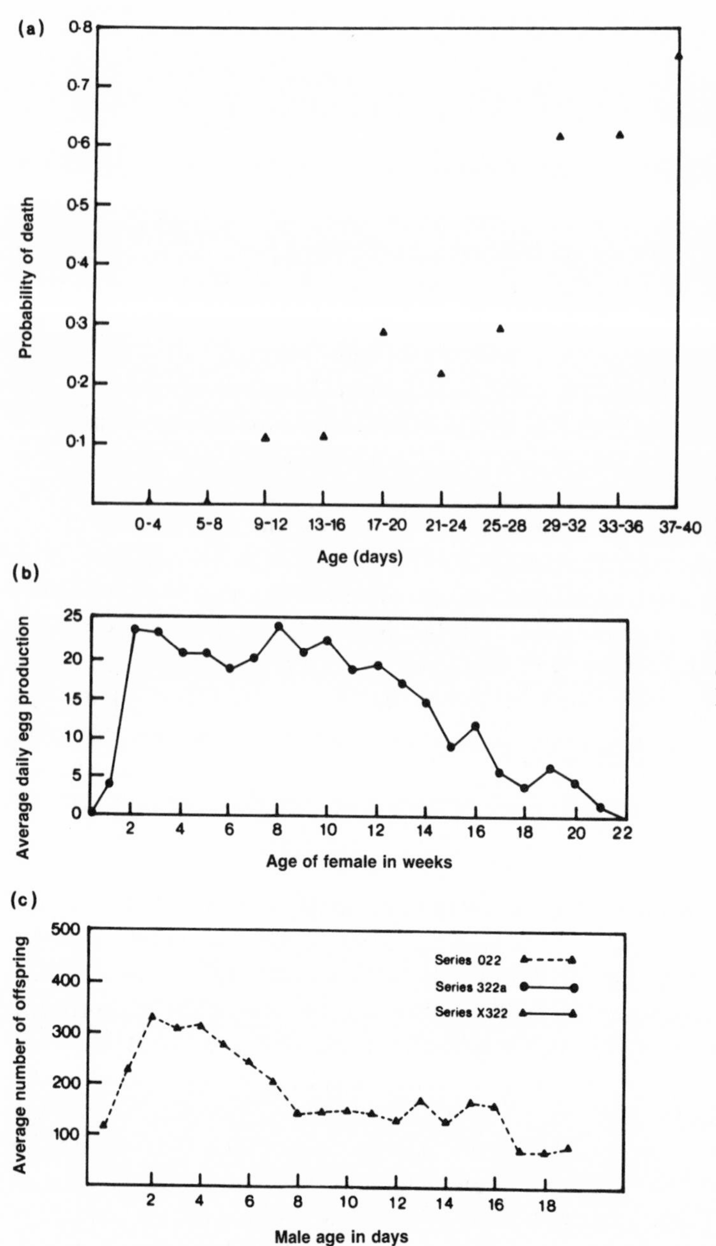

Figure 2.1: Some examples of changes in survival and reproductive rate with age in *D. melanogaster*. (*a*) The probability of death over four-day periods for a cohort of sixty females kept with males throughout life (Partridge et al., 1986). (*b*) Egg-laying rates of females kept with males throughout life (redrawn from Tantawy and Vetukhiv 1960). (*c*) The average daily offspring production of males supplied with twenty virgin females per day (redrawn from Kvelland 1965).

phenomena are of interest here. First, after the adults eclose from the puparium, there is a period during which reproductive activity is at first absent and subsequently increases to a maximum. The length of this period varies between species. Second, once full reproduction has commenced, reproductive rate and survival probability start to decline in both sexes (see fig. 2.1).

In males, the lag before maximum reproductive rate is reached may be associated both with the filling of the accessory glands, which produce the seminal fluid, and with the maturation of sperm in the testes (Clayton 1957, 1962; Steele 1984). In females, the onset of oogenesis in the ovaries and yolk protein synthesis in the fat body and elsewhere are probably important, and mating increases egg production, especially in females mating for the first time early in life (Manning 1967; Baumann 1974; Bouletreau 1978). The time of onset of sexual receptivity and reproduction often differs between the sexes of the same species (Manning 1967; Donegan 1984). The reasons both for this difference and for the variation between species in the timing of sexual maturity are at present obscure.

The second consistent age-related change in demographic characteristics is the aging or senescence shown by both sexes; fertility, mating success, and survival probability decline later in adult life (e.g., Gowen and Johnson 1946; Charlesworth 1980; Long, Markow, and Yaeger 1980; Partridge and Farquhar 1981; data in fig. 2.1). These changes have been documented only in the laboratory, and there is a general dearth of information on the effects of aging on invertebrates in the field.

In terms of natural selection, senescence must be regarded as maladaptive, because fitness components show a decline with age. Nonetheless, as Medawar (1952) first pointed out, senescence could evolve as a result of a change in the relative importance of genetic drift and natural selection in determining the fate of genetic variants with effects at different ages. Medawar's basic point was that in any population the abundance of cohorts of increasing age will decline as a result of externally imposed mortality caused by predation, disease, accidents, and so on. A decline in effective population size increases the effect of genetic drift in the face of natural selection. It therefore follows that mutants that decrease fitness will be less easily eliminated and advantageous mutants less easily incorporated by natural selection if they exert their effects later in life. This idea has since been refined and modeled (Hamilton 1966; Charlesworth 1980), and the conclusion has been reached that the effects of aging will start to be exerted only once reproduction has commenced. *Drosophila* has been used to test hypotheses about the evolution of aging (e.g., Rose and Charlesworth 1981a,b; Rose 1984; Luckinbill et al. 1984).

2.2 Costs of Breeding and Constraints on Longevity

There is now considerable evidence that in *Drosophila* and other insects, reproductive activity by both sexes has a negative effect on longevity (see

Partridge 1986 for review). The evidence for this statement comes from studies where reproductive activity is experimentally manipulated and the effect on longevity measured.

Reproductive activity of *Drosophila* females has been manipulated in various ways. Several studies have compared the longevity of females kept with males with that of females kept virgin throughout life (e.g., Bilewicz 1953; Malick and Kidwell 1966; data in fig. 2.2*a*). This type of experiment in general shows a reduction in the longevity of the females kept with males. The exact nature of the effect of exposure to males is at

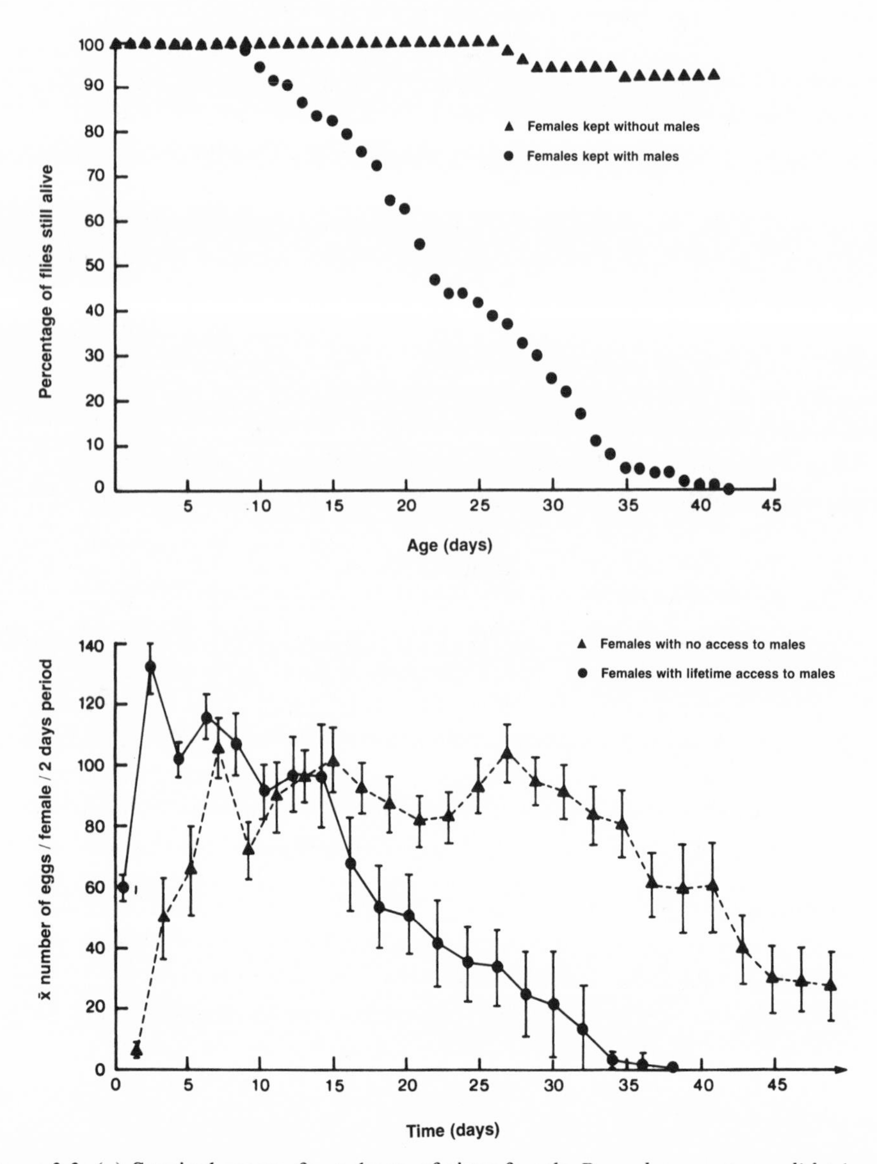

Figure 2.2: (*a*) Survival curves for cohorts of sixty female *D. melanogaster*: *solid triangles*, kept virgin throughout life; *solid circles*, kept with two males throughout life. (*b*) Egg-laying rates (with 95% confidence limits) for the same two groups. From Partridge et al. 1986.

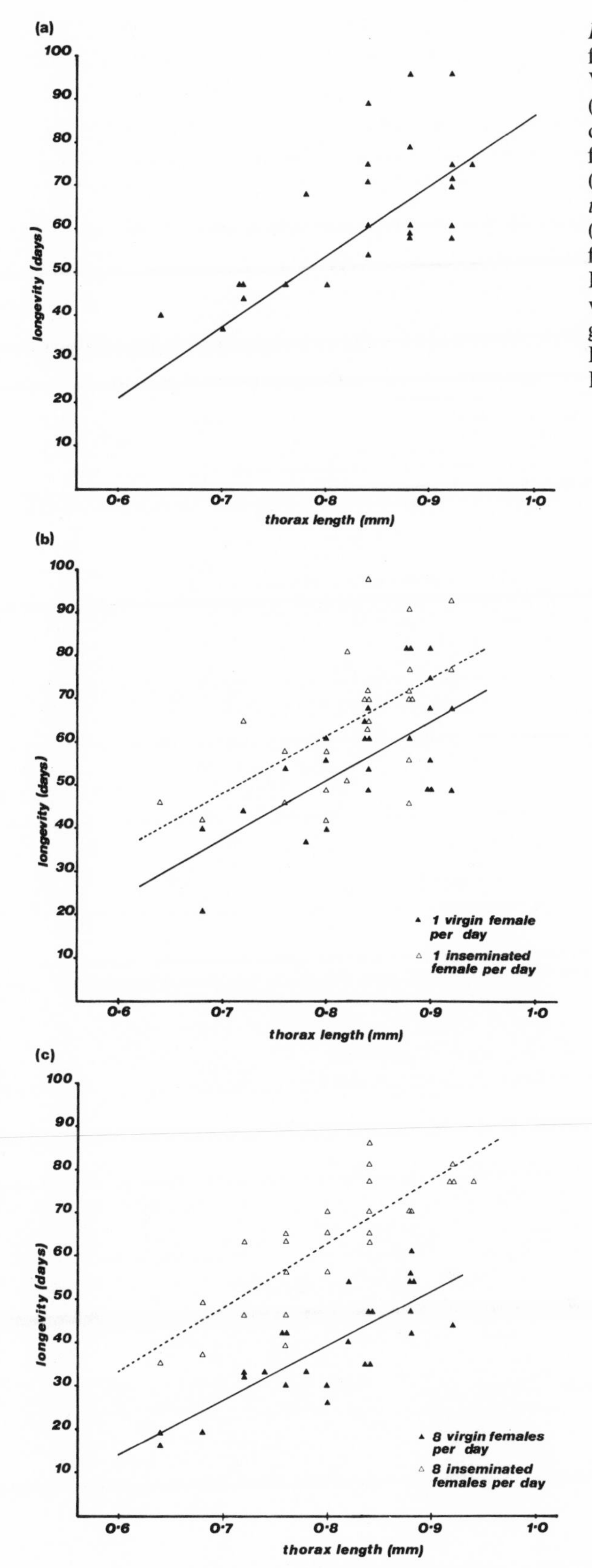

Figure 2.3: Size and longevity for male *D. melanogaster*: (*a*) Virgins; (*b*) supplied with one (*solid triangles*) virgin female or one (*open triangles*) inseminated female per day throughout life; (*c*) supplied with eight (*solid triangles*) virgin females or eight (*open triangles*) inseminated female per day throughout life. In *b* and *c* the differences in elevation between the best-fit regression lines are significant. Redrawn from Partridge and Farquhar 1981.

present uncertain. Although mating has the effect of increasing egg production, especially in young females mating for the first time (David 1963; Manning 1967; Baumann 1974), as the egg-laying period proceeds, egg-laying rates of virgins can come to equal or exceed those of inseminated females (David 1963; Partridge et al. 1986; data in fig. 2.2*b*). Therefore, in addition to any effect on egg laying, males may harm females by, for example, contaminating the food, harassing them, or by some postmating effect (Cohet and David 1966). This is an area where further work is needed. More direct evidence for an effect of egg production comes from experiments where flies are made sterile by high temperatures (Maynard Smith 1958) or radiation (Lamb 1964). These potentially harmful manipulations have the paradoxical result of of increasing longevity, presumably because they destroy ovarian activity. That this is the main explanation is supported by the finding that irradiation does not increase life span in mutant *ovaryless* females (Lamb 1964). This type of experiment may not reveal the full costs of egg production, because yolk protein synthesis is not necessarily turned off by the cessation of oogenesis (Bownes 1983). Some sort of nutritional manipulation might be an appropriate way of manipulating all aspects of egg production.

In males, reproduction has been manipulated by altering the rate at which males are kept supplied with virgin females (Partridge and Farquhar 1981). Longevity is lowest among males supplied with females at the highest rate (see fig. 2.3), and this effect appears to be a direct conse-

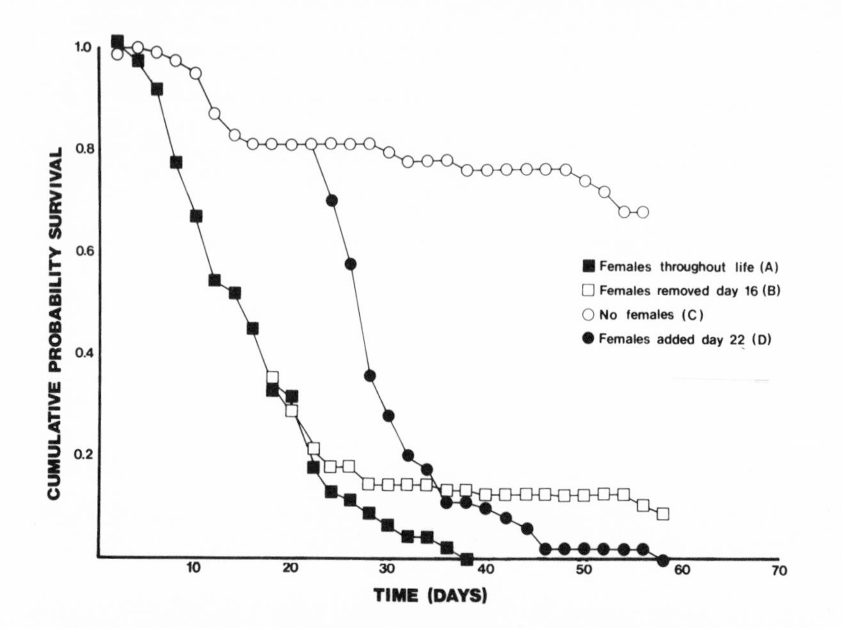

Figure 2.4: Survival curves for a total of 240 male *D. melanogaster*. Group A males died at a significantly higher rate than group C males. Group D males at once adopted a death rate significantly higher than that of group C and statistically indistinguishable from that of group A. Group B males after day 22 died at significantly lower rate than group A males. Redrawn from Partridge and Andrews 1985.

quence of the presence of females (fig. 2.4). It is not clear exactly what aspects of sexual activity are important here. Spermatogenesis, costs of accessory fluid production, and the energetic costs of courtship and mating could all be important. There have been no direct measurements of the energetic costs of courtship, although there is some suggestion that wing vibration causes fatigue (Jallon and Hotta 1979). Accessory fluid seems to be the main limitation on the number of fertile matings a male can perform in a short period (Lefevre and Jonsson 1962; Richmond et al. 1980; Hihara 1981), and since the fluid is composed mainly of protein (Chen 1984), amino acid acquisition may be critical.

The costs of reproduction in the laboratory may not give an accurate reflection of conditions in the field, although the direction of any error is not obvious. Costs could be underestimated because flies in the field are in general in poorer nutritional status (Bouletreau 1978; Steele 1984) and may incur additional costs associated with searching for mates, competition with other members of the same sex, and travel between breeding sites. Flight is known to be particularly costly. For instance, in *D. melanogaster* one hour of tethered flight reduces by an average of sixty the egg production of females over the next two days (Roff 1977). On the other hand, some laboratory manipulations may lead to an overestimate of costs. In the field matings may be less frequent than in the laboratory, and rates of egg production are probably also lower. Laboratory studies of the effects and interactions of these variables would be valuable.

2.3 Proximate Causes of Variation in Reproductive Success

One of the main weaknesses of using *Drosophila*, and indeed many other animals with no parental care, for work on lifetime reproductive success is that it is virtually impossible to estimate preadult mortality in the field. In the laboratory the main source of preadult mortality is larval competition. With increasing numbers of larvae in a fixed quantity of food, the number and size of the emerging adults decline and the variance in size increases (Kearsey 1965; Bos and Scharloo 1974; Robertson 1960; Atkinson 1979; fig. 2.5). For these reasons, studies confined to the adult stage must underestimate individual variation in lifetime reproductive success by ignoring what can undoubtedly be a very large proportion of zero scores.

Confining our attention to the adult stage, we can examine the effect on individual variation in reproductive success of varying the length of time over which the measure is made. There are several reasons for expecting short-term and long-term measures of variation to differ. Because reproductive activity reduces the longevity of both sexes, short-term measures of variance could be overestimates, because flies that reproduced at high levels early on would pay for this by reduced longevity and hence reduced late reproductive success. One the other hand, since life expectancy and fertility decline with age, the benefits of high early reproduction could outweigh its negative effect on longevity, so that a longevity cost of reproduction will not automatically reduce the variation in lifetime success between individuals. Furthermore, because longevity and fertility are often correlated with size in both sexes, short-term measures of variation in reproductive success in a size-variable population could be underestimates because those individuals with higher fertility would also have higher life expectancy. In *Drosophila* in the laboratory, short-term measures of reproductive success in both sexes do show less individual variation than lifetime scores.

In males, mating success has been measured as number of females inseminated (Partridge and Farquhar 1981), and reproductive success has

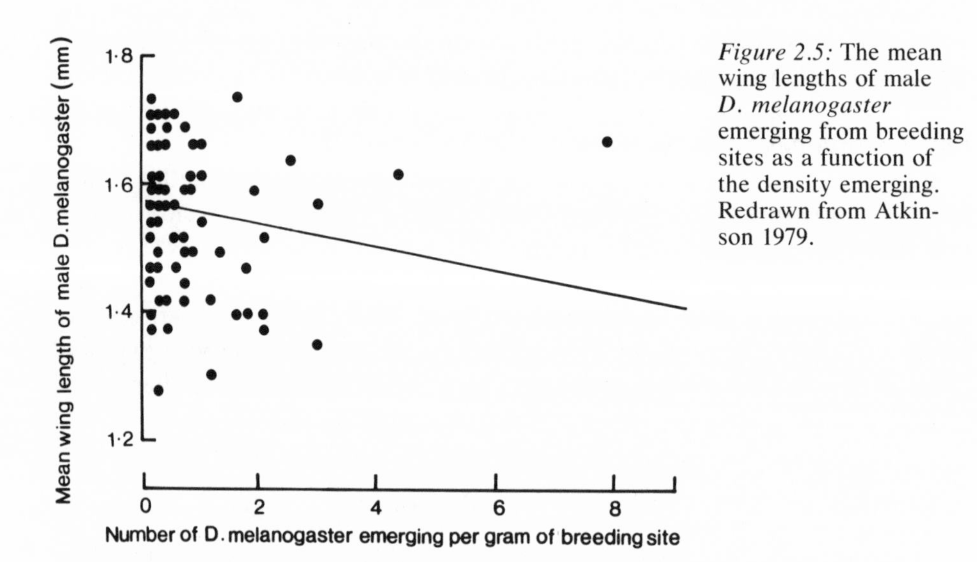

Figure 2.5: The mean wing lengths of male *D. melanogaster* emerging from breeding sites as a function of the density emerging. Redrawn from Atkinson 1979.

Table 2.1 Measures of Variation between Individuals in Lifetime Reproductive Success for Male and Female *Drosophila melanogaster*

Measure	Time over Which Measure Was Made	Standardized Variance ($\sigma^2/\bar{x}^2$)
Males		
Number of females inseminated		
(one virgin per day supplied)	2 weeks	0.007
(n = 25)	lifetime	0.073
	(mean = 56 days)	
Number of females inseminated		
(eight virgins per day supplied)	2 weeks	0.025
(n = 25)	lifetime	0.083
	(mean = 39 days)	
Total adult progeny produced		
(six virgins per day supplied)	1 week	0.076
(n = 16)	lifetime	0.187
	(mean = 24 days)	
Females		
Total adult progeny produced		
(two adult males always present)	1 week	0.136
(n = 65)	lifetime	0.282
	(mean = 19.65 days)	

been measured by counting the total number of adult progeny produced (Partridge et al., n.d.). In the former set of experiments, males were supplied with either eight or one virgin females per day, and the variation between individuals during the first two weeks of life, when all the males in both experiments were still alive, and over the whole life span are shown in table 2.1. The measure of variation used is $\sigma^2/\bar{x}^2$, the standardized variance which gives the potential for selection. Since sample sizes are the same for both measures, the comparison is not biased by mortality effects on sample sizes. Lifetime insemination success shows considerably more variation between individuals than does success during the first two weeks. Adult progeny production also shows greater variation between individuals for the lifetime measure than for the scores from the first week of life, again suggesting that variation in longevity has an important effect on the lifetime score. Female reproductive success was measured by culturing under relaxed competition all the progeny produced by individual females kept with two males throughout life (Partridge et al., 1986). The lifetime scores again show more individual variation than those from the first week of adult life (table 2.1).

These data imply that longevity is a major factor determining the lifetime reproductive success of *Drosophila* under these conditions, and direct analysis confirms this. For instance, several studies have reported a

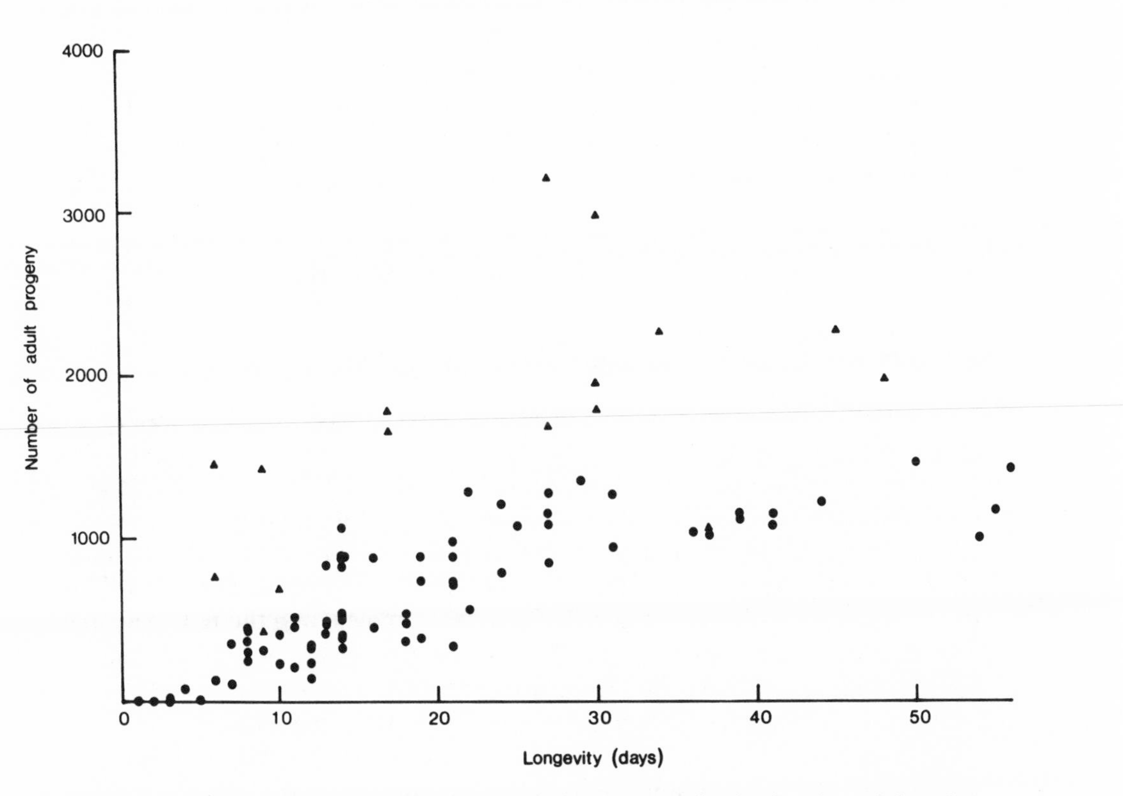

Figure 2.6: Lifetime progeny production in relation to longevity for female (*solid circles*) and male (*solid triangles*) *D. melanogaster*. From Partridge et al. 1986.

strong positive correlation between longevity and lifetime progeny production in females (Kidwell and Malick 1967; Tantawy and Rakha 1964; data in fig. 2.6). Our data for males show a similar picture (fig. 6).

It is probably inappropriate to attempt to partition variation in lifetime reproductive success into the effects of longevity and fecundity using these data. Fecundity of both sexes shows a marked drop with age, so that a figure for mean fecundity that includes data from all ages will not be independent of longevity; long-lived individuals may have a higher proportion of the lower fecundity measures associated with older ages.

One interesting feature of the laboratory data is that females show more individual variation than do males on both measures of progeny production. Since much of the variation in both sexes is attributable to the effects of size, this result suggests that selection for large size in adult females may be particularly intense, which may in part explain why females are in general larger than males in *Drosophila*.

These data probably reflect the maximum progeny production by individuals of the two sexes under conditions of unrestrained access to food and mates. The means (and ranges) were for males 1,699 (426–3,198) and for females 615 (0–1,455). These figures are rather surprising in that they suggest that males are constrained by basic features of their reproductive biology to be able to produce only about three times the progeny of a female. The anisogamy argument (Bateman 1948) has been used to suggest that males are potentially capable of fertilizing the entire reproductive output of many females, and the present results suggest that this effect may not be as large as might be supposed.

All the figures discussed in this section are undoubtedly very different in the field. The range and shape of the size distributions are likely to be very important, and both longevity and fertility are probably reduced by nutritional stress and a greater incidence of externally imposed mortality. While information on the importance of these variables in the field is highly desirable, gathering the relevant data will not be easy.

2.4 Ultimate Causes of Variation in Reproductive Success

The best-documented correlate of reproductive success in both sexes is body size. Size varies between individuals as a result of events during the larval period and does not change once adulthood is reached. Temperature, the nature of the breeding site, larval crowding and genetic factors can all influence body size (Robertson 1954, 1957; Anderson 1973; Sang 1949, 1956; Burnet, Sewell, and Bos 1977; Tantawy 1964; Atkinson 1979). In *Drosophila* females are in general larger than males, and the few available field data do not suggest that the level of individual variation in size differs between the sexes (Atkinson 1979; Sokoloff 1966). A possible evolutionary contribution to the size dimorphism has already been mentioned.

In males kept with unconstrained access to females and food, lifetime progeny production, number of females inseminated in the first two weeks of life, number of progeny produced in the first week of life, and longevity are all positively correlated with size (Partridge and Farquhar 1981, 1983; fig. 2.7). Large males also have higher mating success in the field in *D. melanogaster* and *D. pseudoobscura* (Partridge, Hoffman, and Jones 1987). The reasons for the greater success of larger males under these conditions are probably numerous. Their longevity advantage may be related to improved homeostatic efficiency as a result of a relatively small body surface, and this advantage may allow them to channel more energy into replenishment of accessory glands and testes after each mating and to devote more energy to searching for mates and courtship. Large males do court more than small both in the laboratory (Partridge, Ewing, and Chandler 1987) and in the field (Partridge, Hofman, and Jones 1987). In the laboratory they also produce more courtship song (fig. 2.8).

Fighting may also be important in determining male mating success in the field. Fighting has been observed in the field among male *D. pseudoobscura*, and the victor in a fight is usually larger than the loser (Partridge, Hoffman, and Jones 1987). Fights occur on the surface of fruit during the peak mating period, and their higher rate of wins may explain the greater mating success of the larger males in these circumstances. Fighting has been observed in *D. melanogaster* in the laboratory (Jacobs

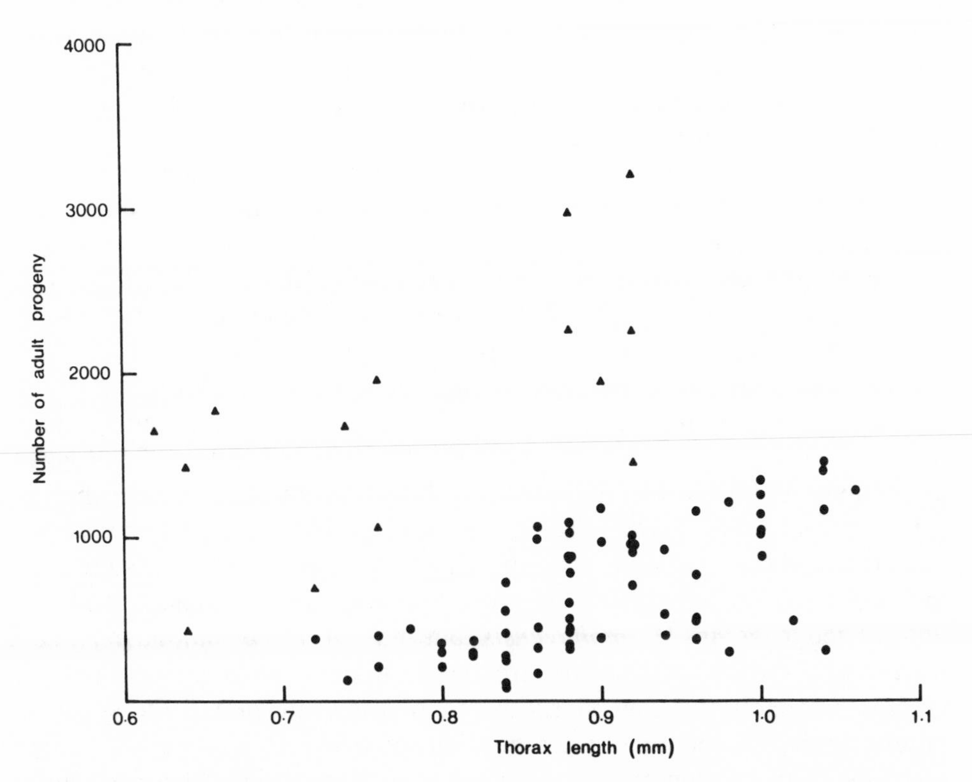

Figure 2.7: Lifetime progeny production in relation to size for male (*solid triangles*) and female (*solid circles*) *D. melanogaster*. From Partridge et al. 1986.

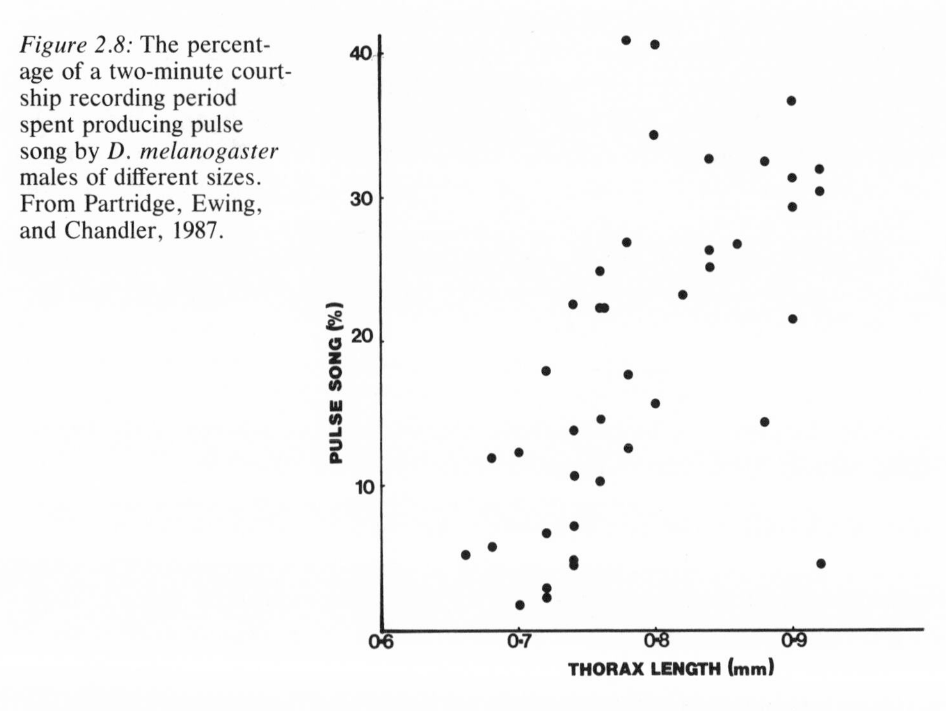

Figure 2.8: The percentage of a two-minute courtship recording period spent producing pulse song by *D. melanogaster* males of different sizes. From Partridge, Ewing, and Chandler, 1987.

1960; Dow and von Schilcher 1975), and larger males win in aggressive encounters (Partridge and Farquhar 1983). The role of fighting in this species in the field is at present uncertain. Female choice may also be involved in producing the mating advantage of larger males, but this has not yet been demonstrated.

Female fertility and longevity and hence lifetime reproductive success are also correlated with body size. In *D. melanogaster*, *D. simulans*, and *D. pseudoobscura*, both phenotypic and genotypic correlations have been demonstrated (Robertson 1957; Tantawy and Rakha 1964; Tantawy and Vetukhiv 1960). Figure 2.7 shows data for phenotypic variation in females taken from the same cultures as the males in that figure.

In some *Drosophila* species, females receive nutrients from the male at the time of mating via the seminal fluid. This type of effect has previously been found in other insects (e.g., Friedel and Gillot 1977; Boggs and Gilbert 1979) but has only recently been investigated in *Drosophila* (Markow and Ankney 1984). Their data showed that mating a female to a male that had been fed on food containing radioactive isotopes resulted in the appearance of the label all over the body of the female and in unfertilized eggs in *D. mohavensis* but not in *D. melanogaster*. *D. mohavensis* remates much more frequently than *D. melanogaster* under laboratory conditions, and Markow and Ankney interpreted the species differences as a reflection of their differing ecologies, *D. mohavensis* being more likely to encounter periods of food shortage. Since then, a male-derived contribution of methionine to yolk proteins in somatic tissue has been demonstrated in both *D. melanogaster* and *D. pseudoobscura*, but its nutritional significance is at present uncertain (Bownes and Partridge, n.d.).

There are as yet no data showing that either short-term fecundity or lifetime reproductive success is increased by multiple mating in *D. mohavensis*. Some data from *D. pseudoobscura* do suggest a nutritional effect of continued exposure to males. Turner and Anderson (1983) compared the progeny production and longevity of females either singly mated at the beginning of the egg-laying period and then kept without access to males or kept with males throughout life. They found that, despite their lower longevity, lifetime production of progeny was higher in the second group, and the effect was particularly marked if the females were nutritionally stressed. Since these underfed females produced fewer progeny than well-fed females, shortage of sperm is not a likely explanation of the effect. The results could also be a consequence of courtship feeding, which has been documented in the closely related *D. subobscura* by Steele (n.d. a,b), who also showed that female egg production in the period immediately after mating is greater if the female was fed during courtship.

2.5 Summary

The chapter summarizes work on lifetime reproductive success in *Drosophila* species. Nearly all of this work has been done in the laboratory, and the danger of producing unrealistic conditions and the need for field studies are emphasized. Two main age-related changes in reproductive activity in *Drosophila* are an increase to a maximum early in adult life and senescence after the peak is reached. Reproductive activity results in lowered longevity in both sexes. Despite this cost of early reproduction, variation in reproductive success between individuals is greater for lifetime measures than for short-term ones, indicating the large contribution of variation in longevity to lifetime measures. Large flies of both sexes have higher longevity and reproductive rates. In some species, multiple mating by females may increase lifetime reproductive success because of nutritional contributions from the male.

3 Sources of Variation in Lifetime Reproductive Success in a Nonterritorial Damselfly (Odonata: Coenagrionidae)

Ola M. Fincke

FOR FOUR YEARS, I have studied the reproductive behavior of *Enallagma hageni*, a clear-winged "bluet" damselfly, common to ponds and small lakes in the eastern United States and Canada (fig. 3.1). Adults provide no parental care, and the sexes are only moderately dimorphic. Females, which can be either green or blue bodied, are slightly larger than the bright blue males. Darwin (1871) noted that many sexual dimorphisms appeared to result from sexual selection (competition for matings or female choice). The potential for sexual selection in species such as *E. hageni*, whose males compete by searching for mates (i.e., "scramble competition"), has been predicted to be intermediate between that of territorial and monogamous species (e.g., Trivers 1972; Emlen and Oring 1977).

Theoretically, one could test the prediction above by comparing the potential opportunity for selection, I (variance/mean^2; Crow 1958; Wade and Arnold 1980), on mating and fertilization efficiencies over the lifetime of a generation of *E. hageni* with those of territorial or monogamous species. However, because of practical problems in applying field data to theoretical models of variance partitioning, conclusions based on a comparison of I values (which lack confidence intervals) are somewhat subjective. The relative standardized variance that is attributed to sexual selection depends upon which selective episodes are used to partition total variance in lifetime reproductive success (LRS), how covariance components (which may be statistical artifacts) are treated, and the extent to which sexual selection on mated males can be detected (Fincke 1986a).

I values indicate a potential opportunity for selection. To determine how the potential opportunity for sexual selection is realized, one needs to identify phenotypic traits that covary with mating or fertilization efficiency. Although sexual selection is not the only cause of sexual dimorphism (e.g., Selander 1966), in general one should expect to find a greater degree of morphological or behavioral sexual dimorphism, or both, in species with a higher potential opportunity for sexual selection than in species with a lower potential.

Figure 3.1: Pair of *Enallagma hageni* in copula (marked male is above female.

In this chapter, I first examine the reliability of standardized variances in major selective episodes of LRS, and the correlations between episodes, across two generations and within a single reproductive season. I then identify demographic, ecological, and seasonal factors that contribute to variation in LRS independent of phenotypes and present evidence that the potential opportunity for selection on mating success is realized as stabilizing selection on body size. Finally, I compare the potential opportunity for sexual selection and salient features of the mating system of *E. hageni* with those of other odonates for which the variation in some aspects of LRS has been measured.

3.1 Measuring Lifetime Reproductive Success in *E. hageni*

The reproductive behavior of *E. hageni* has made it an exceptionally useful study insect. The study population at East Point Pond (100 m perimeter, <1 m depth), near Pellston, Michigan, USA, was univoltine, localized, and isolated (the closest conspecific population was 4 km away). After overwintering as aquatic larvae, adults emerged from early June to early July and were active until early August. The sex ratio at emergence (determined by sexing exuvia) did not differ from 1:1 (113 males:124 females). Because body size (i.e., wing length rather than weight) of the adults was fixed at emergence, age and size effects on reproductive success could be clearly separated.

Newly emerged tenerals matured in an adjacent field where breeding adults also fed and spent the night. Time to sexual maturation was determined by resighting individuals collected within hours of emergence, held in an outdoor insectary until the body and wings had hardened, then color coded to day by a spot of paint on the thorax (marking wings of tenerals can damage them) and released the same day. Although males attained full mature coloration within one day (females required 3 to 7 days), individuals were not considered sexually mature until their first return to the pond to breed. I assumed that mature adults either remained in the area until dying or were unsuccessful in reproducing if they left the site. The reproductive season of the population was short but rather variable (e.g., 3 to 6 weeks in 1980–83); 1981 was an "abnormal" year, in that the population was reproductive for only three weeks. This was the same year in which I marked teneral adults on the day of emergence. Because other populations of *E. hageni* in the same county remained active two weeks after my study population disappeared, I do not know if the restricted season resulted from a marking effect on tenerals or from parasitism, disease, or other natural catastrophes limited to my study pond.

When sexually mature individuals were first sighted at the pond, we marked them by writing a number on the forewing and released them within ten minutes of capture. In 1980, wing length was recorded as a measure of body size. "Life span" refers to the span between first and last sightings of mature adults. Handling and marking had an initial effect in that the probability that a mature individual would be resighted after the day of marking was considerably lower (.48) than subsequent daily survivorship ($\bar{x}$ = .83; Fincke 1986a). In 1980, individuals were sighted at the pond only; in 1982 they were also censused in the adjacent field. Unless otherwise noted, all data refer to sexually mature adults, and means are given ± standard errors. I used ln wing length for all partial regressions, and $(\ln \text{wing} - \bar{x} \ln \text{wing})^2$ for those involving quadratic equations. Because both sexes returned repeatedly to the pond to mate and lay eggs during a predictable time each day (1000–1800 h), I could record lifetime mating success (LMS) of most of the individuals in the population on eighteen days of a twenty-three-day span in 1980 (Fincke 1982). In 1982 I measured lifetime reproductive success (LRS) in terms of eggs laid by females or fertilized by males for a subset of the population on seventeen days of a twenty-eight-day span (details in Fincke 1986a). Rainy or overcast days (when the damselflies were inactive) were excluded.

Although damselflies mature eggs continuously, females visited the breeding site and were receptive to mating only after accumulating 200–500 eggs, which they laid as a "clutch" on a given day (individuals were credited with only one visit per day at the pond). In the morning, males searched the banks for arriving females. By midafternoon, this supply of unpaired females was depleted, and many males still without mates then waited at oviposition sites for females resurfacing from bouts of underwater egglaying. A male mating by either of these two tactics showed no obvious courtship behavior but simply pounced on a female, clasping her thorax with his abdominal appendages (tandem position) and releasing her if she was unreceptive (Fincke 1985). Most tandem pairs could be noted because male arrival was gradual throughout the morning, only a moderate number (10–145) visited the pond on a given day, and mated pairs spent on average, 22 min in copula, plus an additional 50 min in tandem before females submerged in one of three oviposition areas (each about 10 m^2) of sparsely clumped sedges and *Potamogeton*. The water in oviposition areas was less than 0.3 m deep and so clear that I could read the wing numbers of submerged females as they inserted eggs into plant stems. Eggs laid were estimated as a function of oviposition duration.

Upon resurfacing, females either flew to shore and left the pond or were taken in tandem, either by mates that had been guarding above water or by lone males, which would subsequently guard if resurfacing females mated with them. Because sperm displacement in this species is high (Fincke 1984b), a male was credited with 100% of the fertilizations from subsequent oviposition bouts by his mate unless she remated with another male between bouts of submerging, in which case the intercepting male gained a portion of her clutch. Guarding duration was measured as the time from the initial submergence of a mate until the male left the oviposition area (i.e., the area circumscribed by a 1 m radius from the submergence stem) for more than 3 min. Many males continued to "guard" even after their mates resurfaced undetected.

3.2 Yearly and Seasonal Variation in Components of Reproductive Success of Females and Males

Males were about three times as variable in both the number of zygotes produced (fig. 3.2) and the number of matings (tables 3.1 and 3.4) as were females. In both years, males were more variable in all aspects of reproductive success except life span and visits/day alive (table 3.4). Standardized variances in total clutches and total mates of females measured in 1980 and 1982 varied little between the two years (table 3.1). The shorter life span of 1980 females was an artifact of my sampling procedure. Adding the mean interclutch interval to the mean life span in 1980 resulted in an average life span of 7.3 days, similar to that in 1982 when I censused females both in the feeding areas and at the pond.

Variation in mating efficiency (mates/clutch) was inconsequential to the variation in female LRS because only 3 of the 447 mature females sighted at the pond in 1980 and 1982 were not seen to mate. The number of matings/clutch by a female was a consequence of both the number of oviposition bouts/clutch ($r = .27$, $n = 50$, $.1 > p > .05$), and the frequency of interception by a second male. Most (50%, $n = 194$) guarding males were successsful in preventing any interceptions and usually did not remate (see below). Although oviposition duration of females mating multiply ($\bar{x} = 28.9 \pm 2.0$ min, $n = 48$) was not greater than that for those

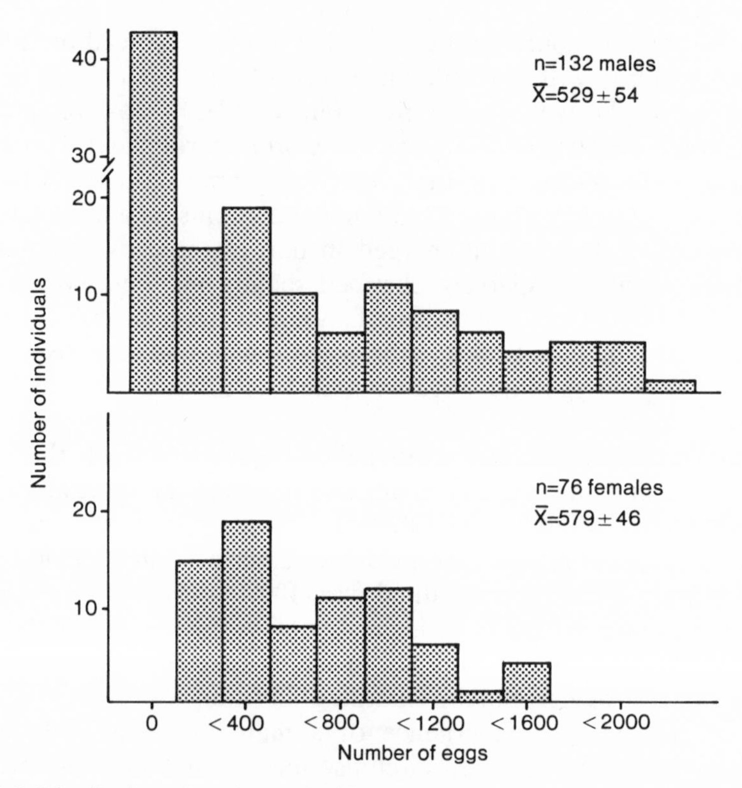

Figure 3.2: Distribution of total eggs fertilized by males and laid by females in 1982.

mating once/day ($\bar{x}$ = 24.7 ± 2.1 min, n = 84, t-test, n.s.), there was a positive correlation between eggs/clutch and matings/clutch for females (table 3.2). Most females (114/210) laid all their eggs in one bout, and only 9% submerged more than two times. I interpret these results to mean that mating did not stimulate increased oviposition but, rather, a female carrying more eggs took longer to oviposit and was consequently more likely to resurface prematurely and remate than those carrying few eggs.

Table 3.3 summarizes the results of partitioning the total variance in the number of eggs laid by females and fertilized by males in 1982 (methods detailed in Fincke 1986a). Both the variance partitioning method and r^2 values (see table 3.2) indicated that variation in total clutches (a function of life span and interclutch interval) and in eggs/clutch contributed about equally to the total variance in LRS of females. The interclutch interval, measured directly from females seen more than once, was not significantly correlated with total eggs laid (r = −.26, n = 44) and was positively correlated with life span (r = .61). Because females seen only once were those at the pond, each had a value of one for clutches/day, the maximum possible. This resulted in a spurious and large negative covariance between clutches/day and life span (table 3.2), and consequently I did not further partition the number of clutches into these two subcomponents.

Sexual selection could potentially act on differences in the mating efficiency of males and, among mated males, on differences in fertilizations/mating other than those arising from differences in female fertility (table 3.3). Among mated individuals visiting the pond on a given day, the average number of eggs laid or fertilized per mating was identical for males and females, but the variation in fertilizations/mating of males ("daily" I_{fm}; table 3.4) was greater than the variation in eggs/clutch of females ("daily" I_o; table 3.1). The absolute difference between daily I_{fm}, and I_o multiplied by R_1, the sex ratio given that a male fertilized his mate's entire clutch, represented the potential opportunity for sexual selection on fertilization efficiency of mated males (I_{fe}) (see Fincke 1986a). For lifetime data, I estimated a relative opportunity for sexual selection on males of 51% ($I_s/I_m = (I_{fe} + I_e)/I_m$), of which 7% was attributable to I_{fe}.

Because males not present at the breeding site did not have the opportunity to mate, matings per visit to the pond was a better measure of mating efficiency than was matings averaged over days alive (matings/day, as distinct from "daily" success above). Variation in matings/day included variation both in mating efficiency and in the proportion of a male's life spent at the pond (visits/day), which should theoretically reflect intervisit

Table 3.1 Year-to-Year Variation in Components of Female LRS in *Enallagama hageni*

Variable	Symbol	Year	*n*	Mean	Variance	I
Total eggs[a] (LRS)	I_f	1982	76	44.5	961.00	0.49
		(daily)	114	27.8	289.0	0.37
Eggs/clutch	I_o	1982	76	24.2	228.01	0.39
		(daily)	114	25.4	179.56	0.28
Total clutches (visits)	I_c	1980	358	1.31	0.39	0.23
		1982	89	1.78	1.14	0.36
Life span (days)	I_l	1980	358	2.15	8.35	1.8
			84[b]	5.88	17.2	0.50
		1982	89	6.17	43.82	1.15
			39[b]	11.08	35.64	0.29
Clutches/day alive	I_p	1980	358	0.90	0.06	0.07
		1982	89	0.71	0.136	0.26
Total mates (LMS)	I_t	1980	358	1.57	1.07	0.44
		1982	89	1.84	1.17	0.34
Interclutch interval (days)		1980	69	5.1	33.8	1.30
		1982	73	5.2	21.16	0.78
		(females)	44	5.7	20.88	0.65
Eggs/mate		1982	76	24.2	156.25	0.27
		(daily)	110	25.4	179.56	0.28
Mates/clutch	I_e	1980	358	1.21	0.34	0.24
		1982[c]	76	1.04	0.21	0.20
Eggs/day alive		1982	76	16.4	278.9	1.04

[a] Throughout table, statistics for "eggs" are given in oviposition duration.
[b] Females seen only once are excluded. In 1980, females were sighted only at the pond.
[c] These are only matings for which oviposition durations were known, not total matings.

Table 3.2 Correlation between Components of LRS in Male and Female *E. hageni*

	Males										
Females	*Eggs/ Clutch*	*Total Eggs (LRS)*	*Total Mates (LMS)*	*Eggs/ Mating*	*Mates/ Visit*	*Mates/ Day*	*Visits/ Day*	*Life Span*	*Visits (Clutches)*	*Date First Seen*	*Size (In wing)*
Total eggs (LRS)	.56 (76) ***		.89 (151) ***	.62 (88) ***	.62 (151) ***	.24 (151) *	−.34 (151)	.46 (151) ***	.58 (151) ***	.02 (151) n.s.	—
Total mates (LMS)	.78 (76) ***	.80 (76) ***		.10 (88) n.s.	.67 (509) ***	.42 (509) ***	−.40 (509) ***	.59 (509) ***	.67 (509) ***	.001 (151) n.s.	−.003 (489) n.s.
Eggs/mating	.63 (76) ***	.54 (76) ***	−.001 (76) n.s.		.05 (88) n.s.	−.08 (88) n.s.	−.21 (88) n.s.	.15 (88) n.s.	.12 (88) n.s.	−.04 (88) n.s.	—
Mates/visit	.56 (76) ***	.13 (76) n.s.	.30 (76) *	−.17 (76) n.s.		.82 (156) ***	.03 (509) n.s.	−.05 (509) n.s.	.03 (509) n.s.	−.05 (509) n.s.	.02 (498) n.s.
Mates/day	.19 (76) n.s.	−.36 (76) ***	−.27 (76) *	−.24 (76) *	.58 (76) ***		.58 (151) ***	−.22 (509) *	−.03 (509) n.s.	−.01 (509) n.s.	.04 (489) n.s.
Visits/day	.10 (76) n.s.	−.57 (76) ***	−.51 (358) ***	−.22 (76) n.s.	.06 (358) n.s.	.48 (358) ***		−.83 (509) ***	−.59 (509) ***	.15 (509) ***	.11 (489) n.s.
Life span	.01 (76) n.s.	.60 (76) ***	.56 (358) ***	.10 (76) n.s.	−.004 (358) n.s.	−.60 (76) ***	−.92 (358) ***		.85 (509) ***	−.14 (509) *	−.07 (489) n.s.
Visits (clutches)	−.10 (76) n.s.	.70 (76) ***	.69 (358) ***	.14 (76) n.s.	−.02 (358) n.s.	−.55 (76) ***	−.80 (76) ***	.79 (358) ***		−.12 (509) ***	−.05 (489) n.s.
Day first seen	−.14 (76) n.s.	−.35 (76) ***	−.22 (76) n.s.	−.23 (76) *	.12 (76) n.s.	.22 (76) n.s.	.31 (76) *	−.32 (76) *	−.26 (76) *		.03 (489) n.s.
Size (ln wing)	—	—	−.03 (298) n.s.	—	.009 (298) n.s.	−.02 (298) n.s.	−.04 (298) n.s.	−.02 (298) n.s.	−.09 (298) n.s.	−.13 (298) n.s.	

Note: There were no differences in the sign or significance of correlations between 1980 and 1982 (1980: n = 509 males, 298 females; 1982: n = 151 males, 88 mated males, 76 females).

*p = < .05.

***p = < .001.

interval. Visits/day alive covaried negatively and spuriously with life span (table 3.2), because males seen only a few times had artificially high values for visits/day. In contrast, for males seen more than once in 1982, the interval between visits to the pond, measured directly, was not significantly correlated with LRS ($r = -.1$, $n = 62$) but was positively correlated with life span ($r = .34$; Fincke 1986a).

As seen in table 3.4, the average total matings per male was higher in 1982 than in 1980 because in 1980 I concentrated on recording all males at the pond, whereas in 1982 I followed tandem pairs and thus underestimated the number of lone males at the pond. My ability to detect individuals at the pond appeared to be consistent between years, as evidenced by the similarity between years in means and variances for life span and the number of total visits. In both years, variation in survivorship ($r^2 = .28$–$.35$), and mating efficiency ($r^2 = .36$–$.45$) were the main sources of variation in lifetime mating success (LMS; tables 3.2, 3.3).

In contrast to the relatively similar year-to-year variation in components of LRS, daily reproductive success of individuals at the pond fluctuated unpredictably from day to day (Fincke 1986a) and was more variable than LRS (tables 3.1, 3.4). Moreover, because more females partitioned a clutch among several males later in the season than did so

Table 3.3 Components of the Total Opportunity for Selection in *E. hageni* and the Selective Episodes Used to Estimate Them

Percentage of Total	Symbol	Selective Episode Measured	Possible Sources of Variation in Episode
Females			
	I_f	Eggs laid	
79	I_o	Eggs/clutch	Female fertility
69	I_c	Total clutches	Survivorship Interclutch interval
Males			
	I_m	Eggs fertilized	
21	I_{fm}	Fertilizations/mating	Female fertility Sperm displacement[a] Mate guarding[a] Alternative tactic[a] Mate choice[a] Male-induced fertility[a]
44	I_e	Matings/visit	Mating efficiency[a] Chance encounters
34	I_p	Visits/day alive	Visits and intervals between visits
	I_l	Life span	Survivorship

Note: $I_f = I_o + I_c$, and $I_m = I_{fm} + I_e + I_p + I_l$ (modified from Wade and Arnold 1980). Percentages were calculated from 132 males and 76 females for which all selective episodes were known. Because correlations between selective episodes were not significant, covariance components were ignored (see Fincke 1986a).

[a] Variation potentially attributable to sexual selection.

Table 3.4 Year-to-Year Variation in Components of Male LRS in *E. hageni*

Variable	Symbol	Year	*n*	Mean	Variance	I	I_m/I_f
Total fertilizations[a] (LRS)	I_m	1982	132	40.7	2323.24	1.40	2.9
		(daily)	451	11.30	393.63	3.08	8.3
Fertilizations/mate[a]	I_{fm}	1982	88	25.4	193.21	0.30	1.1
		(daily)	182	25.7	292.41	0.44	1.6
Mates/visit	I_e	1980	509	0.40	0.16	1.00	4.2
		1982	156	0.70	0.24	0.49	2.5
Mates/days alive		1980	509	0.30	0.132	1.49	
		1982	156	0.50	0.213	0.84	
Life span	I_l	1980	509	4.34	17.39	0.92	0.5
			311[b]	6.47	16.81	0.40	0.8
		1982	156	5.46	32.04	1.07	0.8
			85[b]	9.19	28.09	0.33	1.1
Proportion of life at pond (visits/day)	I_p	1980	509	0.76	0.07	0.13	1.9
		1982	156	0.69	0.11	0.23	0.9
Total mates (LMS)	I_t	1980	509	0.96	1.13	1.21	2.8
		1982	156	1.48	2.16	0.99	2.9
Visits to pond		1980	509	2.38	2.56	0.45	2.0
		1982	156	2.16	2.19	0.47	1.3
Days between visits		1980	211	2.5	8.44	1.35	1.0
		1982	146	2.7	10.45	1.43	1.8
		(males)	62	3.1	9.30	0.99	1.5
Fertilizations/days alive		1982	132	10.0	262.44	2.62	2.5

[a] For total fertilizations, multiply mean by 13 eggs/min.
[b] Males seen once are excluded.

early in the season, the opportunity for sexual selection on the fertilization efficiency of mated males increased dramatically over the reproductive season in 1982, even though the total *relative* opportunity for sexual selection (I_s/I_m) actually decreased late in the season (table 3.5). Because the average reproductive life span was roughly only a quarter of the breeding span of the population, this meant that males emerging later in the season faced a greater selective pressure to guard mates than did males earlier in the season.

3.3 Phenotype-Independent Factors Affecting Reproductive Success

Female Choice

After one oviposition bout, about half (48%, n = 172) of the females taken in tandem refused to mate, whereas only 3% (n = 7/215) refused to mate before having oviposited that day (Fincke 1985). Tandem females refusing to mate a second time had been underwater longer ($\bar{x}$ = 19.5 ± 3.0 min, n = 37) than those that accepted a second mate ($\bar{x}$ = 10.3 ± 2.1, n = 31). Moreover, the first males mating with the seventy-one females that mated with two males on the same day did not differ in size ($\bar{x}$ wing length = 17.6 + 0.06 mm) from the second mates ($\bar{x}$ = 17.4 ± 0.06, *t*-test n.s.). Nor was mating assortative by body size (r = .04, n = 298 pairs; Fincke

1982). These results suggest that a female's "decision" to mate was contingent upon the number of eggs she carried. Thus the variance in male fertilization and mating efficiency resulted from competition among males for fertilizations and mates, or simply from chance encounters with females, rather than from active or passive female choice based on male phenotype.

Demographic and Seasonal Effects

For *E. hageni*, differential investment in reproduction by the two sexes was reflected in sex differences in the time required to reach sexual maturity and in the intervals between periods of sexual receptivity. These differences resulted in a male-biased operational sex ratio that generated the potential opportunity for sexual selection. Females required about three days longer to mature ($\bar{x}$ = 8.8 ± 1.1 days, n = 21) than did males (x = 6.1 ± 0.6 days, n = 29), and consequently more male tenerals (34%, n = 85) than female immatures (20%, n = 105) marked at emergence returned to the pond as adults. Although this male bias in returning individuals could have resulted from disproportionate dispersal of imma-

Table 3.5 Changes in the Daily Opportunity for Sexual Selection on Male *E. hageni* (I_s) at the Pond, Summed over Days of the Reproductive Season

					Percentage of I_m		
	n	Mean	Variance	I	I_e	I_{fe}	I_s
Early season (days 1–10) (R_1 = 1.1)					60	2	62
Fertilizations[a]	252	8.9	392.04	4.95			
Total mates (I_e)	256	0.31	0.276	2.96			
Fertilizations/mating (I_{fm})	66	30.0	408.04	0.45			
$\bar{x}$ mates/clutch (females)	70	1.19	0.084	0.06			
Eggs/clutch (I_o)	51	35.0	637.23	0.52			
Midseason (days 11–19) (R_1 = 1.4)					59	16	75
Fertilizations	138	10.7	334.89	2.93			
Total mates	138	0.47	0.381	1.72			
Fertilizations/mating	107	23.7	204.49	0.36			
$\bar{x}$ mates/clutch	65	1.40	0.112	0.06			
Eggs/clutch	54	28.9	490.49	0.59			
Late season (days 20–28) (R_1 = 1.0)					20	35	55
Fertilizations	61	22.7	384.16	0.75			
Total mates	56	1.07	0.176	0.15			
Fertilizations/mating	52	22.2	201.64	0.41			
$\bar{x}$ mates/clutch	49	1.29	0.256	0.15			
Eggs/clutch	40	28.0	524.17	0.67			

Note: Eggs/clutch were calculated from marked and unmarked females mating with the males for which data are given. Fertilizations/mating were calculated for mated males only. $I_{fe} = I_{fm} - R_1 I_o$ (see text).

[a] Fertilization and egg data are given as oviposition duration in minutes.

ture females, daily mortality (of 0.2) during the slightly longer maturation period of females could more parsimoniously account for the difference. The initial male bias in the mature adult population was counterbalanced by the slightly longer life span of females (tables 3.1, 3.4). Average daily survivorship of mature adults after the first day was constant and was similar in both sexes (1982: $\bar{x}$ = 0.85 ± 0.03, n = 89 females: $\bar{x}$ = 0.80 ± 0.03, n = 156 males; Fincke 1986b). Assuming that the unmarked males (24% of tandems) and females (22% of tandems) seen mating with marked individuals, and the twenty-two unmarked pairs that were excluded from the analysis, were resighted as often as marked individuals ($\bar{x}$ = 2.6 and 1.8 times for mated males and females, respectively), I estimated an absolute sex ratio for 1982 of 1.3:1 (179 males:133 females) or, assuming unmarked females were unique, 179:168.

Females visited the pond on average once every six days. Males visited the pond nearly twice as frequently (table 3.4), resulting in an average operational sex ratios of 3:1 (1980) and 2.6:1 (1982). For both years, there was an inverse relationship between the operational sex ratio and the proportion of males present at the breeding site that mated (r = .87, n = 18 days, 1980). Although the densities of both males and total individuals of both sexes were low at the beginning and end of the season, density had no significant overall effect on the proportion of males mating (Fincke 1982). This was because the activity of both sexes was higher on warm, sunny days, and the numbers of males and females visiting the pond on a given day were correlated (r = .83, n = 18 days).

A seasonal shift in female oviposition behavior accounted for the increase in the relative opportunity for sexual selection on mated males from early to late season (e.g., table 3.5). During mid- and late season, preferred stems became full of eggs, and females had to walk underwater to others or use the macrophytic algae *Chara*, on which they had to move around more to lay the same number of eggs. Consequently the submerged but buoyant females were more likely to lose their hold, and twice as many "popped up" prematurely as did earlier in the season. Lone males responded to this increase in mating opportunities by spending more time at the water in search of resurfacing females in mid- and late season (Fincke 1985).

"Costs" of Reproduction

For males, the act of searching for a potential mate at the pond, regardless of success, appeared to be equally "costly" in terms of survivorship. Both guarding and lone males at the water were preyed upon by spiders, gomphid dragonflies, and frogs, *Rana catesbiena*. In the oviposition areas, the density of the latter was much higher than that of a similar predator, *R. pipiens*, in the adjacent field (pers. obs.). Mated males spent more time away from the pond than did males that visited the breeding site but did not mate (Fincke 1982), suggesting that copulation, guarding, and flying in tandem were energetically costly, that mating males needed time to replenish reproductive materials, or both. Because neither fertilization nor mating efficiency was significantly correlated with life span (table 3.2), I conclude that for males there was no evolutionary trade-off betweeen reproductive effort and survivorship.

The consequence of egg laying to female survivorship, on the other hand, was considerable and could be estimated directly (Fincke 1986b). About a fifth of the females floated in the water after resurfacing, where they risked (.07) drowning or being eaten by an aquatic predator if they were not pulled from the water either by mates or by lone males with whom they "exchanged" fertilizations for vigilance service. Of ovipositing females whose fate was known, 2.6% (n = 232) died as the result of submerging. Few females realized their reproductive potential: the average mature female lived long enough to mature only one clutch, and yet females kept in an insectary produced up to a thousand eggs, enough for two or three clutches (pers. obs.).

3.4 Individual Differences versus Random Factors

Although life span was a good predictor of reproductive success for males and females (fig. 3.3), longevity appeared to be influenced by random factors. Constant survivorship (see above) indicated age-independent mortality. Furthermore, the maximum life span in the field (twenty-five days) was nearly five times the average observed. Mating rates when young (1 to 2 days after sexual maturity) did not differ significantly between short-lived (less than four days) and long-lived (more than seven days) males, suggesting that males that lived longer were not more efficient in mating than were short-lived males. Likewise, longer-lived females were not more fertile per clutch (though they did produce more clutches) than were shorter-lived females (table 3.2).

For males seen more than one day, the mating distributions of males whose visits to the pond were equal, or whose life spans were equal, were statistically indistinguishable from the Poisson distributions expected if

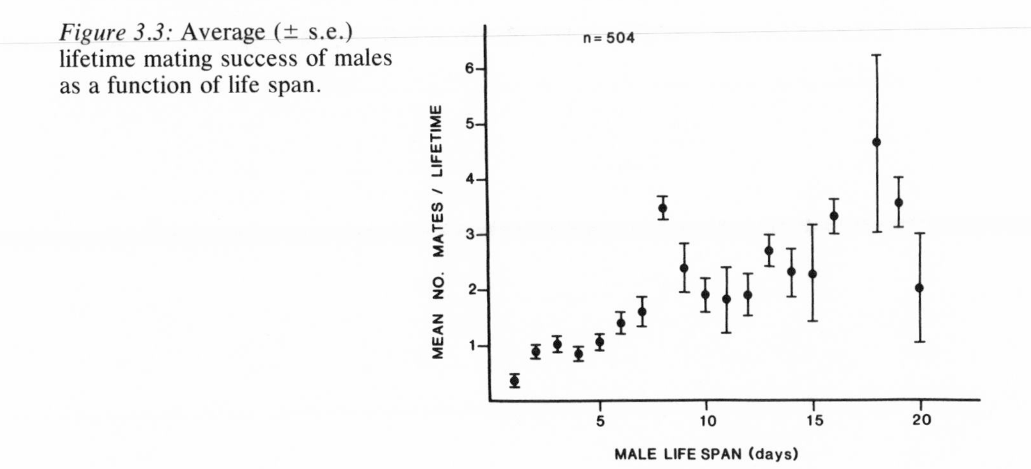

Figure 3.3: Average (± s.e.) lifetime mating success of males as a function of life span.

matings were independent events (Fincke 1986a). This suggested that unless phenotypic differences were found between successful and unsuccessful males, the observed variation in mating success resulted from random encounters with females. Males mating many times did not fertilize more eggs per mating than those mating a few times (table 3.2). However, had I artificially assigned a value of zero for eggs/mating for unmated males, thereby making the correlation betweeen fertilizations/mating and mates/visit for all males significant ($r = .55$, $n = 156$, $p < .05$), I would have incorrectly concluded from this result that certain individuals were consistently more successful than others (e.g., McVey, this volume, Chap. 4). In fact, mated males as a group were phenotypically different from unmated males, but as shown below, the difference was a subtle one. In addition, the fertilization efficiency of mated males depended upon their pre- and postcopulatory behaviors (see below).

Date of Emergence

As seen in table 3.2, the date a mature individual first returned to the pond, a function of the day on which it emerged as an adult (which in turn was a consequence of its span as a larva), was not significantly correlated with fertilizations/mating, mating efficiency, or LRS of males, or with

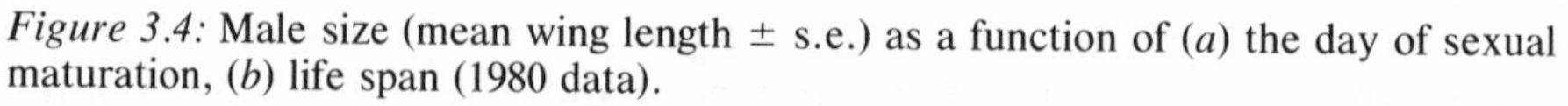

Figure 3.4: Male size (mean wing length ± s.e.) as a function of (*a*) the day of sexual maturation, (*b*) life span (1980 data).

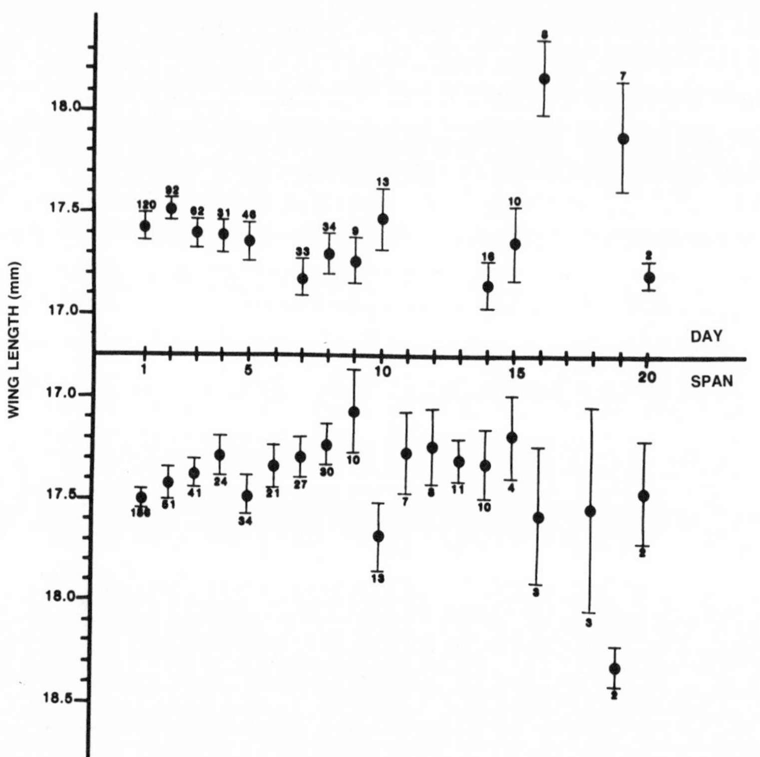

eggs/clutch of females. For both sexes there was a slight but significant negative correlation between the day of first return and life span (table 3.2). Females that emerged late had less time to mature successive egg clutches, as shown by the negative correlation between day of first return and LRS of females. The corresponding correlation for males was not significant, because males returned to the pond more frequently than did females and returned more frequently when younger than when older. Although the correlation between day of first return and wing length for the entire sample was not significant (table 3.2), males maturing early (days 1–7) in the season (79% of total males) showed a slight but steady decrease in wing length ($r = .90$, $n = 6$ days; fig. 3.4*a*). Multiple regression analysis, controlling for life span, revealed a slight but significant negative effect of day first seen on female size ($\beta \pm$ s.d.$= -0.0009 \pm 0.0004$, $p <$ .05). The effect was significant for males only if males seen once were excluded from the analysis ($\beta \pm$ s.d. $= -0.001 \pm 0.0005$, $p < .05$).

The day of first return plotted against size gave a significant fit with a positive quadratic (for squared component, $\beta \pm$ s.d $= 1.00 \pm 0.40$, $n =$ 487, $p < .05$), indicating that intermediate-sized males were those with the earliest day of first return. Nearly three-fourths (72%) of the males were first seen within the first five days of the study in 1980 (fig. 3.4*a*), suggesting that emergence was temporally clumped. Individuals emerging any earlier were likely to face large and unpredictable fluctuations in daily temperatures characteristic of late spring in northern Michigan (pers. obs.).

Size Effects

In neither sex was body size significantly correlated with LMS, mating efficiency, number of visits to the breeding site, or survivorship (table 3.2, fig. 3.4*b*). Body size of males was not significantly correlated with any of the variables above during early season (days 1–4), or midseason (days 5–13), but it was negatively correlated with life span ($r =$ .47, $n = 56$) and consequently with total mates ($r = -.36$) (but not with mating efficiency) during late season (days 14–25, 1980). The average body size of adults first seen in midseason was smaller ($\bar{x} = 17.3 \pm 0.05$, $n = 135$) than that in either early ($\bar{x} = 17.4 \pm 0.03$, $n = 305$) or late season ($\bar{x}= 17.5$ $\pm$ 0.1, $n = 56$; t-test, $p < .05$). Controlling for day of first return, I found no evidence for indirect selection (see Lande and Arnold 1983), on body size with respect to survivorship (standardized partial $\beta \pm$ s.d $= -0.06 \pm$ 0.045, $n = 487$ males, n.s.; std. $\beta = -0.03 \pm 0.06$, $n = 295$ females, n.s.), LMS of males (std. $\beta = -0.001 \pm 0.045$), or clutches for females (std. $\beta =$ -0.082 ± 0.06, n.s.).

For males, selection on body size was stabilizing with respect to lifetime mating success. Mated males were significantly less variable in size than were unmated males (F-test, $p < .02$; Fincke 1982). Males of intermediate size obtained more matings and visited the pond more often than those at the extremes (fig. 3.5*a*,*b*). There was no direct evidence for

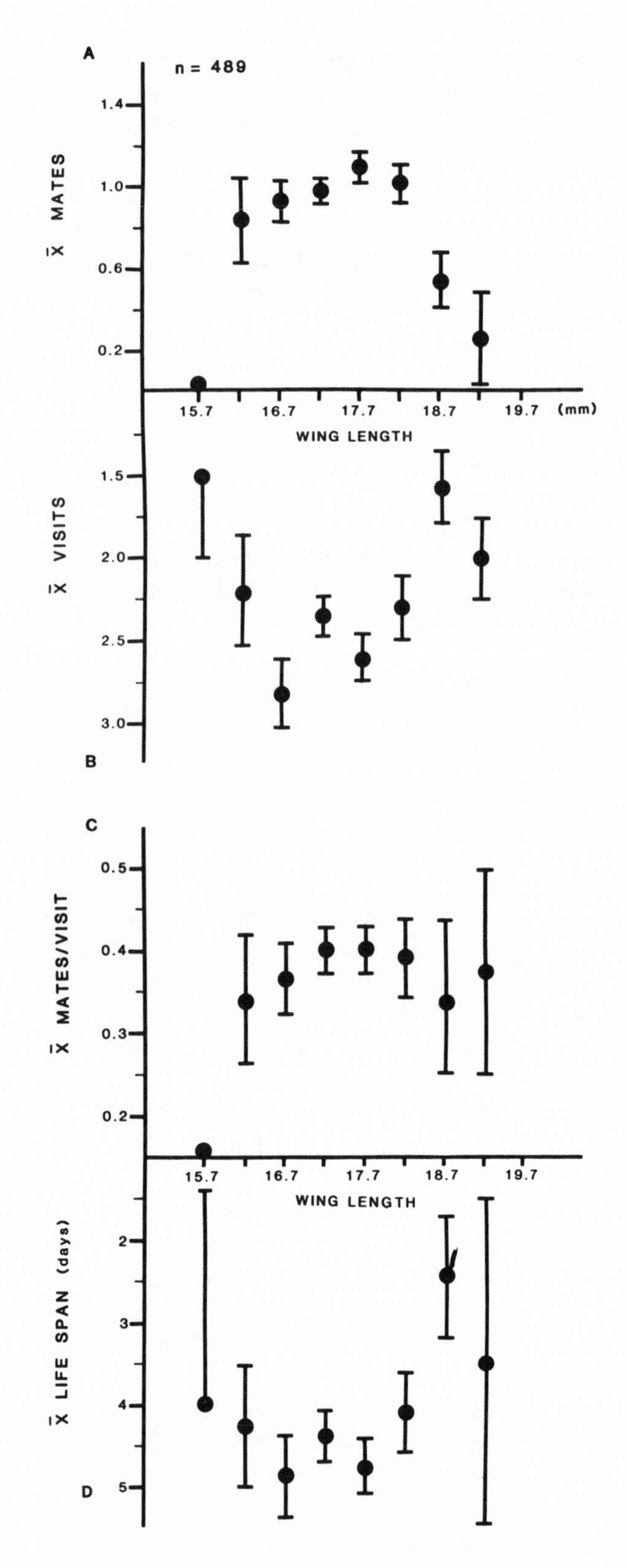

Figure 3.5: Means + s.e. of fitness correlates as a function of male size. Tests for stabilizing selection for size on males with respect to (*A*) mating success ($\beta \pm$ s.d $= -76.0 \pm 29.2$, $t = -2.61$, $p < .05$); (*B*) total visits, $\beta \pm$ s.d. $= -110.5 \pm 44.1$, $t = -2.51$, $p < .05$); (*C*) mating efficiency, $\beta \pm$ s.d. $= -17.2 \pm 10.7$, $t = -1.6$, n.s.; (*D*) survivorship, $\beta \pm$ s.d. $= -173.5 \pm 114.6$, $t = -1.5$, n.s. Betas are partial regression coefficients for the squared component of ln wing.

stabilizing selection on either mating efficiency or life span (fig. 3.5*c*,*d*). The average male size (wing length = 17.4 ± 0.03 mm) appeared to result from the interaction of two selective forces. As indicated by the peaks of the curves in *a* versus *b* of figure 3.5, selection on total matings (a function of mating efficiency and life span) favored slightly larger males than selection on total visits (a function of life span and time between visits). These results suggested that small size was more advantageous for survivorship than for mating efficiency.

No evidence for stabilizing selection on size with respect to total clutches or life span was found for females (partial β ± s.d. for squared predictor = 0.07 ± 0.06; −0.12 ± 0.29, n = 298 clutches and life span respectively). Unfortunately I did not measure body size in 1982 when I measured total eggs laid, but for 1980 data the correlation between female body size and the number of clutches was not significant (table 3.2). Moreover, few females lived long enough to realize their full reproductive potential. In laboratory-reared insects, clutch size has been shown to be positively correlated with body size (Hegmann and Dingle 1982), and I assume this is true for *E. hageni* females as well, though females varied only slightly in body size (fig. 3.6). I suspect that much of the variation in

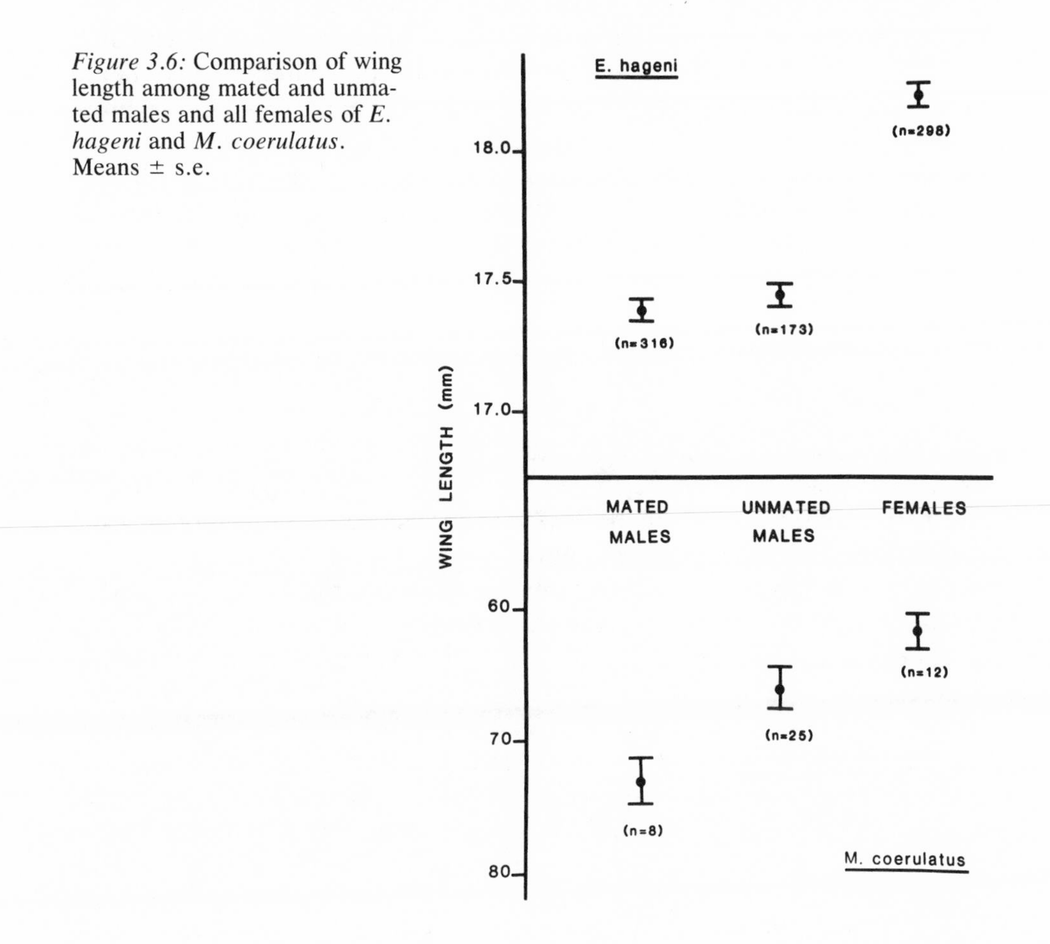

Figure 3.6: Comparison of wing length among mated and unmated males and all females of *E. hageni* and *M. coerulatus*. Means ± s.e.

my estimates of the number of eggs laid/visit resulted from differential success in oviposition rather than from large differences in clutch size.

Age Effects

Aging negatively affected the mating success of males by increasing the intervals between their visits to the breeding site and thereby decreasing the average mates/day alive of long-lived males (mates/day for males seen more than three times was negatively correlated with life span, $r = -.19$, $n = 105$). Mates/visit increased from 0.45 for one-day-old mature males to a high of 0.78 for three-day-old males, then leveled out to an average of 0.42/visit for males older than three days (Fincke 1982). The initial increase may result from experience, but if so, it was quickly countered by aging effects on mating rates after day 3.

Color Morph

There was no difference in the average number of matings, life span, or size between heteromorphic green females (74% of total) and andromorphic blue females (26%) (Fincke 1982). Although individual males mating more than once showed no preference for mates of one morph over the other (Fincke 1982), when experimentally presented with tethered females of each morph, males took heteromorphic females in tandem more quickly than they did andromorphic females (Fincke, pers. obs.; D. Paulson, pers. comm.). Because tandem pairs were harassed regardless of the color morph of the female (Fincke 1986b), my results do not support the conclusion of Robertson (1985) that by "mimicking" males, andromorphic females avoid male harassment but mate less frequently than do heteromorphs.

Behavioral Effects

The proportion of his mate's clutch that a male fertilized (fertilization efficiency; Fincke 1986a) depended on his guarding duration, his use of alternative mating tactics, and his ability to displace sperm (though I did not measure individual variation in the latter). By failing to guard, a male risked losing potential fertilizations to intercepting males via sperm displacement (38% chance), or owing to death of his mate before she laid all her eggs (3% chance). However, a male had only a 3% chance of finding another receptive female on a given day (Fincke 1986b). Few males (10.5%, $n = 194$) abandoned submerged mates before 30 min had passed. Contrary to the popular assumption that males should always be ready to mate, in a field experiment, guarding males did not mate with additional females presented to them within 30 min of submergence by their mates. This latency to remating, which corresponded closely to the average oviposition duration/clutch of females, was characteristic of guarding males and not exhibited by males mating multiply in an insectary (Fincke

1986b). Not surprisingly, guarding duration was positively correlated with the number of eggs a male fertilized ($r = .34$, $n = 163$ first mates; $r = .61$ $n = 40$ second mates; Fincke 1986b)

A male that found a female on the bank of the pond before she had oviposited that day (76% of all matings) could potentially fertilize her entire clutch, whereas if he caught her at the oviposition site, after she resurfaced from one or more bouts of egg laying, he could at best fertilize only a partial clutch. (Fewer than 1% of the females arrived at oviposition sites unaccompanied by a male.) In 1982, 32 males (28%) mated two or more times in the same day, of which 63% used both tactics. For these males, the average number of mates ($\bar{x} = 3.4 \pm 0.3$) and the average oviposition duration by mates ($\bar{x} = 92 \pm 8.4$ min), were about one standard deviation above the mean values of all mated males ($\bar{x} = 2.3 \pm 0.15$ mates; $\bar{x} = 58.0 \pm 5.0$ min). Although the use of alternative tactics was independent of male size, age, or success or previous tactic used (Fincke 1985), the males above could have differed phenotypically from less successful males in ways I did not measure.

3.5 Comparison of *E. hageni* with Other Odonates

A comparison of *E. hageni* with other odonates tentatively supports the hypothesis that as males are better able to control access to oviposition sites required by females, the potential for sexual selection on males increases. I argue that in odonates this potential sexual selection has been realized in the form of male guarding behavior, sexual differences in wing and body coloration that are used in displays to signal an individual's sex, and male-biased dimorphism in body size.

In contrast to *E. hageni*, in which males are the smaller sex, males of the Neotropical giant damselfly *Megaloprepus coerulatus* are significantly larger than females and are distinguishable from them by differences in wing coloration that they use for sexual recognition. Males defend territories for up to three months, around water-filled tree holes where females come to mate and oviposit (Fincke 1984a). Selection on mating success favors large males (fig. 3.6), because larger males are less often displaced by intruders than are smaller males and are passively chosen as mates by females, who mate only with territory holders (98% of the matings; Fincke, n.d.). Estimates of heritability of body size were very low (<0.22; Fincke, n.d.), as might be expected if body size has been under strong selection (Fisher 1930). Although I_e/I_t could not be accurately estimated, the opportunity for selection on male mating success (I_t) over three months of a seven month season was 5.1, nearly five times the analogous measure for both *E. hageni* and *Erythemis simplicicollis* (I_t = 1.07), a less sexually dimorphic territorial dragonfly (McVey, this volume, chap. 4). *M. coerulatus* is long-lived (maximum life span of marked individuals = seven months), and reproductive adults remain dispersed throughout the forest. Consequently, individuals have the opportunity to establish their superiority over competitors at tree-hole sites, making random encounters with females less likely to contribute to variation in reproductive success than in *E. hageni*.

Like *Megaloprepus*, some territorial males of *E. simplicicollis* were consistently more successful in mating (McVey, this volume, chap 4), although specific phenotypic correlates of mating efficiency were not obvious and have yet to be identified. Given McVey's method of estimating I_s after Wade and Arnold (1980) ($I_s = I_m - R\ I_f$), her most appropriate ratio to use for comparisons is that in which I_m was calculated from average eggs fertilized per day alive (eggs/day). This measure of I_s/I_m was 65%–79% for *Erythemis* and only 47% for *Enallagma*. Although I_m/I_f (where I_m was calculated from the total eggs fertilized per male) was lower in *Erythemis* (0.9) than in *Enallagma* (2.9), this ratio is not appropriate for comparing the relative importance of sexual selection between species because it confounds variation in survivorship, a component of natural selection, with variation in mating efficiency, a component of sexual selection.

Because the number of fertilization opportunities for males is ultimately determined by female mating patterns, among nonterritorial species the relative opportunity for selection on mating efficiency should increase as females partition their egg clutches among a greater number of males (Fincke 1986a). Nonterritorial coenagrionids exhibit a similar degree of sexual dimorphism in body coloration and size (females are larger than males) but differ in postcopulatory behavior. The potential for sexual selection was lower for species in which males remained in tandem with ovipositing mates (contact guarding) than in *E. hageni*, in which males defended the perches on which their mates submerged (noncontact guarding). For *C. puella*, a contact guarder, Banks and Thompson (1985) calculated that variation in mating efficiency (matings/day alive) contributed to only 22% of total variation in LMS (using $1 - r^2$, where r is the correlation coefficient between life span and LMS). This is about a third of the comparable value for *E. hageni* ($1 - .35 = .65$). Relative to noncontact guarding, contact guarding limits mating opportunities of males because a second male is less likely to intercept a female and fertilize part of her clutch. It is thus not surprising that Hafernik and Garrison (1986) found that the standardized variance in male LMS of *Ischnura gemina*, in which females mated only once per clutch, was lower than that for either *C. puella* or *E. hageni*. I predict that the lowest potential for sexual selection among odonates will be found in species such as *Ischnura verticalis*, in which females mate only once per lifetime and oviposit in a broad range of substrates, thereby precluding male control over their reproduction. In *I. verticalis* sexual recognition is primarily behavioral, males do not associate with females after copulation, and mating encounters are opportunistic, away from oviposition sites (Fincke 1987).

3.6 Conclusions and Summary

Although components of LRS for males and females varied relatively little between the two apparently "normal" study years, the opportunity for sexual selection on mated males increased considerably within a reproductive season. Differences among males in the rate of encounters with females and in life span were the two main sources of variation in male LRS. Although I found no phenotypic predictor of mating efficiency or survivorship per se, males of intermediate size had the highest number of mates and visits per lifetime. The potential opportunity for selection on mating success was realized as stabilizing selection in this population.

Its relatively small size, coupled with high density of males around concentrated oviposition areas at the pond, made mate searching a more efficient way to obtain matings for *E. hageni* than defense of an oviposition area. Considerable synchrony in female receptivity, coupled with benefits to submerging females of mutiply mating, made noncontact mate guarding advantageous to mated males and simultaneously favored unmated males that switched to an alternative mate-finding tactic. Because mating opportunities were very low (28% and 41% of the males failed to mate in 1982 and 1980, respectively), a male's ability to maximize the number of fertilizations/female encountered was more advantageous than his ability to obtain multiple matings/visit (though the most successful males were those that did both). Although the heritability of such traits as sperm removal, guarding propensity, and the use of alternative mate-finding tactics may prove difficult to measure, I have shown that in *E. hageni* these traits function in the context of sexual selection, even though the current population may not be able to respond to selection on these traits. A comparison of *E. hageni* with odonates for which comparable data were available indicated that territorial species exhibited greater sexual dimorphism related to sexual displays and correspondingly higher potentials for sexual selection on males than did nonterritorial species.

3.8 Acknowledgments

I am grateful to T. Clutton-Brock and two anonymous reviewers for critical comments on the manuscript and to A. Grafen for advice on data analysis. I thank R. Drapcho and C. Sagrillo for excellent field assistance. The University of Michigan Biological Station provided both financial aid and logistical support of the fieldwork. Additional funding was provided by a Teaching-Research Fellowship and a Neuro-Behavioral Traineeship from the University of Iowa and by a NATO Postdoctoral Fellowship at the University of Oxford.

4 The Opportunity for Sexual Selection in a Territorial Dragonfly, *Erythemis simplicicollis*

Margaret E. McVey

SEXUAL SELECTION IS TRADITIONALLY viewed as selection arising from variance in mating success, whereas natural selection stems from variance in other components of fitness (Darwin 1859, 1871; Ghiselin 1974; Arnold and Wade 1984a; Wade 1979; Lande 1980; Wade and Arnold 1980; Arnold and Houck 1982; Arnold

Figure 4.1: Adult male *Erythemis simplicicollis* eighteen days after emergence. Most of the thorax and abdomen is a light powdery blue. Females and prereproductive adult males, on the other hand, have a green thorax and green-and-black-striped abdomen. Mean hind wing length of males = 30.9 mm (± 1.45 s.d., n = 669), and that of females = 31.7 mm (± 1.63 s.d., n = 116).

1983a). Selection and the evolutionary response to selection, however, are different phenomena (Fisher 1930; Haldane 1954; Arnold and Wade 1984a), the latter requiring genetic contributions to phenotypes (Lande 1980; Arnold and Wade 1984a,b). Nonetheless, the opportunity for selection can be measured as the variance in relative fitness (Crow 1958).

In promiscuous species, seasonal variance in reproductive success (RS) of males tends to exceed that of females. Because males can cycle more quickly between matings than females can between clutches, in a panmictic population without sexual selection, variance in male RS could exceed variance in female RS because of chance alone. Consequently, measuring the "opportunity" for sexual selection from relative variances (Wade and Arnold 1980) is insufficient. One must also demonstrate that individual attributes can influence relative RS. If there are no phenotypic characteristics that influence reproductive success, the "opportunity" for sexual selection is simply the opportunity for genetic drift.

This chapter investigates the opportunities for sexual and natural selection in a dragonfly, *Erythemis simplicicollis* (Say), a species in which males practice resource-defense polygyny (Emlen and Oring 1977). Females lay their eggs on algal mats that are present only over portions of some ponds or lakes. Most males of the species (fig. 4.1) are territorial, defending sections of algae where they attempt to capture approaching females. Some males, called satellites, perch on or near the territories of other males and attempt to intercept females. Females begin to lay eggs within a minute of copulation, and all males attempt to guard their mates by hovering within 10–20 cm. Takeovers do occur, and some females mate with up to seven males in one day.

This study covered a single population in Florida where adults emerge steadily from April through October. The resulting age structure of the population is similar to that of long-lived avian populations if one equates dragonfly days postemergence to avian years. Following emergence, the young adults forage away from the ponds for about two weeks before attempting to reproduce (McVey 1985). The breeding life span of individuals averaged between eight and ten days in two years of study, the maximum being forty-five for both males and females (McVey 1981).

The goal of this report is to estimate the opportunities for sexual selection on a daily and lifetime basis and to examine the covariance between traits contributing to reproductive fitness for males facing two different competitive situations. In addition, the contribution of male attributes to reproductive success is examined experimentally.

4.1 Methods

The lifetime RS data reported here are from males marked between 12 April and 8 July 1979 at or around two ponds in the Austin Cary Memorial Forest, eight miles northeast of Gainesville, Florida, USA. These were in a series of nine ponds and one lake isolated from other suitable bodies of water more than 1 km of pine forest in all directions. Until about 28 May, only pond 9 supported the algal mats required by the species (situation 1). The algae was artificially restricted to a 2-m strip around the perimeter of the pond. Between 27 and 29 May, pond 1 developed an edging of surfaced *Hydrilla verticillata* acceptable to females (situation 2). Finally, in late July ponds 4 and 5 developed surface vegetation and were used for territory removal experiments (McVey 1981), and in late August the lake developed some algae.

Males were individually marked with small dots of enamel on two wings on their first day at the pond. All were aged upon capture using thoracic color patterns (McVey 1985). For a given color pattern, the standard deviation of postemergence age was ±1.2 days. Females were also marked but could not be aged.

In situation 1, with all males confined to the single pond 9, females outnumbered males 1.9 to 1. In the second situation, with two ponds available, average male densities at pond 1 were less than those at pond 9 (0.15 and 0.26 males/m shoreline respectively) and at pond 1 males outnumbered females 2.6 to 1.

Males were selected for inclusion in the two sequential competitive situations as follows. After 28 May, males could move between ponds 1 and 9, 60 m apart. The date of first breeding was checked for all males that switched ponds at some point in their lives. The earliest such arrival date was considered the cutoff for inclusion of newly arriving males in situation 1 (only pond 9 available). All males arriving after that date (22 May) had an opportunity to switch ponds if they lived more than six days. The same procedure (searching for the earliest arrival date for males which switched to ponds 4 or 5) was applied to determine the cutoff date for males included in situation 2.

Twelve observers (including myself) were responsible for recording all copulations, durations of ovipositions, escalated agonistic behaviors, and territorial boundaries at half-hour intervals and noting perching locations for all animals present on all occupied ponds between 0900 and 1800 h daily. I checked observer reliability in reporting copulations and the duration of ovipositions by comparing their records with simultaneous records I took on the same pond sections (41 h testing, 3–5 h per assistant). At pond 9, assistants reported 99% of copulations (n = 78). Of the total seconds of oviposition performed by females following mating with the male hovering nearby and by solitary females, 95% and 73% were recorded respectively (out of 2,760 sec for 51 mated and 726 sec for 54 unmated females). At pond 1, observers recorded 65% and 39% respectively (out of 1,810 sec for 35 mated and 200 sec for 16 unmated females). Because unmated females accounted for only 20% of oviposition, the reduced observer reliability on this measure is relatively unimportant. These values and others determined for ponds 4 and 5 were used to correct the data before analysis.

We checked the unobserved ponds twice daily between 1100 and 1500 h for nonbreeding males and removed any scraps of floating vegetation to prevent females from ovipositing in these areas. Although we should have complete lifetime records for males breeding in the study site, two lines of evidence suggest that some young males probably did emigrate after one day at the pond(s). First, the daily probability of mortality was a constant ($p = .11$) for all days following initial capture except the first, when it was markedly higher (.20 to .55 depending on the month; McVey 1981). Second, in late August we found three marked males, all of which had disappeared after one day from our site, defending recently formed algae on the lake 300 m distant. The probable number of emigrants equals the difference between the *y*-intercept of the regression of the log number "surviving" (excluding the first day) and the initial population size, or 2.7 and 5.7 males in situations 1 and 2, respectively.

With more than 1 km of woods surrounding our ponds, immigrants that had successfully bred elsewhere before arrival should have been very rare. In Ohio, Currie (1961) found limited dispersal of males during the prereproductive period and even lower dispersal of breeding males. Only 11.7% (28/239) of prereproductive and 2.5% (3/118) of reproductive males seen more than once crossed a 0.1-km-wide stretch of woods separating two ponds. For more details on methods, see McVey (1981).

4.2 General Background and Measuring Reproductive Success

More males and females attended the ponds in the hours around solar noon than earlier or later (fig. 4.2). Females remained at the pond only for as long as required to lay the clutch of the day, ranging from a few seconds to a maximum of 15 min for females that mated several males. Most male residents arrived by 0930 h and stayed until 1600 h, with few or no breaks for foraging.

Some males (56%) established territories and defended them throughout an entire day. Other males were permanent satellites (12%), and the rest (32%) would switch between resident and satellite tactics. Switches occurred if an early resident was evicted by a newly arriving male or if a

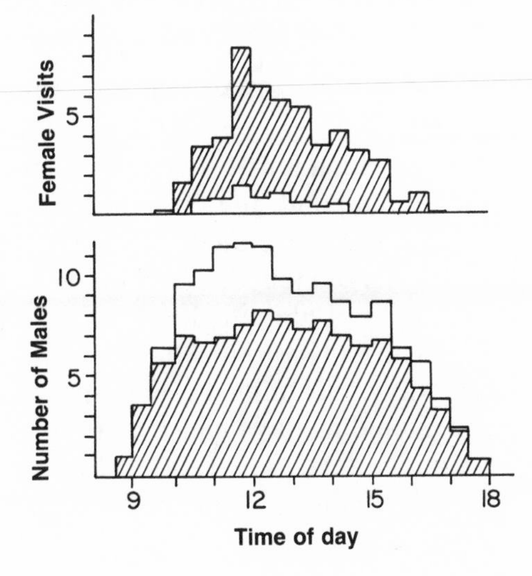

Figure 4.2: Top: Average number of female visits resulting in copulation per half-hour interval for seven days between 28 April and 10 June at pond 9. Hatched and open sections represent resident and satellite copulations, respectively. *Bottom*: Average number of males present as residents (*hatched*) and satellites (*open*) per half-hour interval, same samples. S.e. of mean ranged from 0.54 males (1100 to 1500 h) to 1.0 (earlier and later).

satellite successfully drove a resident from all or part of his territory. The defeated male usually remained at the pond as a satellite. Males in either role captured and mated females. Males were present for some period on 99.4% of the nonrain days.

I report reproductive success for females as eggs laid and for males as eggs fertilized. Differences in hatching or nymphal success within and between ponds were not known. For females, I estimated the number of eggs laid from the duration of oviposition, an estimate of female abdominal temperature, and the relation between temperature and egg flow rate as described by McVey (1981, 1984, and n.d.c). For the males, reproductive success had two components: immediate fertilizations of the eggs that a female laid within minutes after copulation, before remating or leaving the pond; and fertilizations of eggs in the female's next clutch the following day.

On average, females laid 80% of their eggs immediately after mating. Using the sterile-male technique, McVey and Smittle (1984) demonstrated that the last male to mate fertilizes virtually 100% of the eggs a female lays within 5 min of copulation, which is the longest uninterrupted period of time required to lay the clutch for the day. Females laid the remaining 20% of their eggs on the following day before remating. Sperm precedence by the female's last mate for these eggs was 50% or higher in nine out of ten matings. I estimated the number of eggs a male fertilized the day following a copulation using the proportion of females for which a given male was probably the last mate; the number of eggs laid the next day by unmated females; and a sperm precedence range of 50%–100%.

Rain days are eliminated from all analyses (unless stated otherwise) because both males and females generally remained quiescent awaiting better weather before visiting the ponds (McVey 1981). In situation 1, rain/total male days = 50/491 = 10%; in situation 2, rain/total male days = 20/1010 = 2%.

4.3 Variation in Female Reproductive Success

Most females deposited a clutch of eggs ranging from 50 to 2,200 daily except on rainy days (McVey 1981; McVey and Smittle 1984). Marked females were seen at pond 9 on 76% of the nonrain days. On the 24% of days in which females were missing, however, it is not clear whether they visited the pond; the ratio of days missing to total longevity for individual females was correlated with the conspicuousness of female wing markings. Thus, marked females might have been present on most of their "missing" days but not identified.

The daily variance in female reproduction was therefore calculated in two ways, including days missing as if the female did not produce any eggs and excluding days missing, assuming that she had oviposited but had not been identified. The ratio of the variance to the mean squared for female daily fecundity (I_f) was between .440 and .833 (table 4.1).

4.4 Variation in Male Daily Fertilization Success

Competitive Situation 1, Residents

Male mating success depended in part on territory location. At pond 9 the algal ring was relatively homogeneous in appearance, but twice as many females visited some areas (per meter of shoreline) as visited others (McVey 1981). In addition, male mating success increased with increasing territory size (McVey, n.d.b). During the hours around solar noon, a male gained on average 0.52 matings/h per additional meter of territory defended over territories ranging from 2 to 11 m (partial correlation coefficient holding density constant = .39, d.f. = 2,53, $p < .002$).

Table 4.1 Opportunity for Sexual Selection and Variance in Components of Lifetime Reproductive Success for *Erythemis simplicicollis*

	Mean	σ^2	I	I_m/I_f	I_s	I_s/I_m
Daily fecundity[a]						
Situation 1 (pond 1)						
Females including days missing	650	35.2	0.833			
				2.55	1.68	.79
Males	1420	428.5	2.125			
				4.83	1.89	.89
Females excluding days missing	860	32.4	0.440			
Situation 2 (ponds 1 and 9)[b]				3.47	2.23	.77
Males	990	282.2	2.88			
				6.55	2.54	.88
Pond 1 only[b]				7.47	4.04	.65
Males	370	87.8	6.22			
				14.10	5.06	.81
Longevity (days)						
Males	8.76	58.3	0.760			
Situation 1				0.83		
Females	6.97	44.4	0.914			
Situation 2				0.81		
Males	9.61	68.1	0.736			
Total lifetime RS (× 100)						
Males	130.5	21,910	1.29			
Situation 1				0.936	.53	.42
Females	39.7	2,126	1.35			
Situation 2				1.41	.75	.39
Males	103.4	20,278	1.90			

Note: Female data represent partial lifetimes. Female survivorship equalled male survivorship (McVey 1981), hence the average female longevity should be 1.8 to 2.6 days longer than listed, and the mean lifetime RS should be adjusted upward accordingly. Given a constant probability of surviving over all ages, however, the ratio of the variance to the mean squared (I) remains unaffected by the age at which females were first captured.

[a] Mean daily fecundity for females and males is computed as the mean of all female-days or all male-days, respectively.

[b] The same forty-seven females are used for all comparisons, assuming that their fecundity was unaffected by which pond they would use. $I_s = I_m - R \times I_f$ (Wade and Arnold 1980), where I_s is the potential intensity of sexual selection and I_m and I_f are the total selection potentials on males and females, respectively. See table 4.2 for sex ratios (R = males/female) and sample sizes (female-days and male-days) of the groups listed.

Not only did males gain more matings with larger territories, they also gained 9.3 fertilizations per mating per additional meter of algae up to territories of 9 m, after which fertilizations per mating remained constant (partial correlation coefficient holding density constant = .32, d.f. = 2,159, $p < .001$). Females laid more eggs on larger territories because they could wander longer before encountering interference from neighboring residents. On territories larger than 9 m in length, females could usually lay their entire clutches before being encountering neighboring residents, hence no further increase in the duration of oviposition with increasing territory size was possible.

Competitive Situation 1, Satellites

Satellites on average mated as often as residents holding territories 3.5 m in length—0.68 copulations per hour (in the four hours around solar noon; McVey n.d.b). If a female laid eggs following copulation with a satellite, she would have to deposit them in an area that was defended by a resident male. Although satellites attempted to guard their mates, they succeeded on average for only 14 sec. Resident males holding a 1.8-m territory had equal difficulty in guarding their females. Males did not defend territories smaller than 2 m (McVey, n.d.b).

Females refused to oviposit after 27% (51/189) of copulations with satellite males, but after only 7% (121/1,682) with residents (two-tailed χ^2 test, $p < .05$). Following "refusals," females flew away from the pond and then reentered within a few minutes, whereupon they were captured by other males. Because sperm precedence by the last mate is 100%, a female can effectively cancel one mating by accepting another, which is probably easier than refusing to copulate in the first place. The few females that were unable to bend their abdomens, a necessary precopulatory act, were clasped by their "mates" for up to 7 min, far longer than the 10–60 sec required for copulation.

Females did not discriminate among males by physical characteristics; instead, they used behavioral clues to a male's status. In a typical resident copulation, the pair would perch on top of the algal mat or on a twig or stick near the center of the male's territory. Satellites sometimes perched on algal mats for copulation, but if chased by residents they might fly in copula for a minute or two and/or fly to the ground a meter or more from the pond edge. Females that received such treatment refused to oviposit 44% of the time (26/59). These same males would sometimes fill territorial vacancies as they arose for a few minutes or hours until evicted by another male. Females refused to oviposit after mating with these "temporary" residents only 11% of the time (4/44), which was not significantly different from the refusal rate for the full-time residents (two-tailed χ^2 test, $p > .05$; McVey 1981). Females probably discriminated against satellites because of the inevitable disturbance of a resident male fighting with the satellite, usually causing delays in oviposition and sometimes injuring the female (McVey 1981).

Competitive Situation 2

In addition to the preceding considerations, when two ponds were available males' fertilization success depended upon which pond they attended. Females preferred the thick algal mats in pond 9 to the surfaced fronds of *Hydrilla* on pond 1, laying on average a total of 421 eggs daily per meter of shoreline at pond 9 compared with a total of only 41 eggs daily per meter at pond 1. Although pond 1 (96 m shoreline) was larger than pond 9 (52 m shoreline), similar numbers of males attended each pond daily (14.4 ± 1.8 s.d. and 13.7 ± 1.9, respectively). As a result, males fertilized on average 5.8 times as many eggs at pond 1 as at pond 9 (table 4.2).

Daily Potential for Sexual Selection

On a daily basis male reproductive success varied greatly even within a single pond (table 4.1). Within each pond, male fertilization success improved with increasing territory size and the amount of time spent defending algae. The potential for sexual selection compared with total selection on males (I_s/I_m; table 4.1 notes) was 8% to 16% higher at the more competitive pond 9 (.79 to .89) than at pond 1 (.65 to .81; table 4.1) and ranged from 65% to 89% of the total daily opportunity for selection. Male body size had no significant effect on daily RS (McVey 1981).

4.5 Daily Success and Longevity

Individual male RS varied somewhat between days owing to fluctuations in the operational sex ratio (Emlen and Oring 1977), weather and perhaps male condition. Despite within-male variability, individual differences in

Table 4.2 Correlations between Daily Reproductive Success and Breeding Longevity in *Erythemis simplicicollis*

Group	n	Total Days	Mean Daily RS	Mean Longevity	Sex Ratio	Correlations		
						b	r^2	p
Males								
Situation 1	56	441	1344	7.9	0.54	.559	.315	.02
Situation 2	105	990	934	9.4	0.78	.238	.187	.05
Pond 9	46	382	1600	8.3	0.24	.608	.331	.02
Pond 1	47	417	277	8.9	2.63	.134	.334	.02
Switch ponds	12	191	943	15.9	0.77	.279	.386	.21
Both situations	161	1431	1076	8.9	0.68	.309	.307	.01
Females								
Including days missing	47	295	650	7.0		.027	.048	.69
Excluding days missing	47	223	860	7.0		.219	.237	.11

Note: Mean daily reproductive success (RS) is expressed as the mean number of eggs fertilized or laid across all male-days or all female-days (total days), respectively, and serves to describe the pond situation. Longevity excludes rain days, and the sex ratio equals the average number of males/females alive daily. Each correlation is based on the longevity and average daily RS (total lifetime RS divided by number of breeding days) of n individual animals.

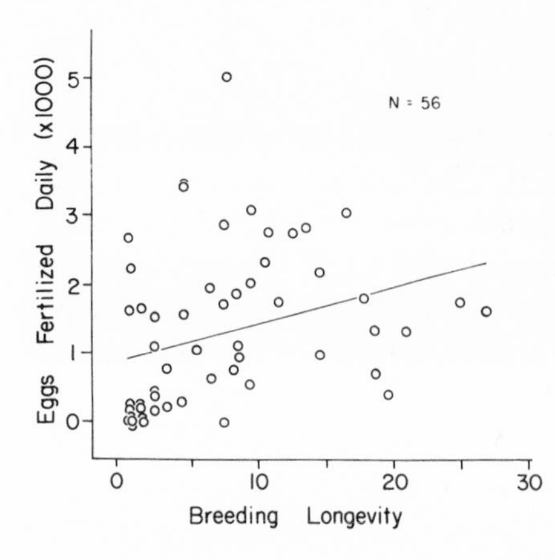

Figure 4.3: Correlation of average daily RS with breeding longevity of males in situation 1 (pond 9 only). Daily RS ($\times$ 100) = .559 $\times$ longevity + 9.03, s.e. of b = 11.4.

performance were consistent ($p < .001$, $H = 151.21$, Kruskal-Wallis one-way ANOVA, $n = 381$ days from 37 individuals, situation 1, excluding males with longevities < 4 days). The reason for consistency of individuals' performances was that the more successful residents retained the same territories from one day to the next, whereas part- and full-time satellites tended to remain as such (McVey 1981). Moreover, in situation 2 animals rarely switched between ponds (only 12/105 switched at all, and these switched on average only 1.3 times per lifetime).

Given that some males achieved consistently high daily RS while others were consistently unsuccessful, the possibility existed that the former group was shorter-lived than the latter because they protected more females and defended larger or higher-quality territories. Nonetheless, in situation 1 daily RS was positively correlated with longevity (fig. 4.3 and table 4.2). Thus, within a single competitive landscape, the more successful males did not appear to suffer higher mortality risks.

One might still expect survivorship differences between males operating under different overall levels of competition, as between ponds 1 and 9. The males at pond 1 engaged in fewer contests than those at pond 9 because there were fewer mates to defend and lower males densities. If animals were more likely to die while attending pond 9 than pond 1, on comparing all males, including those that switched ponds, daily reproductive success would be negatively correlated with survivorship. Instead, daily reproductive success was again positively correlated with survivorship in situation 2 (n = 105; table 4.2). For females no correlation existed between daily RS and longevity (table 4.2).

The positive correlations between daily RS and longevity suggest that some male attributes or correlated attributes contributed positively both to competitive success and to survivorship. Two other factors, however, might have caused the positive correlation, making this conclusion unwarranted. The first is the confounding effect of age on reproductive success, and the second is the possibility of differential emigration of animals of differing competitive skills. Both possibilities are suggested by the fact that the correlation of average daily RS with longevity depended upon the poor performance of the short-lived males. If males that disappeared before reaching their fourth breeding day are eliminated from the data, the correlations between daily RS and longevity disappear in all data sets (pond 9 only, $p = .68$, $n = 35$; both ponds, $p = .50$, $n = 73$). I will therefore examine each possibility in turn.

One could obtain a positive correlation between average daily RS and longevity if reproductive success improved with age. If mortality was random with respect to male performance and age, the daily average RS of males that die young will include only the low "subadult" success rates. Young male *Erythemis* (twelve to eighteen days postemergence) were not as successful as middle-aged (nineteen to twenty-seven days) males (fig. 4.4). Young males that attempted reproduction fertilized fewer eggs daily than did middle-aged males (fig. 4.4, top), and many did not even attempt to breed until an older age (fig. 4.4, bottom). The modal age for males first arriving at the ponds was 16 days, range 12–26, with 85% of males arriving by 19 days postemergence.

The positive correlation of average daily RS with longevity did not result from age-dependent RS in this population, however. Short-lived males (disappearing before their 4th breeding day) fertilized significantly

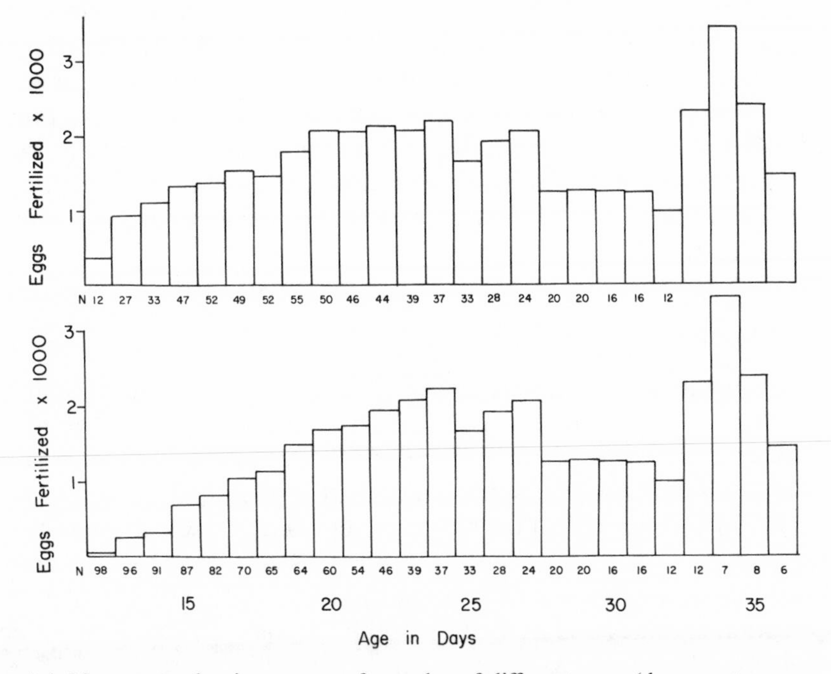

Figure 4.4: Mean reproductive success for males of different ages (days postemergence). *Top*: Only males present at pond 9 are included, hence the sample size first increases and then decreases with age. *Bottom*: Includes males that had not yet arrived at the pond as males with zero success. The number of males dying betweeen twelve and twenty-six days after emergences that never appeared at the ponds is unknown. Within age-classes, daily RS approximated a zero-truncated log-normal distribution. One standard deviation above the mean ranged from 900 to 3,000 eggs.

fewer eggs in their first 1–3 breeding days than did longer-lived males on their first 3 days despite the virtually identical age distribution of the two groups on the first day (daily RS averaged 590, 430, and 500 for short-lived males (n = 34, 19, 8) and 1,150, 1,520, and 2,360 for the longer-lived males (n = 64, 64, 59) on days 1, 2, and 3, respectively; p = .054, .003, and .018, Mann-Whitney U test, days 1, 2, and 3).

To reduce the possibility that I included males who had emigrated after one day rather than died, before data analysis I eliminated three and six males from each situation (see methods) that were the youngest and doing the least well on their one day at the pond to help bias the data against finding a positive correlation between average daily RS and longevity. Thus the slight positive correlation does suggest that male characteristics that promote short-term reproductive success correlate with those aiding male survival. The effect depends primarily on noncompetitive males dropping out of the population quickly.

4.6 Potential for Sexual Selection over Lifetimes

When averaging males' daily RS over their lifetimes, the higher successes of some days were canceled by the lower successes of others. In addition, mortality was independent of daily RS for long-lived males. Consequently, the ratio of sexual selection to total selection (I_s/I_m) averaged over lifetimes was about half that computed on a daily basis (table 4.1).

In estimating variance in reproductive success, I include only those individuals that reached reproductive maturity and attempted to breed on at least one day. If the population is on average stable, then more than 99.9% of eggs laid fail to contribute to the next generation (McVey 1981; Benke and Benke 1975). Consequently, natural selection is most intense on the nymphal stage.

4.7 Partitioning Selection

The previous analysis confounds the variance in female survivorship and daily clutch size in one factor, I_f. To examine the separate contributions of survivorship and average daily fecundity to lifetime reproductive success (LRS) and the potential for sexual and natural selection, I use Arnold and Wade's (1984a,b) paradigm for analyzing episodes of selection (table 4.3). For simplicity, I have restricted the analysis to situation 1. I consider survivorship as the first episode that accounts for 57% of the total potential for selection in males and 68% of the total in females. I chose to define the next episode for males as daily mating success, a component of sexual selection. The final episode, defined as the number of eggs fertilized per mating, depended on how long a male guarded his mate and consequently was also considered as a component of sexual selection (dependent on male mating tactics and on territory size). The product of w_2 and w_3 was equal to daily fecundity.

Table 4.3 Partitioning of the Total Opportunity for Selection in Male *Erythemis simplicicollis*, Situation 1 (Pond 9 Only)

Source of Variance in Relative Fitness	Contribution to Total Opportunity for Selection		
	Symbol	*Value*	*%*
Males			
Natural selection (breeding longevity w_1)	I_1	0.729	57
Sexual selection (mating success/day w_2)	I_2	0.448	35
Cointensities and change in covariance (12)	*[a]	−0.117	−9
Subtotal (matings/lifetime = w_1w_2)	I_{12}	1.071	83
Sexual selection (eggs/mating w_3)	I_3	0.365	28
Cointensities and change in covariance (123)	*[a]	−0.150	−12
Total selection (eggs fertilized per lifetime = $w_1w_2w_3$)	I_{123}	1.286	100
Females			
Natural selection (breeding longevity w_1)	I_1	0.914	68
Fecundity including days missing (w_2)	I_2	0.343	25
Fecundity excluding days missing (w_2)	I_2	0.330	24
Cointensities and change in covariance (12)	*[a]	0.093	7
or	*[a]	0.106	8
Total selection (eggs laid per lifetime w_1w_2)	I_{12}	1.350	100

Note: Male daily fertilization success, I_{23}, = 0.786 while $I_2 + I_3 = 0.813$, meaning that a cointensity (*) of −.027 or −2% exists between matings/day and fertilizations/mating. Male daily RS therefore accounted for about 61% (0.786/1.286) of the total opportunity for selection, and a cointensity (*) of −18% exists between longevity and daily RS.

[a] Cointensities (*) and changes in covariance include the unweighted covariance before selection, the weighted covariance after selection, and the difference in covariance that results from the previous selection episode; for (12) = COI(1,2) + COI(12 | 2) + (COI(12,2 | 1) − COI(12,2)) (see Arnold and Wade 1984a).

Despite the positive correlations between breeding longevity and daily fecundity (table 4.2) and between matings/day and eggs/mating (r^2 = .371, $p < .001$, $n = 56$) for all males weighted equally, the cointensities and changes in covariance between each episode were negative (table 4.3). This probably resulted for two reasons. First, the positive correlations depend heavily on the males with short life spans and few matings per day. Second, the animals with short life spans also contributed more to the variance than did the longer-lived males, because daily RS was averaged over fewer days. For females the cointensities and changes in covariances were slightly positive. The variance in longevity contributed more than twice as much to variance in LRS in females as in males.

4.8 Male Attributes and Lifetime Reproductive Success

Erythemis adults exhibit a constant probability of dying per unit time for most or all of the breeding adult life span, which yields a high variance in adult survivorship. Much of this mortality is probably due to accidents and predation in such a way that phenotypic characteristics are of little importance. Even for male *Erythemis*, in which some attribute appears to contribute positively to both survivorship and to longevity, most of the longer-lived males' mortality risks are independent of daily RS.

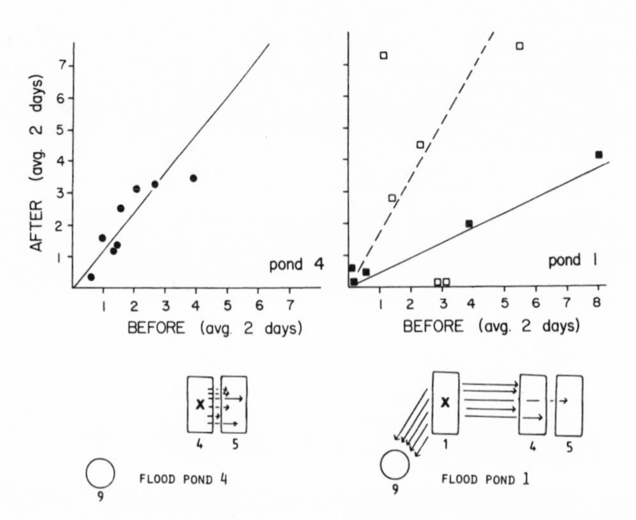

Figure 4.5: Territory removal experiments. The correlation of the value of a males' territory holdings for two days before and after they deserted the flooded pond. Values are averaged as the percentage of the total female oviposition at all ponds. *Left*, all animals moved from pond 4 to pond 5 (Spearman's rank correlation, $p < .001$). *Right*, solid squares represent males moving to ponds 4 or 5 (SRC, $p > .05$) and open squares represent the males moving to pond 9 (SRC, $p < .05$); the solid and dashed lines are the predicted ratios for these two groups, respectively.

Of the variance in average daily breeding fecundity among males, what proportion resulted similarly from chance events over which individual males had no control? Much of male fecundity depended upon the territory defended, and males tended to retain the same territory for several days. If territory occupancy was largely a result of chance acquisition, then male phenotype would be of little importance, and the apparent potential for sexual selection could have little influence on genetic evolution.

To test whether male attributes influenced territory acquisition and maintenance, I removed territories from males, forcing them to reestablish in entirely different ponds (McVey 1981, n.d.a). In these two experiments, conducted in late July and early August after ponds 4 and 5 became available, all territories at each of two ponds were eliminated by increasing water depth by 15 cm, thus submerging the surface vegetation (rooted to the bottom) gradually over a 1-day period (fig. 4.5). Without oviposition sites to defend, males moved to the remaining ponds. Before both experiments, other males at the target ponds defended contiguous territories so that no "vacancies" existed.

If male attributes contributed significantly to the size of territory obtained and the timing of residency with respect to the probability of female visits, then the value of a male's holdings (expressed as expected RS) before and after each experiment should have been positively correlated. If males assessed ponds before settling, one might predict that the ratio of the value of males' holdings before and after the experiment would be 1.0. If males simply moved to the first pond they encountered and remained there, one would predict that the ratio of territory values before and after the experiment would depend upon the ratio of females to males at the experimental (algae removed) and target ponds.

The relative numbers of females and males did differ between some ponds (table 4.4). If males did not sample different ponds before settling, one would predict that the ratio of males' territory values before and after the experiment would equal the average male fertilization success (as a percentage of total oviposition) at the target pond divided by that at the original pond. For males moving from pond 1 to pond 9, the predicted ratio of male territory values would be 3.6 (percentage of total oviposition at target pond 9)/2.3 (percentage of total oviposition at original pond 1) = 1.6. For males moving from pond 1 to pond 4 or 5 the predicted ratio would be 1.3/2.3 = 0.57, and for males moving from pond 4 to pond 5 it would be 2.1/2.0 = 1.1.

Individual male attributes did significantly influence the value of territory obtained in the experiment (fig. 4.5). Furthermore, the males leaving pond 1 supported the idea that males did not sample different ponds but stayed at the first one they encountered. At least seven males obtained their territories by evicting established residents from part of all of their territories for part or all of each day.

4.9 Summary

In situations 1 and 2, males that adoped the satellite tactic for large portions of each day not only achieved a lower daily RS, but tended to live shorter lives than predominantly resident males. Because the satellite

Table 4.4 Territory Removal Experiments: Male and Female Distribution before and after Each Experiment

Pond	Seconds of Oviposition[a]	% of Total Oviposition	Number of Males[b]	% per Male	Seconds of Oviposition[a]	% of Total Oviposition	Number of Males[b]	% per Male
Flood 1	*Before: 22 July to 24 July*				*After: 26 July to 28 July*			
Pond 9	1,830	48.6	13.7	3.55	3,136	66.9	16.3	4.10
Pond 1	1,329	35.3	14.4	2.45	—	—	—	—
Ponds 4, 5	606	16.1	24.6	0.65	1,552	33.1	30.0	1.10
Total	3,765	100.0	52.7		4,688	100.0	46.3	
Flood 4	*Before: 31 July to 2 August*				*After: 3 August to 5 August*			
Pond 9	2,773	57.8	16.3	3.55	1,184	35.0	13.2	2.65
Pond 4	797	16.6	9.7	1.71	—	—	—	—
Pond 5	1,229	25.6	22.3	1.15	2,199	65.0	29.8	2.18
Total	4,799	100.0	48.3		3,383	100.0	43.0	

[a] Three-day average of oviposition seconds. Eggs were laid at a rate of between eight and eleven per second (McVey 1981, 1984). Individual females averaged eighty-four seconds of oviposition daily during periods of good weather (n = 26 marked females, 114 female-days, July only).

[b] Average number of different individuals present for three days. The % per male is the percentage of total seconds of oviposition at all ponds.

tactic did not compensate for its low returns by extended longevity, it cannot be considered part of a mortality-fecundity trade-off as first envisioned by Gadgil (1972). It appears to be part of a strategy conditional on relative male competitive ability, the threshold for switching being the point (length of territory defended) at which the two tactics are equally rewarding per unit time (see Dominey 1984).

A large contribution to variation in lifetime RS was the variation in survivorship. In situation 2, where one might expect males at pond 9 to expend more energy and to take more risks than at the less competitive pond 1, no survivorship differences existed. Most mortality appears to result from random events, and the "opportunity" for this aspect of natural selection to affect changes in genotypes across generations appears to be slight.

The opportunity for sexual selection over lifetimes represented 39% and 42% of the total potential for selection on males in two very different situations (table 4.1). A more detailed partitioning of selection into episodes revealed that although sexual selection appeared to account for 43% (100% − 57%) of the total variance in LRS in situation 1, it actually accounted for 61% after weighting individuals by their longevities (table 4.3).

The potential for sexual selection was not simply the spurious result of rapid cycling of males relative to females and chance acquisition of territories, but was the result, in part, of the variance in male competitive abilities. Which aspects of phenotype were important is not yet known. Some males might have carried parasites, others might have been inefficient foragers, and still others might have had muscle or wing injuries that prevented vigorous flight. Although the variance in competitive abilities might have depended largely on environmental effects during ontogeny, the potential for sexual selection to cause genetic changes exists.

4.10 Acknowledgments

I am very grateful to Peter Marler, Steven Green, and Donald Kroodsma for their support and council throughout the study. The investigation was funded by a National Institute of Health National Research Service Award no. 7524 from the National Institute of General Medical Science. Many thanks to Sheree McFarland, Janice Manjuck, Greg Divine, Linda Soar, Kim Iseman, Debbie Brooker, Chris Smith, Janet Land, Andy Lane, Bob Koff, and David Elmer for their excellent field assistance.

5 Mating Success and Fecundity in an Ant-Tended Lycaenid Butterfly

Mark A. Elgar and Naomi E. Pierce

THE AUSTRALIAN LYCAENID BUTTERFLY *Jalmenus evagoras* Donovan exhibits an unusual mating system that derives in part from its close association with ants. Both larvae and pupae of *J. evagoras* secrete food rewards for workers of several species of *Iridomyrmex*. In return the ants protect the larvae and pupae against predators and parasitoids. Populations of *J. evagoras* deprived of their attendant ants cannot survive (Pierce 1983; Pierce et al. 1987), and females of the species even use ants as cues in ovipositions (Pierce and Elgar 1985).

The larvae of *J. evagoras* aggregate and pupate in clusters on the upper branches of their *Acacia* host plants. Adult males search for mates by regularly investigating trees containing juveniles of the species. They hover around the trees, sometimes tapping a pupa with their antennae, perhaps thereby "tasting" its age and sex. When a pupa is about to eclose (emerge as an adult) as many as twenty males may gather around it, forming a "mating ball." The males engage in a frenzied scramble as the pupa ecloses and copulation takes place before a teneral female has even had time to expand her wings (see fig. 5.1). Pairs remain mating on a tree for several hours. Females mate only once, and mated females vigorously reject further advances by males. Although mating in *J. evagoras* does not always involve the formation of a visually dramatic mating ball (eclosing pupae are often found by single males), females are almost always mated before their wings have hardened, and virtually every mating is readily and unmistakably observable because of the conspicuous location of pupae and the lengthy copulation time. It is therefore possible to obtain a comprehensive record of matings in a field population of this butterfly.

In this chapter we describe several of the cues that males use to search for females. These include males' response to the presence or absence of ants, their interest in male and female pupae of different sizes and ages, and their attraction to clusters of conspecific adults. We then

Figure 5.1: Male *Jalmenus* surround an eclosing pupa.

analyze the lifetime mating success of males, which is an important component of their lifetime reproductive success. We assess whether particular morphological and behavioral traits are correlated with the components of a male's lifetime mating success. Using a combination of field and laboratory observations, we examine the importance of size and longevity for female fecundity. Finally, we discuss how associations with ants may have shaped the evolution of the mating system of *J. evagoras*.

5.1 Natural History and Study Site

J. evagoras is a multivoltine butterfly found along the east coast of Australia from Melbourne in the south to Gladstone in the north. Although widespread in its distribution, where it occurs *J. evagoras* forms discrete, highly localized populations. Males and females show almost no sexual dimorphism in wing color or pattern (Common and Waterhouse 1981, Pierce 1984). Although there are no differences in adult wing length, female pupae are larger than male pupae, and adult females are 60% heavier than adult males (see table 5.1). Males of *J. evagoras* eclose several days before females. According to our field estimates, males also live longer than females (although this may simply reflect lower emigration rates).

Our study site was around the village of Mount Nebo in Queensland, approximately 10 km west of Brisbane (152° 47′ E, 27° 23′ S). Here *J. evagoras* feeds predominantly on the foliage of young plants of *Acacia irrorata* and *A. melanoxylon*, which come up as second growth after land has been cleared. The larvae and pupae are tended by ants in the

Iridomyrmex anceps group (sp. 25, Australian National Insect Collection) that form large polygynous and polydomous colonies. Populations of *J. evagoras* overwinter as eggs that hatch in early spring, usually by mid-November. There are a minimum of three broods per season. The general biology of *J. evagoras* has been described in Kitching (1983), Pierce (1983, 1984, 1985, 1987), Pierce and Elgar (1985), Pierce and Young (1986), and Pierce et al. (1987). Our observations were made in a garden approximately 20 m by 20 m, from January through March 1984. All correlations described below are Spearman rank coefficients unless otherwise specified.

5.2 Methods and Results

Male Mate-Searching Behavior

Response to Ants

Methods. The methods used in this field experiment are a modification of those described by Pierce and Elgar (1985). Four fifth-instar larvae were placed on each of twelve potted plants of *A. irrorata* that had been

Table 5.1 Comparison of Mean Body Size, Longevity, and Eclosion Date for Males and Females of *J. evagoras*

	Females	Males	t_s
Juvenile body size			
Pupal length (mm)	13.9 (1.2) n = 46	12.8 (1.2) n = 46	4.35**
Pupal width (mm)	5.4 (0.7) n = 46	4.8 (0.7) n = 46	4.39**
Median day of eclosion[a]	14	8	
Adult body size			
Forewing length (mm)	21.5 (2.3) n = 116	21.0 (2.0) n = 126	1.79
Body length (mm)	16.2 (1.7) n = 116	16.1 (1.6) n = 126	0.56
Body weight (mg)	72.2 (30.4) n = 42	45.7 (17.3) n = 52	5.30**
Estimated longevity (days)	3.1 (3.5) n = 45	6.9 (5.9) n = 35	3.59*
Proportion of observed life span[b]	0.63 (0.26) n = 45	0.73 (0.21) n = 29	

Note: With the exception of body weight, all data were collected from individuals caught in the field. Numbers in parentheses are standard deviations.

[a] First pupa eclosed on day 1.

[b] Number of days observed per life span.

* $p < .05$; ** $p < .01$.

Table 5.2 First Approaches of Males of *J. evagoras* to Plants Containing Larvae with and without Ants

	Approaches		
	Previous Days	*Swap Day*	*Subsequent Days*
Plants with ants	98	80	90
Plants without ants	52	79	60
χ^2	13.50**	0.00	5.61*

*$p < .05$; **$p < .01$.

arranged in a circular arena about 4 m in diameter. This spacing of plants was within the range of distances found between plants in natural populations. Ants from nests in the garden were allowed to tend the larvae on six adjacent plants in the arena but were excluded from the other six plants. This arrangement was left for four days, and on the latter two days, for about three hours each, we recorded the first plant visited by individual males as they entered the arena (Pierce and Elgar 1985). There were at least nine males active in the study site during the experiment, and individual butterflies were observed more than once during the five days of observation. Our data reflect the behavior of most of these males rather than .the behavior of just one or two individuals. After two days we removed all the plants in the arena and replaced them with new plants that had not been previously infested with larvae and ants. The positions of plants with and without ants were swapped so that those positions where ants had been excluded now had ants and vice versa. Larval density was kept constant on each day during the entire seven-day experimental period.

Results. The presence of ants influenced the mate-searching behavior of the males (table 5.2). Males preferentially approached and landed on plants with larvae and ants. However, they became confused when the positions of these plants were changed (on the swap day) and were equally likely to approach and land on plants with and without ants on that day. Thus males of *J. evagoras* can learn the positions of plants containing larvae and ants.

Response to Pupae

Methods. We arranged twelve potted food plants (*A. melanoxylon*) in a circular arena with a diameter of 4 m. A single pupa was measured (length and width) and then hung onto each plant. All pupae were tended by ants. We observed individual males that entered the arena and recorded their approaches and landings on the plants (see above). We also noted the eclosion date and sex of each pupa and made several recordings each day of the number of ants tending it. When a pupa eclosed it was replaced with another one. The experiment was conducted over a period of three weeks, during which time thirty-one pupae eclosed (seventeen males and fourteen females). The time each pupa was on a plant before eclosing averaged three days (s.d. = 1.7 days).

Results. There was no correlation between the proportion of times a pupa was visited by males and the number of days before it eclosed (see fig. 5.2*a*), indicating that males apparently have to approach a pupa before they can assess its state. However, there was a significant negative correlation between the proportion of times a pupa was landed on and the number of days before it eclosed (fig. 5.2*b*). Thus, males visit pupae randomly with respect to age but land on them more frequently when they are about to eclose.

We examined whether males prefer certain types of pupae by analyzing the residuals about the regression curve (see fig. 5.2*b*). For example, if males spend more time with female pupae, the positive residuals should represent mostly female pupae. There was no relationship between the sex of the pupa and whether it was preferentially landed upon ($\chi^2 = 0.039$, d.f. = 1, $n = 31$). Males could use size as an indicator of sex, since female pupae are larger than male pupae (table 5.1), but there was no evidence that males preferentially landed on larger pupae ($\chi^2 = 0.017$, d.f. = 1, $n = 31$). There was a tendency for males to spend more time with pupae that were tended by more ants on the day before it eclosed (Fisher's exact probability = 0.06, $n = 31$), although there was no association between the sex of the pupa and the number of ants tending it on that day

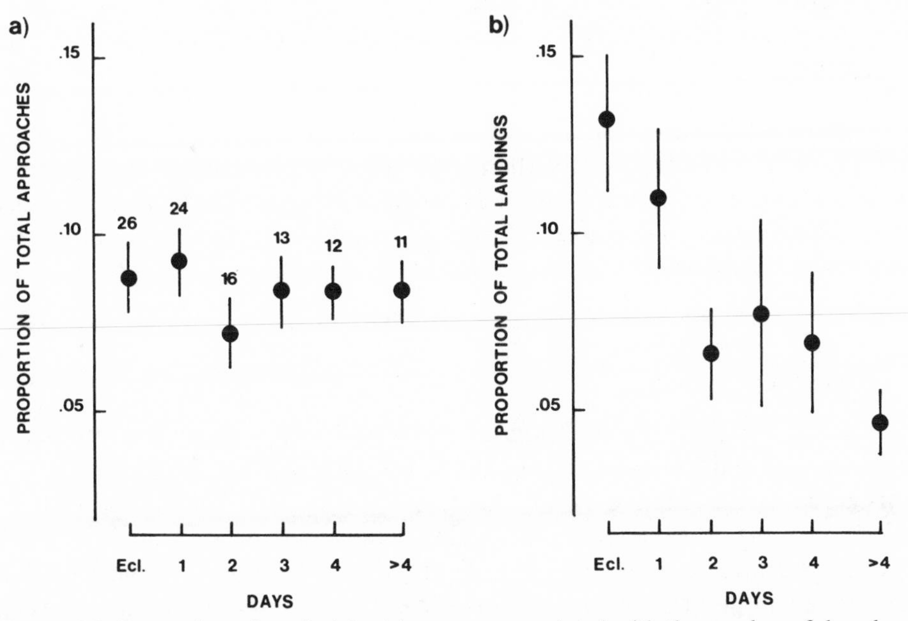

Figure 5.2: Proportion of total visits (*a*) was not correlated with the number of days before eclosion, but the proportion of total landings (*b*) was negatively correlated with the number of days before eclosion ($r_s = -.305$, $p < .001$, $n = 105$). Bars represent standard errors, with sample sizes above.

Table 5.3 Frequency of Visits and Landings by Males of *J. evagoras* to Plants with Varying Numbers of Pinned Adults

	Number of Trees	Total Visits	Total Landings
No pupae, no adults	3	54	0
One pupa only	3	65	2
One pupa and one pinned adult	3	71	5
One pupa and three pinned adults	3	108	22
χ^2		22.08*	41.76*

*$p < .001$, d.f. = 3.

(Fisher's exact probability = 0.32, n = 31). In summary, males are sexually indiscriminate in their mate-searching behavior, and this often results in attempted copulations with eclosing males.

Response to Adults

Methods. In this experiment we used the same twelve-plant arena described above, but instead of placing single pupae on each plant, we employed four treatments. The plants contained either one pupa and three dead conspecific adults pinned next to the pupa; one pupa and one pinned adult; one pupa only; or nothing. Each treatment had three replicates, randomly assigned to different plants. Again, we observed the approaches and landings of males for about six hours over a two day period. The treatments on different plants were changed on the second day to control for possible pupal maturity or tree-position effects. None of the experimental pupae eclosed during the course of the experiment or for two days afterwards.

Results. Males visited and landed on trees with pinned adults significantly more frequently than on trees without pinned adults (table 5.3). Males did not interact with the pinned adults but simply landed beside them. Therefore males use conspecific adults as cues in their mate-searching behavior.

Male Reproductive Success

Methods

Each pupa on trees bearing juveniles of *J. evagoras* in the study site was marked and followed. To the best of our knowledge the individuals in this garden were isolated by a distance of at least 1 km from other colonies of *J. evagoras*. Five emigrant males from our marked sample were retrieved from a second study site slightly over 1 km away, suggesting that the actual population of *J. evagoras* we were studying was larger than simply those individuals contained in the garden. However, the extremely high recapture rate of our marked individuals and the comparatively low rate of immigration into our study site indicated that we were sampling most of the breeding individuals in a colony that was isolated by distance from other individuals.

After a male had eclosed and his wings had expanded and hardened (about 30 min), we captured him by either encouraging him to walk off the plant onto our hands or simply picking him up by the thorax and wing between thumb and forefinger. It was never necessary to use a butterfly net. We measured his forewing, hind wing, antenna, and body length, then wrote an identifying number on his right forewing using an enamel-based marker pen. The age of two males that were not caught immediately after eclosion was estimated by wing wear. Females were captured, measured, and marked in a similar manner approximately one hour after they had eclosed, while they were still in copula.

Continuous observations were made each day from about 0630 h until midafternoon for the entire twenty-nine-day breeding period. Individuals that were present at the field site and the outcome of all mating tournaments were recorded for each day. Pairs remained in copula for at least two hours ($\bar{x}$ = 4.32 h, s.d. = 1.80, n = 41); hence we were able to observe every mating in our study site.

Male Survival

Forty males that eclosed were marked and observed over a period of four weeks. Male survivorship is shown in fig. 5.3. A quarter of the "eclosed" male population disappeared within the first day, but there was

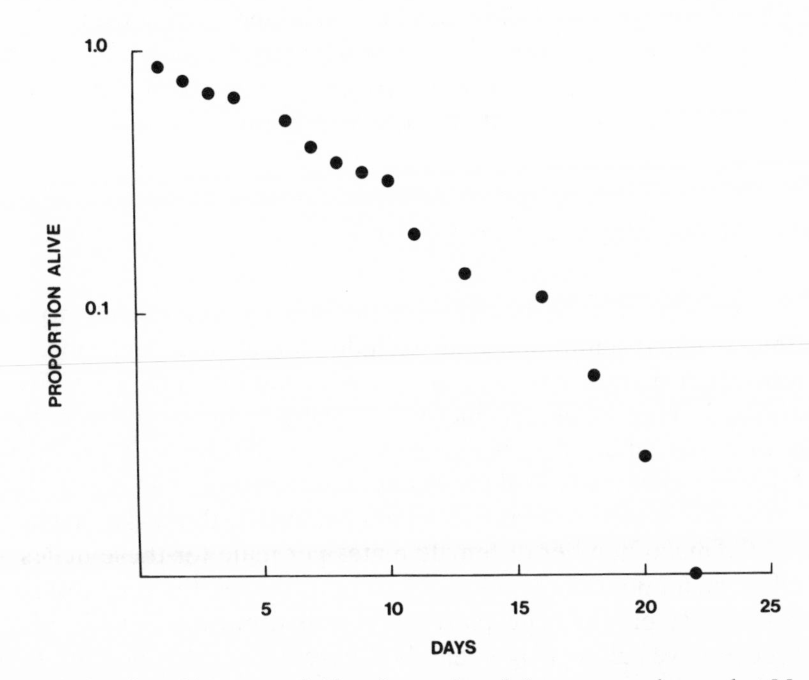

Figure 5.3: Survivorship curve of thirty-five males of *J. evagoras* observed at Mount Nebo, Queensland, during February and March 1983. The proportion alive is given on a $\log_{10}$ scale.

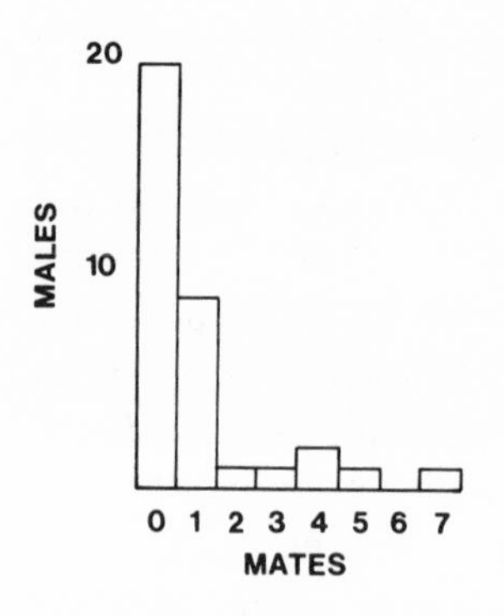

Figure 5.4: Frequency distribution of male lifetime mating success in *J. evagoras*.

a steady rate of disappearance over subsequent days. Since recapture rates were extremely high (table 5.1), we assumed that disappearance within the first day was due to both emigration and mortality, whereas later disappearances primarily reflect mortality. The basis for this assumption rests on our observations of five known emigrants that were observed at another field site, about 1 km away. All these emigrants left their original site within twenty-four hours of eclosing. These five emigrants were excluded from the analysis of male mating success (see below). Furthermore, on the basis of wing wear, there were no "old" male immigrants into our study site during the observation period. We saw numerous birds attempt to eat adults of *J. evagoras*, including willie wagtails (*Rhipidura leucophrys*), pied butcherbirds (*Cracticus nigrogularis*), Lewin honeyeaters (*Meliphaga lewinii*), and kookaburras (*Dacelo gigas*). Most of these attempts were made while the butterflies were on the wing or sitting in the grass, and we never saw a male taken while it was in a "mating ball" or sitting beside a pupa on a plant.

Male Mating Success

Our analysis of male lifetime mating success refers only to those males that reached adulthood and excludes larval and pupal mortality. Therefore our measures of the variance in male lifetime mating success are overestimated. There was considerable variability in the lifetime mating success of males of *J. evagoras*: the most successful male mated with seven females, whereas 57% of the males failed to mate at all (fig. 5.4). The overall variance in lifetime mating success among the thirty-five males was 2.65, and the mean number of female mates per male for these males was 0.97, reflecting an unbiased overall sex ratio (although the operational sex ratio was almost purely male biased). The distribution of male lifetime mating success was almost significantly different from that derived from a Poisson probability function (Kolmogorov-Smirnov test: $D = 0.203$, $.1 > p > .05$), suggesting that lifetime mating success may not be a random process.

Components of Lifetime Mating Success

We have identified three components of male lifetime mating success for *J. evagoras*: longevity, encounter rate, and mating efficiency. Thus, male lifetime mating success in *J. evagoras* is expressed as LMS = LS × ER × ME, where for each male LS = male lifetime (estimated from the number of days from first to last sighting); ER = encounter rate (total number of females that eclosed during the male's lifetime per male lifetime); ME = mating efficiency (number of mates per total number of females that eclosed during the male's lifetime).

Out of thirty-five males, twenty-nine were alive on days when females eclosed and therefore had an opportunity to mate. These males will be referred to as "breeding" males. The mean, variance, and standardized variance of the components and products of the components of lifetime mating success among breeding males are shown in table 5.4. The percent contributions of the components, derived using the method of Brown (this volume, chap. 27) are shown in table 5.5. The 94% of the variance in male lifetime mating success due to breeding males can be broken down into 12.3% attributable to variation in life span, 9.6% attributable to variation in encounter rate, and 40.4% attributable to variation in mating efficiency.

Most of the variance in male lifetime mating success lies in longevity and mating efficiency and their covariances. The considerable simultaneous independent variation and covariation between longevity and mating efficiency is due to the highly significant correlation between these two variables (see below). This high degree of covariance makes the analysis

Table 5.4 Mean and Variance of the Components of Lifetime Mating Success of Breeding Males of *J. evagoras*

Component	Original		Standardized Variance
	Mean	*Variance*	
LS	7.45	36.76	0.66
ER	1.61	1.34	0.52
ME	0.06	0.01	2.78
LS ER	10.69	63.36	0.55
LS ME	0.81	2.50	3.81
ER ME	0.09	0.02	2.47
LS ER ME	1.05	3.02	2.74

Note: LS = lifetime; ER = encounter rate; ME = mating efficiency (see text for details). Standardized variance is the variance divided by the square of the mean.

Table 5.5 Percentage Contribution of the Components of Lifetime Mating Success to Variation in LMS in Breeding Males of *J. evagoras*

Component	Longevity	Encounter Rate	Mating Efficiency
Longevity	13.08		
Encounter rate	−14.61	10.24	
Mating efficiency	157.69	−21.91	43.03
3+	−87.51		

difficult to interpret biologically, because it is impossible to determine which is the most important component of a male's making success. The analysis of male lifetime mating success excludes nonbreeding males, many of which survived less than one day (see fig. 5.2). The importance of longevity might increase if this mortality were included.

Determinants of Male Lifetime Mating Success

Male forewing length was significantly correlated with both longevity (r_s = .370, p = .014, n = 35) and mating efficiency (table 5.6) but not with the encounter rate of males with eclosing females (r_s = .227, p = .100, n = 35), and encounter rate was not correlated with longevity (r_s = .153, p = .200, n = 35). However, there was a significant correlation between eclosion date and encounter rate; males that eclosed earlier had a higher encounter rate than males that eclosed later in the breeding period (r_s = .421, p = .006, n = 35).

Since body size (measured by forewing length) was correlated with longevity, it was necessary to derive partial correlation coefficients in order to establish whether body size or longevity had a greater influence on mating efficiency. This analysis revealed that mating efficiency was

Table 5.6 Relationship between the Absolute Forewing Length, Longevity, and Mating Efficiency of Thirty-five Males of *J. evagoras*

	Mating Efficiency	
	r_s	*Partial Correlation Coefficient*
Forewing length	0.385*	0.292
Longevity	0.497**	0.395*

*p < .01; **p < .001.

Table 5.7 Mean Forewing Length, Longevity, Age, and Number of Mates of Males of *J. evagoras* Present for Each One-Week Period during the Twenty-eight-Day Season

	Week				
	1	*2*	*3*	*4*	$F_{3,60}$
Number of males present during the week	18	24	12	11	
Forewing length	22.6	21.1	20.6	21.1	2.607
	(1.8)	(2.6)	(2.3)	(1.9)	p = .06
Days present	3.4	3.8	5.5	3.5	2.946
	(2.1)	(2.3)	(2.3)	(1.6)	p = .04
Age	2.6	4.8	8.3	10.6	7.171
	(1.2)	(3.2)	(5.1)	(7.2)	p < .001
Number of mates	0.3	0.7	0.5	0.4	0.928
	(0.6)	(0.1)	(0.8)	(0.5)	p = .43

Note: Several males were observed for more than one week. The *F*-statistic is derived from one-way ANOVA, and the age data were transformed before analysis. Standard deviations are given in parentheses.

directly correlated with longevity, whereas it was correlated with body size only through the effects of longevity (table 5.6).

The Effects of Relative Body Size and Age

Although absolute body size was not directly correlated with mating efficiency (table 5.5), it was still possible that an individual's relative size affected his mating ability. This is because the average size of males varied during the course of the breeding period (table 5.7), and thus each individual's relative size also varied. We analyzed the importance of a male's relative size to his mating efficiency in two ways.

The first approach was to examine the relationship between male relative size and mating efficiency. For every male that contested for a female, we calculated the proportion of other males during each contest that were smaller than he was. We calculated relative size per contest rather than relative size per day because a male's relative size alters when males in copula are effectively removed from the population. The median relative size per contest for each male during his lifetime was significantly correlated with mating efficiency (fig. 5.5). However, relative size was also correlated with male longevity (r_s = .381, p = .02, n = 35), and male longevity was correlated with mating efficiency (see table 5.6). Partial correlation analysis reveal that mating efficiency was still correlated with both relative body size (r = .356, $p < .05$) and longevity (r = .388, $p < .05$).

The second approach was to look at a male's mating success over a week. The mating season was divided into four weeks. Each male was assigned to one of the four weeks, depending upon which week was most representative of his life. Where a male overlapped two whole weeks, he was assigned to the week that adjoined the maximum number of additional

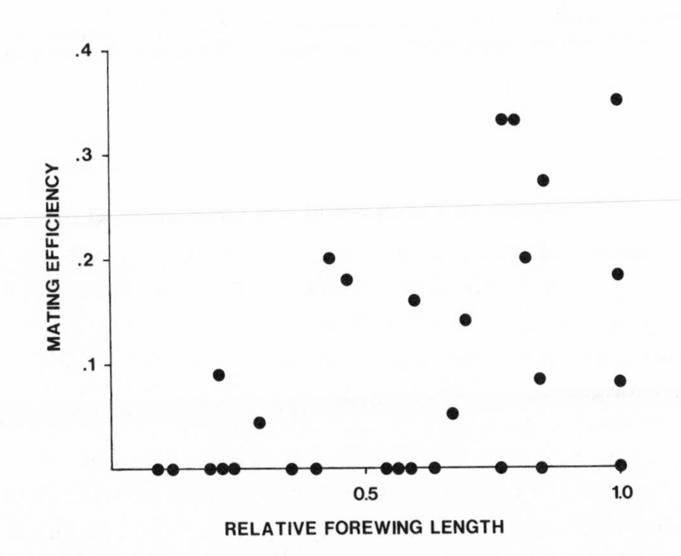

Figure 5.5: Relationship between mating efficiency and relative forewing size of breeding males (r_s = .475, $p < .005$, n = 29).

days he was alive. The average male size, number of mates, days present, and age were calculated for each of the four weeks (table 5.7), and we then derived the deviation from the average for each male by simply subtracting the average score from his score for each variable.

Deviations in size, mates, and days were significantly correlated with each other (deviation in size with deviation in mates, $r_s = .485$, $p < .005$, $n = 35$; deviation in size with deviation in days, $r_s = .369$, $p < .02$, $n = 35$; deviation in days with deviation in mates, $r_s = .604$, $p < .001$, $n = 35$). Partial correlation analysis revealed that deviation in number of mates was still correlated with deviation in size ($r_s = .354$, $p < .05$) and deviation in days present ($r_s = .523$, p $< .05$). Age at time of mating was not an important variable determining mating efficiency. Deviation in age at time of mating was correlated with deviation in mates ($r_s = .389$, $p < .01$), but this was primarily due to the effects of longevity. Deviation in days present and deviation in age were also correlated ($r_s = .625$, $p < .001$), and the partial correlation of deviation in age on deviation in number of mates, controlling for deviation in days present, was not significant ($r = .122$, $p > .2$). Thus there was no evidence for age-specific mating success among males.

Determinants of Female Fecundity

Methods

Females were reared in the laboratory on *A. irrorata* and tended by colonies of *Iridomyrmex* sp. 25 ants (see Pierce 1983). After eclosion, females were weighed, measured, and allowed to mate with males of known age and size. They were then placed in individual oviposition cages containing a small cutting of *A. irrorata* and lengths of scored wooden dowling upon which they could lay eggs. Females were fed three times daily with a 3:1 mixture of water and honey. We recorded the total number of eggs each female laid during her lifetime.

Results

Females were observed in the field site for a much shorter period than males (see table 5.1). We do not know whether this is because females have a shorter life span than males or because they emigrate more frequently and at all ages. Unlike males, there was no correlation between female body size and longevity in the field (forewing length: $r_s = .187$, $p > .10$; body length: $r_s = .158$, $p > .15$, $n = 45$). In the laboratory, females survived between ten and twenty-two days (mean = 14 days, s.d. = 1.8, $n = 16$) and laid between 55 and 455 eggs with a mean of 237 (s.d. = 131, $n = 16$). There was a highly significant correlation between female weight and the number of eggs she laid per day (fig. 5.6). The total number of eggs a female laid was not correlated with either the number of days she lived in the laboratory ($r_s = .090$, $p > .35$, $n = 16$) or the weight of her male mate ($r_s = .075$, $p > .35$, $n = 16$).

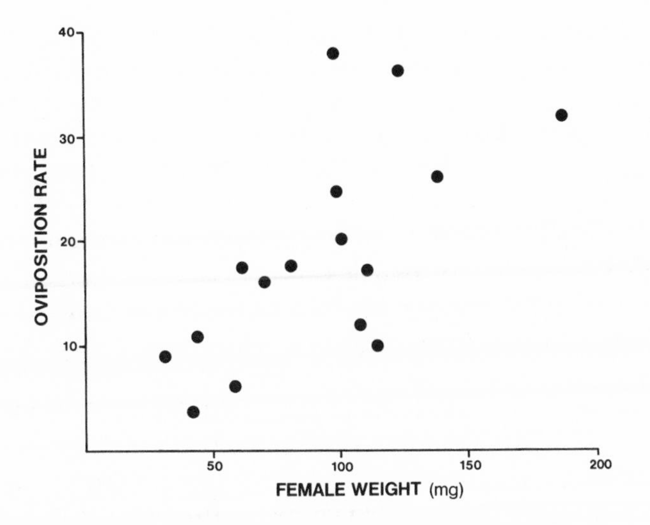

Figure 5.6: Relationship between female body weight and oviposition rate in *J. evagoras* under laboratory conditions (r_s = .641, p = .004, n = 16). Oviposition rate is the total number of eggs laid per total number of days alive.

Size-Selective Mating

Since larger females have a higher fecundity than smaller females and larger males have a competitive advantage, we might expect to find size-selective mating in *J. evagoras*. We found no evidence of this; relatively larger males did not mate with larger females (r = .046, $p > .4$, n = 31). The lack of linear size-selective mating may be related to a male's expectations about potential future matings. In our study site, the probability that at least two females of *J. evagoras* would eclose on any day during the breeding period was .34. It seems unlikely that a male could increase his reproductive success by rejecting a smaller female and waiting for a larger one.

5.3 Discussion

Two types of mate-locating behavior are commonly described for butterflies: perching (or territorial behavior), in which a male alights in a characteristic location and investigates passing butterflies that might be potential mates; and patrolling, in which males fly almost continuously in search of females (see Scott 1972, 1974, 1975; Rutowski 1982, 1984; Silberglied 1984). In both cases, a male's searching behavior is dependent upon the unpredictable arrival of adult females. In *J. evagoras*, the location of female pupae *is* predictable. Males locate plants with conspecifics and ants, learn their positions, and trapline from one plant to the next in search of eclosing females. Although they may remain on one plant for several hours, they will readily leave that position for another (for example, to join a mating ball); they are not territorial as are other lycaenids (e.g., Powell 1968; Scott 1972; Douwes 1975; Suzuki 1976; Alcock 1983).

It is not surprising that males have incorporated ants into the set of cues they use in locating conspecifics, because of the obligate association between *J. evagoras* and its attendant ants. Healthy larvae and pupae are never found in the field without attendant ants, and females use workers of *Iridomyrmex* sp. 25 as cues in oviposition (Pierce and Elgar 1985). Males are capable of learning the positions of plants with and without ants, and this may decrease energetic costs and the risks of predation by reducing the time spent searching for females.

Males also used pupal age as a cue in their mate-searching behavior. The cue the pupa emits is probably a volatile pheromone that the males detect through olfaction, because if an observer crushes a late-stage pupa, his fingers also become attractive to males. Males of birdwing butterflies *Ornithoptera priamus caelestis* (Borch and Schmid 1973) and *O. p. poseiden* (A. Hiller, pers. comm.) as well as several species of *Heliconius* (Bellinger 1954; references in Brown 1981; Boppré 1984) gather around pupa that are about to eclose. In certain Heliconiinae, males can apparently detect the difference between male and female pupae (Gilbert and Longino, cited in Boppré 1984). This is not the case for *J. evagoras*, which is surprising, since there is considerable competition for mates and selection for sex discrimination might be expected. One possible explanation for the lack of sex discrimination by males of *J. evagoras* is that if the probability of two or more pupae eclosing simultaneously is very low, then there simply may not be a cost to waiting for a male pupa to eclose.

The presence of males sitting in a group around a pupa is a good indicator that a pupa is about to eclose, and males of *J. evagoras* use each other as cues in mate searching. Attraction to conspecific adults is quite common among butterflies, including lycaenids (Douwes 1975), although the context in which attraction occurs is not always distinguished (see discussion in Silberglied 1984).

Body size in *J. evagoras* is clearly an important feature of male reproductive success. This is because male body size is correlated with both longevity and mating efficiency, two components of male lifetime mating success. It is interesting to note that although variation in male lifetime mating success does not greatly exceed a random model, it does not necessarily indicate that mating is in any sense random (cf. Sutherland 1985a). Our data show that larger males are generally more successful, and hence the assumptions of a Poisson distribution are not met.

One important result from our observations of *J. evagoras* concerns the relationship between absolute and relative body size. We found that mating efficiency was more strongly correlated with relative size than with absolute body size. Relative size was a more appropriate measure than absolute size once longevity was controlled for in these mating contests, because the average size of individuals changed over the course of the season. This may be a common feature of insect mating systems, and future analyses of male mating success in insects should consider relative as well as absolute size. Interestingly, there was no evidence for age-specific mating success; relative mating success was not correlated with male age at time of mating except through the effects of longevity.

Little evidence exists for a relationship between body size and mating success in other butterflies. Territorial defense in the black swallowtail butterfly *Papilio polyxenes* depends largely on the length of tenure (Lederhouse 1982), and the outcome of disputes over territories in the speckled wood butterfly *Parage aegeria* is usually resolved by an ownership convention (Davies 1978). Both these studies implied that mating success was positively correlated with territory ownership. Wickman's (1985) study of the small heath butterfly *Coenonympha pamphilus* is the first to confirm this assumption.

Female fecundity was strongly correlated with body size, suggesting that body size is an important component of female lifetime reproductive success in *J. evagoras*. The relationship between body size and fecundity has been commonly found in the Lepidoptera (e.g., David and Gardiner 1961; Baker 1968; Labine 1968; Marks 1976; Suzuki 1978; Lederhouse 1981; Hayes 1981; but see Boggs 1986) and in other insects (see Thornhill and Alcock 1983 for review). Although we do not have direct field measurements of female lifetime fecundity, our field results indicate that female longevity is probably quite short. It seems likely that selection would favor females that lay most of their eggs within the first few days after mating (see also Boggs 1986).

Our results also provide evidence for a possible selective advantage of early male emergence in this species. The eclosion of males before females (often referred to as "protandry"; see Thornhill and Alcock 1983 for review) is a widespread characteristic of butterflies and other insects. Although protandry is commonly regarded as a mechanism that increases a male's encounter rate with females and thereby increases his reproductive success (Wiklund and Fagerström 1977; Singer 1982; Wiklund and Solbreck 1982), there have been no quantitative studies of the effects of protandry on mating success. Males of *J. evagoras* that eclosed earlier in the season encountered more females than males that eclosed later. This result was not confounded by male longevity or size, since neither variable was correlated with encounter rate. The analysis of the components of male lifetime mating success indicated that some of the variation can be explained by encounter rate, suggesting a selective advantage for protandry. This benefit may impose an upper limit on male body size; if there is a positive correlation between developmental time and body size, larger males may encounter fewer females and hence experience a lower mating success (Darwin 1871; Lederhouse, Finke, and Scriber 1981; Singer 1982; Partridge and Farquhar 1983).

The close association that larvae and pupae of *J. evagoras* have with ants may have influenced the mating system of this butterfly in several important ways. The propensity of myrmecophilous lycaenids to aggregate and to occur in highly localized populations is likely to be the result of relying upon attendant ants for defense (Pierce 1983; Pierce and Elgar

1985). Because of their effective ant guard, the larvae of species such as *J. evagoras* are able to pupate openly in clusters on their host plants. The dense, localized populations of *J. evagoras* and its conspicuous pupation sites would have the effect of promoting intense competition among males for females. Males are able to investigate regularly every plant in an area bearing conspecific pupae and engage in active tournaments for eclosing females. This would not be possible for a butterfly species that was widely dispersed or whose pupae were concealed and difficult to find.

Since attendant ants also play a role in male mate-searching behavior, selection may have favored those females that are able to maintain a retinue of attendant ants. Males are attracted to plants with workers of *Iridomyrmex* sp. 25 and investigate pupae with more ants more frequently than pupae with fewer ants. By generating competition among males, a female may end up mating with a male of higher quality or better competitive ability. Therefore, rather than excluding the possibility of female choice, the intense competition among males of *J. evagoras* may in fact allow females to make a passive mate choice (see Halliday 1983; Partridge and Halliday 1984). One observation that conflicts with this idea is the absence of sex discrimination by males of *J. evagoras*. If it were advantageous for females to maximize competition among males, then selection would favor any mechanism that allowed males to recognize female pupae.

Finally, attendant ants may exert a direct effect on the reproductive success of both males and females of *J. evagoras*. Our study has shown that body size is an important correlate of both male mating success and female fecundity. However, the presence of attendant ants places a considerable limitation on body size. In laboratory experiments examining the effect of ants on development of *J. evagoras*, Pierce et al. (1987) found that the pupae and adults of tended larvae were significantly smaller than the pupae and adults of their untended counterparts, presumably because of energy lost in feeding attendant ants, and this effect was especially true for females. Of course, in the field, the larvae and pupae of *J. evagoras* cannot survive without ants (Pierce 1983; Pierce et al. 1987). However, for *J. evagoras*, one of the costs of associating with ants is levied in the final adult size, and this cost is surely reflected in the reproductive success of these butterflies.

5.4 Acknowledgments

We thank Steve Albon, Anthony Arak, Steve Austad, Tim Clutton-Brock, Nick Davies, John Endler, Alan Grafen, Paul Harvey, Martin Taylor, and Per-Olof Wickman for their comments and criticisms. Pete Atsatt and John Smiley captured our five male emigrants, and Kim Benbow provided field assistance. Tony Hiller first showed us the populations of *J. evagoras* at Mount Nebo, and Roger Kitching has been a continuing source of support and encouragement. We are indebted to the hospitality of Charmaine

Lickliter, who not only allowed us to invade her paddock but also relieved arduous days with numerous cups of tea. Pierce was supported by a Fulbright Postdoctoral Research Fellowship, and Elgar by a Christ's College Research Scholarship and by travel grants from Christ's College, Cambridge, and the Cambridge Philosophical Society.

6 Reproductive Success and Group Nesting in the Paper Wasp, *Polistes annularis*

David C. Queller and Joan E. Strassmann

Eusocial insects are characterized by overlapping generations of individuals that live together and cooperate in caring for the brood of a few reproductively active individuals (Wilson 1971). Eusociality is restricted to wasps, bees, ants, and termites. In the most complex species, morphologically distinct castes perform specialized behavioral tasks (Wilson 1971). The primitively eusocial insects, like highly social vertebrates (Wilson 1975), lack sterile castes. Although many individuals forgo most or all of their potential offspring production, all are capable of reproducing at some time in their lives. The mixing of progeny of different individuals within a common nest complicates the measurement of reproductive success, but it also provides a special incentive for making the effort. It is only when individuals can produce offspring that we can investigate how they are selected to forgo that option.

Of all the primitively eusocial insects, paper wasps of the genus *Polistes* are among the easiest to study because they build exposed paper nests that permit easy observation (fig. 6.1). There have been a number of studies of reproductive success in *Polistes* (e.g., West-Eberhard 1969; Gibo 1978; Noonan 1981; Klahn 1981). Our study of *Polistes annularis* offers the advantage of several years of data on a large sample of nests in a natural situation (not on buildings, and not artificially protected from predation). *P. annularis* is also particularly interesting because it has an unusually high degree of group nesting among spring females.

Every spring, new nests are founded by mated females of *P. annularis*. These females, known as foundresses, begin nests within a few meters of their natal nest, often in association with other females who emerged from the same nest the previous year (Rau 1931; Strassmann 1979; Strassmann 1983). Association sizes range from one to twenty-two foundresses. One female, the queen, lays most of the eggs. The others help in building the nest, as well as in feeding and defending the brood, until the first workers emerge. Then they cease all foraging and remain on the nest.

Figure 6.1: Polistes annularis foundresses on their nest.

The first brood to emerge from the nest consists of worker females and, on some nests, a few males. Workers remain associated with the nest and help to rear brood. Though a worker has a low probability of becoming an egg layer, she can mate and become a queen if the foundresses die. After at least one generation of workers, reproductive females emerge, followed by males. Whether a female becomes a worker or a reproductive female that will mate, hibernate, and then nest the following spring is probably not fixed at emergence but depends on conditions at that time (Strassmann et al. 1984). Nests cease brood production at dates that vary from 19 July to 22 October. In autumn females gather nectar and store it in concentrated form in the nest for later consumption. Females return to their nests on warm winter days to feed on their honey and to interact with nestmates. Mating takes place at hibernacula in early January.

We will not deal with some important aspects of reproductive success, such as sexual selection and the effects of age structure, that are treated in many other chapters of this book. Instead we will focus on the reproductive success of foundress females and particularly on group nesting. We begin by analyzing the variance in foundress reproductive success and then examine selection for group nesting, using both quantitative genetic measures and inclusive fitness methods.

6.1 Methods

This population of *Polistes annularis* is found on a west-facing cliff with a large overhang that overlooks Lake Travis, about 30 miles west of Austin, Texas, USA. Active nests, which number in the hundreds in any given year, occur only along a 200 m section of cliff face that is out of direct sunlight for most of the day. Studies like this are possible because the entire colony cycle takes place on or in the cliff, from the building of daughter nests within meters of the parental nest to hibernation in cracks in the cliff. The analyses here focus on data collected in 1977 and 1978, since these years had the most complete information. Data on many of the points discussed here are also available for 1976, 1979, and 1980. These data show that 1978 was a fairly typical year for wasp reproduction, whereas 1977 was a poor year. Weather records for Austin, Texas, indicate that 1978 was normal in both temperature and rainfall but 1977 included a summer drought that lasted several months. (Earlier in 1977 a flood of the reservoir below the cliff destroyed some nests, but these were excluded from the fitness analyses.) The 1977 data, though atypical, do show reproductive success in a crunch year, and such years could be important in the evolution of foundress associations.

The number of foundresses on nests was assessed in two ways. On some nests, an exact count was obtained by marking all foundresses seen on the nest. On others, the number of foundresses was scored as the maximum number present at any one census. This method is unlikely to miss many foundresses, since censuses were typically done at night or early morning when all females are likely to be on the nest. It is also unlikely to count foundresses twice, since they rarely move between nests (Strassmann 1983). Censuses were conducted at about three-day intervals before worker emergence and every four to ten days thereafter. To obtain an estimate of what foundresses achieved on their own (without workers) we counted the number of pupae just before the emergence of the first brood (28 April and 1 May in 1977, 1 May in 1978).

The reproductive success of a colony was scored as the number of reproductive females reared from the nest. They were counted (and marked with enamel to obtain data on overwintering success) by etherizing all nest occupants when no developing brood remained. Reproductive females can be distinguished from workers at the end of the season because they do not forage extensively and so have unworn wings. In addition, their abdomens are full of thick, white fat, which can be seen through the intersegmental membrane (Eickwort 1969; Strassmann et al. 1984). Males were also counted, but they are excluded from most of the analyses because they may leave their natal nests a few days after emergence, which would lead to undercounting, and because they sometimes move to the vicinity of other nests, which would lead to misattribution.

Reproductive success was computed as the probability of surviving to reproduction times the mean reproduction of those that survive, but first a sampling bias had to be corrected. The first component was scored for all nests, but only a subset of successful nests was counted and marked because this required etherization (although etherization did not in fact seem to disorient or otherwise cause harm). Therefore failures were overrepresented relative to successful nests that were scored for number of reproductives. We arrived at unbiased statistics (means, variances, covariances) on reproductive success of foundresses by weighting foundresses on failed nests by that proportion of foundresses from successful nests that were scored for reproductives (.905 in 1977 and .418 in 1978).

Because of nonnormality, we do not report significance levels for our covariance measures of selection. Instead, we report significance levels for a nonparametric measure of association, Kendall's coefficient of rank correlation. To correct for the weighting problem, this was calculated after randomly eliminating failed nests in the proportions given in the paragraph above.

Data on the reproductive success of foundresses were of two kinds. For some analyses, it was sufficient to divide the colony reproductive success equally among foundresses, that is, use the between-group component of individual fitness. Data on the relative contributions of foundresses within nests was obtained by a total of 212 hours of observations of egg laying by individually marked foundresses on eighteen nests (Strassmann 1981b). Analyses exclude nests in which fewer than two eggs were observed to be laid.

If all the foundresses on a nest die, a mated worker becomes queen, so our estimates of foundress fitness sometimes count granddaughters rather than daughters. Reproductives emerge from a nest for about thirty days before it ceases to contain developing brood. Subtracting the egg-to-adult development time of about forty days (Strassmann and Orgren 1983) from this interval roughly estimates the period when reproductive eggs are laid. In 1977, 41% of nests that reared reproductives lost all foundresses before this period, and 27% lost them during it (n = 47 nests). In 1978 these figures were 44% and 51%, respectively (n = 39). Our estimate of fitness is therefore unconventional from the standpoint of selection equations, which typically consider one generation at a time. But there is no reason, for example, why a selection differential (difference between unselected and selected means) cannot be calculated over two generations if care is used in its interpretation. The extent to which this selective difference is heritable will be influenced by the fact that it sometimes involves a second episode of genetic transmission and recombination, and the selective response will be diminished accordingly.

6.2 The Opportunity for Selection

Within and between Groups

The opportunity for selection (Arnold and Wade 1984a), sometimes called the intensity of selection (Crow 1958), is the variance in individual relative fitness. Since relative fitness is scaled by mean fitness (2.36 reproductive females per foundress in 1977, 5.82 in 1978), the opportunity for selection serves as a useful measure for comparing reproductive variance in different species. It measures the amount of change in fitness due to selection within a generation and, if fitness were completely heritable, the change between generations. As a consequence, it also describes the upper limit of change for any phenotypic character (Arnold and Wade 1984a).

In a social species such as *P. annularis*, individual fitness can be written as $p_{ij}W_i$, where W_i is the number of reproductives produced by

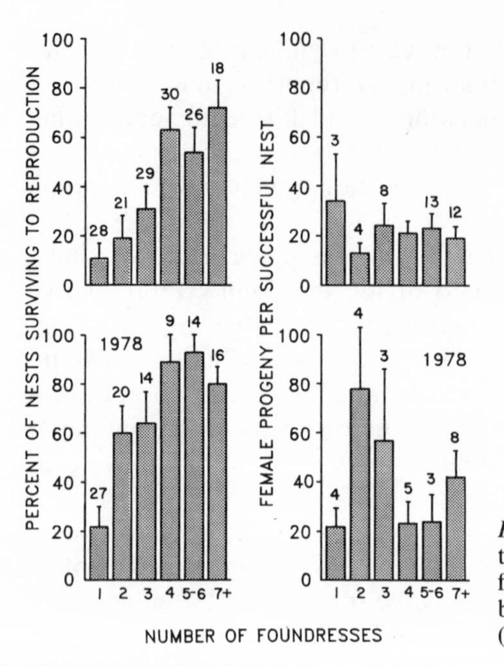

Figure 6.2: Colony survival and reproductive success of surviving colonies for different foundress association sizes. Error bars denote standard errors. Sample sizes (colonies) are given over the bars.

colony i, and p_{ij} is the fraction of them descended from the jth foundress. We have considerable information on colony fitness (fig. 6.2), and some on how colony reproduction is divided among foundresses (table 6.1), but since the latter set of nests was destroyed by flooding, the two sets of data do not overlap. However, the required estimates of the mean and variance of $p_{ij}W_i$ can be synthesized from the two separate data sources using the following approximations (Bury 1975):

$$\overline{p_{ij}W_i} \approx \bar{p}_{ij}\overline{W}_i$$

$$V(p_{ij}W_i) \approx \overline{W}_i^2V(p_{ij}) + \bar{p}_{ij}^2V(W_i) + 2\bar{p}_{ij}\overline{W}_i\text{cov}(p_{ij},W_i),$$

where the mean and variance of W_i are weighted by the number of foundresses in the group. The selection opportunity or intensity, I, is therefore approximately

$$I = \frac{V(p_{ij}W_i)}{\overline{p_{ij}W_i}^2} = \frac{V(p_{ij})}{\overline{p_{ij}}^2} + \frac{V(W_i)}{\overline{W}_i^2} - \frac{2\text{cov}(p_{ij},\ W_i)}{\overline{p_{ij}}\,\overline{W}_i}, \tag{1}$$

which consists of separate intensities due to p_{ij} and W_i and a cointensity (see Arnold and Wade 1984a for the components missing from this approximation).

Calculation of the values in eq. (1) is straightforward except for the covariance, which poses a problem because p_{ij} and W_i are obtained from different data sets. But $\text{cov}(p_{ij},\ W_i) = \text{cov}(p_{i.},W_i)$, where $p_{i.}$ is the mean p_{ij} for a nest, which is the reciprocal of foundress number. Since foundress numbers are known for all nests with W_i data, the covariance can be calculated. (Note that the utility of this method is not limited to data sets with no overlap; it also allows one to take advantage of having many more values for W_i owing to its being easier to sample.)

For 1977, the three components of I in eq. (1) are 4.38, 2.46, and −0.43, respectively, and assuming that the 1977 observations on p_{ij} are valid for 1978, the 1978 components are 4.38, .846, and −0.42. Their sums, the total opportunities for selection, are reported in table 6.2. These estimates could be slightly biased because the p_{ij} estimates in table 6.1 come from a sample of nests with somewhat more than the average number of foundresses. In addition, since queens usually survive longer than all their subordinate cofoundresses (thirty-four of thirty-nine nests), our p_{ij} values probably somewhat underestimate queen fitness relative to subordinates.

Though they are useful for calculating I, the components of eq. (1) are less interesting than the decomposition of I into components due to the mean fitness of members of group i and to the deviation of individual

Table 6.1 Egg Laying by Foundresses in *Polistes annularis*

Nest Number (i)	Number of Females	Number of Eggs Laid	Egg Layer (j)	Proportion of Eggs Laid (p_{ij})
1	3	7	1	.571
			2	.286
			3	.143
2	4	4	1	.750
			2	.250
3	15	6	1	.333
			2	.500
			3	.167
4	3	2	1	1.000
5	6	2	1	.500
			2	.500
6	8	5	1	.800
			2	.200
7	23	11	1	.364
			2	.364
			3	.182
			4	.091
8	7	2	1	1.000
9	6	2	1	1.000
10	2	2	1	1.000
11	12	5	1	.200
			2	.400
			3	.400
12	9	3	1	.333
			2	.666

Note: The queen, determined by behavioral observations (Strassmann 1981b), is foundress 1 on each nest. Foundress 2 on nest 11 became queen during the course of the study.

Table 6.2 Opportunity for Selection on Foundresses in *Polistes annularis*

	1977	1978
Within groups	1.87	1.53
Between groups	4.54	3.28
Total	6.41	4.81

fitness from the group mean. This is an analysis of variance in relative fitness, into between-group and within-group components. This partition is of interest because of the view that the evolution of an altruistic social trait depends on positive between-group selection overriding negative within-group selection (Wilson 1980; Wade 1985).

The variance of the group mean fitness, taken over all individuals, is calculated as $\Sigma_i N_i(w_{i.} - 1)^2/(N - 1)$, where N_i is the number of foundresses on nest i, N is the total number of foundresses, $w_{i.}$ is the mean individual relative fitness of group i, and 1 is the population mean fitness. Subtracting this component from eq. (1) gives an estimate of the variance of individual deviations from the group mean. These values are reported in table 6.2. The within-group component and the total may be somewhat underestimated because some sources of within-group variance have not been included, particularly that due to varying times of foundress death.

What do these measures mean? The view that selection between groups is group selection while selection within groups is individual selection can be misleading (Grafen 1984; Nunney 1985), and one feature of our partition illustrates this. Single-foundress nests affect only the between-group component, because a single foundress's fitness cannot deviate from its group mean. Yet it hardly seems appropriate to attribute selection on single foundresses entirely to group selection.

Nevertheless, this partition of I is relevant to social selection, whatever we choose to call the components. It shows that about one-third of the variation in reproductive success results from individuals' outcompeting members of their own colony. It also shows the maximum amount that individual relative fitness can change owing to that cause compared with other causes.

Sequential Selection Episodes

For the between-group component of individual relative fitness, we can further partition the opportunity for selection, I, into parts due to different episodes of selection, specifically into parts due to survival to worker emergence, survival to reproduction, and number of reproductive females reared. Since it is the between-group component that is being partitioned, each foundress within a group is assigned the same fitness value for each component. If the colony survives a given selection episode, each foundress gets a value of 1 for that fitness component; if the colony fails, each foundress gets a value of 0. Similarly, the number of reproductives reared on a nest is divided equally among its foundresses.

The reason for concentrating on the between-group component of the selection opportunity is partly practical; we do not have sufficient data on within-group fitness. But it is also true that the between-group component is of special interest for a question to be considered later, selection on foundress association size. Since association size is a phenotypic trait that has the same value for all members of a group, there is no within-group component of selection on it. By partitioning the between-group component of I, we will later be able to see to what extent the selection opportunity is actually realized in selection on association sizes.

The fitness model is multiplicative: total fitness = (survival to worker stage) × (survival to reproduction) × (number of reproductives). For multiplicative fitness models, Arnold and Wade (1984a) have derived a partitioning of I into a set of intensities, one for each episode of selection, and a set of cointensities of selection between different episodes. This partition has its uses, but for models like ours it has a drawback in the way it treats induction of zero fitness values by early selection episodes. If, for example, a colony fails to survive to worker emergence, its foundresses must be assigned zero values for later selection episodes, regardless of how successful they might have been had they actually experienced those later episodes. The Arnold and Wade partition incorporates such strings of zero values into the cointensities, but it seems preferable, at least for sequential episodes, to attribute this effect solely to the early episode where the selection actually occurs.

This property is possessed by an alternative partition of I that can be derived from another of Arnold and Wade's results. The selection differential is the difference between mean values of a phenotypic trait before and after selection. It can be written as S = cov(z,w), where z is individual phenotypic value and w is relative fitness (Robertson 1966; Price 1970). For a multiplicative fitness model with k components or selection episodes, the selection differential can be partitioned into additive components as

$$S = \mathrm{cov}(z,w) = \Sigma_k\, \mathrm{cov}(z,w_k), \qquad (2)$$

where w_k is the individual's relative fitness in the kth selection episode (Arnold and Wade 1984a). In each episode, an individual's w_k is weighted by the cumulative fitness from previous episodes. For example, failure to survive the first episode ($w_1 = 0$) would give zero weight to that individual in calculating the covariance for the next episode. Relative fitness can be viewed as a phenotypic trait with the selection differential, cov(w,w) = var(w) = I, that is, the selection differential on relative fitness is the selection opportunity. From eq. (2), I can therefore be written as $\Sigma_k\, \mathrm{cov}(w,w_k)$.

This partition of I has the minor drawback that its components are not themselves selection opportunities (variances in relative fitness), but retaining the form of the whole seems less important than retaining its properties. Each of the covariances between relative fitness in the kth episode and total relative fitness measures the amount of selection on total

Table 6.3 Components of the Selection Opportunity (I_k) and the Selection Differential (S_k) on Foundress Association Size, Owing to Three Selection Episodes

Episode (k)	Sample Size (n)[a] (1)	I_k (2)	% of $\Sigma_k I_k$ (3)	S_k (4)	% of $\Sigma_k\|S_k\|$ (5)
1977					
Survival to worker stage	510	.33*	7.2	.47*	23.8
Survival to reproduction	384	.60*	13.2	.27	13.6
Number of female reproductives	266	3.62*	79.6	−1.24*	62.6
Total	—	4.55	100.0	−.50	100.0
1978					
Survival to worker stage	189	.10*	3.1	.63*	12.8
Survival to reproduction	171	.15*	4.6	−.09	1.8
Number of female reproductives	151	3.02*	92.3	−4.22*	85.4
Total	—	3.27	100.0	−3.68	100.0

[a] To correct for a sampling bias, foundresses on failed nests are counted as .905 units in 1977 and .418 units in 1978 (see methods).

*The associations represented by these covariance values are significant at $p < .01$, as assayed by the nonparametric Kendall's rank correlation.

relative fitness in the kth episode, thereby retaining in miniature a useful property of the whole opportunity for selection.

Table 6.3 reports estimates of cov (w,w_k) for the three multiplicative fitness components. The two survival components composed a relatively small fraction of the selection opportunity. In 1977 colony mortality was 38% before worker emergence and an additional 21% before reproduction. For 1978 these figures are 25% and 12%. In each of these periods, 80%–100% of the failures were due to the death or disappearance of all resident females. The other nests either were knocked down by birds or fell down with part of the crumbly limestone cliff. These nest mortality values are quite low for *Polistes* (Yoshikawa 1954; Gibo 1978; Strassmann 1981a), perhaps because of the protected location or the large colony sizes. By far the greater part of the opportunity for selection on foundresses was due to the number of reproductives produced by successful colonies.

6.3 Selection on Foundress Association Size

While the selection opportunity tells us about the potential for selective change, most biologists really want to know about selection on actual traits. The most pressing problem posed by *Polistes* and other eusocial insects is how selection can cause individuals to aid others at the expense of their own reproduction. In *P. annularis* this question is relevant both to subordinate foundresses and to workers. We will focus on the former, primarily because foundresses sometimes adopt the alternative of nesting alone, allowing estimation of selection parameters.

How does selection act on foundress grouping? We can first address the question from a purely phenotypic standpoint. The number of foundresses associating can be treated as a phenotypic (in this case behavioral) trait of each of the foundresses in the group. For example, each of the wasps that form an association of five is assigned a phenotypic value of 5. Since *Polistes annularis* foundresses rarely switch between groups (Strassmann 1983), each can be assigned a single fixed, phenotypic value. This value is meant to reflect the wasps' early-season choice of association size, so it does not change when subsequent mortality of group members reduces the actual association size.

This trait is subject to selection like any other. Since we have fitness estimates for associations of different sizes (fig. 6.2; table 6.4), we can calculate phenotypic measures of selection on the trait. These are reported in table 6.3. As noted earlier, the selection differential, which measures the average difference between unselected and selected phenotypes, is the covariance between individual relative fitness and the phenotypic value. We have used formula (2) to calculate this in parts corresponding to the same multiplicative fitness components used in the analysis of the selection opportunity. Again, the calculations have been performed using the between-group component of individual fitness, a simplification that is permitted because there is no within-group component of the phenotypic trait.

Table 6.3 shows that survival to worker emergence exerts directional selection for larger associations. Survival from that time until rearing of reproductives has a smaller effect, the direction of which differs in the two years. The most pronounced effect is due to the number of reproductives produced by surviving colonies, which exerts strong directional selection for smaller foundress associations. Note that while larger associations

Table 6.4 Inclusive Fitness in *Polistes annularis*

	Number of Foundresses per Nest					
	1	*2*	*3*	*4*	*5–6*	*7+*
1977						
1. Mean number of subordinates (n)	0	1	2	3	4.5	7.1
2. Mean colony success (C_n)[a]	3.61	2.48	7.45	13.53	12.30	13.54
3. Minimum r needed if x =0	—	none	none	none	none	none
4. Minimum r needed if x =.088	—	none	none	none	none	none
5. Maximum A_s possible if r =.5, x =.088	—	none	1.24	2.12	1.35	1.06
6. Maximum A_s as % of C_0	—	none	34	59	38	30
1978						
7. Mean number of subordinates (n)	0	1	2	3	4.4	10.5
8. Mean colony success (C_n)[a]	4.94	47.25	36.43	20.97	21.98	38.02
9. Minimum r needed if x =0	—	.11	.31	none	none	none
10. Minimum r needed if x =.088	—	.03	.17	none	none	none
11. Maximum A_s possible if r =.5, x =.088	—	23.06	9.23	3.40	2.62	2.43
12. Maximum A_s as % of C_0	—	466	187	69	53	49

Note: r =relatedness; x =each subordinate's direct reproduction relative to the queen's; A_s = the expected fitness of subordinates if they nested alone; C_0 = fitness of lone foundress.

[a] Calculated from data in figure 6.3.

sometimes outproduce small ones (fig. 6.2), the effect is neither consistent nor very pronounced, and the per capita measures of selection we have calculated indicate a clear advantage to being in smaller groups.

Column 5 of table 6.3 shows what percentage of the total selective change (Σ_k $|cov(z,w_k)|$) is due to each component. Comparison with the percentages in column 3 shows that in both years, the first episode causes more change than expected on the basis of its contribution to the selection opportunity, and the last episode causes less change than expected. That is, more of the selection opportunity was realized in the early episode. This may be because foundresses were still present in approximately their original numbers. By the last episode most foundresses had died, and the original foundress association number would affect fitness primarily in an indirect and attenuated way (e.g., through the workers they reared).

Figure 6.3 depicts the selection differentials for each episode of selection and the changes they are predicted to produce in the value of the mean association size. The actual changes owing to the third component, the number of reproductives produced, will depend on the genetic transmission of the trait as well as on phenotypic selection. If heritability of this component is zero, it will effect no change. If heritability is 1, then it should change mean association size by the full amount of the selection differential. The shaded areas in figure 6.3 show the range encompassed by these extremes, and the mean size of the next year's associations is predicted to fall in this range. The observed value for 1979 falls within the predicted range, though the range is so large that this is not surprising.

However, the observed value for 1978 is far higher than the upper end of the predicted range. The only fitness component missing from this analysis is the overwintering success of the reproductive progeny. Its selection differential was not reported in table 6.3 or figure 6.3 because it seemed more likely to be a trait of these progeny rather than of the spring foundresses. This tends to be supported by the small values of the selection differentials on foundress association size for this component, −.173 for 1977 and −.219 for 1978 (based on overwintering data from thirty-four and thirteen colonies). Including these values would shift the predicted ranges to the left. This only increases the discrepancy between observed and predicted values for 1978.

The discrepancy might be due to some environmental difference between years that caused the genes affecting foundress association to be expressed differently; 1977 was a poor year in terms of the numbers of reproductives reared, presumably because of the drought. If those wasps that were reared were generally in poor condition, then we can hypothesize that they may have been more likely to join together in large groups in 1978. But as we discuss below, there is no evidence that *P. annularis* foundresses in poor condition form larger groups (Strassmann 1979; Sullivan and Strassmann 1984).

Even in the absence of environmental differences between years, this kind of analysis has some important limitations. It provides an adequate description of phenotypic selection, which may tell us something useful

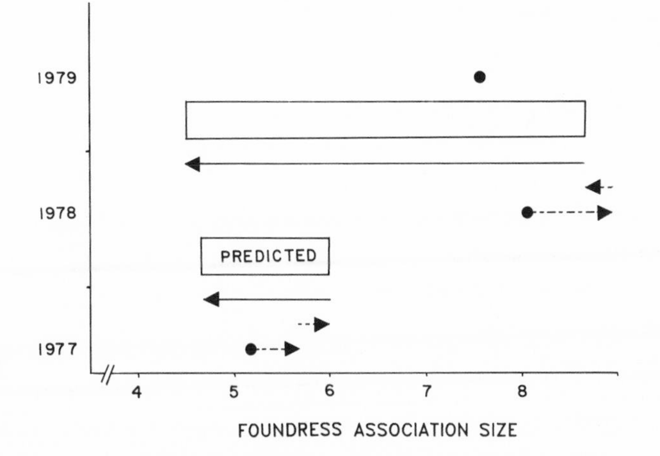

Figure 6.3: Selection on foundress association sizes in 1977 and 1978. Solid circles represent mean springtime association sizes, averaged over individuals. The arrows represent components of the selection differential as follows: selection due to survival to worker emergence—dashed arrows from circles; selection due to survival to reproduction—the other dashed arrows; selection due to number of female reproductives reared—solid arrow. Predicted ranges for spring associations in 1978 and 1979 are given.

about genetic selection when the relationship between genes and phenotypes is reasonably straightforward. But this relationship is more complex than usual for a social trait such as association size. First, the expression of the trait depends on other individuals. Consider a foundress who, given a complete range of choices, would join in a group of five relatives. Her actual choices will be limited by the number and behavior of her relatives. For example, she may have only one group of six relatives to join, or perhaps two groups of size two and nine. Second, individual environmental effects might be important. For example, wasps in poor condition might be more likely to join as subordinates, a possibility discussed below.

There is a third reason the relationship between genes and association size is complex. The transmission of genes is primarily through the direct reproduction of the dominant queen. But the size of the association is probably determined by the behavior of the joining subordinates (if subordinates are selected to join, queens are selected to accept them, but not vice versa). Problems involving this kind of separation between phenotypic traits and genetic transmission are generally analyzed by inclusive fitness methods, to which we now turn.

6.4 Inclusive Fitness

An individual's inclusive fitness is its production of offspring, stripped of all components that are due to the behavior of other individuals and augmented by the individual's contributions to the reproduction of others, with each such contribution multiplied by its relatedness to the individual affected (Hamilton 1964). Although the definition seems complex, its application to the evolution of social traits can be quite simple, because a behavior is predicted to be favored by selection if it results in a net increase in inclusive fitness, that is, when

$$\Delta w_a + r\Delta w_r > 0, \tag{3}$$

where r is the mean relatedness of the actors to the relatives affected, and Δw_a and Δw_r are their respective average fitness changes due to the behavior. Assumptions underlying formula (3), known as Hamilton's rule, have been discussed by Seger (1981), Michod (1982), Grafen (1984), and Queller (1984).

Polistes annularis foundresses often face several options in joining a colony, and as noted above, these may vary for different individuals depending on the number and grouping behavior of their relatives. Our analysis will focus on one particular kind of option, the choice between nesting alone and joining a group of relatives. One reason for this focus is pragmatic; this kind of choice is available to almost all foundresses, excepting only those that have no surviving natal nestmates to join. More important, this is the choice that is most germane to the theoretical question whether it pays individuals to forfeit their own direct reproduction in order to benefit others.

For each colony, let n be the number of foundresses subordinate to the queen (association size = n + 1) and let C_n be the mean number of reproductives produced by colonies with n subordinates. If each subordinate produces x of these reproductives for every one produced by the queen, the direct reproduction of each subordinate is $xC_n/(nx + 1)$, and the direct reproduction of the queen is $C_n/(nx + 1)$. Finally, let A_s and A_q represent the number of reproductives produced by a subordinate and queen respectively if they had nested alone. Hamilton's rule (eq. 3) can now be written

$$\left[\frac{x\,C_n}{nx + 1} - A_s\right] + \frac{r}{n} \cdot \left[\frac{C_n}{nx + 1} - A_q\right] > 0. \tag{4}$$

The 1/n in the second term is necessary because each of the n subordinates in a colony is responsible for that fraction of the queen's gain from having subordinates.

Generally, A_s and A_q have both been estimated by C_0, the reproductive success of singly founded (zero subordinates) colonies (Metcalf and Whitt 1977b; Gibo 1978; Noonan 1981). We will follow this procedure, but it is important to point out the limitations it imposes on the analysis. Since both strategies—join or found alone—are present in the population, we cannot predict a priori which should have the higher payoff, that is, whether Hamilton's rule should be satisfied. If foundresses are not constrained in their choices, the presence of both strategies leads to the expectation that their payoffs are equal. If one has a lower payoff, its presence in the population is probably due to a constraint that prevents some individuals from adopting the better strategy. For example, founding alone might persist as a strategy, even if it did not pay as well as joining, if some individuals had no relatives to join. Our analysis, and others like it, should therefore not be construed as a test of Hamilton's rule. Rather, Hamilton's rule is used as a probe for the existence of constraints that, by limiting options, contribute to variation in inclusive fitness.

What Relatedness Is Required?

Formula (4) can be used to ask several different questions. First, what value of r is required for joining as a subordinate to pay at least as well as founding alone? Because sex determination is haplodiploid in the Hymenoptera, relatedness could be as high as .75 for full sisters. Solving eq. (4) for r yields

$$r > \frac{n[(nx + 1)C_0 - xC_n]}{C_n - (nx + 1)C_0},$$

with the sign of the inequality reversed if the denominator is negative. Rows 3 and 9 of table 6.4 list the minimum required relatedness for associations of different size classes on the assumption that subordinates have no direct reproduction ($x = 0$), while rows 4 and 10 report similar values using an estimated value of $x = .088$ (each subordinate lays 8.8% of the number of eggs laid by the queen), calculated from the data in table 6.1 (since the regression of colony values of x on foundress number was not significant, the same value of x was used for all association sizes).

In 1977, subordinates in all association sizes do worse than single foundresses for all possible values of r ($r \leq .75$). The same holds for associations of four or more in 1978, but joining smaller associations is better than nesting alone for some of the possible values of r. Though we do not have a precise measurement of r, it is possible to obtain a rough estimate.

How Closely Related Are Foundresses?

Relatedness can be estimated by using recursion methods on at least one generation of pedigree information (Seger 1977; Murray 1985). Here we seek a recursion equation specifying r_s', the relatedness of spring foundresses, in terms of r_s, the relatedness of foundresses in the preceding spring. An equilibrial value can then be obtained by setting $r_s' = r_s$.

Spring foundresses typically associate with their natal nestmates of the preceding fall, and if they do so at random with respect to relatedness, then r_s' is equal to the relatedness of fall nestmates, r_f. Female nestmates in the fall include one or more groups of sisters, each the progeny of a different egg layer. Let the average relatedness within sister groups be r_{fw}

and the average relatedness between nestmates from different sister groups be r_{fb}, and designate the relative frequencies of these two kinds of relationship as $freq(r_{fw})$ and $freq(r_{fb})$. The average relatedness among female fall nestmates is:

$$r_f = r_{fw} \times freq(r_{fw}) + r_{fb} \times freq(r_{fb}). \tag{5}$$

Electrophoretic studies of *Polistes metricus* have shown that the daughters of any given female are generally full sisters (Metcalf and Whitt 1977a), so we will ignore the effect of the small amount of multiple mating and assign r_{fw} as 3/4. To be conservative in our estimate of r_{fb}, we will assume that all of the sisterships on a nest are daughters of different workers and that each of these workers is the daughter of a different spring foundress. If foundresses and workers both mate at random, this assumption implies that $r_{fb} = r_s/16$. Substituting this, as well as $r_f = r_s'$, into eq. (5) gives the desired recursion equation: $r_s' = 3 \times freq(r_{fw})/4 + r_s \times freq(r_{fb})/16$. Assuming the population is at equilibrium, the relatedness of spring foundresses will be the same in successive years. Substituting $r_s = r_s'$ and solving yields the equilibrium:

$$r_s' = \frac{3 \times freq(r_{fw})}{4 - freq(r_{fb})/4}. \tag{6}$$

The relative frequencies of within-sistership and between-sistership relations can be estimated from the spring egg laying data in table 6.1, provided we assume that the summer eggs that will develop into reproductives are divided among egg layers in the same manner. This assumption ignores two basic features of the *Polistes* colony cycle: that there are more potential egg layers (mated workers) as the season progresses, and that the dominance of queens tends to become more pronounced (Röseler et al. 1984). Though these two factors will likely cause some error in our estimates, the errors will be in opposite directions and may approximately cancel out. Let N_i be the number of reproductive females produced by the ith nest, and let p_{ij} be the fraction of these reproductives who are members of the jth sistership on the nest (for any i, $\Sigma p_{ij} = 1$). The numbers of full sisters and other female relatives of a female in sistership j of nest i are therefore approximately $p_{ij}N_i$ and $(1 - p_{ij})N_i$, respectively. This includes the individual herself among the full-sister relationships, but this error will be small for the large values of N_i in *P. annularis* (fig. 6.2). Multiplying these values by the number of sisters in the sistership, $p_{ij}N_i$, and summing over all sisterships in the population gives the total number of r_{fw} and r_{fb} relationships: $\Sigma\Sigma p_{ij}^2 N_i^2$ and $\Sigma\Sigma p_{ij}(1 - p_{ij})N_i^2$. Because of the flood, we do not have N_i values associated with the p_{ij} values in table 6.1, but by assigning to each successful 1977 nest the p_{ij} values of a randomly chosen nest in table 6.1, we obtain the sums 31,665 and 17,088, respectively. These correspond to values of .649 for $freq(r_{fw})$ and .351 for $freq(r_{fb})$, which, substituted into eq. (6), gives $r_s' = .498$. Following this procedure for 1978 nests gives frequencies of .703 and .297 and an r_s' value of .537.

Given all the assumptions required to obtain these estimates, they should be used with some caution. But a value of approximately .5 is accurate enough to perform the task required of it here. Table 6.4 shows that in 1978, subordinates in associations of size 2 and 3 require relatedness to be at least .029 and .170, easily exceeded by the estimated value of .5, and that all other associations require values above the maximum, .75. This means that 86% of subordinates in 1978, and 100% in 1977, seem to have joined associations that offer them a lower mean reproductive return than the alternative of nesting alone.

Is Joining the Best of a Bad Situation?

The inclusive fitness analysis does not alter the earlier conclusion that foundresses fare better on small colonies. As a rule, single foundresses outreproduce subordinates on other nests. It remains to examine three possible biases that could have exaggerated the advantage of small associations. First, if larger colonies produced proportionately more males, then their relative success will have been underestimated by counting only females. Second, it would not be surprising if smaller colonies had a higher incidence of worker takeover owing to the death of all foundresses, resulting in a dilution of foundress genes in the reproductive progeny. Third, small colonies may be less effective at preventing usurpation of the nest by a foreign female (Gamboa 1978).

The data on these points were not complete enough to include in the fitness estimates, but they are sufficient to suggest that these potential biases are not important in *P. annularis*. The data on males show that single foundresses do not differ significantly from others in the sex ratio of reproductives reared, although the number of single-foundress nests was small (Mann-Whitney U test, $p \gg .05$; In 1977 $n = 48$ with 3 singles; in 1978 $n = 22$, 3 singles). Nor was the sex ratio reared correlated with foundress association size (Kendall's rank correlation, $p \gg .05$). The same two tests proved negative ($p \gg .05$) for worker takeovers before or during the reproductive egg-laying period (1977, $n = 45$, 2 singles; 1978, $n = 16$, 1 single). Finally, out of 141 nests observed, only 10 were usurped by foreign females, and these were distributed evenly among association sizes (foundress numbers = 1, 2, 4, 4, 6, 6, 8, 8, 9, 10).

Thus there is no apparent bias, and it appears that subordinates really do fare worse than single foundresses. Why then do they join? As noted earlier, this may be an indication that their options are limited by their environment (in quantitative genetic terms, there may be a gene-environment interaction, with genes for single nesting being expressed only under certain conditions). This could be because there are not enough good nesting sites for all foundresses to nest alone. Or some foundresses may simply be poorly suited to nesting alone because of a poor genetic makeup or, more probably, because of an environmental insult such as poor feeding (West-Eberhard 1969). Indeed, parents (or workers) may have evolved feeding schedules that force some of the colony's progeny into choosing subordinate roles (Alexander 1974). If this argument is to explain joining, the environmental insult must affect the ability to found alone without similarly decreasing the ability to provide aid as a subordinate (Craig 1983).

How poor would subordinates have to be at nesting alone to account for their preference for joining? Solving eq. (4) for A_s, the expected fitness of a subordinate if she nested alone, yields

$$A_s < \frac{nx(C_n - rC_0) + r(C_n - C_o)}{n(nx + 1)},$$

if we continue to assume that queens nesting alone would not differ from single foundresses ($A_q = C_0$). Using the estimates of $r = .50$ and $x = .088$, the threshold (maximum) values of A_s that would justify joining are given in rows 5 and 11 of table 6.4. These are also listed as percentages of the mean fitness of actual single foundresses (rows 6 and 12). In 1977 it did not pay to be a subordinate in an association of two even if the alternative was death ($A_s = 0$). But aside from this case, being a subordinate will pay for foundresses who, by nesting alone, would be less than 30%–70% as successful as those who actually did found alone.

A female's condition, and hence her ability to nest alone, may be correlated with her size, as measured by wing length, dry weight, or fat weight. In this population of *P. annularis* there was no evidence that small size was an important cause of large foundress associations (Strassmann 1983; Sullivan and Strassmann 1984). Though queens were often the largest individuals on their nests, they were often smaller than subordinates on other nests. There was no tendency for smaller females to join larger foundress associations. The variance in wing length among females on a new nest was not different from that of all females from the same natal nest, suggesting that smaller individuals are not joining larger ones preferentially. Finally, most of the variance in the size of autumn reproductive females was between nests rather than within them. This means that if parents or workers were manipulating feeding in a way that ensured some optimal distribution of progeny around a critical size threshold, that threshold must differ considerably among natal nests, and this does not seem likely.

These data strongly suggest that poor condition does not control joining, but they do not prove it because the measures used may not be good enough indicators of condition. However, a clearer indicator is available. Experimentally depriving some females of their winter honey stores lowers their probability of survival and causes survivors to build smaller nests than undeprived controls (Strassmann 1979). Despite this

Table 6.5 Population Density and Association Size in *Polistes annularis*

Year	Total Number of Foundresses	Association Size (Mean ± s.e.) Averaged	
		Over Nests	*Over Foundresses*
1977	775	3.8 ± .2	5.2 ± .1
1978	667	4.7 ± .3	8.2 ± .2
1979	845	4.7 ± .3	7.6 ± .2
1980	557[a]	4.9 ± .2	6.1 ± .1

[a] May be an underestimate.

evidence of relatively poor condition, the deprived females did not form larger associations than the controls.

Females could be forced into nesting in groups because there are not enough good nesting sites. The cliff face on which the study nests were built is unusual in that it has a large overhang that protects most of the nests from the sun until after 1500 h. Nesting ends abruptly at the end of the overhang, and nests near the edge tend to have daughter nests toward the center of the colony rather than expanding the edges. Extensive searches of other cliffs around the lake failed to turn up other similarly suitable cliffs. The only other site where nests were found was a tree-shaded cliff. *P. annularis* is generally reported to nest in large aggregations near water (Rau 1929). For nest site constraints to cause large foundress associations, it is not enough that the site in general be restricted; it is also necessary that nests could not be more tightly packed within the site. Density of nests is great at the study site, approaching one nest every 2 or 3 m^2. It is not clear whether they could be even denser in some areas. Perhaps internest aggression limits the proximity of nests. It appears that nests are denser along ridges or cracks in the cliff, perhaps because sight lines between neighboring nests are blocked, which would reduce internest aggression. Also, the crack or ridge may provide a visual landmark that aids workers in finding their own nest (workers do sometimes join the wrong nest and stay with it). But on average a nest could expect to gain as many workers as it lost, so this may not be a major factor favoring nest spacing. A more detailed analysis of the effects of nest location on foundress number will be presented elsewhere.

That nest sites may be limiting in *P. annularis* is indicated by the fact that 1%–7% of the nests built in a given year were reuses of the natal nest of the preceding year (1977–80). Nest reuse is generally rare in *Polistes* (West-Eberhard 1969). It is a poor option because the cells are dirty and infested with organisms that eat the meconium deposited in the bottom of the cell. Even if the cells are not used over, but a new nest is built protruding from the old one, the entire structure is more likely to fall.

However, if nest sites are limited, we would expect that years with large total numbers of foundresses present would also have large association sizes. No such pattern emerges for 1977–80, regardless of whether association sizes are averaged over colonies or over individuals (table 6.5).

6.5 Discussion

In *P. annularis*, the opportunity for selection was greater between groups than within them, so we have concentrated more on the between-group component. It is worth noting however that the within-group component is not negligible. Behavioral interactions among *Polistes* foundresses on a nest strongly reflect reproductive competition (e.g., Pardi 1942; West-Eberhard 1969; Noonan 1981; Strassmann 1981b; Pfennig and Klahn 1985).

Both the selection differentials and the inclusive fitness analysis suggest that selection favors smaller association sizes in *P. annularis*. This result holds true for a "typical" year and also for an unusually dry year, in which reproduction was depressed. It is therefore difficult to argue that the apparent disadvantage of joining as a subordinate might be balanced by an advantage accruing to multiple nesters in unusually difficult years.

Other inclusive fitness studies of *Polistes* have shown similar results, though not as extreme. In two separate studies of *P. fuscatus*, subordinates in associations of two outperformed single foundresses, but those in larger associations did not (Noonan 1981; Gibo 1978). Subordinates in *P. metricus* seem to have an inclusive fitness advantage over single foundresses, but associations of more than two were not studied (Metcalf and Whitt 1977b). In each of these studies it was advantageous to be a subordinate on the smallest and most common kind of multiple association, and any joining of disadvantageous, larger associations could conceivably be explained as relatively uncommon mistakes or as the suboptimal tail of a generally adaptive distribution. But this is quite unlikely for *P. annularis*, since single foundresses outperformed subordinates on all association sizes in 1977 and outperformed 86% of subordinates in 1978.

Despite selection for smaller associations, there was no detectable decrease in mean association size over the course of the study (table 6.5). Perhaps there was no heritability for this trait. More likely, subordinates were not really experiencing as large a cost as we estimated, since they may not have been as productive as single foundresses even if they had nested alone. Foundresses surely differ in condition and body size, and the latter is an important factor in determining queenship in this and other species (e.g., Turillazzi and Pardi 1977; Noonan 1981; Sullivan and Strassmann 1984; Dropkin and Gamboa 1981). Yet we could find no evidence that body size or condition exerts a strong influence on association sizes. In contrast, single foundresses in *P. fuscatus* are larger than subordinates, and in one out of two years studied, body size of single foundresses was correlated with reproductive success (Noonan 1981). There is little evidence bearing on whether parents or workers manipulate some of their reproductive progeny into being subordinates by limiting their feeding as larvae. If this were true, reproductive females should be quite variable in size, but Haggard and Gamboa (1980) found them no more variable than workers in *P. metricus*.

Of course, highly eusocial insects do determine castes by larval feeding (Wilson 1971). But primitively eusocial insects may be, in some respects, more like vertebrates with comparable levels of cooperation. Nest-site limitation is an attractive possibility for explaining grouping in *P. annularis*, because similar ecological constraints seem to be important in selecting for cooperative breeding in birds (Emlen 1984; Woolfenden and Fitzpatrick 1984). The limited evidence is not strongly either for or against nest-site constraints, and the hypothesis deserves further study. It is known that group nesting can be influenced by the kind of nest site. In both *P. nimpha* (Cervo and Turillazzi 1985) and *P. exclamans* (Strassmann and Hughes, unpublished data), nests on natural substrates are usually singly founded, while those on buildings may have a high incidence of group nesting.

Although we are left without a definite answer as to why foundress groups are so large in *Polistes annularis*, we are left with a clear direction. Valuable as it is, measuring reproductive success is not enough. We must also investigate the reproductive options that are presented to different individuals by their environments, including their social environments. Inevitably, this must lead to experimental manipulations aimed at forcing individuals to choose options they would otherwise not take or allowing them to choose options that would otherwise not be permitted. In the first category, it would be very useful to measure the success of subordinate foundresses forced to nest alone, perhaps by removing their associates. In the second, the importance of nest site constraints could be assessed by artificial increases in nest sites or decreases in foundress population densities.

6.6 Summary

In *Polistes annularis* foundress females establish new nests, alone or in association with relatives, from which they rear one or more broods of workers followed by reproductives of both sexes. Foundress reproductive success, as measured by the number of reproductive females reared, averaged 2.36 in 1977 and 5.82 in 1978. For these two years the selection opportunity, or variance in relative fitness of foundresses, was 6.53 and 4.95. Most of this variance was due to differences between groups in mean success of foundresses, but a substantial part, about 30%, was due to differences within foundress associations. A further partition of the between-group component of the variance showed that relatively little was due to nest survival and most was due to differences in per capita productivity of successful nests.

Calculations of both selection differentials and inclusive fitness seemed to indicate selection for smaller association sizes. But there was no decrease in mean association size during the study, suggesting that subordinates in large groups may have been unable to nest alone. One hypothesis, that poor condition prevented subordinates from nesting alone, was contradicted by the available data. There was insufficient evidence to judge a second hypothesis, that a shortage of good nest sites limits single founding.

6.7 Acknowledgments

For help with nest censuses and observations we thank Diana Crowell, Scott Davis, Robert Matlock, Dana Meyer, Beverly Strassmann, and Richard Thomas. For nest observations we also thank Christine Becker, Paul Bruton, Doug Crosby, Madeline Daigle, Juan Ibarra, Tony Jones, Tony Mitchell, Diane Stallings, and Christi Steinbarger. Larry Gilbert, Alan Templeton, Bill Mueller, and John Smiley provided much useful advice and encouragement during the field study. We thank Colin Hughes and Pekka Pamilo for careful comments on the manuscript.

2 Amphibians

IN HIS CHAPTER, HOWARD describes the extent and causes of variation in reproductive success in two species of frogs with contrasting mating systems: bullfrogs (*Rana catesbiana*), in which males compete for territories in areas where females deposit their eggs over an extended breeding season; and woodfrogs (*R. sylvatica*), in which males do not defend territories and rarely engage in direct combat, competing indirectly for access to females during a very short breeding season. In bullfrogs, males and females have similar age-specific growth rates, whereas in woodfrogs females are significantly larger than males.

In both species, standardized variance in the number of zygotes produced per season was greater for males than for females, though in bullfrogs the inclusion of individual differences in the survival of progeny to hatching reduced differences in variance between the sexes. Although Howard was unable to follow members of either species throughout their reproductive lives, computer simulations indicated that in both species variance in lifetime reproductive success was substantially higher than variance within seasons.

Howard's study is the only one in this book where it is possible to make a direct comparison of the effects of body size on fitness in males and females in species showing contrasting patterns of sexual dimorphism. In both bullfrogs and woodfrogs body size and reproductive success increase with age, in females as a consequence of increased numbers (and size) of eggs produced, in males on account of increased numbers of matings and average size of mates. Comparisons between the species showed that the intensity of selection on size in bullfrogs is similar in the two sexes, whereas in woodfrogs size exerts a stronger effect on the fitness of breeding females than on that of males.

7 Reproductive Success in Two Species of Anurans

Richard D. Howard

ADVANCES IN SEXUAL SELECTION theory during the past decade (e.g., Trivers 1972; Emlen and Oring 1977) have stimulated a proliferation of studies on the reproductive behavior of individuals in nature. Investigations of anuran mating behavior, in particular, have also been aided by developments in techniques for marking and identifying individuals in the field (Emlen 1968) and by a comprehensive review of anuran social behavior (Wells 1977). During the past six years, the reproductive behaviors of more than thirty species of anurans have been investigated (Howard and Kluge 1985). These studies illustrate several advantages of working with frogs and toads in nature, including their ease of capture, visibility, and suitability for field experimentation. However, the greatest asset of anurans is that their external mode of fertilization allows unambiguous quantification of the reproductive success of individuals. Unlike the other species discussed in this volume, observation of copulations for most anuran species provides complete certainty of both male and female parentage.

Here I report estimates of seasonal reproductive success (RS) and predictions of lifetime reproductive success (LRS) for two species of ranid frogs: bullfrogs (*Rana catesbeiana*) and woodfrogs (*Rana sylvatica*). Despite being congeners, these two species show little similarity in ecology, life history, or mating behavior. Bullfrogs are the largest North American ranid (fig. 7.1). At the latitude of my investigation, bullfrogs often require two years to complete metamorphosis (Collins 1975); individuals become sexually mature one year (males, some females) or two years (females) after metamorphosis; and adult life spans range from five to eight years past metamorphosis. The prolonged life in the tadpole stage and mostly aquatic adult stage necessitates that bullfrogs use only permanent bodies of water throughout their life cycle. Breeding occurs during a two-month period in summer. Males are highly aggressive and defend territories that are used as egg deposition sites. Females initiate mating by approaching and touching territorial males. The possibility of mate choice

Figure 7.1: Male bullfrogs aggressively defend territories used as egg deposition sites. Fights are typically wrestling matches in which the larger male prevails.

by females is supported by my observations of females approaching and then rejecting some males only to approach and solicit copulation from other males.

Woodfrogs are among the smallest ranid species (fig. 7.2). Young complete metamorphosis within four months of egg deposition; individuals become sexually mature one or two years after the egg stage (males) or two or three years after the egg stage (females); and adult life spans probably do not exceed four years, with most individuals only living to reproduce during one breeding season. Woodfrogs spend most of their adult life in woodland areas and use temporary ponds and marshy areas only for reproduction and larval development. Their breeding season may last one week, but given favorable weather, all breeding can occur within twenty-four hours. Breeding occurs during early spring once ponds have thawed. Males do not defend territories or engage in male-male combat; however, unpaired males will attack amplexed pairs before egg deposition to take possession of females clasped by these males. Observations and experiments provide no support for female choice in woodfrogs (Howard and Kluge 1985). Females deposit their clutches at communal egg deposition sites.

Additional differences between these two species include sex ratio and number of adults at the breeding site. During four years of investigation, the breeding sex ratio of bullfrogs was essentially 1:1; and the population contained 26 to 58 males and 23 to 55 females (Howard 1981). During an equivalent number of breeding seasons, the breeding sex ratio of woodfrogs ranged from 1.85 to 4.00 males/female, and the population contained 4,161–6,151 males and 1,041–3,142 females (Howard and Kluge 1985).

7.1 Measures of RS

Data on bullfrogs were collected during 1975–78; methods used have been described elsewhere (Howard 1978a, 1983). Data on woodfrogs were obtained during a four-year study (1980–83) in collaboration with A. G. Kluge (see Howard and Kluge 1985 for a more complete description of methods and results). Both studies occurred in southeastern Michigan, USA; the populations were approximately 39 km apart.

Bullfrogs

Three measures of RS were collected on bullfrogs during each breeding season: number of matings, number of zygotes, and number of hatchlings produced (females) or sired (males) (table 7.1). Mating success

Figure 7.2: Male woodfrogs compete for mates by attempting to displace already paired males from their females before egg deposition begins. In this instance a large male is displacing a marked small male that has clasped a large female. This interaction is occurring in a communal egg deposition site (note egg masses beneath individuals). The larger male eventually took possession of the female and fertilized her eggs.

Table 7.1 Standardized Variance in Seasonal Reproductive Success for Male and Female Bullfrogs

Year	Number of Matings		Number of Zygotes		Number of Hatchlings	
	$\sigma^2/\bar{x}^2$	*Ratio*	$\sigma^2/\bar{x}^2$	*Ratio*	$\sigma^2/\bar{x}^2$	*Ratio*
1976						
Males (n = 38)	1.38	12.39	1.99	9.12	2.34	5.52
Females (n = 22)	0.11		0.22		0.42	
1977						
Males (n = 26)	1.50	7.45	1.47	3.72	1.52	3.96
Females (n = 26)	0.20		0.40		0.38	
1978						
Males (n = 29)	1.83	12.04	3.06	10.96	3.18	9.86
Females (n = 23)	0.15		0.28		0.32	

was determined by direct observations of copulations. Number of zygotes was estimated from photographic slides taken immediately after egg deposition for forty-four different egg masses. Every egg that could be discerned was counted. Clutch size of forty-nine other egg masses could not be estimated in this way because no photographic record was possible; in these cases clutch size was estimated using regression equations predicting clutch size given female size (Howard 1983). For females, total number of zygotes produced per year equaled the sum of eggs in all clutches deposited; for males, number of zygotes produced per year equaled the sum of eggs in all clutches fertilized. Number of surviving hatchlings per individual equaled the number of zygotes produced by each individual multiplied by embryo survival probabilities estimated for each egg mass. Survival estimates were determined from samples of embryos taken every ten to twelve hours from each egg mass until the time of hatching. In total, 93 egg masses were sampled and 67,354 embryos were examined (see Howard 1978b).

Standardized variance in these RS measures ($\sigma/\bar{x}^2$, after Wade and Arnold 1980) was from four to twelve times greater in males than in females (table 7.1). However, ratios of standardized variance in RS between the sexes declined by as much as 53% when the RS measures were ranked from least complete (number of matings) to most complete (number of hatchlings). The relative contributions of number of matings (M), average number of zygotes per mating (Z), and average percent hatching success (H) to total variation in RS were estimated in an earlier analysis using stepwise multiple regression (Howard 1983). However, this analysis can be improved upon in two ways. First, M, Z, and H interact multiplicatively to produce total RS rather than interacting additively as modeled by the regression procedure used; this problem can easily be overcome by taking log transformations of all the variables. Second, the relative contribution of each of the three independent variables to variation in log (total RS) would be best expressed after removing the influence of the other two independent variables from the independent variable being analyzed. To do this, I calculated the squared second-order semipartial correlation for each of the independent variables (after Kerlinger and Pedhazur 1973). As in the earlier analysis, mating success accounted for most of the variation in RS for males. Of the ninety adult males in the population during 1976–78, forty-three (47.8%) did not mate. Among the successfully breeding males, squared semipartial correlations indicated that 38.0% of the variation in log (total RS) was explained by log M; 33.9% of the variation was explained by log Z; and 22.1% of the variation was explained by log H.

All sixty-eight females in the population during 1976–78 mated once or twice. Squared semipartial correlations revealed that 51.5% of the variation in log (total RS) was explained by log H; 43.9% of the variation was explained by log Z; and 20.4% of the variation was explained by log M. Note that the sum of the squared semipartials exceeds 100%; this results from having both positive and negative correlations between the independent variables (Thorndike 1978). Although the relative importance of the three variables in this analysis differs somewhat from that obtained when using nontransformed data in a stepwise multiple regression (Howard 1983), both types of analysis clearly indicate that data on all three variables are necessary to encompass variation in total RS, particularly for females. Consequently, data on all these variables had to be obtained for males to compare relative variation in RS between the sexes (Howard 1983).

Excessive mortality of marked individuals owing to winterkills during the last two years of the study precluded estimation of LRS. However, I devised a computer simulation that predicts standardized variance in LRS based on number of hatchlings produced, using field data on age-specific and sex-specific survivorship and fecundity (Howard 1983) (table 7.2). Survivorship probabilities were calculated using the proportion of individuals of each age-sex category that survived between 1975 and 1976. Fecundity estimates considered the proportion of each age-sex category that successfully mated in 1976, the average number of hatchlings produced by these successful breeders, and the amount of variation in number of hatchlings they produced. A simulation "run" consisted of a cohort of one hundred one-year-old males and one hundred one-year-old females in which the particular yearly outcome of survival and RS was randomly chosen from the appropriate age-sex distribution calculated from field data. I set maximum life span at five years of age; however, given the low annual rates of survival in this species, fewer than 3% of the individuals in a cohort reached this age, on average. The simulation predicted standardized variance in LRS within a cohort for each sex. I then used data from one hundred such cohorts to obtain average standardized variance in male

Table 7.2 Survivorship and Fecundity Data for Male and Female Bullfrogs

Age x (yr)	Probability of Survival (age $x-1$ to age x)	Probability of Successful Mating	Number of Hatchlings Produced ($\bar{x}$ ± s.d.)
Males			
1		0.11	2732 ± 1289
2	0.42	0.55	3581 ± 2671
3	0.42	0.69	6402 ± 3986
4	0.52	0.83	17511 ± 8687
5	0.23	1.00	19346 ± 10035
Females			
1		0.10	2007 ± 1097
2	0.42	1.00	3372 ± 4650
3	0.42	1.00	7228 ± 4159
4	0.69	1.00	10238 ± 6241
5	0.26	1.00	11147 ± 7517

Note: Survivorship based on mark-recaptures of known individuals between 1975–76; fecundity data were obtained in 1976. These data were also used in a computer simulation to predict variation in LRS within and between the sexes (see text).

LRS, in female LRS, and in the ratio of male to female LRS. Simulations predicted that the average standardized variance in LRS was 5.91 ± 0.08 for males and 4.35 ± 0.04 for females. These predicted values exceeded their seasonal equivalents (table 7.1), mostly because of the effect of variation in survivorship on LRS (Howard 1983). In contrast, the average ratio of standardized variance in LRS between the sexes equaled 1.37 ± 0.03, which was well below all seasonal ratios obtained. In this case both age-specific fecundity and age-specific survivorship influenced the ratio of LRS between the sexes.

In separate analyses, I also considered how much the standardized variance is influenced either by the inclusion of nonbreeding individuals in the analysis or by the mortality of breeding and nonbreeding individuals. Much of the variation in LRS resulted from the inclusion of individuals that obtained no matings. If only successfully breeding individuals were considered, simulations predicted that standardized variance in LRS would be 1.45 ± 0.02 for males and 1.46 ± 0.01 for females. These values represented 74% and 65% reductions, respectively, from the predicted values of standardized variance in LRS based on all individuals in a cohort. The ratio of standardized variance in LRS between the sexes was 1.02 ± 0.02. Thus, when unsuccessfully breeding individuals are excluded from the analysis, male and female bullfrogs became even more similar in LRS variation.

The effects of mortality on standardized variance in LRS were removed by restricting the analysis to those individuals that lived to an advanced age. For example, for all those males that lived through their fourth year, average standardized variance in LRS equaled 0.31 ± 0.08; for four-year-old females, it equaled 0.30 ± 0.08 (n = 10 cohorts). The average ratio of standardized variance in LRS between the sexes for these cohorts was 1.13 ± 0.47. Thus, standardized variance in LRS within each sex for these four-year-olds was considerably lower than that calculated for the entire cohort of one hundred individuals, indicating a strong effect of eliminating mortality; however, the variation between the sexes was roughly similar to that of the entire cohort, revealing the effect of sex-specific differences in age-specific fecundity.

Woodfrogs

Two measures of RS were collected on woodfrogs: number of matings and number of zygotes produced (females) or sired (males) (table 7.3). All females that arrived at the breeding site mated once and only once; hence they showed no variation in mating success. In contrast, standardized variance in male mating success ($\sigma^2/\bar{x}^2$) equaled 2.02. Males were also 22.48 times more variable than females in zygote production. However, only a portion of the standardized variance among males in mating success or zygote production could be directly attributed to sexual selection (Howard and Kluge 1985). In the year these data were collected, 58% of the males did not mate because of the heavily male-biased sex ratio. After sex-ratio effects were taken into consideration, the amount of standardized variance in zygote production among males equaled 0.22, and the ratio of standardized variance in RS between the sexes was only 2.66.

LRS was predicted for woodfrogs using the simulation procedure described earlier for bullfrogs. As with bullfrogs, the survivorship and fecundity parameters used for woodfrogs were based on field data (table 7.4). Given the large number of males sampled, I could partition age-specific matings probabilities more thoroughly for male woodfrogs than for male bullfrogs. In particular, at each age, males had separate probabilities of mating once and of mating twice. The $\bar{x}$ ± s.d. number of zygotes that a successful male could sire in one mating was based on the overall distribution of clutch sizes produced by a sample of 110 females. Because year-to-year survival was low (12.5% for males and 8.0% for females), the

Table 7.3 Standardized Variance in Seasonal Reproductive Success for Male and Female Woodfrogs

	Number of Matings		Number of Zygotes (raw data)		Number of Zygotes (adjusted for sex ratio)	
	$\sigma^2/\bar{x}^2$	*Ratio*	$\sigma^2/\bar{x}^2$	*Ratio*	$\sigma^2/\bar{x}^2$	*Ratio*
Males (n = 524)	2.02		1.88		0.22	
		n.a.		22.48		2.66
Females (n = 110)	0		0.08		0.08	

Table 7.4 Survivorship and Fecundity Data for Male and Female Woodfrogs

Age x (yr)	Probability of Survival (age $x-1$ to age x)	Probability of Successful Mating[a]	Number of Zygotes Produced ($\bar{x}$ ± s.d.)
Males			
1		0.23	600 ± 174
		(0.03)	600 ± 174
2	0.125	0.32	600 ± 174
		(0.05)	600 ± 174
3	0.125	0.32	600 ± 174
		(0.05)	600 ± 174
4	0.125	0.32	600 ± 174
		(0.05)	600 ± 174
Females			
1		0.00	0 ± 0
2	0.08	1.00	614 ± 171
3	0.08	1.00	741 ± 138
4	0.08	1.00	741 ± 138

Note: Survivorship based on mark-recaptures of known individuals between 1980–82. Male fecundity data obtained in 1981; female fecundity data obtained in 1980–82 (Howard and Kluge 1985). These data were also used in a computer simulation to predict variation in LRS within and between the sexes (see text).

[a]For males, once or twice.

initial cohort size of one-year-old animals was increased to one thousand individuals. As with bullfrogs, one hundred different cohorts were followed, then reproductive statistics were averaged.

Some demography of the woodfrog population is uncertain. In particular, it was not clear whether males became sexually mature at one or two years of age and whether females mature at two or three years of age. I assumed that males mature at one year of age in the simulation; however, this assumption was of little consequence because the male age at sexual maturation was the beginning point of the simulation and demographic events before this age were never considered. In nature, sexual maturation in most, if not all, females occurred one year later than in males (Berven 1981; Howard and Kluge 1985); however, it was unknown whether these immature females differed from adult females in their annual rates of survivorship. Here I will initially assume that no one-year-old females bred and that their annual survivorship was the same as for older females; I then evaluate some departures from these assumptions.

Simulations predicted that the average standardized variance in LRS in a cohort of one thousand individuals was 3.39 ± 0.01 for males and 13.99 ± 0.05 for females. These predicted values greatly exceeded their seasonal equivalents, particularly for females (table 7.3). The reason for this difference was similar to that for the bullfrog results: survival differences among individuals have a strong influence on the amount of variation in LRS. Interestingly, the average ratio of LRS between the sexes was 0.24

± 0.01. Thus, in contrast to seasonal data, computer simulations predict that females will be more variable than males in terms of LRS.

As with bullfrogs, I also considered how the standardized variance in LRS was affected by nonbreeding individuals and by mortality. Including nonbreeding individuals had an even stronger affect on variation in LRS for woodfrogs than for bullfrogs. If the analysis is limited to successfully breeding individuals, average standardized variance in LRS equaled 0.25 ± 0.001 for males and 0.18 ± 0.004 for females (n = 100 cohorts). These values represented 93% and 99% reductions, respectively, in the average standardized variance in LRS predicted when both successful and unsuccessful breeders are considered. The average ratio of standardized variance in LRS between the sexes equalled 1.46 ± 0.04.

The effect of mortality on variation in LRS was removed by analyzing only those individuals that have lived through their third year of life. Standardized variance in LRS for these three-year-olds averaged 1.40 ± 0.20 for males and 0.084 ± 0.01 for females (n = 10 cohorts); the average ratio of standardized variance in LRS between the sexes equaled 16.90 ± 3.28. Males were more variable in LRS than females in this analysis because, on average, 48.7% of the males in each cohort of three-year-olds did not breed owing to the low mating probabilities that result from a biased sex ratio (table 7.4). By contrast, all females had bred twice by this age. When only successfully breeding males were considered, the average standardized variance in their LRS was 0.22 ± 0.04 and the average ratio of standardized variance in LRS between the sexes equaled 2.64 ± 0.35. The values obtained in both these analyses closely matched those of their seasonal counterparts (table 7.3).

Other simulations were performed in which 10% of the one-year-old females bred (a generous estimate given all available field data). These simulations predicted an average standardized variance in female LRS of 5.90 ± 0.02 and an average ratio of standardized variance in LRS between the sexes of 0.56 ± 0.01. Standardized variance in female LRS decreased linearly with the percentage of one-year-old females breeding. However, the sexes were not predicted to be similar in standardized variance in LRS until at least 20% of all females breed as one-year-olds. In another set of simulations, I considered what would happen if one-year-old females did not become sexually mature but had a yearly survivorship equal to that of adult males (12.5% survive) rather than that of adult females (8% survive). In this case the average standardized variance in female LRS would equal 8.18 ± 0.02 and the average ratio of standardized variance in LRS between the sexes would be 0.41 ± 0.01. That is, female standardized variance in LRS would still exceed that of males.

7.2 The Effect of Age on RS

Unlike most birds and many mammals, individuals in many anuran species grow appreciably in body size during adulthood. For both bullfrogs and woodfrogs, the major influence of age on variations in RS results from differences in body size; thus I will consider age effects and size effects simultaneously. The age-size relationship I use for bullfrogs is based on an eight-year capture-mark-recapture program initiated by Collins (1975) on the bullfrog population that I studied. I also use the age-size relationship Collins (1975) obtained for woodfrogs; however, our study populations differ.

Bullfrogs

For both males and females, age affected the probability of mating, the average number of hatchlings produced by successfully breeding individuals, and the amount of variation in their RS (table 7.2). These relationships existed for females because larger, older females laid larger clutches ($r^2 = .78$; $p = .0001$; $n = 32$), and in most years studied, only larger, older females deposited two clutches during a single breeding season (Howard 1983). Larger female bullfrogs produced larger eggs in their first clutch of the season ($r^2 = .58$; $p < .01$; $n = 16$) but not in their second clutch ($r^2 = .12$; $p > .05$; $n = 14$). However, second clutches of individual females consisted of smaller eggs overall than their first clutch ($t_{11} = 2.07$; $p = .03$). Unfortunately, the effect of egg size on either offspring survival or RS was unknown.

Bullfrog females actively selected their mates (Howard 1978a; 1983). Larger, older females usually mated with larger males; smaller, younger females appeared to mate at random with respect to male body size. However, the positive size assortment among mated pairs was weak (see below); as a result, female size was not correlated with offspring hatching success (Howard 1983).

Cues that females used to evaluate mate or territory quality are unknown. Furthermore, temperature changes caused larval survivorship to vary considerably from day to day, thus reducing the predictability of "good" egg deposition sites. Because of this unpredictability, it would be best to consider differences among the specific males and their territories that each female visited before mating and use these alternatives to evaluate actual mating choices. Relevant data are scarce because the male visitation patterns of particular females must be known and the quality (in terms of percent hatching success of offspring) of each territory of the males visited must be estimated. Data meeting these requirements were collected in 1978. In this year, thirteen of seventeen females mated in the highest-quality territory of the territories they sampled ($p = .025$; binomial test). Only ten of these females mated with the largest male that they visited, and this was not significant ($p = .315$). However, I do not know whether larger, older females tended to select the "best" option available more often than younger females.

Larger males achieved more matings each year (1976: $r_s = .59$; $p = .0002$; 1977: $r_s = .79$; $p = .0001$; 1978: $r_s = .30$; $p = .054$) and usually mated with larger, more fecund females (1976: $r_s = .34$; $p = .042$; 1978: $r_s = .30$;

$p = .051$ but not in 1977: $r_s = .21$; $p = .24$). In one year of the three years sampled, the offspring sired by larger males also had increased hatching success ($r_s = .66$; $p = .0008$; $n = 20$ males). In this year, hatching success primarily resulted from how well the young survived leech predation (Howard 1978b). In the other two years, offspring sired by males of various sizes had high hatching success. Presumably the low density of territorial males in these two years may have allowed all males to obtain territories of high quality (Howard 1983).

Age influenced the type of mating tactic male bullfrogs employed. Larger, older males always defended territories. Small, one-year-old males either were territorial or employed a ‘‘satellite’’ behavior, in which they remained close to a larger territorial male and attempted to intercept females that he attracted. Territorial males had a greater chance of mating than satellite males but also suffered greater changes of predation. However, the tendency of young males to employ satellite behavior rather than territoriality did not appear to result from the increased risks of mortality associated with territoriality (Howard 1984). Instead, the chance of successful mating by these young males was low because they usually controlled suboptimal breeding territories, and/or were discriminated against as mates by females.

Woodfrogs

Older individuals of both sexes achieved greater RS. For females this relationship resulted from a significant clutch size/body size relationship ($r^2 = .61$; $p = .0001$; $n = 110$ females); Larger female woodfrogs also produced larger eggs ($r = .53$; $p < .05$; $n = 18$; Collins 1975). For males, larger individuals had an increased chance of mating ($r = .66$; $p < .05$; 12 male size categories) and tended to mate with larger females ($r = .17$; $p < .01$; $n = 571$) (Howard and Kluge 1985).

All observations and experiments indicated that initial pairing of male and female woodfrogs was random with respect to male size and age (Howard and Kluge 1985). Pairs did not begin egg deposition until several hours after pairing, however, and during this time large males obtained an advantage through male-male competition. In particular, small males were more often deposed from their mates than were large males when paired with females of similar size (Howard and Kluge 1985). Large males were also more capable of usurping females paired to other males. Thus, in contrast to bullfrogs, male-male competition in woodfrogs occurred after amplexus was achieved, not before.

Preliminary data also suggested that a male's size influenced his spatial position in a breeding chorus. Males in and around an established communal egg deposition site tended to be larger than males a few meters away from the site ($t_{45} = 1.75$; $p = .044$). Females that males encountered at the egg deposition site were almost always paired; thus larger males might have been positioning themselves in the best area to take possession of females by dislodging an already paired male from his mate.

7.3 Sexual Selection and Sexual Dimorphism

Body size influences RS for both sexes in bullfrogs and woodfrogs. Yet in bullfrogs the sexes are similar in body size at a given age and have similar size-specific growth rates (Howard 1981), whereas in woodfrogs females are significantly larger than males (Wright and Wright 1949; Howard 1980) and grow more rapidly at a given size (Bellis 1961; Howard and Kluge 1985). Fortunately, the body-size patterns in these two species can be readily interpreted on the basis of the relative effects of size on RS for each sex.

For bullfrogs, body size influences average RS in a similar fashion in males and females (fig. 7.3). An analysis of covariance indicates that regression lines calculated separately for the sexes do not differ in slope ($F_{1,6} = 2.68$; $p = .15$) or intercept ($F_{1,7} = 0.74$; $p = .41$). Because the slope of a trait versus RS regression is equivalent to the univariate selection gradient (Lande 1979; Arnold and Wade 1984a,b), the similarity of regression slopes for the sexes here means that male and female bullfrogs should at present be experiencing similar selective pressures with respect to body size.

For woodfrogs, average RS increases more with body size in females than in males (fig. 7.4). An analysis of covariance indicates that the male and female regression lines differ significantly in slope $F_{1,23} = 24.948$; $p <$ .001) (Howard and Kluge 1985). Thus the sexes differ in the selection gradient on body size; in particular, a given increment in body size confers

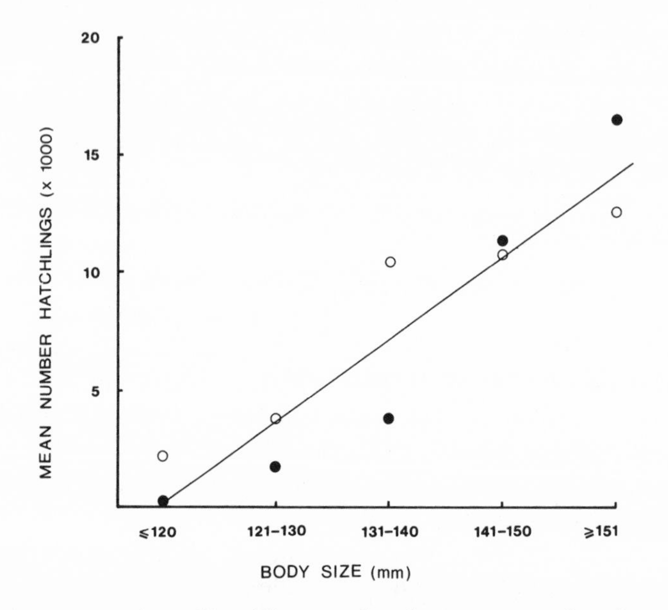

Figure 7.3: Average number of hatchlings produced per year as a function of body size for male (*solid circles*) and female (*open circles*) bullfrogs. Data are from 1976. An analysis of covariance indicates that the same positive relationship exists between these two variables for both the sexes. Common regression equation is $Y = 350.05x - 41656.60$ ($r^2 = 0.86$; $p < .01$).

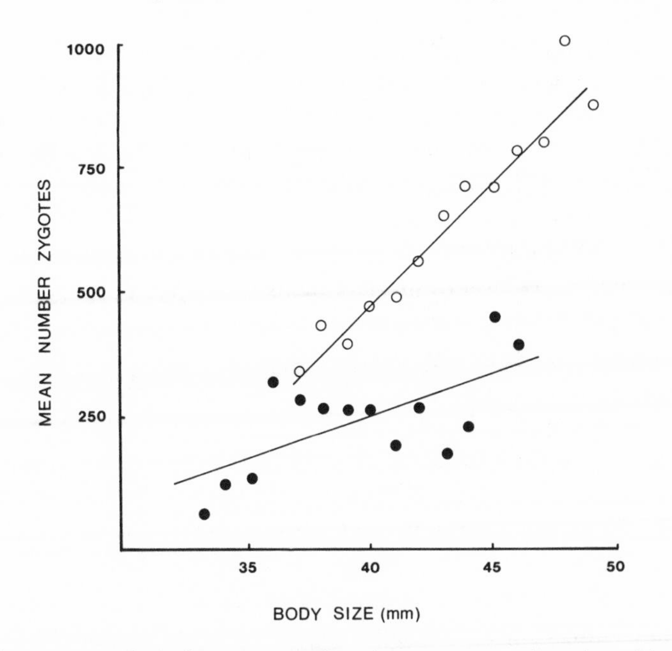

Figure 7.4: Average number of zygotes produced per year as a function of body size for male (*solid circles*) and female (*open circles*) woodfrogs. An analysis of covariance indicates that the regression lines differ for the two sexes (see text). Regression equation for males is $Y = 16.55x - 413.3$ ($r^2 = .43$; $p = < .05$) and for females is $Y = 50.86x - 1557.6$ ($r^2 = .95$; $p < .01$).

a greater reproductive reward, on average, to a female woodfrog than to a male woodfrog. Assuming that larger size can be attained only at some cost, males that have lower growth rates than females should be favored, thereby producing a sexual dimorphism.

It is crucial to note that comparisons of the relative effects of phenotypic traits on RS between the sexes requires that RS be measured in the same currency for each sex. In addition, the adequacy of the RS measure used depends on how much of the actual variation in RS it encompasses for each sex.

7.4 Discussion

Differences between the sexes in patterns of parental investment should influence sex-specific differences in the amount of competition for mates and in variation in RS (Trivers 1972). The sex providing less investment should have greater variation in RS than the sex providing more investment. The relative amount of variation in RS between the sexes is further increased if the spatial and/or temporal distribution of the more heavily investing sex provides some individuals of the opposite sex an opportunity to monopolize mates. Such monopolization can result from controlling either individuals of the heavily investing sex directly or the resources they require for successful reproduction (Emlen and Oring 1977).

As with most species, females provide more parental investment than do males in both woodfrogs and bullfrogs. The relative energy commitment of the sexes in a clutch is apparent in the duration of time between matings. I have watched male bullfrogs mate with two females in one night, then seek additional copulations. However, females take 24.7 ± 5.2 days n = 13; 1978 data) between their first and second clutches. Male woodfrogs can obtain two or even three matings within a twenty-four hour period, but the short breeding season in this species precludes females from mating than once a year.

The body size–fecundity relationship in female bullfrogs and woodfrogs is common for species in which maternal care is not provided, because selection appears to have favored the number of eggs a female can produce rather than the number of young she can rear. Presumably, lack of maternal (and paternal) care in most anurans results because differences in life-style between tadpoles and adults make parents ineffective in reducing predation on young or providing further nourishment for them (Howard 1978b). A consequence of the clutch size–body size relationship is that females can vary appreciably in quality from the male's perspective. Size-assortative mating increases the amount of variation in male RS, particularly if the males mating with the larger, more fecund females also obtain a disproportionate share of the matings. Both species show weak size-assortative mating, but for different reasons: in bullfrogs it results from active female choice; in woodfrogs it results from male-male competition for females. In both species it is the larger, older males that tend to mate with larger females and to mate more often. I have never observed male bullfrogs or woodfrogs discriminate among females as mates; however, Berven (1981) reported male mate choice in a woodfrog population he studied.

When data on seasonal standardized variance in RS are extended to predict standardized variance in LRS, it becomes clear that survivorship affects standardized variance in LRS much more than fecundity does. The strong effect of survivorship on LRS makes standardized variance in LRS an insensitive index of the opportunity for sexual selection. Most of the annual mortality of adult bullfrogs in the population I studied occurs outside the breeding season (e.g., winterkills) and is similar for the sexes, hence unrelated to sexual selection. Relatively little mortality is directly related to reproduction: older territorial males incur higher risks of predation.

Arnold and Wade (1984a,b) provide a promising way of separating the effects of components of RS such as survivorship and fecundity to allow a more precise estimation of the opportunity for sexual selection. However, their methodology has so far been applied only to data on seasonal RS, not LRS. For bullfrogs, use of such seasonal data may overestimate the opportunity for sexual selection because age differences among individuals have such a strong effect on RS (table 7.2).

A major goal in documenting present-day variation in RS or LRS between the sexes is to determine the relationship between RS and phenotypic outcomes of past selection across species—that is, to interpret patterns of sexual dimorphism. Various statistics related to RS variation have been used in recent comparative studies, including the socioeconomic sex ratio (Clutton-Brock, Harvey, and Rudder 1977) and mean or maximum harem sizes (Alexander et al. 1979). However, the sex-specific differences in selection gradients I provide here should be a more precise predictor of sexual dimorphism. This measure considers both sex-specific differences in RS variation and how this variation relates to phenotypic differences. Given the proliferation of reproductive studies recently done on taxa such as anurans, we should be soon able to compare the effects of differential sex-specific selection gradients across a number of species and develop a more rigorous understanding of the forces affecting sexual dimorphism.

7.5 Summary

Although congeners, bullfrogs and woodfrogs have few life-history and behavioral characteristics in common. Proximate mechanisms for increased RS for females in both species include attaining a large body size to increase capacity both for more eggs and for larger eggs per clutch; however, mate choice can be demonstrated in female bullfrogs but not in female woodfrogs. Male bullfrogs defend territories used as egg deposition sites, and all male combat occurs before amplexus. Initial pairing of males and females appears to be random in woodfrogs. Aggression among male woodfrogs primarily occurs when unpaired males attempt to displace already paired males and take possession of their females. Thus in bullfrogs sexual selection occurs before pair formation, whereas in woodfrogs it occurs after pair formation.

For both species, standardized variance in seasonal RS is greater in males than in females. However, computer simulations using field data on demographic parameters predict that the sexes will be fairly similar in standardized variance in LRS for bullfrogs and that females are more variable in LRS than males in woodfrogs.

Although body size influences RS for the sexes of both species, RS in male and female bullfrogs appears to increase with size in a similar fashion, whereas in woodfrogs RS increases more with body size for females than for males. Selection gradients obtained from the body size versus RS regressions are consistent with the degree of sexual dimorphism in size in these two species. Male and female bullfrogs are similar in size; female woodfrogs are significantly larger than male woodfrogs.

3 Birds

MONOGAMY IS THE PREDOMINANT breeding system among birds (Lack 1968), and the thirteen chapters in this part reflect this. The species represented fall into four groups: four short-lived, primarily monogamous passerines where partners usually remate each year (great tits, *Parus major*; song sparrows, *Melospiza melodia*; house martins, *Delichon urbica*; and pied flycatchers, *Ficedula hypoleuca*); five long-lived monogamous species where partners commonly mate for life (European sparrowhawks, *Accipiter nisus*; Bewick's swans, *Cygnus columbianus*; snow geese, *Anser caerulescens*; kittiwakes, *Rissa tridactyla*, and fulmar petrels, *Fulmarus glacialis*); one lek-breeding species (black grouse, *Tetrao tetrix*); and two social breeders (groove-billed anis, *Crotophaga sulcirostris*, and Florida scrub jays, *Aphelocoma coerulescens*). In four of the monogamous species, feeding territories are defended during the breeding season, predominantly by males in great tits and song sparrows and by both sexes in sparrowhawks and Bewick's swans. In anis, feeding territories are defended seasonally by all group members, and in scrub jays a permanently mated pair and their helpers defend a communal territory throughout the year. In contrast, in house martins, pied flycatchers, snow geese, kittiwakes, and fulmar petrels only the area of the nest is defended.

Unlike the studies of invertebrates, most studies in this section concentrate primarily on females. Comparisons emphasize the extent to which differences in the stage at which reproductive success is calculated can affect estimates of the opportunity for selection. With the exception of sparrowhawks (chap. 13), standardized variance in the number of young fledged by breeding adults was higher (mean for eight studies, 0.75) than standardized variance in eggs produced (mean for six studies 0.47), and standardized variance in the number of recruits was higher still (mean for four studies, 2.23).

These results suggest that individual differences in the survival of offspring are one of the most important components of variation in total fitness among breeding females, and analysis of the components of variation in lifetime success confirms this (see chaps. 8, 9, 10, and 20). In most species, individual differences in clutch size are slight. The contributions of breeding life span and the proportion of nonbreeders in the population varied between studies, but at least part of this variation was caused by differences in the definition of nonbreeders: some studies included only nonbreeding adults in this category (see chaps. 9, 10, 11, 14, and 20) while others included individuals that died between fledging and adulthood (see chaps. 13 and 16).

In most species, age had an important effect on breeding success. Many components of reproductive success initially improve with age (see chaps. 9, 10, 11, 13, 14, 16, 19, and 20), in some cases for as long as ten years (see chap. 17). Factors responsible for this increase include maturation, the individual's own experience of breeding (chap. 12), the age or experience of its mate (chap. 9), the duration of the pair-bond (chap. 14), dominance rank (chap. 19), and the number of resident helpers (chap. 20). At the other end of the life span, components of reproductive success or survival declined with age in both short- and long-lived species (see chaps. 9, 10, 13, 14, 16, and 20), though these effects appear to be more pronounced in the latter. In most cases the causes of this decline are still unknown.

Changes in climate or food availability between years cause pronounced fluctuations in breeding success that account for much of the variation in lifetime reproductive success in all four short-lived species (see chaps. 8–10). Effects of this kind may easily obscure relationships between phenotypic variation and fitness when animals from different cohorts are included in the same analysis. However, they do not necessarily indicate that selection is unimportant. In longer-lived species, similar differences exist between breeding seasons, but because individuals breed in several years, they are a less important cause of individual differences in lifetime breeding success, and the effects of phenotypic variation may be easier to demonstrate.

Seven studies investigated the effects of morphological or behavioral variation on breeding success. Components of breeding success were related to body size in house martins and sparrowhawks (chaps. 11 and 13), to laying date in fulmars (chap. 17), to the duration of the pair-bond in Bewick's swans (chap. 14), to breeding unit size in anis (chap. 19), and to the number of resident helpers in scrub jays (chap. 20). The potential complexity of selection pressures is clearly demonstrated by the extensive studies of selection on body size in song sparrows (chap. 10) and on clutch size in great tits (chap. 8). In neither study was lifetime reproductive success consistently related to phenotypic factors. However, in both there was evidence of alternating directional selection and of opposed selection pressures at different stages of the life history.

Five of the studies of monogamous birds examined variation in the success of males as well as females (chaps. 9, 10, 11, 14, and 16).

Comparisons of the extent and causes of variation in breeding success between the sexes in the monogamous species show that they are generally similar. This is not always the case, however. For example, in house martins (see chap. 11) higher mortality of females generates a male-biased sex ratio among adults and an increase in the opportunity for selection among males compared with females. And in Bewick's swans, the social rank and breeding success of females are more strongly affected by the dominance and body size of their mates than by their own size and rank (see chap. 14).

Black grouse (chap. 18) provide an interesting contrast with the monogamous species. Competition for mates is intense, and standardized variance in lifetime mating success among breeding males is high. The effective reproductive life spans of males are shorter than those of females, and male success depends on the ability to attract females.

The studies of anis (chap. 19) and scrub jays (chap. 20) show the extent to which an individual's social environment can affect its fitness in group-living species. In both species the size of the breeding group exerts a strong effect on the fitness of its members. In anis, individuals of both sexes adopt different reproductive roles that affect their survival and their reproductive rate. Not only may individuals switch roles several times during their lifetime, but the costs and benefits of particular roles are likely to differ between individuals. While complexities of this kind underline the importance of longitudinal records of individual life histories, like the study of *Polistes*, they emphasize the need to analyze the short-term payoffs of particular reproductive decisions.

8 The Great Tit, *Parus major*

A. J. van Noordwijk and J. H. van Balen

The great tit has been studied intensively since 1933, when Kluyver started to ring all adults and nestlings in the population in Oranje Nassau's Oord, an estate near Wageningen, the Netherlands, in order to follow the fate of all individuals in the population (Kluyver 1951). Because all parents were identified on the nest, Kluyver was able to relate the survival of fledglings to properties of the parents and thus to follow the reproductive success of individuals over a complete lifecycle, from breeding adults to breeding adults. This is important, since there are opposite trends in successive stages of the life cycle, leading to false conclusions if data from partial life cycles are extrapolated. For example, the number of fledglings is positively correlated with clutch size, but fledgling survival is negatively correlated with clutch size (van Noordwijk, van Balen, and Scharloo 1981a). In this chapter, which is based on data from a relatively undisturbed population, we present an analysis of variability in lifetime reproductive success (LRS), together with a discussion of the relationship between inbreeding and variance in offspring number, and of the effects of genetic variation for reproductive traits on the variation in reproductive success.

The Great Tit

The great tit is a common songbird of deciduous woodlands, weighing about 17–19 g (female) or 18–20 g (male). It is a hole-breeding species that will readily breed in nestboxes (fig. 8.1). In the Netherlands it is normally a resident territorial species. Its diet consists of insects and other arthropods throughout the year, supplemented in winter with seeds, preferably beechmast. In very cold winters with prolonged snow cover the territorial system may break down completely, and many individuals will migrate to villages (van Balen 1980; Drent 1983), which is the normal pattern in boreal regions.

From 1955 onward, Kluyver's original population study (Kluyver

1951) was continued and replicated in two small and two larger populations in different parts of the Netherlands. The two main study sites are the Hoge Veluwe near Arnhem and the island of Vlieland in the north of the country. Throughout the study nestlings have been ringed and parents have been identified for most nests. Only a small proportion of the breeding population has been missed every year, mostly in nests that failed before the moment of ringing and catching the adults. In the Hoge Veluwe, about 50% of the breeding birds have been ringed as nestlings in the study area, whereas on Vlieland this amounts to 80%. Basic information about the breeding biology has been recorded throughout the study, including the onset of laying, clutch size, hatching rate, and breeding success. In addition, data on egg size and body size have been collected for periods of several years in connection with specific projects.

Because the population on Vlieland is almost closed, it is ideal for investigations on the density dependence of survival. A recent reanalysis of these data (Tinbergen, van Balen, and van Eck 1985) shows that experimental reduction of the number of fledglings per nest leads to an increase in the annual recovery rates of the remaining fledglings. In

Figure 8.1: Great tit on nest. Photo by Lukas Keller.

contrast to earlier interpretation (Kluyver 1971; Klomp 1980), the adult recovery rates are not found to be density dependent. Instead, the recovery rates are positively related to the seed supply in winter and negatively related to the number of fledglings raised annually in years with a low winter seed stock.

8.1 Variation in Reproductive Success and Its Components

Methods and Samples

The population reported on is from a study area in "the Hoge Veluwe," described by van Balen (1973, 1980). For the analysis of LRS, we selected a period (1969–83) in which no experiments were performed during the breeding season. There were two discontinuities during this period: heavy storm damage necessitated an alteration of the study area in 1972–73, and during the first seven winters (up till 1975–76) seed food was supplied on six or seven sites.

We have analyzed the data on all females breeding for the first time in the years 1969–80, using the breeding data up till 1983. This implies that data on a few females in the last cohorts are incomplete, because they were still alive and breeding at the end of the period considered (± 10% of the last cohort). A few birds were omitted when their clutches had been manipulated or when breeding parameters were not completely known. All clutches were included, whether hatched or not.

For the remaining birds (n = 635) we compiled a set of breeding parameters, including the number of reproductive years, number of clutches, number of eggs laid, number of young fledged, and number of offspring recruited as breeding birds into the same population. The data were calculated separately for all clutches of a female, all first clutches, and all late (repeat and second) clutches. In addition, recruitment was calculated separately for male and for female offspring.

The total number of breeding offspring, or recruits, per female (LRS) was split into five components: life span of the female (LS, defined as number of reproductive years), number of clutches per year (CY), number of eggs per clutch (EC), number of fledglings per egg (FE), and number of recruits per fledgling (RF), so that LRS = LS × CY × EC × FE × RF. To these data we applied the partitioning of variance following the method by Brown (this volume, chap. 27). In the total sample there were, by definition, no nonbreeders. In the subsamples of recruits from first clutches, females that did not produce a known first clutch were treated as nonbreeders. Females without late clutches were treated similarly.

The analysis is restricted to local recruitment, neglecting recruitment into breeding populations outside the study area. About half the breeding birds are immigrants; that is, they were not ringed as nestlings in the study area. There is considerable emigration from the study area in summer (Drent 1984) and also in other parts of the year. The discussion addresses the important question to what degree the conclusions are likely to be affected by differential dispersal.

The most extensive data on the occurrence and effects of inbreeding in a natural population are those from the island of Vlieland. About 80% of the breeding birds have been locally ringed as nestlings, and hence pedigrees are fairly complete. Common ancestors of the female and the male of a breeding pair can be identified. The degree of relatedness is expressed in Wright's coefficient of inbreeding F, which is equal to the proportion of the genome in the offspring that consists of two identical copies of genes in the common ancestor. The inbreeding coefficient allows a quantitative analysis of the effects of inbreeding.

Results

The frequency distributions of LRS and its main components are given in figure 8.2. By far the most variable component in the life cycle is the variation in the postfledgling survival, or local recruitment rate (RF) of the offspring (table 8.1). A first analysis was made for all recruits (tables

Table 8.1 Basic Statistics on Variation in Lifetime Reproductive Success in Great Tit Females

	Original		Standardized
Component	*Mean*	*Variance*	Variance
LS	1.6236	0.9543	0.3620
CY	1.2500	0.1411	0.0903
EC	8.5608	3.7773	0.0515
FE	0.5873	0.0832	0.2412
RF	0.0823	0.0135	1.9865
LS CY	2.1024	2.4863	0.6036
LS EC	13.9572	80.5438	0.4169
CY EC	10.6320	13.8801	0.1212
LS FE	0.9810	0.6052	0.6657
CY FE	0.7199	0.1475	0.2736
EC FE	5.1701	7.8039	0.3088
LS RF	0.1507	0.0476	2.6663
CY RF	0.1029	0.0200	1.8858
EC RF	0.7204	1.0397	2.0930
FE RF	0.0555	0.0064	2.7525
LS CY EC	17.9874	195.6811	0.6482
LS CY FE	1.2532	1.2868	0.9058
LS EC FE	8.6122	48.6334	0.7299
CY EC FE	6.2898	12.2093	0.3091
LS CY RF	0.1941	0.0816	2.9242
LS EC RF	1.3180	3.7262	2.8454
CY EC RF	0.8953	1.5167	1.9539
LS FE RF	0.0991	0.0194	3.1459
CY FE RF	0.0679	0.0086	2.3669
EC FE RF	0.4858	0.5011	2.9245
LRS	1.0914	2.3576	3.3407

Note: N = 635 breeding females, 693 recruited offspring. Data from Hoge Veluwe, 1969–83.
LS = life span of female; CY = number of clutches per year; EC = number of eggs per clutch; FE = egg and nestling survival, fledglings/eggs; RF = survival of offspring from fledging till first breeding; LRS = lifetime reproductive success (number of locally recruited offspring over lifetime).

8.2 and 8.3). The result is that variation in LRS is mainly due to variation in RF. Life span (LS) and egg and nestling survival (FE) are the next largest contributors to the variation in LRS. Similar results were obtained in the separate analyses of female and male recruits (tables 8.4 and 8.5). The other components in these analyses were identical to the previous one. For each of the separate sexes the variance in RF makes an even larger contribution to LRS than the variance for the combined data. Summarizing: a high reproductive success is due primarily to a high recruitment rate and secondarily to a long life span and a high nesting success.

Simultaneous variation of variables also contributed significantly to

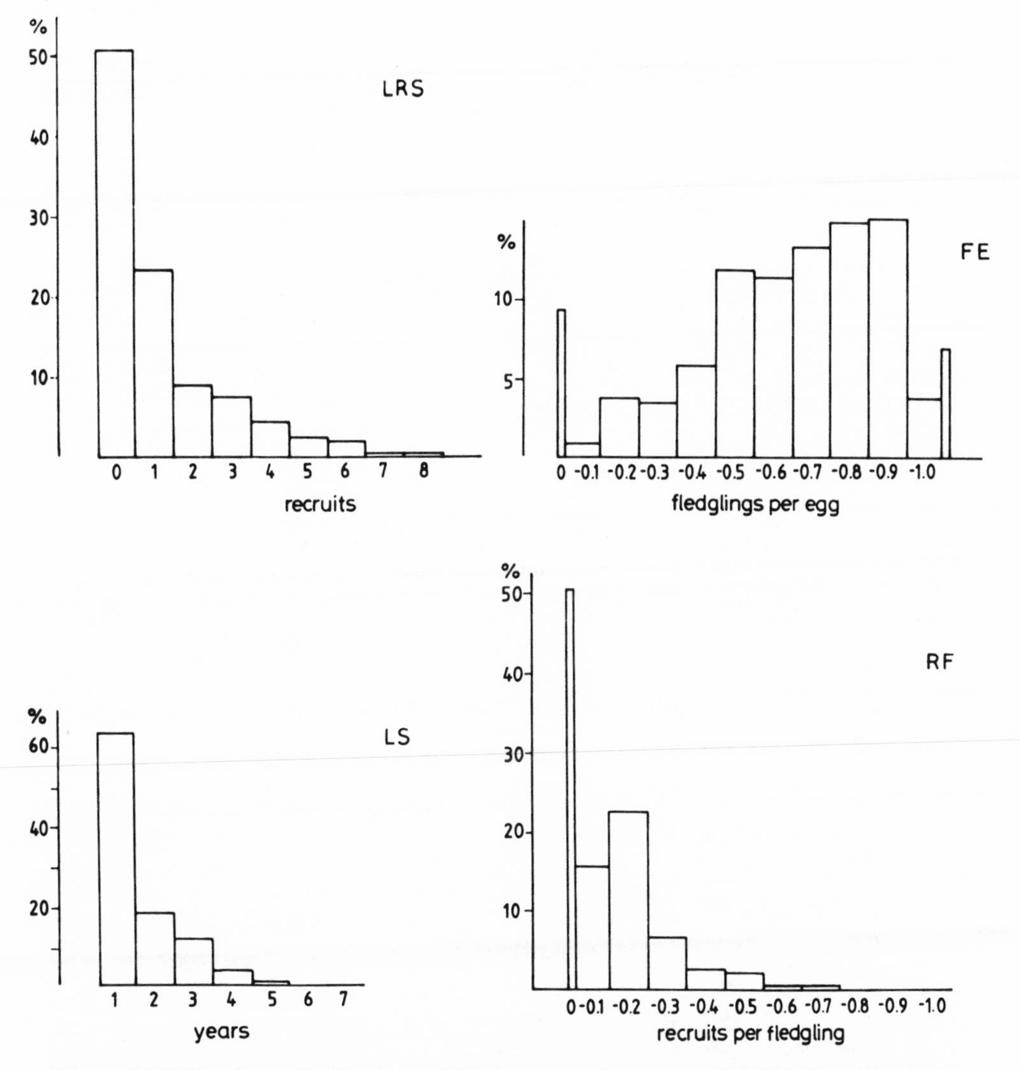

Figure 8.2: Frequency distributions of lifetime reproductive success (LRS) and its main components: life span (LS), egg and nestling survival (FE), and local recruitment rate (RF). Data from 635 females breeding in 1969–83 in Hoge Veluwe.

Table 8.2 Partitioning of Variance in Lifetime Reproductive Success in Great Tit Females: Percentages of Total Variance

Component	LS	CY	EC	FE	RF
LS	10.84				
CY	4.53	2.70			
EC	0.10	−0.62	1.54		
FE	1.87	−1.73	0.48	7.22	
RF	9.51	−5.72	1.65	15.71	59.46
Three- and four-way contributions, −7.55					

Note: Data from Hoge Veluwe, 1969–83.
LS = life span of female; CY = number of clutches per year; EC = number of eggs per clutch; FE = egg and nestling survival, fledglings/eggs; RF = survival of offspring from fledging till first breeding.

Table 8.3 Partitioning of Variance in Lifetime Reproductive Success in Great Tit Females: Off-Diagonal Entries

Component	LS	CY	EC	FE	RF
LS		3.55	−0.46	−0.74	−12.01
CY	0.98		−0.76	−2.38	−11.09
EC	0.56	0.14		0.11	−1.42
FE	2.61	0.65	0.37		1.37
RF	21.53	5.37	3.06	14.34	

Note: Data from Hoge Veluwe, 1969–83. Independent variation below the diagonal, correlation above the diagonal.
LS = life span; CY = number of clutches per year; EC = number of eggs per clutch; FE = egg and nestling survival, fledglings/eggs; RF = survival of offspring from fledging till first breeding.

Table 8.4 Separate Analyses for Female and Male Recruits in the Partitioning of Variance in Lifetime Reproductive Success in Great Tit Females: Percentages of Total Variance

Component	LS	CY	EC	FE	RF
Females					
LS	6.07				
CY	2.54	1.51			
EC	0.06	−0.35	0.86		
FE	1.05	−0.97	0.27	4.04	
RF	9.66	−5.91	0.84	10.70	70.84
Three- and four-way contributions, −1.21					
Males					
LS	8.24				
CY	3.44	2.06			
EC	0.08	−0.47	1.17		
FE	1.42	−1.32	0.36	5.49	
RF	−1.56	−4.48	3.05	20.52	72.34
Three- and four-way contributions, −10.34					

Note: Data from Hoge Veluwe, 1969–83.
LS = life span of female; CY = number of clutches per year; EC = number of eggs per clutch; FE = egg and nestling survival, fledglings/eggs; RF = survival of offspring from fledging till first breeding.

Table 8.5 Separate Analyses for Female and Male Recruits in the Partitioning of Variance in Lifetime Reproductive Success in Great Tit Females: Off-Diagonal Entries

Component	LS	CY	EC	FE	RF
Females					
LS		1.99	−0.26	−0.42	−15.98
CY	0.55		−0.42	−1.34	−12.31
EC	0.31	0.08		0.06	−2.81
FE	1.46	0.37	0.21		−6.39
RF	25.65	6.40	3.65	17.09	
Males					
LS		2.70	−0.35	−0.56	−27.75
CY	0.74		−0.58	−1.81	−11.02
EC	0.42	0.11		0.08	−0.68
FE	1.99	0.50	0.28		3.07
RF	26.19	6.53	3.73	17.45	

Note: Data from Hoge Veluwe, 1969–83. Independent variation below the diagonal, correlation above the diagonal.

LS = life span of female; CY = number of clutches per year; EC = number of eggs per clutch; FE = egg and nestling survival, fledglings/eggs; RF = survival of offspring from fledging till first breeding.

variation in LRS. Table 8.3 shows that the largest contributions are made by simultaneous independent variation in LS and RF and in FE and RF. The same is true for the separate analyses of female and male recruits (see table 8.5).

Next, separate analyses were run for first clutches and for late clutches. This was done because there are large differences in the local success of first and late clutches (e.g., Dhondt and Hublé 1968). In the analysis of recruits from first clutches there are fifty-six nonbreeders (birds for which only late clutches are known). The results (tables 8.6 and 8.7) are similar to those of the first analysis, in terms of both the contributions from single variables and the contributions from combinations of variables. As expected, the contribution of nonbreeding individuals is negligible.

In the analysis of recruitment from late clutches (table 8.6), more than half the females were nonbreeders, but only 12% of the variation in LRS was due to nonbreeding, that is, to the failure to produce a late clutch. Again, among the contributions from single variables those from recruitment rate (RF), life span (LS), and egg and nestling survival (FE) are predominant. There are also large contributions from simultaneous independent variation of LS and RF and of FE and RF. In addition, there are contributions from variation in RF and CY (number of clutches per year) and a negative contribution of higher-order combinations of variables.

Regarding the covariances, the striking feature is their positive sign in most analyses. This implies that a long life span of a female is associated with a high annual number of fledglings produced and also with a high recruitment rate of the offspring. Thus no trade-offs between life span, fertility, and offspring survival were found in this comparison of LRS of different females (but see Tinbergen, van Balen, and van Eck 1985). This

Table 8.6 Separate Analyses for First and Late Clutches in the Partitioning of Variance in Lifetime Reproductive Success in Great Tit Females: Percentages of Total Variance

Component	LS	CY	EC	FE	RF
First clutches					
LS	10.20				
CY	−0.72	0.43			
EC	0.24	−0.01	1.23		
FE	2.12	0.14	0.39	6.91	
RF	10.93	1.30	−0.46	12.24	54.86
Three- and four-way contributions, −3.97					
N = 579; 56 nonbreeders					
Proportion of animals included, 0.912					
Overall variance, 2.006					
Percentage of variance due to excluded animals, 4.17					
Remainder, 95.83					
Late clutches					
LS	6.09				
CY	−4.92	2.80			
EC	0.71	−0.24	1.53		
FE	1.96	−0.67	−0.14	7.46	
RF	10.01	−10.66	−8.56	129.82	92.94
Three- and four-way contributions, −140.13					
N = 301; 334 nonbreeders					
Proportion of animals included, 0.474					
Overall variance, 0.243					
Percentage of variance due to excluded animals, 12.02					
Remainder, 87.98					

Note: Data from Hoge Veluwe, 1969–83.
LS = life span of female; CY = number of clutches per year; EC = number of eggs per clutch; FE = egg and nestling survival, fledglings/eggs; RF = survival of offspring from fledging till first breeding.

implies that the differences between females are greater than the differences resulting from different life-history tactics (van Noordwijk and de Jong 1986). These differences between females may reflect either properties of the individuals or properties of the environment. The great variation in recruitment rates between years (see fig. 8.3) shows that there is a large environmental component in this variation (also see sec. 8.6).

The general conclusion from these analyses on subsets of the data is that variation in the recruitment rate is always the most important factor in determining the overall reproductive success. We will therefore examine the annual variation in recruitment rate in more detail.

Annual Variation in Recruitment Rate

The recruitment rate varies considerably between years (fig. 8.3). Apart from the early years, when the population was in a build-up phase, the number of recruits per female varied between 0.20 and 1.2. On average the local recruits form about one-quarter of the population, another one-quarter being immigrant yearling birds and one-half consisting of surviving older birds. This means that the extremes in the recruitment rate are associated with a halving or doubling of the population size, since the numbers of local recruits and immigrants show a strong positive correlation and adult survival and recruitment rate are also positively correlated.

Over the years there have been many studies of the survival rates of great tits. The one factor that has always been significant in explaining the differences between years is the beech-crop index (Perrins 1966). Even in areas without beeches, the beech-crop index is strongly associated with overwinter survival (see Tinbergen, van Balen, and van Eck 1985). Unfortunately the compilation of this index stopped after 1976. Our own estimates since then indicate that in 1979 there was a poor to moderate beech crop, but much less than 1976, while 1977, 1978, 1980, 1981, and 1982 were very poor years. All seven autumns with a high beech-crop index (>4; see fig. 8.3) were preceded by a below-average breeding density, which makes it hard to distinguish between food availability and population density as the decisive factor (see Perrins 1966). The main

Table 8.7 Separate Analyses for First and Late Clutches in the Partitioning of Variance in Lifetime Reproductive Success in Great Tit Females: Off-Diagonal Entries

Component	LS	CY	EC	FE	RF
First clutches					
LS		−0.87	−0.20	−0.34	−8.53
CY	0.15		−0.02	0.04	0.48
EC	0.44	0.02		0.09	−2.80
FE	2.45	0.10	0.30		−0.94
RF	19.45	0.82	2.35	13.18	
Important three-way contributions:					
LS EC RF 4.86 *.958					
LS FE RF −4.48 *.958					
Four-way contributions, −3.94 *.958					
Late clutches					
LS		−5.81	0.23	−0.42	−19.57
CY	0.89		−0.47	−1.77	−24.31
EC	0.49	0.23		−0.74	−16.04
FE	2.38	1.10	0.60		93.42
RF	29.58	13.56	7.48	36.40	
Important three-way contributions:					
EC FE RF −57.22 *0.88					
LS CY RF −51.54 *0.88					
CY FE RF −34.02 *0.88					
Four-way contributions, −41.57 *0.88					

Note: Data from Hoge Veluwe, 1969–83.
LS = life span of female; CY = number of cluthes per year; EC = number of eggs per clutch; FE = egg and nestling survival, fledglings/eggs; RF = survival of offspring from fledging till first breeding.

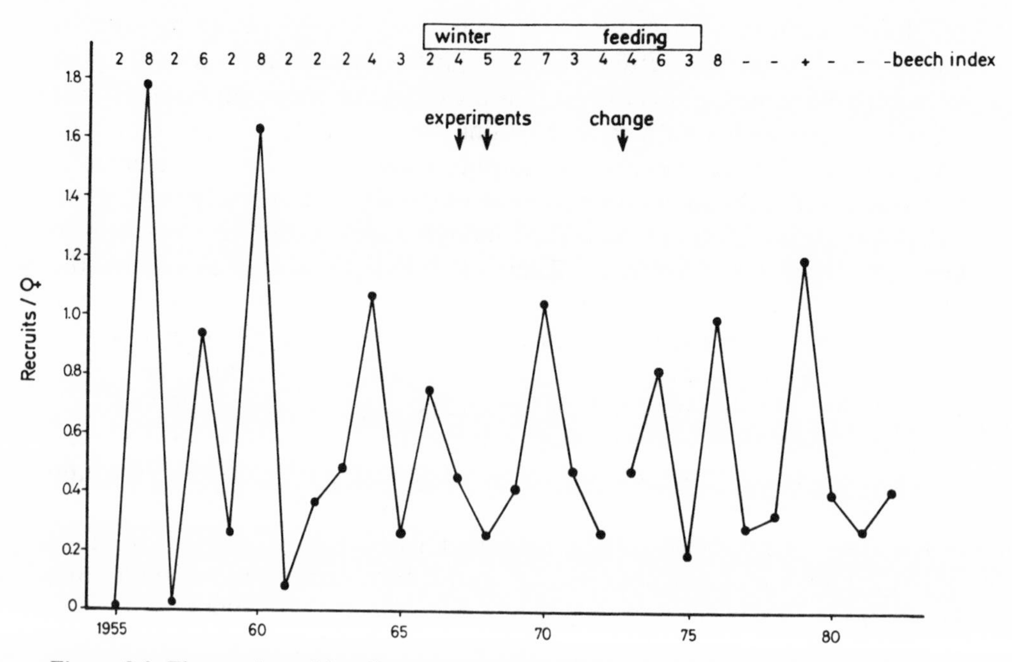

Figure 8.3: The number of locally recruited offspring per female in the Hoge Veluwe 1955–82. Beech-crop index (scale 1–10) and period of winter feeding are indicated along the top. In 1967 and 1968 the numbers of fledglings were experimentally reduced and between 1972 and 1973 the study area was rearranged.

arguments that attribute the variation in recruitment rate to some variable associated with the beech-crop index rather than to density are that the coefficient of rank correlation of the recruitment rate with the beech-crop index is higher than that with density, and that the variation in recruitment rate in years with a poor beech crop is small compared with the difference between years with a poor beech crop and years with a rich one. This is also true when only years with a low population density are analyzed.

8.2 Inbreeding

Inbreeding is related to variation in reproductive success in two ways. On the one hand, the occurrence of inbreeding will increase in every random-mating population with greater variance in number of offspring produced. On the other, inbreeding depression may cause a related pair to produce fewer offspring and thus enhance variation in reproductive success.

The Occurrence of Inbreeding

It is difficult to evaluate the occurrence of inbreeding because there is no generally accepted null hypothesis. A previous statement (van Noordwijk and Scharloo 1981) that the occurrence of inbreeding on Vlieland is at least three times the expectation in an ideal population of the same size is based on wrong assumptions (M. G. Bulmer, pers. comm.). Population geneticists generally use the random-mating population as a reference. A recent analysis of the largest subpopulation on the island of Vlieland shows that the average degree of relatedness of partners is equal to that of neighbors and also equal to that of randomly chosen birds of opposite sex (van Noordwijk et al. 1985). Thus neither dispersal nor other inbreeding-avoidance mechanisms have a noticeable effect on the occurrence of inbreeding. This implies that, apart from population size, the variance in offspring number is the main determinant of the occurrence of inbreeding.

The Effects of Inbreeding on Reproductive Success

The most extensive data on the effects of inbreeding in a natural population are those from the island of Vlieland. Inbreeding has two different effects on the hatchability of eggs (van Noordwijk and Scharloo 1981). If the laying female is inbred (if its parents are related) the probability that the clutch will contain eggs that fail to hatch is significantly higher than in outbred females, and this probability is positively related with the degree of inbreeding (fig. 8.4*a*). There is no relation between the number of eggs failing to hatch and the degree of inbreeding. This is as expected, because the effect must be due to the genotype of the female laying these eggs, which either does or does not contain genes to make the eggs more susceptible to environmental conditions. If the embryo in the egg is inbred (if the laying female is related to its male partner), there is

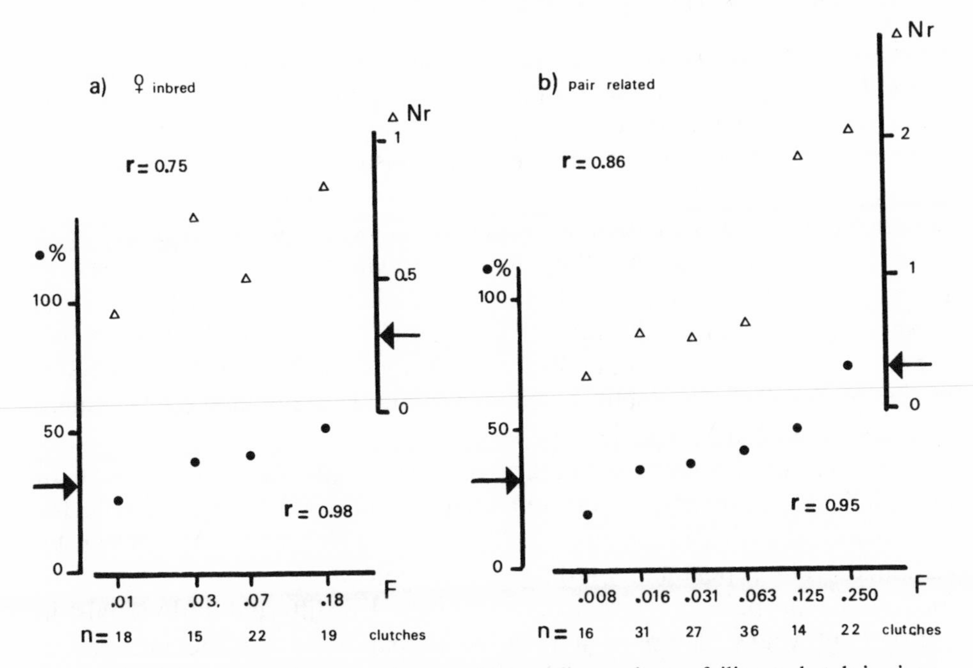

Figure 8.4: The relation between degree of inbreeding and eggs failing to hatch is given as the proportion of all clutches in which eggs failing to hatch were found (*solid circles*) and as the mean number of eggs failing per clutch (*open triangles*). In *a* the laying female is inbred (and the eggs have a "normal" genotype); in *b* the female is related to its male partner, and therefore the embryo's genotype has a higher degree of homozygosity.

also a reduced hatching rate (fig. 8.4*b*). In this case there is a positive relation between the mean number of eggs failing to hatch and the degree of inbreeding as well as a positive relationship between the proportion of clutches containing any eggs failing to hatch and the degree of inbreeding. The degree to which the male is inbred has no effect on the hatching rate of the eggs he sires. Inbreeding is a major cause of hatching failure in the population on Vlieland; at least half of all eggs failing to hatch may be attributed to inbreeding depression.

Inbreeding affects the variance in the number of nestlings produced. However, it turns out that the lower hatching rate is more than compensated for by higher recruitment. If one combines the reproductive success of the related pair and that of the inbred offspring, closely related pairs produce twice as many breeding grandchildren as unrelated pairs. This could be because related pairs occur mainly in large families (the offspring of successful individuals) simply because the proportion of potential mates that are relatives is dependent on family size. If the properties that made the ancestors successful are partly inherited, the individuals involved in inbreeding are then expected to have an above-average reproductive success as well (for a more detailed discussion see van Noordwijk and Scharloo 1981). The main problem with this hypothesis is that important heritable variation in reproductive success is not a priori likely.

8.3 Heritability of Traits Affecting Reproductive Success

Heritability of reproductive success is difficult to measure in wild populations. In theory, one could compare the reproductive success of parents with that of their offspring, but this is not possible for the unsuccessful parents. A comparison of sibs has the disadvantage that sibs share environmental conditions during growth. Thus, even if there is a resemblance between sibs in their reproductive success, it is not possible to draw conclusions about whether this resemblance has a genetic basis. In natural populations the possibility that resemblance between parents and offspring is environmental rather than genetic is very real but smaller than for sib comparisons (Falconer 1981; van Noordwijk 1984a). Furthermore, genetically and environmentally caused family resemblance can be separated (see van Noordwijk 1984b).

So far, only some preliminary results for the direct estimation of heritability of number of recruited offspring produced are available for data from 1980 till 1982 on Vlieland. A first rough approximation can be obtained from repeatability—that is, the constancy of performance of individuals or pairs or territories occupied by different individuals in subsequent years. There is a positive repeatability for the number of recruits produced, but this is likely to be an effect of the environment rather than an effect of the birds, since the reproductive success of different birds occupying the same nestbox is also positively correlated (Koelewijn, pers. comm.).

Table 8.8 Parent-Offspring Resemblance in Number of Recruits Produced

Year	Regression ± s.e. (n)	p
Daughters		
1980	0.15 ± 0.09 (41)	<.10
1981	−0.04 ± 0.20 (44)	>.5
Sons		
1980	0.23 ± 0.11 (44)	<.06
1981	0.05 ± 0.10 (32)	>.05

Note: Data from Vlieland, 1980–82. Parents and daughters are compared for the number of recruited sons, parents and sons are compared for the number of recruited daughters.
ANCOVA table for 1980 data:
Variance between slopes: $F_{1,81} = 0.30$, n.s.
Variance between intercepts: $F_{1,82} = 5.22$, $p < .05$.
Pooled regression: slope = 0.19 ± 0.07 (85), $p < .01$.

The problem of having no parents without offspring can be circumvented by comparing parents and daughters for the number of male offspring at breeding age and comparing parents and sons for the number of female recruits. The results for the Vlieland data (table 8.8) indicate that a significant resemblance ($p < .01$ for the pooled regression) was present for offspring born in 1980 but not for those born in 1981. Caution is necessary in interpreting these results, since the number of recruits shows a Poisson distribution. Analyses for data from other years have not yet been undertaken. At present it is too early to say whether this resemblance has a (partly) genetic basis.

Instead of regarding reproductive success as a single trait, it is more promising to consider the extent to which some well-defined traits are inherited and to what extent they contribute to variation in reproductive success. Below, we will review the evidence for heritability in the following traits: clutch size, egg size, number of clutches per year, and longevity.

Clutch Size

About 40% of the variation in the size of first clutches is genetic (table 8.9). In these analyses the clutch size in later clutches has not been investigated, and the variation caused by differences in environmental conditions between years has also been largely eliminated (Perrins and Jones 1974; van Noordwijk, van Balen, and Scharloo 1981a). However, the variation between years is small compared with the phenotypic variation within years. Clutch size in second clutches is positively correlated with clutch size in first clutches of the same female but negatively correlated with nest success in first clutches. There are too few data on parents and offspring both having second clutches to investigate a family resemblance in the size of second clutches.

Table 8.9 Heritability and Repeatability Estimates for Reproductive Traits in the Great Tit

Population	Repeatability	Heritability	Reference
Clutch size			
Wytham	51 (696/267)	48 ± 10 (256/–)	Jones 1973
Hoge Veluwe	36 (1,277/480)	34 ± 13 (1,242/336)	van Noordwijk, van Balen, and Scharloo 1981a
Vlieland	54 (509/189)	46 ± 14 (1,327/362)	van Noordwijk, van Balen, and Scharloo 1981a
Liesbos	30 (422/178)	25 ± 21 (–/139)	van Noordwijk, van Balen, and Scharloo 1981a
Oosterhout	46 (165/62)	50 ± 30 (–/57)	van Noordwijk, van Balen, and Scharloo 1981a
Egg volume			
Wytham	72 (–/84)	72 ± 22 (–/81)	Jones 1973
Oulu	58 (421/–)	86 ± 29 (–/45)	Ojanen, Orell, and Väisänen 1979
Hoge Veluwe	70 (118/59)	66 ± 24 (93/60)	van Noordwijk et al. 1981
Vlieland	61 (84/42)	72 ± 30 (110/54)	van Noordwijk et al. 1981

Note: The estimates are in percentages of the total phenotypic variance. The number of clutches observed and the number of individuals in the repeatability and the number of daughters in the heritability estimates are given in parentheses.

Egg Size

About 70% of the phenotypic variation in egg size is genetic (table 8.9; Jones 1973; Ojanen, Orell, and Väisänen 1979; van Noordwijk et al. 1981). In the great tit all eggs usually hatch within a single day in first clutches. Small eggs produce small hatchlings, and initial size differences may lead to a size hierarchy within the brood, so that when food is scarce eggs that are small relative to the clutch mean have a lower chance of producing fledglings (Schifferli 1973). The importance of absolute egg size, the variation between clutches, probably lies in the length of the period that hatchlings can survive starvation.

Number of Clutches per Year

Three types of clutches can be distinguished: first clutches, repeat clutches following a first clutch that failed, and second clutches following a successful first clutch. The proportion of second clutches in the population is highly variable, and there is no evidence that the same females tend to produce second clutches in subsequent years (Kluyver 1951). The questions of why and when second clutches are produced, and by which individuals, are not yet fully understood (Kluyver 1951; den Boer-Hazewinkel 1980).

8.4 Reproductive Success and Evolutionary Change

The significance of the genetic variation in the traits listed above depends on the time scale being considered. In some years there are great differences in reproductive success associated with, for example, clutch size or onset of laying (van Noordwijk, van Balen, and Scharloo 1981*a*, *b*). However, these selective differences are not unidirectional, and over a twenty-year period, which is equal to some ten generations, there is no *net* selection for these traits. To take clutch size as an example, the number of fledglings is always positively correlated with clutch size and the recruitment per fledgling is always negatively correlated. The slopes are variable. In "good" years nearly all eggs result in fledglings and the recruitment rate is almost independent of clutch size, resulting in a net advantage for large clutches. In poor years the opposite is true—there is a net advantage for small clutches (van Noordwijk, van Balen, and Scharloo 1981a). The amount of genetic variation in clutch size is sufficient to allow the mean clutch size to change by as much as two eggs in a decade.

These annual variations in environmental conditions are presumably responsible for maintaining substantial genetic variance in many life-history traits. The net result is that there is no clear advantage for any particular phenotype if we combine data gathered over many years. A change of environmental conditions could lead to a rapid adaptation. If one were to analyze the variation in reproductive success during a transitional period of this kind, one could find a considerable effect of clutch size on reproductive success. The results in Tables 8.2–8.5, indicating a small contribution of clutch size to variation in reproductive success, are consistent with what is known about natural selection for clutch size, but the conclusions depend critically on the time scale chosen.

8.5 Discussion

The average number of locally recruited offspring produced by the 635 females for which we analyzed the data is 1.09 (table 8.1). To maintain population size, every female should on average produce two offspring, one male and one female. Immigration compensates for the difference between required and realized local recruitment. We argued above that part of the recruits settle outside the study area and are therefore excluded from our analysis. A first point of discussion must therefore be how reliable the conclusions are likely to be—in other words, how selective emigration is likely to be. We will not discuss the behavioral mechanisms involved in dispersal in any detail (see Drent 1984) but will concentrate on the general pattern. First, we have shown (tables 8.4 and 8.5) that the importance of the various components that explain recruitment is very similar for male and female recruits. Because the emigration rates differ between the sexes, this is an indication that migration has no overriding effect on the relative importance of the components. Second, the immigration rate and the recruitment rate show a strong positive correlation in our data as well as in the Wytham data (Webber 1975).

With respect to the variance in reproductive success, however, we can only say that there is just one piece of evidence for an effect of emigration on recruitment rates, whose quantitative importance is usually limited. In years with a high population density, there may be a small group of "guest pairs" or intruders, birds that breed in a nestbox on the territory of another pair (Eyckerman 1974; Dhondt and Schillemans 1983; Drent 1983). A subgroup of these guest breeders has a territory elsewhere, without a suitable nest site. The guest pairs tend to leave the study area with their offspring immediately after fledging, and the probability that they produce recruits into the study population is low. Under special conditions regarding the distribution of favorable foraging sites and the origin of the parents, this phenomenon can be of quantitative importance for local recruitment (Drent 1984).

Our results show all the weaknesses inherent to correlative data analysis. Nevertheless, there is a simple explanation that is consistent with all results. The most important aspect in the population dynamics is that there are "good" years and "poor" years. In the good years population size is below average, the beech-crop index is high, the recruitment rate of offspring born is high, and the survival of the breeding birds to the next breeding season is also high. In poor years the opposite is true except that population density may either be large or small. Given that no two subsequent years can be good, at least in terms of beechmast, and that the differences between good and poor years are considerable, we can explain the observed results—that recruitment rate is the major determinant in LRS—since high recruitment rates usually occur only once in a bird's lifetime.

One aspect of the results is especially useful for discriminating between alternative explanations, namely, the relatively small contribution of life span to the total variation in reproductive success. If the data for single years are analyzed, there is only a small effect of age on reproductive success. If the environmental conditions are constant, one therefore expects the reproductive output to be proportional to the reproductive life span. In fact, the contribution of life span is largely shared with the contributions of the interactions of life span with fecundity and with local recruitment. One might paraphrase and say that life span is especially important when fecundity and recruitment rate are high, or in "good" years.

We have already stressed the importance of the time scale of the analysis. Our results reflect the net effects over a decade. On this time scale, the annual fluctuations in recruitment rate, presumably caused by differences in environmental quality, are the single important determinant of the population structure. There are no consistent advantages correlated with phenotypic properties of individuals, all traits that were analyzed are selectively neutral on this time scale, and hence the existence of genetic variation in these properties is not important in explaining the observed variation in reproductive success. This implies only that these processes have little *net* effect over a period of a decade, roughly five generations.

8.6 Acknowledgments

We gratefully acknowledge the assistance of J. Visser in carrying out the computations. We thank H. P. Koelewijn for his kind permission to reproduce his unpublished results on the inheritance of reproductive success.

9 Lifetime Reproductive Success of the Great Tit, *Parus major*

R. H. McCleery and C. M. Perrins

THE GREAT TIT IS perhaps the most-studied small bird in the world. Its wide distribution, together with its abundance and the readiness with which it will nest in boxes, makes it a very convenient subject for ecological and behavioral study. Long-term projects in progress include those described here in Wytham woods in the United Kingdom (Lack 1966; Perrins 1979) and those in the Hoge Veluwe forest and the Frisian Islands in the Netherlands (van Noordwijk and van Balen, this volume, chap. 8) and near Antwerp in Belgium (Dhondt 1971).

The great tits in Wytham live mainly in mixed deciduous woodland at a density of about one pair per hectare, in contrast to the populations described in chapter 8, which are mainly in coniferous woodland at a rather lower density. Great tits are short-lived, monogamous birds; only a small proportion of the fledged young reach breeding age, and more than half of those that breed do so only once. Second broods within a season are almost unknown in Wytham (in contrast to some of the Continental populations), though pairs that lose their clutches or broods through predation often lay replacement clutches.

9.1 Data and Analysis

In this chapter we examine a number of the parameters that affect the reproductive output of individual great tits during their lifetime. During the period of study examined here (1960–84), roughly one thousand nest boxes were available to the tits each year over some 230 ha of mixed deciduous woodland. The nest boxes are visited at regular intervals throughout the nesting season, and the nestlings are ringed before fledging. From these visits we know the date when each clutch was started, the number of eggs laid, the number of young that hatched, and the number of young that fledged.

Up to 1964 we caught some of the females during the day, while they were incubating, or at night, while they were roosting with their chicks; no males were caught. Since 1965 we have attempted to catch both sexes of the breeding birds while they were feeding their young. For the most part this is successful; over 90% of the breeding birds that raise young are caught each year. The parents of those broods that fail for one reason or another are not usually caught, however, so if those that succeed are in any way different from those that fail, there is the possibility of bias. Nest failure due to predation from weasels, *Mustela nivalis*, was common until the early 1970s (Dunn 1977), at which point the nest boxes were changed to ones that are virtually predator-proof; hence the proportion of nests lost to predators decreased markedly about this time and the proportion of breeding adults caught increased.

Trapping the breeding birds also gives us the opportunity to measure the survival rates of birds raised in the boxes the year before. However, this figure differs from those on other aspects of nesting success in that it is not complete. Many young birds will have survived and emigrated from the wood, and so we do not encounter them. We presume that the sample of birds surviving in the wood reflects the success of the broods concerned. It is extremely rare for us to recapture a bird we have missed for two breeding seasons. Hence, for those breeding up to and including 1982, we can be reasonably certain of their subsequent fate; since this is not true for the two years 1983 and 1984, the data on survival and recruitment of young for adults breeding in these two years have been omitted.

Approximately half the breeding attempts in any year are made by birds that have not themselves been reared in Wytham. The detailed relationship between the study population and birds in neighboring areas is not known, but it does not seem simply to be buffered against fluctuation by immigration from the surroundings, since there is a positive correlation between the number of immigrants and the number of Wytham birds breeding in any season. The total breeding population has remained fairly constant over the study period, though it can double or halve between years (Perrins 1979), so we have assumed where necessary that the population is stable.

The great tit starts to breed at different times in different years, and there are major effects of year on breeding success, especially on the number of young that survive after fledging. This makes it difficult to make direct comparisons between years. To overcome this, we have usually centered the data so as to produce tables showing deviations from the norm of that particular group in that particular year, calculated by subtracting the mean for the year from each score. For example, a laying date of day 30 (day 1 = 1 April) in 1964, when the mean laying date was 34.3, would be scored as $30 - 34.3 = -4.3$. We found that dividing the difference from the mean by the standard deviation for the year to produce *z*-scores (as in Perrins and McCleery 1985) made little difference to the results but made the tables much harder to understand, so we have not done so in this chapter. In calculating lifetime reproductive success (LRS) we have not centered the data in this way but have simply taken the total number of recruited young produced by an individual in its lifetime. Although this leaves the problem of between-year variation unsolved, it avoids some other difficulties. In practice most individuals produce zero recruits in a season, a substantial minority one, and very few two or more, so the correction factor for any year is always small (about 0.1) relative to the differences in reproductive output between individuals.

Over half of the adult great tits die between one breeding season and the next (see table 9.3). As a result we have far more data on one-year-old birds than two-year-olds, and the sample rapidly diminishes with age. A weakness of the study is that we are unable to sex the nestlings before fledging. As a result, we cannot determine whether there is any sex-related differential mortality before or after fledging.

Age and Reproductive Success

One of the most important factors affecting the reproductive success of any individual great tit is its age. This has been the subject of several previous studies (e.g., Kluijver 1951; Perrins 1965; Perrins and Moss 1974; Harvey *et al.* 1979). Females tend to lay about 0.5 fewer eggs on average

Table 9.1 Two-Way Analysis of Variance Separating Male and Female Age Effects on Lay Date, Clutch Size, Number Fledged, and Number of Offspring Recruited to the Breeding Population in the Great Tit

Source of Variation	Sum of Squares	d.f.	Mean Square	F	p
Lay date					
Mean	171.43	1	171.43	3.23	.07
Female age	666.42	2	313.21	5.90	.003
Male age	114.76	2	57.38	1.08	.34
Female × male	18.76	4	4.69	0.09	.99
Error	83,868.32	1,581	53.05		
Clutch					
Mean	8.73	1	8.73	3.89	.05
Female age	22.77	2	11.39	5.07	.006
Male age	7.79	2	3.89	1.73	.18
Female × male	8.43	4	2.11	0.94	.44
Error	3,550.94	1,581	2.25		
Number fledged					
Mean	0.01	1	0.01	0.00	.95
Female age	2.25	2	1.22	0.30	.74
Male age	20.50	2	10.25	2.70	.07
Female × male	21.21	4	5.30	1.39	.23
Error	6,010.82	1,581	3.80		
Number of recruits					
Mean	0.80	1	0.08	0.11	.74
Female age	0.55	2	0.27	0.38	.68
Male age	4.73	2	2.36	3.28	.04
Female × male	2.01	4	0.50	0.70	.60
Error	1,140.17	1,581	0.72		

Table 9.2 Means of Standardized Scores of Lay Date, Clutch Size, Number Fledged, and Number of Offspring Recruited Broken down by Age of Male and Female Parents in the Great Tit

	Female Age			
Male Age	*1*	*2–4*	*5+*	*Marginal*
Lay date				
1	1.00	−1.64	0.00	
2–4	−0.39	−2.99	−1.13	
5+	0.78	−1.95	−1.47	−0.54
Clutch size				
1	−0.19	0.36	0.05	
2–4	0.03	0.33	0.71	
5+	−0.29	0.25	0.34	0.06
Number fledged				
1	−0.10	0.18	−1.00	
2–4	0.29	0.24	0.00	
5+	−0.06	−0.12	0.49	0.07
Number of recruits				
1	−0.03	−0.09	−0.31	
2–4	0.13	0.18	−0.07	
5+	0.14	0.01	0.22	0.03

when aged one year than in their subsequent breeding seasons, though the age of the female seems to have little effect on the recruitment rate of offspring into the breeding population. Pairs where the male is a first-year bird lay later, lay fewer eggs, fledge fewer young, and produce fewer recruits than pairs in which the male is older than one year, but the lay-date and clutch-size effects are due to the age of the female in these pairs, not to that of the male (Perrins and McCleery 1985). Mated pairs tend to be of the same age; this appears to be due to the age distribution of unmated individuals at the start of the breeding season rather than to any selective mating (Perrins and Mcleery 1985), but it means that the age of one parent is not statistically independent of the age of the other. To examine the effects of age on fecundity more closely, we took all pairs for which the ages of both parents were known, broke them down by age of male and female, and used two-way analysis of variance (BMDP2V; Dixon 1981) to look for effects of male age on measures of reproductive success (table 9.1). This appears to confirm that female age affects lay date and clutch size independent of the age of the male but has little effect on the fledging success or recruitment rate of offspring into the breeding population. On the other hand, male age considered independent of female age has little effect on lay date and clutch size, but it does affect fledging success and recruitment rate, with older males generally being slightly more successful (table 9.2). With recruitment rate the effect is not straightforward. The trend in recruitment rate for females of increasing age is different depending on the age of the male (see also Perrins and McCleery 1985).

There is a slight bias in this analysis toward more successful pairs

Table 9.3 Life Tables for Female and Male Breeding Great Tits in Wytham

	l_x	m_x	$l_x \cdot m_x$	$x \cdot l_x \cdot m_x$	v_x
Females					
1	1.000	0.359	0.359	0.359	0.624
2	0.387	0.370	0.143	0.286	0.686
3	0.171	0.401	0.068	0.206	0.715
4	0.076	0.518	0.039	0.157	0.706
5	0.037	0.328	0.012	0.061	0.386
6	0.014	0.154	0.002	0.013	0.154
7	0.004	0.000	0.000	0.000	0.000
8	0.001	0.000	0.000	0.000	0.000
GRR = 2.130, R_0 = .624, T = 1.082					
Males					
1	1.000	0.326	0.326	0.326	0.673
2	0.425	0.392	0.167	0.333	0.817
3	0.213	0.425	0.090	0.271	0.847
4	0.109	0.580	0.063	0.253	0.826
5	0.044	0.293	0.013	0.064	0.610
6	0.023	0.383	0.009	0.053	0.606
7	0.008	0.643	0.005	0.036	0.643
8	0.002	0.000	0.000	0.000	0.000
GRR = 3.041, R_0 = .673, T = 1.336					

Note: Age-related survival (l_x) is derived from the life spans of all recorded breeding birds of known age. Age-related fecundity (m_x) is derived only from pairs where the ages of both parents were known.

than average, shown by the marginal cell values in table 9.2; since the measurements are deviations from the seasonal mean, the marginal cell totals should be zero in an unbiased sample. Pairs with one parent or both of unknown age tend to be slightly less successful on all four measures of reproductive success than the average for the whole population, owing to our not trapping and aging the parents of breeding attempts that fail at an early stage.

In spite of the difficulties of interpretation, we have calculated the mean number of surviving young produced by males of different ages, using untransformed scores (table 9.3).

The Survival Rate of Adult Great Tits

Figure 9.1 shows the log-linear survivorship curves for 3,127 breeding great tits in Wytham. The same data are shown as the proportion surviving to each age in table 9.3. To estimate mortality we have included all the breeding adults whose date of birth was known and that have not been seen for at least two breeding seasons, assuming that such birds have died rather than emigrated. There is slightly higher mortality among first-year birds, though this is not very apparent in the log-linear plot. The drop in survivorship of females aged seven or more is not significant (standardized residual = 1.88, $p > .05$); in any case it is based on a total of only nine individuals.

The log survival curve suggests that for some reason males survive better to ages four and six than might be expected. There is a slight but persistent tendency for males to survive better than females ($\chi^2 = 17.87$, d.f. $= 5$, $p < .01$).

In table 9.3 we combine the information on age-related fecundity and age-related mortality into life tables for male and female great tits. The proportions surviving to each age are derived from the same data set as figure 9.1, based only on birds that have bred at least once (since we cannot sex the birds as nestlings). The age-related fecundities are the numbers of offspring recruiting to the breeding population, divided by two since each mated pair must produce two viable offspring to replace themselves (e.g., Pianka 1978). In compiling these figures we used only pairs in which both parents were known. If all recorded pairs were used there would be a bias toward less successful breeding attempts by females, because in the early years of the study we were more likely to catch the female than the male when a brood failed because of predation or desertion. If we ignore the small numbers of birds aged seven or more, there is a slight tendency for mean fecundity to peak at age four in both sexes, though as we have seen (tables 9.1 and 9.2), in the case of females this may be due to the age of the mate. The decline in the fecundity of older birds may be partly due to the increasing probability that they have lost their original mate, since pairs breeding together for the first time tend to be less successful than pairs of the same age that have previously bred together (Perrins and McCleery 1985).

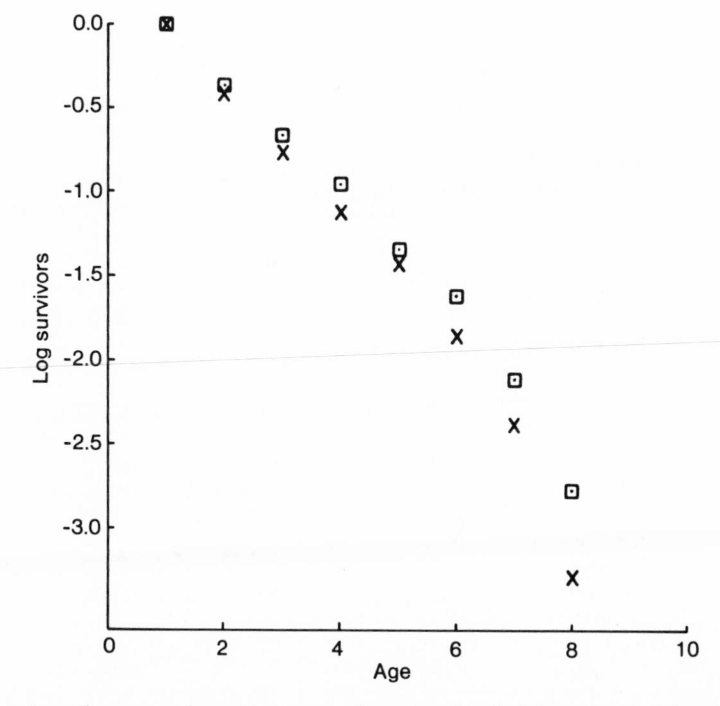

Figure 9.1: Log-linear survivorship curves for great tits that have bred at least once in Wytham. Males (*dotted squares*): 1,192 individuals; females (*X's*): 1,935 individuals. Birds seen breeding in 1983 or 1984 and those whose date of birth is unknown are omitted.

The reproductive value in both sexes is highest when the birds are aged two to four and slightly lower for first-year birds and those aged five or more. The consistently higher reproductive value of males of given age compared with females depends mainly on the lower mortality of first-year males. Males aged two to four have greater "fecundity" than females of the same age, though first-year males have lower "fecundity" than first-year females.

Relation between Subsequent Survival and Reproductive Output

There is no evidence that raising a large number of young has any deleterious effect on the parents' chances of survival. Table 9.4, panel A shows the relation between the subsequent survival of birds of each sex and their breeding parameters for birds that fledged at least one chick. We had thought that the opposite might be the case, that birds more successful at breeding might also have higher chances of survival, but there is no evidence that this is so either. However, if we examine the same trends including those birds that fail to fledge any young, we find a slightly different pattern for females. Those birds that failed to survive to breed in a subsequent season also tended to have laid later, fledged fewer young, and produced fewer recruits, though their clutches were no smaller (table 9.4, panel B). Part of the reason for this is that many of these failed broods were taken by predators, and in some cases the females were also known to have been taken.

Table 9.4 Comparison between Birds That Survived to Breed the Following Season and Those That Did Not, Using Four Measures of Reproductive Effort

Measure	Died	Survived	p[a]
A. Fledged at Least One Chick			
Males			
Clutch	.00	.00	n.s.
Fledged	.01	−.02	n.s.
Recruits	−.01	.01	n.s.
Lay date	.05	−.10	n.s.
Females			
Clutch	.02	−.03	n.s.
Fledged	−.01	.01	n.s.
Recruits	−.01	.02	n.s.
Lay date	.20	−.31	n.s.
B. All Breeding Birds			
Females[b]			
Clutch	.01	−.03	n.s.
Fledged	−.72	.03	<.001
Recruits	−.08	.03	.001
Lay date	.26	−.88	<.001

Note: Data centered by subtracting the mean for the year.
[a] Probability that the means differ under the null hypothesis, using a t-test.
[b] Only females are shown; for males the result is the same as in panel A.

It is difficult to point to which is cause and which is effect; late breeding is known to result in higher nestling mortality and poorer survival after fledging (Perrins 1965); Late breeding also causes the adult to be late in completing molt and to have a shorter wing length when molt is completed (Dhondt 1981), both of which may influence overwinter survival.

Lifetime Reproductive Success

In the sample of data collected between 1960 and 1982, males produced more young during their lifetime than females. A total of 1,449 females that bred in the nest boxes produced 1,010 young that are known to have survived to breed. The mean reproductive success for females is thus 0.697 surviving young. The standard deviation is 1.109, giving a standardized variance ($\sigma^2/\bar{x}^2$; Wade and Arnold 1980) of 2.53. Since 881 of these females raised no young that survived to breed, the 1,010 young were produced by only 528 (36%) of the females, an average of 1.91 each. The comparable figures for males are: 924 males raised 860 young, giving a mean reproductive success of 0.931 (s.d., 1.281), standardized variance ($\sigma^2/\bar{x}^2$) of 1.89; 461 males raised no surviving young, so 463 (50%) raised 860, or 1.86 each. Three factors, none of which we are able to quantify, may produce this apparent imbalance:

1. We do not know the sex ratio at fledging.
2. There could be higher mortality of males than females between fledging and breeding.
3. There have been claims (e.g., Bulmer and Perrins 1973) that some one-year-old males do not breed; this was thought to happen because of a slight imbalance in the sex ratio of breeding birds, so that some males were unable to get a mate.

Any one (or a combination of two or all) of these factors might explain why those males that do breed have a higher reproductive output than females.

Variance in Reproductive Success

Individual differences in lifetime reproductive success are large, with most individuals failing to produce any surviving young (fig. 9.2). If the probability that an offspring survived was random and independent of whether the parent had produced any other surviving offspring, these distributions would be Poisson, but this hypothesis may be rejected since both for females ($\chi^2 = 139.53$, 3 d.f., $p < .001$) and for males ($\chi^2 = 156.04$, 4 d.f., $p < .001$) the distributions depart significantly from Poisson. In both sexes more individuals than expected produce no recruits and fewer than expected raise one. For males fewer than expected raise two, but more than expected raise larger numbers in both sexes, suggesting some clumping in the production of recruits—that is, that individuals that produce one are likely to produce another.

Figure 9.3 shows, for males and females separately, the distributions of (*a*) lifespan (LS); (*b*) mean clutch size (EC); (*c*) number of chicks hatched per egg (HE); (*d*) number of chicks fledged per hatchling (FH);

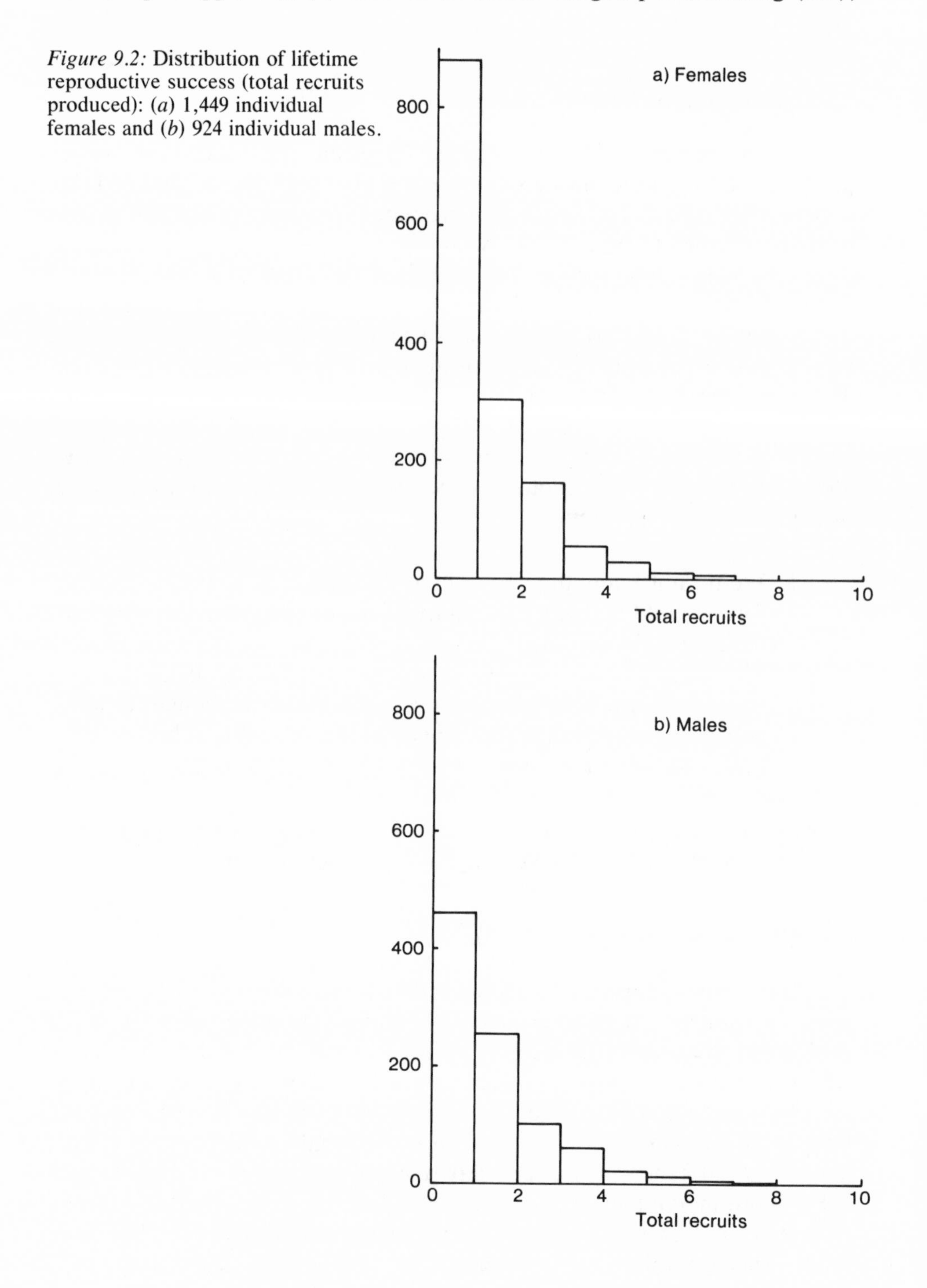

Figure 9.2: Distribution of lifetime reproductive success (total recruits produced): (*a*) 1,449 individual females and (*b*) 924 individual males.

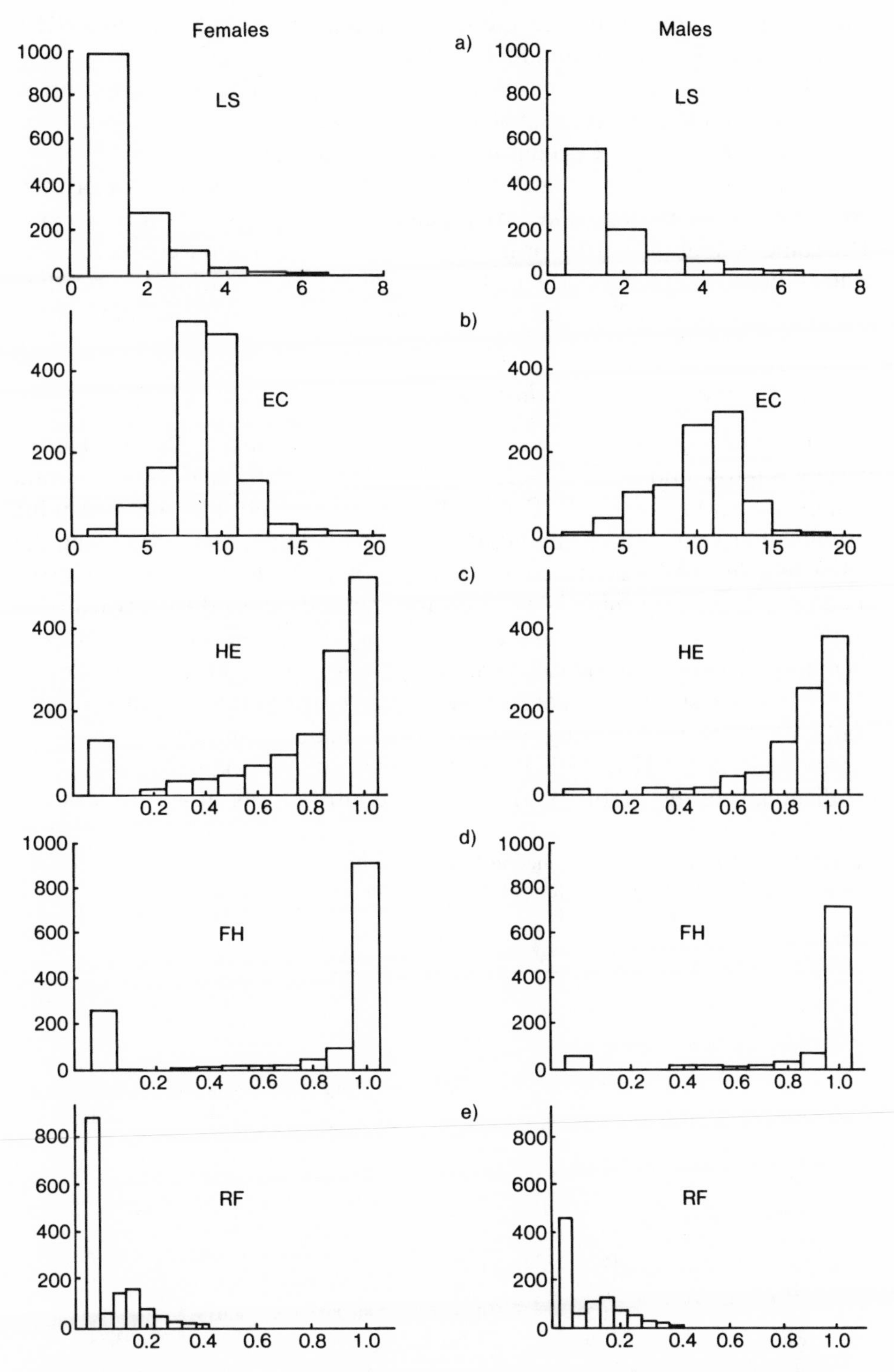

Figure 9.3: Distributions of (*a*) life span (LS); (*b*) mean clutch size (EC); (*c*) number of chicks hatched per egg (HE); (*d*) number of chicks fledged per hatchling (FH); (*e*) number of recruits to breeding population per fledgling (RF), shown for females and males separately.

and (*e*) number of recruits to the breeding population per fledgling (RF): *b–e* are based on each individual's total lifetime production.

The number hatched per egg and number fledged per hatchling can be seen to be bimodal in females (see McCleery and Perrins 1985). The main reason for this is again predation; those nests that are not attacked by predators tend to have a high success, hatching and raising most of their young, whereas in those taken by predators all the eggs or young are lost. The same is doubtless true for males, though it is less apparent from these figures because most of the males are caught later during the nestling period; hence the males tend not to be identified at nests that fail.

Components of Reproductive Success

The lifetime reproductive success of an individual is the product of variables *a* to *e* above: LRS = LS × EC × HE × FH × RF. Since second broods are rare in Wytham (Perrins 1979), we need not include a term for attempts per year, unlike van Noordwijk (this volume, chap. 8). Using the 1,449 females and 924 males for which complete lifetime reproductive success can be determined, we used the method devised by Brown (this volume, chap. 27) to partition the variance in lifetime reproductive success owing to variations in the different components of the breeding cycle.

Table 9.5 shows the percentage variance attributable to the components of LRS in breeding females. The major contribution is from recruitment rate (RF), with some influence of life span (ca. 8%). The off-diagonal terms (table 9.5) indicate contributions that are due to

Table 9.5 Partitioning of Variance in Lifetime Reproductive Success in All Breeding Great Tit Females and Males: Proportion of Variance Attributable to Each Component of LRS

Component	LS	EC	HE	FH	RF
Females					
LS	7.83				
EC	−2.11	1.66			
HE	−0.79	−0.23	3.20		
FH	2.01	−0.16	−0.43	5.26	
RF	7.90	5.58	9.66	22.58	50.51
3+	−12.48				
Males					
LS	24.41				
EC	−17.57	6.26			
HE	0.67	−0.26	2.26		
FH	1.23	0.09	−0.29	4.03	
RF	21.01	7.86	−4.56	7.83	93.38
3+	−46.35				

LS = Life span; EC = mean clutch size; HE = number of chicks hatched per egg; FH = number of chicks fledged per hatchling; RF = number of recruits to breeding population per fledgling. 3+ = three-way and higher-order interactions.

Table 9.6 Partitioning of Variance in Lifetime Reproductive Success in All Breeding Great Tit Females and Males: Breakdown of Off-Diagonal Terms

Component	LS	EC	HE	FH	RF
Females					
LS		−2.75	−2.01	0.00	−11.37
EC	0.63		−0.49	−0.58	1.49
HE	1.22	0.26		−1.25	1.78
FH	2.01	0.43	0.82		9.62
RF	19.28	4.09	7.87	12.96	
Males					
LS		−20.43	−0.36	−0.61	−21.66
EC	2.86		−0.52	−0.38	−3.08
HE	1.03	0.26		−0.46	−8.51
FH	1.84	0.47	0.17		0.78
RF	42.68	10.94	3.94	7.05	

Note: Independent variation below the diagonal, correlation above the diagonal.
LS = Life span; EC = mean clutch size; HE = number of chicks hatched per egg; FH = number of chicks fledged per hatchling; RF = number of recruits to breeding population per fledgling.

simultaneous variation in two components, and this can be broken down into that due to correlations between the components and that due to independent variation (table 9.6). In the case of females the off-diagonal components are noticeable in three cases, HE × RF and LS × RF (both mainly due to independent variation), and FH × RF, due in about equal measure to correlation and independent variation. These results are extremely similar to those of van Noordwijk (chap. 8) for great tits in the Netherlands, and taken at face value they suggest that there is a positive relation between individual differences in fledging success and recruitment rate and between life span and recruitment rate. However, the sample contains the females who failed to raise any fledglings and is therefore subject to the problem that nests suffering predation have zero reproductive output and females involved with such nests have short lives.

In tables 9.5 and 9.6 we also present an identical analysis for the great tit males for which we have complete data. The picture is very similar with two exceptions. There is a negative interaction (reduction in total variance) attributable to the combination of EC and LS, mainly owing to negative correlation between these components, and there is a positive contribution from EC × RF, mainly owing to independent variation.

We have already pointed out that most breeding birds fail to leave any surviving offspring, giving rise to a bimodal shape for the distribution of RF, with a large peak at zero and a more or less normal-looking distribution of about 0.15 recruits per fledgling. This suggests that it might be fruitful to consider the breeding birds as being of two kinds, those producing at least some offspring and those producing none. Using Brown's method, we assigned all individuals with RF of zero—those that produced no surviving young—to the "nonbreeder" class, allowing us to give the proportion of total variance in LRS attributable to birds that failed

Table 9.7 Partitioning of Variance in Lifetime Reproductive Success in Great Tit Females and Males, Excluding Animals That Produced No Recruits into the Breeding Population: Proportion of Variance Attributable to Each Component of LRS

Component	LS	EC	HE	FH	RF
Females					
LS	58.46				
EC	−9.45	9.02			
HE	−13.68	−1.54	5.25		
FH	−7.64	−0.86	0.43	2.29	
RF	−49.34	2.05	−6.18	−1.28	55.92
3+	−4.73				
Proportion of animals included, 0.3920					
Overall variance, 1.2306					
Percentage of variance due to excluded animals, 61.28					
Males					
LS	76.70				
EC	−49.39	17.90			
HE	−1.35	−0.12	3.62		
FH	−0.68	−0.87	0.12	2.22	
RF	−39.48	−4.05	−11.72	−1.77	66.10
3+	−9.81				
Proportion of animals included, 0.5011					
Overall variance, 1.6420					
Percentage of variance due to excluded animals, 52.59					

LS = Life span; EC = mean clutch size; HE = number of chicks hatched per egg; FH = number of chicks fledged per hatchling; RF = number of recruits to breeding population per fledgling. 3+ = three-way and higher-order interactions.

Table 9.8 Partitioning of Variance in Lifetime Reproductive Success in Great Tit Females and Males, Excluding Animals That Produced No Recruits into the Breeding Population: Breakdown of Off-Diagonal Terms

Component	LS	EC	HE	FH	RF
Females					
LS		−12.82	−15.63	−8.50	−70.20
EC	3.36		−1.85	−0.99	−1.17
HE	1.96	0.30		0.35	−8.05
FH	0.85	0.13	0.08		−2.10
RF	20.85	3.22	1.87	0.82	
Males					
LS		−57.23	−2.93	−1.65	−68.40
EC	7.83		−0.49	−1.10	−10.80
HE	1.58	0.37		0.08	−13.08
FH	0.97	0.23	0.05		−2.61
RF	28.92	6.75	1.36	0.84	

Note: Independent variation below the diagonal, correlation above the diagonal.
LS = life span; EC = mean clutch size; HE = number of chicks hatched per egg; FH = number of chicks fledged per hatchling; RF = number of recruits to breeding population per fledgling.

Table 9.9 Regression Analyses: Total Recruits, Total Fledglings, and Total Recruits for Successful Birds Only, on Life Span

Independent Variable	Coefficient of Life Span	Intercept	*t*-Ratio for Coefficient	$\overline{R}^2$
Total recruits	0.385	0.114	13.05***	.11
Total fledglings	4.609	3.670	27.40***	.57
Total recruits (successful birds only)[a]	0.313	1.220	7.74***	.09

[a]Excludes individuals who produced no recruits.
***p <0.001.

to raise any offspring and to partition the remaining variance among components of reproductive output as before.

More than half the variance (53% and 61% for males and females, respectively) is attributable to those animals that bred but failed to produce any offspring (tables 9.7 and 9.8). The most noticeable feature of the matrices is that for breeders with surviving offspring LS has almost as much importance as RF as a source of variance in LRS; this is true for both males and females. There is also a more noticeable contribution from clutch size (EC). The negative correlation between LS and RF seen in both sexes does not appear to be of biological significance but arises from the selection of birds with one or more surviving offspring. Both the total number of fledglings and the total recruits produced by an individual have, not surprisingly, a significant positive relation with life span (table 9.9). When the regression of total recruits on life span is calculated only for successful breeders, the slope is similar but the *y*-intercept is 1.22, since the minimum number of recruits produced by any bird is one. The predicted RF can be calculated by taking the ratio between recruited young and total fledglings predicted for each life span from these equations. When this is done for the curve derived from successful breeders, predicted RF declines slightly as life span increases. The decline appears to be an arithmetic artifact owing to the constraint that successful birds have at least one recruited offspring by definition. If the exercise is repeated using the regression for all breeders, where the intercept is about zero, the predicted RF is almost independent of life span. We feel this illustrates one of the dangers of interpreting the off-diagonal contributions to the total variance in LRS.

The regression analysis of table 9.9 gives a description of the relation between life span and reproductive success. Each additional breeding season produces on average about 4.6 extra fledglings and 0.3 of a recruit into the breeding population. However, the coefficients of determination are small, and we have already shown that life span is much less important than recruitment rate of the offspring in accounting for variance in LRS among all breeding birds (tables 9.5 and 9.6).

Individual Differences

We looked at the first and second breeding attempt of each individual to see whether birds producing one or more surviving offspring in their first attempt were more or less likely than chance to repeat their success in the following year. Using a two-by-two contingency analysis, we found no such effect in either sex ($\chi^2 = 0.46$ for females, $\chi^2 = 0.33$ for males, 1 d.f.). There is, however, some lack of independence in the relationship between life span and total young produced (table 9.10), which might cause the observed relationship between LS and RF seen in tables 9.5 and 9.6. In both sexes a contingency analysis between life span and number of surviving young (classified as zero, one, or more than one produced) shows a departure from the null hypothesis that number of surviving young and life span are independent. In both sexes there are more individuals than expected with a life span of one year and no surviving young and fewer than expected living to age three or more and producing no offspring. This appears to indicate that birds that manage to keep breeding for three or four seasons are likely to produce a surviving chick eventually, and that those which have several surviving young are likely to have produced them over several seasons rather than all at once. The latter effect may explain the relative importance of life span in accounting for variance in lifetime reproductive success among successful breeders in contrast to its lesser importance when unsuccessful breeders are included.

There are grounds for thinking that body size may be related to reproductive success, since survival of young is correlated with tarsus length, and tarsus length shows a heritability of about .75 in this population (Garnett 1981). We have not so far been able to relate body size measures to LRS, but further work on this is in progress.

Table 9.10 Contingency Analysis of the Relation between Life Span and Total Number of Recruits Produced

Number of Recruits	Life Span (years)						
	1	*2*	*3*	*4*	*5*	*6*	*Total*
Females							
0	682	139	39	11	5	5	881
	601.3[a]	172.1	68.7	21.9	9.7	7.3	
1	193	66	31	11	1	2	304
	207.5	59.4	23.7	7.6	3.4	2.5	
2+	114	78	43	14	10	5	264
	180.2	51.6	20.6	6.6	2.9	2.2	
Total	989	283	113	36	16	12	1,449
Total $\chi^2 = 137.34$; d.f. $= (3 - 1) \times (6 - 1) = 10$.							
Males							
0	324	79	33	15	6	4	461
	277.4[a]	97.3	41.9	27.4	8.5	8.5	
1	145	68	25	11	2	4	255
	153.4	53.8	23.2	15.2	4.7	4.7	
2+	87	48	26	29	9	9	208
	125.2	43.9	18.9	12.4	3.8	3.8	
Total	556	195	84	55	17	17	924
Total $\chi^2 = 80.01$; d.f. $= (3-1) \times (6-1) = 10$.							

[a]Expected frequencies are given below observed frequencies.

Cohort Effects

For the population as a whole, both recruitment rate of young and overwinter survival of adults vary significantly between years. Since these parameters account for most of the variation in LRS between individuals, it seems likely that the LRS of any individual depends on the season of first breeding, particularly since most individuals breed only once or twice. The calendar year of first breeding accounts for a significant proportion of the variance in LRS (females $F = 4.96$, d.f. = 21,1411, $p < .001$; males $F = 4.05$, d.f. = 18,898, $p < .001$). More interesting perhaps, in view of the distinction between successful and unsuccessful breeders, a contingency analysis of successful versus unsuccessful breeders broken down by year departs from the null hypothesis for both females ($\chi^2 = 129.23$, d.f. = 21, $p < .001$) and for males ($\chi^2 = 89.32$, d.f. = 18, $p < .001$), showing that the proportion of birds starting to breed in any year that are successful differs between years.

9.2 Discussion

Age and Reproductive Success

Breeding success increases with age up until about age four, and then it decreases somewhat in females but not in males, though the changes in success are relatively small. An increase in the probability of not breeding at all with increasing age has been shown by Dhondt (1985).

The number of recruits produced per breeding attempt is influenced by the male's age but not, apparently, by that of the female when her age is considered independently of that of her mate. There is a relationship between hatching success, fledging success, and size of the area surrounding the nest (correlated with territory size), probably owing to density-dependent effects of predation (McCleery and Perrins 1985; Dunn 1977). At present we do not know whether older males are able to hold larger territories and hence reduce predation risk through increased spacing, but in any case spacing has much less of an effect on the number of surviving young than the events that occur between fledging and recruitment. Possibly older males are better at securing food for their newly fledged offspring.

Besides the age of the mate, experience in breeding together also influences seasonal reproductive success. Birds that change mates between seasons (either through the death of the partner or through divorce) do less well than other birds of the same age (Perrins and McCleery 1985).

There are several reasons for expecting a relation between reproductive output and adult survival. On the one hand, the theory of life-history strategies predicts that increased reproductive effort in one season may reduce the chances of survival to the next. Nur (1984a) showed that for experimentally manipulated broods of blue tits (*Parus caeruleus*) an increase of brood from three to fifteen was associated with a decrease in female survival to the next breeding season of 54%, and for great tits on Vlieland (Netherlands) Tinbergen et al. (1985) showed that in years where the winter food supply was poor there was a negative correlation between the mean fledglings per adult and the survival rate of the adults. However, a number of workers have reported a positive relation between reproductive success and survival rate owing to differences in the quality of individuals (Coulson 1968; Smith 1981b) or their territories (Hogstedt 1980). Our results support neither of these ideas. In a direct comparison, birds that survive to breed again seem to be neither more nor less successful breeders than those that do not (table 9.4). However, our analysis does not constitute a rigorous test for a trade-off between reproductive effort and survival, for if different individuals are of inherently different quality, and if they can assess their own quality, as seems likely (Perrins and Moss 1974), then each individual solves its own trade-off between survival and reproduction, and the population as a whole will not show any relation between these quantities unless it is manipulated experimentally (cf. Hogstedt 1980).

Lifetime Reproductive Success

We have shown that variance in the recruitment rate of the offspring seems to be the most important component of variance in LRS between individuals. We know some of the factors that affect whether a young bird is recruited into the breeding population (Perrins 1965). These include the date in the season when the chick fledges (early birds have a greater chance of subsequent survival than those that fledge later) and the weight at which it leaves the nest (the heavier the chick, the greater its chance of survival). The importance of both these features seems to be mediated through the effect they have on the young bird's chances of establishing itself in the population in the late autumn (Garnett 1981). We do not know the mechanism by which these factors relate to attributes of individual parents. It is perhaps significant that the repeatability of successful reproduction is rather low, suggesting that attributes of individual parents do not reliably lead to successful reproduction. This is perhaps surprising, since we know that individual birds tend to be consistent in their behavior between seasons. For example, they tend to lay at the same time and do not usually shift territory unless their previous nest suffered predation (Harvey, Greenwood, and Perrins 1979).

The number of young that each parent produces during its lifetime is also affected by the number of years it manages to survive and breed. Therefore any attribute of the individual that affected its probability of survival would affect its lifetime reproductive success, though as we have shown, this seems to be less important than the recruitment rate of the offspring. However, we cannot predict which adult birds are most likely to survive from one year to the next. Age does not appear to be important, apart from the fact that first-year breeders have a lower probability of survival than older birds.

Our results suggest that there is a slight but consistent difference in the numbers of young raised during their lifetimes by males and females. This comes about mainly because males tend to live longer than females. Since each offspring has precisely one male and one female parent, we can only suppose that this means fewer males than females actually enter the breeding population. This might happen for a number of reasons (see above and Perrins and McCleery 1985).

It seems clear that by lumping all our data we have obscured some important differences between cohorts. The proportion of birds starting to breed in any year that are successful varies from year to year. It is quite possible that such cohort differences explain the absence of age-related effects, since the optimum trade-off between survival and reproductive effort varies in an unpredictable way, though some of the variation between years should have been removed by the standardization of the data against the mean for the year. In the case of LRS we used untransformed data, so variation between years is large; while it does not appear to be random, we have not so far discovered the causes, which could include predation, weather, and food supply. We looked in a preliminary way at the influence of the beech-mast index on mean LRS for a cohort, but we found no significant result. Since both recruitment rate and adult survival are density dependent, there is probably a complex interaction between successful breeding and population dynamics, giving rise to cohort differences. The analysis of cohort differences is complicated by the fact that birds with different life spans live through different sets of years. Ideally, one should break down each cohort by life span; even with our data set this leads to very small numbers in each group.

9.3 Summary

In summary, LRS is highly variable between individual great tits, although—as one would expect of a short-lived monagamous animal—males and females show a similar distribution of LRS. Most of the variance between individuals is attributable to variation in the recruitment rate of offspring, much of it caused by the difference between producing no surviving offspring and producing at least one. Life span seems to be somewhat less important. It is clear that cohort differences contribute a significant proportion of the variance between individuals, though the causes are not yet known and we cannot yet say which (if any) attributes of individuals differentiate the successful and unsuccessful individuals. Further work is in progress on both these questions.

10 Determinants of Lifetime Reproductive Success in the Song Sparrow

James N. M. Smith

DIFFERENTIAL REPRODUCTIVE SUCCESS IS the simplest and most practical measure of the operation of natural selection (Fisher 1958). Most available estimates of reproductive success, however, have referred to only part of an organism's lifetime (LeBoeuf 1974; Howard 1983), and the subsequent fates of any offspring produced are usually unknown. Though some recent studies (McGregor, Krebs, and Perrins 1981; Clutton-Brock, Guinness, and Albon 1982; Fincke 1982; Partridge and Farquhar 1983) have measured reproductive success over extended periods, these are all incomplete in some way. They refer either to males only (McGregor, Krebs, and Perrins 1981;

Figure 10.1: Song sparrow. Photo by Ervio Sian.

Fincke 1982), to captive populations (Partridge and Farquhar 1983), or to relatively few individuals with truncated life spans (Clutton-Brock, Guinness, and Albon 1982).

The ideal organism for studying reproductive success can be marked for life at birth, is highly sedentary, is not too numerous, and can be readily censused and observed at all stages in its life, and its reproductive attempts can be documented fully. The song sparrow (*Melospiza melodia*, Emberizidae; fig. 10.1) population inhabiting Mandarte Island, British Columbia, comes close to meeting this ideal. The population is resident throughout the year, all individuals can be marked at birth by colored leg bands, life spans are short (up to four years, occasionally up to eight years), all successful nesting attempts can be followed, and almost all young that recruit as breeders have already been banded as nestlings. Birds usually breed in monogamous pairs (3% of males are bigamous), and many birds occupy territories throughout the year. Clutches are small (usually three or four eggs), and each pair usually raises two broods per breeding season. Earlier accounts of the population, and of aspects of its reproductive success, can be found in Tompa (1964), Smith and Roff (1980), Smith et al. (1980), and Smith (1981a,b).

The aims in this chapter are to describe the lifetime reproductive success and demographic background of four cohorts of birds that hatched between 1975 and 1978; to consider how various measures of lifetime success are related to one another and to some shorter-term measures of success; and to look for relationships between lifetime reproductive success, morphology, and social status.

10.1 Study Area

Song sparrows are resident on Mandarte Island throughout the year. The 6 ha island contains four important habitats for song sparrows. A continuous band of shrubs and small trees (mostly *Symphoricarpos alba* and *Rosa nutkana*) provides breeding and feeding habitat. Grass meadows are used extensively for foraging and rarely for breeding; cliffs and the rocky intertidal zone surrounding the island are used only for feeding. The major vertebrate species that share the habitat with song sparrows are two to ten pairs of fox sparrows, *Paserella iliaca*, a potential competitor; the abundant deer mouse, *Peromyscus maniculatus*, a predator and probable competitor of the song sparrow; and two predators: the northwestern crow, *Corvus caurinus* (ca. twenty pairs), and the glaucous-winged gull, *Larus glaucescens* (ca. two thousand pairs). Song sparrows are sometimes parasitized by a brood parasite, the brown-headed cowbird (*Molothrus ater*). The cowbird has surprisingly little effect on reproductive success in the song sparrow (Smith 1981a), and its effects are ignored in this study. A more detailed account of the study area is given by Tompa (1964).

10.2 Methods

The work reported here covers the ten-year period from 1974 to 1984. Adult birds and independent young of the year were first trapped using mist nets in the autumn of 1974. All breeding birds were banded in 1975 and thereafter, and almost all successful nests were found each year during incubation by observing the female building a nest or by following her to the nest after a foraging trip. Young were individually color marked in the nest at about six days of age, and nearly all of the few young that fledged from nests that were not found were captured in mist nets before they became independent from the parents. Less intensive work was carried out in 1980, when the population dipped to only eleven breeding pairs.

Because all locally hatched birds were marked, immigrants could be readily recognized, except in 1980–81. Five immigrants (four females and one male) were recruited to the population between 1975 and 1980, and just two females immigrated from 1981 to 1984. Young song sparrows leave the nest from ten to fourteen days after hatching and continue to depend on their parents for food for about twenty more days, after which they leave the natal territory to join juvenile flocks. I assessed survival from banding to independence by censusing these flocks and noting the identity of each bird.

The morphology of most individuals was measured after trapping in the late summer and autumn. I measured the weight, tarsus length, and wing length and the beak length, depth, and width as described by Smith and Zach (1979). Most individuals were trapped and measured several times, and mean values of each morphological measurement were used, corrected ($[x - \bar{x}]/s.d.$) to size in the first autumn for each bird (Smith, Arcese, and Schluter 1986) and for slight differences in average size among cohorts (Smith and Zach 1979). The size of the territory each bird occupied was estimated at the end of April each year from the positions of song perches of males and from the locations of boundary disputes between birds of either sex. Territory is defined here as the area of shrubs and trees defended, and I assume that males and females defended identical areas. Territories varied in size in relation to both population density and age. I have therefore also corrected for these differences in my analyses.

In this chapter I focus on the lifetime reproductive success of the four cohorts of young birds hatched between 1975 and 1978. These birds have now all died, and I have complete information for all nondispersing individuals on the following traits: age, life span, number of breeding attempts, mean clutch size, mimimum number of eggs laid, number of young surviving to independence from parental care, number of independent young surviving to breed (including "floater" males), territory size, week of hatching, and morphology. I analyzed data on the week of hatching, because hatch date is known to have a strong influence on the social dominance of juvenile birds (Arcese and Smith 1985). Although some information on reproductive success was available for all birds, some cases (n = 15 for males, n = 9 for females) are omitted here, since they were involved in a temporary removal experiment in 1979 (Smith, Yom-Tov, and Moses 1982) or because they were not captured for morphological measurements. Thus, sample sizes vary slightly for some data sets analyzed here, depending on whether complete data was available for the variable under consideration. The maximum number of missing cases did not exceed 15% of the number of breeding birds. None of the major conclusions differ when these birds are included in the analysis.

Analysis

The analysis depends on the following key assumptions: (1) birds that disappear from the island, either because they die or disperse, have negligible reproductive success; (2) males fertilize all the eggs laid by females who share a territory with them; (3) the few individuals excluded from analysis (see above) are a random sample of the total population of their cohorts. I shall consider the correctness of these assumptions in the discussion. Data were not transformed for most statistical analyses for three reasons. First, morphological data were approximately normally distributed, and transformations did not improve the fit to normality. Second, some variables had modes of zero and could not be normalized by

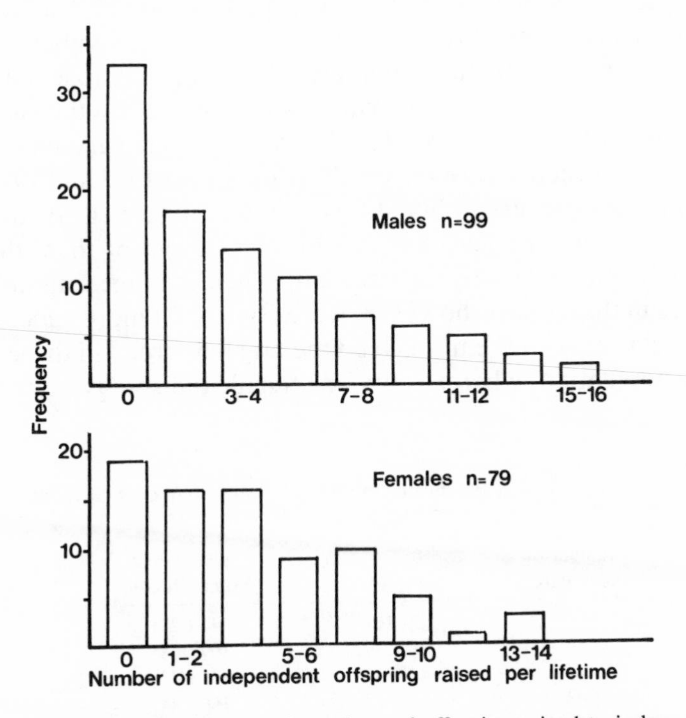

Figure 10.2: Frequency distributions of numbers of offspring raised to independence from parental care (thirty days after hatching) by female (*bottom*) and male (*top*) song sparrows during their lifetimes.

transformation (e.g., fig. 10.2). Finally, sample sizes in this study are moderately large. Under these conditions, parametric tests are fairly robust as long as the data do not deviate too far from the assumptions (Zar 1984). Some of the most critical analyses were repeated using nonparametric methods. Data were, however, standardized ($[x - \overline{x}]$/s.d.) to a mean of zero and variance of one, when making comparisons across age-, sex-, or year-classes, for traits that varied significantly among classes (e.g., territory size, reproductive traits, morphological traits). The effects of correction on morphological variables were slight and did not affect the outcomes of any statistical analyses. Most analyses were carried out using standard parametric statistical tests.

10.3 Results

Variation in Lifetime Reproductive Success

Tables 10.1 and 10.2 show the demography and survivorship of the four cohorts studied. All surviving birds are included here, whether or not they were captured for morphological measurement. Survival from egg to breeding age (one year) was rather low (9%) in 1975 but was fairly similar among the other three cohorts. Adult survival, however, differed strikingly among cohorts because of a severe population crash in 1979–80. This demographic background strongly influenced the mean lifetime reproductive performance of birds in each cohort (table 10.3). This table presents three separate measures of reproductive performance: number of eggs laid (females) or fertilized (males); number of young that reached independence; and number of young surviving to breeding age on the island. Birds in the 1976 cohort were most successful overall, since survivors had three opportunities to breed before the crash. Birds in the 1975 and 1977 cohorts were slightly less successful but still performed quite well. Birds hatching in 1978 were by far the least successful, since most of them died in the crash after breeding in only one season, when average conditions were poor. Males in this cohort, however, were somewhat more successful than females, because six of them survived the crash to breed in the subsequent expanding population, whereas only one female did so. Overall, males and

Table 10.1 Demography of the Four Cohorts of Song Sparrows Studied

Year of Hatching	Number of Female Parents	Eggs Laid	Independent Young Reared	Proportion of Young Surviving to Breeding Age
1975	34	161	44 (.27)[a]	.32
1976	30	229	109 (.48)	.44
1977	44	333	144 (.43)	.35
1978	48	377	166 (.44)	.48

Note: Data from Mandarte Island, British Columbia. Figures refer to the survival of members of each cohort before reaching breeding age.

[a] Proportion of eggs laid is given in parentheses.

Table 10.2 Survivorship of the Four Cohorts of Song Sparrows Studied: Number Alive in May of Each Year

Year of Hatching	Years after Hatching						
	1	*2*	*3*	*4*	*5*	*6*	*7*
1975							
Females	3	1	0	0	0	0	0
Males	11	8	3	1	0	0	0
1976							
Females	23	16	12	1	0	0	0
Males	25	19	12	2	1	0	0
1977							
Females	24	16	3	3	0	0	0
Males	27	18	5	4	3	2	0
1978							
Females	33	1	0	0	0	0	0
Males	46	6	4	2	1	0	0

Note: Data from Mandarte Island, British Columbia. Figures refer to the survival of members of each cohort after reaching breeding age.

Table 10.3 Estimates of Lifetime Reproductive Performance for Seventy-nine Female and Ninety-six Male Song Sparrows

Year of Hatching	*n*	Number of Eggs Laid/Fertilized	Number of Independent Offspring Raised	Number of Offspring Surviving to Breeding Age
1975				
Females	5	12.8 ± 3.09	4.6 ± 1.44	1.6 ± 0.60
Males	11	13.0 ± 3.64	4.2 ± 0.87	0.9 ± 0.39
1976				
Females	21	15.9 ± 1.55	7.2 ± 0.86	2.8 ± 0.40
Males	25	13.5 ± 1.66	5.2 ± 0.76	1.7 ± 0.33
1977				
Females	23	12.9 ± 1.33	4.4 ± 0.58	1.4 ± 0.29
Males	25	11.9 ± 2.29	4.6 ± 1.68	1.5 ± 0.47
1978				
Females	30	6.1 ± 0.51	0.8 ± 0.20	0.1 ± 0.06
Males	35	5.3 ± 0.93	1.6 ± 0.52	0.6 ± 0.29
Total				
Females	79	11.1 ± 0.76	3.8 ± 0.42	1.3 ± 0.18
Males	96	10.1 ± 0.97	3.3 ± 0.31	1.2 ± 0.19
Variance				
Females		45.63	13.94	2.56
Males		90.32	9.22	3.47

Note: Data from birds that bred on Mandarte Island, British Columbia. All measures are means ± s.e. except for the variances at the bottom of the table.

females were about equally successful, although females raised slightly more independent young on average, because none of them failed to obtain mates. The variance in number of eggs laid or fertilized was significantly higher for males (table 10.3, $p = .001$, F-test), but females varied slightly more (table 10.3, $p = .03$, F-test) in the production of independent young.

There was considerable variability in lifetime reproductive performance among individuals that survived to breeding age and remained on the island (fig. 10.2). The modal number of offspring reared to independence

Table 10.4 Mean and Variance of the Components of Lifetime Reproductive Success in Female Song Sparrows

	Original		
Component	*Mean*	*Variance*	Standardized Variance
L	1.633	0.697	0.261
F	6.708	6.163	0.140
S	0.094	0.012	1.403
LF	11.101	45.733	0.381
FS	0.666	0.640	1.612
LS	0.179	0.054	2.280
LFS	1.266	2.658	2.511

Note: N = seventy-nine breeding females.
L = life span; F = fecundity (number of eggs laid); S = survival of offspring to independence.

Table 10.5 Percentage Contribution of the Components of Lifetime Reproductive Success to Variation in LRS in Female Song Sparrows

Component	L	F	S
L	10.41		
F	−0.68	5.45	
S	24.53	2.86	55.87
LFS	1.56		

N = 205; 126 nonbreeders
Proportion of breeders, 0.39
Overall variance (OV), 1.40
Percentage of OV due to nonbreeders, 27.28
Percentage of OV due to breeders, 72.72

L = life span; F = fecundity (number of eggs laid); S = survival of offspring to independence.

Table 10.6 Percentage Contribution of the Components of Lifetime Reproductive Success to Total Variation in LRS in Female Song Sparrows, Including Nonbreeders

Component	L	F	S
L	7.57		
F	−0.50	3.97	
S	17.84	2.08	40.63
LFS	1.14		

L = life span; F = fecundity (number of eggs laid); S = survival of offspring to independence.

Table 10.7 Percentage Contribution of Seven Multiplicative Components of Lifetime Reproductive Success to Variation in LRS in Female Song Sparrows

Component	LS	CY	CS	PH	PF	PI	PB
LS	67.36						
CY	−18.91	12.79					
CS	8.57	0.70	8.71				
PH	8.02	−8.35	0.57	14.15			
PF	2.93	−5.57	0.30	−1.16	13.08		
PI	−18.52	−8.14	−7.12	−4.80	1.63	31.52	
PB	−49.30	19.82	−32.07	−12.29	−19.16	1.98	79.16

Note: Data are for forty-four females that produced at least one offspring surviving to breeding age. L = life span; CY = mean number of clutches laid per year; CS = mean clutch size; PH = proportion of eggs that hatched; PF = proportion of hatching young that reached banding age (ca. six days); PI = proportion of banded young that reached independence (ca. thirty days); PB = proportion of independent young that survived to breeding age.

was zero for both males and females, but several birds of each sex reared ten or more independent young. The higher frequency of total reproductive failure among males (fig. 10.2) results from males that survived as floaters but failed to gain a breeding territory, or from males that gained a territory but no mate. There were no floater females, and females occasionally bred in polygynous trios when there were shortages of territorial males (Smith, Yom-Tov, and Moses 1982).

Lifetime reproductive performance of females was broken down into two sets of multiplicative components, and the contributions of these components to total variation in LRS was examined using the method devised by Brown (this volume, chap. 27). The results of these analyses are displayed in tables 10.4–10.7. First, a simple three-component model was examined, with the variables being life span (in years), fecundity (eggs laid during the lifetime), and survival (proportion of young surviving to breeding age). Among seventy-nine breeding females, the major source of variation in LRS, was survival of the offspring (56%), followed by life span (10%) and fecundity (5%) (table 10.5). Breeders contributed 73% of the overall variance in LRS (table 10.5).

A more complex seven-component model was then examined. Fecundity was subdivided into two subcomponents: mean number of clutches produced per year and mean clutch size. Survival was divided into four subcomponents: proportion of eggs that hatched, proportion of hatching young that reached six days of age (when young were banded), proportion of banded young reaching thirty days of age (when they reach independence from parental feeding), and proportion of independent young surviving to breeding age (one year). The contribution of these variables to the percentage variation in LRS was then examined for forty-four females that raised at least one offspring to breeding age. (Eliminating cases where pairs of variables are zero reduces the possibility of misleading correlations between variables.) The results of this analysis are presented in table 10.7. The three most important variables were, in order: survival from independence to breeding age, life span, and survival from banding age to independence. The contributions of other variables were low, and interactions among variables contributed little to the variance except where they involved the two most important variables. This analysis shows that variation in survival after the period of parental care to the attainment of breeding age was more important than variance in survival during the period of parental care. Within the period of parental care, the contribution to the variance from survival during incubation and the early nestling stage (to six days of age), was about equal to the contribution from survival between six and thirty days of age.

Correlations between Measures of Reproductive Performance

This section describes correlations between different measures of reproductive performance. These data are of interest because most measures of reproductive success available are of single components of fitness, such as clutch size or seasonal breeding success. If these components are not closely correlated with lifetime fitness, they are of limited usefulness.

For each bird, I calculated the number of eggs laid in the first year (females only); the number of eggs laid during the lifetime (females only) or the number of eggs fertilized during the lifetime (males only); the mean clutch size (females only); the number of breeding attempts made in the lifetime; the number of independent young raised in the first year; the number of independent young raised during the lifetime; the proportion of these young that survived to sexual maturity; the life span of the bird (in years); and the number of offspring that reached breeding age.

Table 10.8 Correlation Matrix between Nine Measures of Reproductive Performance in Female Song Sparrows

Measure	EGG1	BREED-ATT	CLUTCH-SIZE	EGG-LIFE	YOUNG1	YOUNG-LIFE	YOUNG-RECRUIT	YOUNG-SURVIVAL	LIFESPAN
EGG1	1.00	—	—	—	—	—	—	—	—
BREEDATT	.58	1.00	—	—	—	—	—	—	—
CLUTCHSIZE	.42	.38	1.00	—	—	—	—	—	—
EGGLIFE	.65	.93	.53	1.00	—	—	—	—	—
YOUNG1	.31	.51	.32	.57	1.00	—	—	—	—
YOUNGLIFE	.32	.77	.39	.80	.82	1.00	—	—	—
YOUNGRECRUIT	.29*	.58	.36	.65	.63	.77	1.00	—	—
YOUNGSURVIVAL	.05[a]	.10[a]	.03[a]	.12[a]	.14[a]	.03[a]	.47	1.00	—
LIFESPAN	.23*	.78	.38	.74	.50	.78	.57	.03	1.00

Note: $N = 79$. EGG1 = eggs laid in first breeding year; BREEDATT = total breeding attempts during lifetime; CLUTCHSIZE = mean size of all clutches laid; EGGLIFE = total eggs laid during the lifetime; YOUNG1 = total independent young raised in first year; YOUNGLIFE = total independent young reared in lifetime; YOUNGRECRUIT = total number of young surviving to breeding age; YOUNGSURVIVAL = proportion of independent young that survived to breeding age; LIFESPAN = number of years lived.

[a]Not significant.

*$p > .01$ (all other correlation coefficients differ significantly from zero at $p < .001$).

Table 10.9 Correlation Matrix between Seven Measures of Reproductive Performance in Male Song Sparrows

Measure	BREEDATT	EGGSFERT	YOUNG1	YOUNG-LIFE	YOUNG-RECRUIT	YOUNG-SURVIVAL	LIFESPAN
BREEDATT	1.00	—	—	—	—	—	—
EGGSFERT	.97	1.00	—	—	—	—	—
YOUNG1	.46	.48	1.00	—	—	—	—
YOUNGLIFE	.81	.84	.64	1.00	—	—	—
YOUNGRECRUIT	.66	.71	.46	.87	1.00	—	—
YOUNGSURVIVAL	.16[a]	.21[a]	.17[a]	.34	.64	1.00	—
LIFESPAN	.85	.85	.27	.80	.69	.18[a]	1.00

Note: $N = 92$. Conventions are as in table 10.8, except EGGSFERT = eggs fertilized during the male's lifetime.

[a]Not significant.

Table 10.8 presents the correlation matrix for all females, and table 10.9 presents that for all males. In females, most measures were strongly intercorrelated ($p < .001$), but most correlations involving the proportion of young surviving to breeding age were not significant. In males, similar relationships were found. Only the proportion of young surviving to breeding age was not strongly correlated to other measures of fitness (table 10.9). Correlations within cohorts resembled those for all birds, save for females in the 1978 cohort. Since only one female in this cohort survived the crash, life span was uncorrelated with other characters, and reproductive success in the first year was virtually synonymous with lifetime success.

Many of these positive correlations follow inevitably from the life history of the song sparrow. In any iteroparous organism with small clutches or litters, reproductive success accumulates steadily with life span, unless total reproductive failure is frequent. In particular, both univariate and multivariate correlation methods (principal component analysis) show that the total number of eggs laid or fertilized, the total number of breeding attempts, the total number of offspring reared to independence, and the total number of offspring reared to sexual maturity are all closely correlated with life span. Although these correlations are high, they are by no means perfect. For instance, the number of offspring raised to independence explains only 59% of the variation in the number of young surviving to breeding age in females and 76% of the variation in males. Other significant positive correlations were much lower, and two were somewhat surprising. The number of independent young raised in the lifetime is correlated with clutch size. This is because average clutch size varied significantly from year to year and was greatest in 1976 and 1977, when individuals in the most successful cohorts raised most of their offspring. The number of offspring raised in the first breeding year was closely correlated with number raised in the lifetime, because over half of all birds (56%) survived for only one breeding season (table 10.2). It is noteworthy that the variable accounting for the largest fraction of total variation in lifetime reproductive success (YOUNGSURVIVAL, = PB in table 10.7) was the most poorly correlated with other measures of success. Annual variation in YOUNGSURVIVAL is closely correlated with population density (unpubl. results).

Is Lifetime Fitness Related to Morphology, Territory Size, and Male Mated Status?

If a trait is correlated with fitness and the trait is heritable, it will be selected for. Selection has previously been shown to operate on morphological variation (Boag and Grant 1981; Price and Grant 1984; Price et al. 1984) and territory size (Price 1984) in passerine birds, but no study to date has tested for selection using lifetime fitness measures.

I used as characters the six morphological measures described above, territory size, and hatch date. The morphological measurements were standarized for differences among sexes and years. Except for body weight, the morphological characters are known to be heritable in the Mandarte population (Smith and Zach 1979; Smith and Dhondt 1980; Schluter and Smith 1986a). To reduce the measurement error and the number of statistical tests employed, I reduced the number of morphological variables to two by extracting the first two principal components from the correlation matrix based on all individuals combined. The first principal component (PC1) is an index of general body size, and PC2 is an index of body shape (the size of the beak relative to the size of the body). Together, size and shape accounted for over 60% of the total variance in measurement (Schluter and Smith 1986b).

Territory size was standarized for differences among sexes, years, and age-classes (older birds tend to have bigger territories). Only males that held territories throughout life were included in the initial analyses. Males of different territorial and mated status are compared later. Hatch week was standardized only for differences among years.

I used the total number of young raised to independence during the lifetime as my measure of fitness and regressed this on the four traits (size, shape, territory size, hatching date) to test for selection (Lande and Arnold 1983; Arnold and Wade 1984a,b). The four traits were used untransformed to test for directional selection, and the squares of the traits were used to test for stabilizing selection. I also used unstandardized morphological measurements to ensure that standardization did not bias the outcome of the tests. None of the characters tested showed any evidence that either stabilizing or directional selection ocurred in males or females in association with variation in lifetime reproductive success. This conclusion was not altered when I used nonparametric correlation tests, and inspection of the data did not reveal any obvious nonlinear effects.

Since lifetime reproductive success was so poorly related to any attribute of the song sparrow that I had measured, I investigated whether lifetime reproductive success was itself a heritable trait. I calculated heritabilities by computing correlations among sibs (Falconer 1981), using the following standardized measures of lifetime success for each sex: number of eggs laid or fertilized, number of breeding attempts, number of independent young raised as a yearling, number of independent young raised in the lifetime, number of offspring reaching breeding age, and life span. Eleven of the twelve calculated heritabilities were very close to zero, and the twelfth was not statistically significant. None of the measures of lifetime reproductive success, therefore, was heritable, even after the major proximate influences on reproductive success had been removed.

One character that had a clear influence on lifetime reproductive success of males was mated status at the start of their first breeding year. Males with territories and mates produced most independent young in their lifetimes; males that were floaters at the end of April in their first breeding season produced by far the fewest young; and males with territories but no mates in their first spring produced intermediate numbers of young (fig. 10.3, top). Eight of the fourteen floaters disappeared after their first breeding season, but six later obtained territories and mates. Males that had mates in their first year had life spans similar to those of territorial males that did not (fig. 10.3, bottom) and were more successful simply because they bred in their first year. However, I could find no

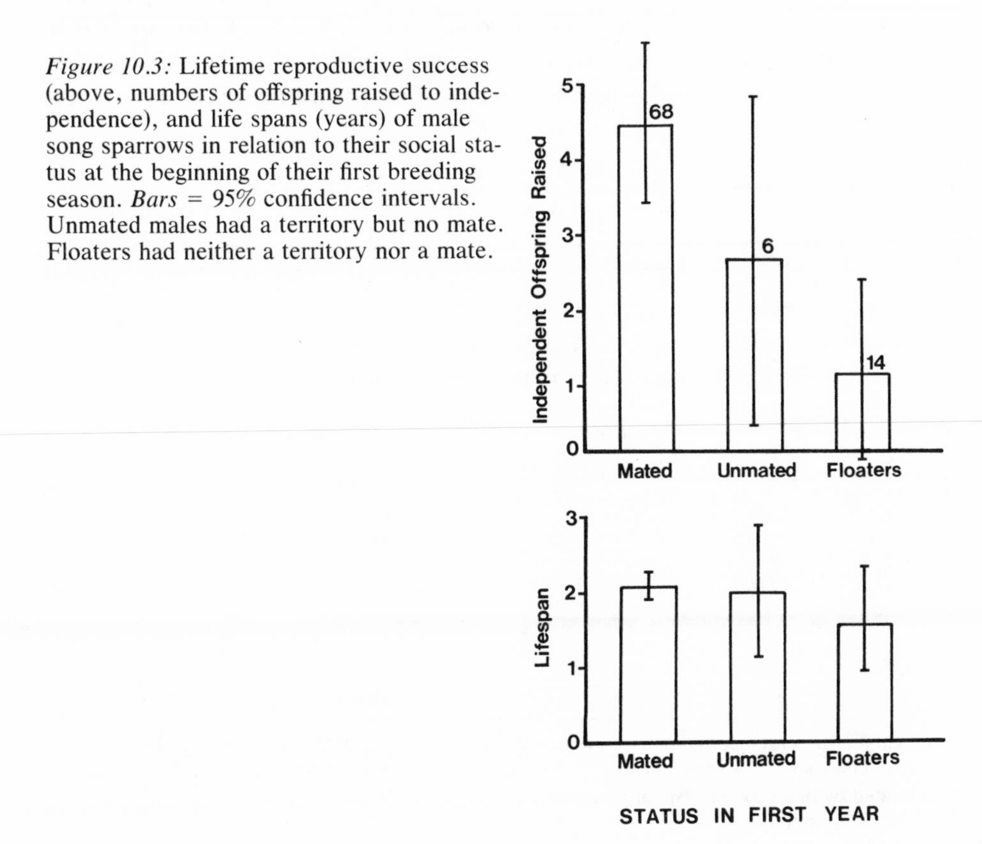

Figure 10.3: Lifetime reproductive success (above, numbers of offspring raised to independence), and life spans (years) of male song sparrows in relation to their social status at the beginning of their first breeding season. *Bars* = 95% confidence intervals. Unmated males had a territory but no mate. Floaters had neither a territory nor a mate.

morphological variable that was associated with mated status among males.

In summary, lifetime fitness in the song sparrow was unrelated to the morphology of individuals, to the sizes of territories they occupied, or to their hatch dates. Among males, however, lifetime fitness was strongly related to their mated status in their first breeding year, but mated status was unrelated to morphology. In addition, lifetime reproductive success was not a heritable trait within the population. These conclusions should be treated with some caution, since they are based on parametric analysis of data that were not always normally distributed. I believe, however, that these results are robust, for the reasons given above.

Effects of Age on Reproductive Success

I have shown above that life span has an important influence on variation in reproductive success. The main reason for this is that longer-lived birds accumulate more breeding attempts, but breeding success also can change with age (see review in Hannon and Smith 1984). Table 10.10 presents the data relating reproductive success to age. The pattern of variation is complex, with considerable changes in average success from year to year. In four of the eight sex/age-classes, there was significant variation in the numbers of independent young raised per year. In all four of these cases, reproductive success increased from age one to age two, although it declined in the third breeding year for the 1976 cohort. This year, 1979, was unfavorable for breeding for all age-classes.

Table 10.10 Variation in Annual Reproductive Success by Cohort and Age in Song Sparrows

		Mean Number of Independent Young Reared						
Cohort	*ANOVA*	*1976*	*1977*	*1978*	*1979*	*1981*	*1982*	*1983*
1975								
Females	n.s.	3.3	1.7	0	—	—	—	—
Males	n.s.	1.2	2.9	1.8	1.0	—	—	—
1976								
Females	***	—	3.3	4.3	1.5	—	—	—
Males	***	—	2.0	3.7	1.1	—	—	—
1977								
Females	n.s.	—	—	3.0	2.0	1.3	—	—
Males	*	—	—	1.8	2.1	1.8	5.0	4.0
1978								
Females	n.s.	—	—	—	0.8	3.0	—	—
Males	***	—	—	—	0.6	4.5	4.0	3.0

Note: Sample sizes are approximately the same as in table 10.2. No data were collected in 1980. ANOVA gives the probability that success varied between years within the cohort, as indicated by one-way analysis of variance.

*$p < .05$; ***$p < .001$.

Examination of differences among three cohorts within years is possible in 1978 and 1979 (table 10.10). In both years, two-year-old males were more successful than both yearlings and older birds ($p < .02$, ANOVA), and this was also true for females in 1979 ($p < .01$, ANOVA). Thus, reproductive success of song sparrows of both sexes tended to increase from the first to the second breeding year but to decline thereafter. Since few birds survived to breed for three years, this pattern of variation in reproductive success tends to add weight to the positive relationship between life span and reproductive success. A fuller account of the effects of age on reproductive performance in the song sparrow on Mandarte Island is given by Nol and Smith (1987).

10.4 Discussion

Violation of the Assumptions of the Study

Although I was able to study the reproduction of virtually the entire population, the results rest upon the validity of several key assumptions laid out earlier. The assumption I have least confidence in is that immmigration and emigration rates are low and in equilibrium. Mandarte Island contains uniformly favorable song sparrow habitat compared with the vegetation on surrounding islands. It is possible, therefore, that Mandarte generates successful dispersers that breed in less favorable habitats elsewhere. If this is true, our estimates of lifetime reproductive success are biased downward and refer only to a selected subset of individuals.

The second assumption that is known to be suspect is that territorial males fertilize all eggs laid within their territories. About 10% of eggs fail to hatch, and some of these are infertile. P. Arcese (pers. comm.) has seen occasional extrapair copulations on the island. Genetic studies, however (Smith and Zach 1979; Smith and Dhondt 1980), suggest that most territorial males do father the young that hatch within their territory.

Some birds were excluded from parts of our analysis because we did not trap and measure them or because we held them temporarily in captivity. I do not feel that this constitutes a serious source of bias, since so few individuals were involved. Finally, a few birds in the four cohorts studied lived during 1980, when I did not measure individual variation in reproductive success. This also is not a serious source of bias, since the breeding population in this year was so low (eleven pairs, from which eight birds had hatched in 1979 and fell outiside the cohorts considered here).

Patterns of Variation in Lifetime Reproductive Success

Three major patterns emerge from this study. First, the lifetime production of young by song sparrows depends strongly on the cohort in which they hatch. Eggs laid in 1975 survived poorly to breeding age, but the resulting adults were successful as breeders. Eggs laid in 1976 to 1978 produced proportionately more breeding recruits than in 1975, and birds in the 1976 and 1977 cohorts had high lifetime success. Birds in the 1978 cohort, however, reproduced and survived poorly as adults. The annual survival rate of adult song sparrows on Mandarte (table 10.2) usually averages above 50%, but it was very low (14%) in 1979–80. The population crash between 1979 and 1980 acted as a guillotine on the production of young by all cohorts represented in 1979. The cause of the crash is not known. In addition, annual reproductive success varied widely among years (table 10.10). The 1978 cohort was virtually eliminated by the population crash of 1979–80, and its only breeding season was the worst of the nine breeding seasons studied to date. It was thus very unsuccessful. The population expanded from thirty-four to sixty-six breeding pairs during 1975–79, and many of the individuals alive during the early stages of the population expansion had high lifetime success, while few birds that recruited to breed at peak numbers did so. This pattern has been repeated since 1981 (unpubl. results).

Lifetime reproductive success is likely to be time dependent in other organisms. Populations of many species fluctuate in abundance, in either an irregular or a cyclic pattern (Lack 1954, 1966; Keith 1963; Krebs 1978). Individuals that breed during population increases may have high lifetime success but may not contribute many of their genes to the next cycle (Chitty 1967). Conversely, animals that breed successfuly during population declines contribute greatly to subsequent generations, even though their average per capita reproductive success may be low. Individual lifetime success in populations exhibiting long-term stability in numbers and age structure is likely to vary less.

Second, variation in survival of offspring explains much of the variation in lifetime reproductive success among individuals. Part of this result stems from the differences among cohorts considered above. Some cohorts had good average survival and produced many surviving offspring, while others performed poorly on both counts. Variation in survival was most marked over the winter, after the young became independent. Little is known of the factors influencing survival after young birds leave the nest compared with what we know of the fates of eggs and nestlings (e.g., Petrinovich and Patterson 1983). In the song sparrow, two important determinants of overwinter survival are dominance status as an independent juvenile, which is in turn determined largely by hatch date (Arcese and Smith 1985), and population density. Juveniles survive less well at high population densities (unpubl. results).

Third, life span was an important influence on lifetime reproductive success. Since most territorial males and all surviving females breed each year, their success inevitably increases as they accumulate breeding attempts. Success rates per nesting attempt are high in this population, compared with those for other open-nesting passerines (Lack 1954, 1966; von Haartman 1957). Some males fail to breed in their first year because they cannnot obtain territories or mates, and these birds perform poorly compared with males that obtain a territory and a mate in their first year (fig. 10.3).

Life span also has an important influence on lifetime reproductive success in male damselflies (Fincke 1982), in female red deer (Clutton-Brock, Guinness, and Albon 1982), and in male fruitflies (Partridge and Farquhar 1983). This pattern is very likely to be a general one.

There were slight sex differences in the variability in reproductive success. Variability in the number of eggs fertilized among males was greater than variability in the number of eggs laid by females (table 10.3). This follows simply from two causes. Some males never reproduced because they failed to gain a territory or a mate, while a very few others obtained more than one mate in one year and reproduced more successfully (Smith, Yom-Tov, and Moses 1982). Females, on the other hand, varied slightly more than males in the production of independent young (table 10.3). This had no obvious explanation, but the effect was weak. Overall, variation in lifetime success among females and males was similar, as expected for a predominantly monogamous species.

Selection, Survival, and Reproductive Success

I found little evidence of selection on lifetime fitness, and I also found that lifetime reproductive success was not a heritable trait. Does this mean that variation in lifetime success is all due to chance or to purely phenotypic effects? We can get some insight into this by considering separate periods of the life cycle in turn.

Variation in lifetime reproductive success in the song sparrow was divided approximately equally among three periods: from fertilization to attaining independence from parental care, from independence to breeding age, and from breeding age to death (tables 10.4–10.7). In the first of these periods, it is difficult to identify the action of selection because survival of the individual depends both on the phenotype of the young and on the phenotypes of its parents. In the second period, survival over the first winter, consistent directional selection operated on females for tarsus length and beak length. D. Schluter and I (Schluter and Smith 1986b) have shown that females with shorter tarsi and longer beaks survived better each year from 1976 to 1979. Among males, directional selection was absent during this period, but there was weak stabilizing selection on body shape. In the third period, adult life, we found no selection on overwinter survival among adults from 1976 to 1979, but we observed strong stabilizing selection on adult body size in both sexes during the population crash in 1979–80. Finally, we found directional selection on annual reproductive success of females in 1979. In this year the traits favored (short beaks and long tarsi) were the very ones selected against among females in their first winter.

Thus, although selection was not seen to operate on lifetime reproductive success, it was frequently noted during shorter time periods. In the case of females, there was antagonistic selection for the same traits at different times within the life span (i.e., a trade-off between survival and reproduction). This is one reason I failed to detect selection over the lifetime. In addition, selection was observed in some years and not in others, and effects of selection over a lifetime could be readily masked by chance events in years when selection was absent.

The traits examined here in relation to lifetime reproductive success were available because they are relatively convenient to measure. The variation in survival, and in annual and lifetime reproduction, must also reflect variation in more complex morphological, behavioral, and physiological traits. On the other hand, chance events such as accident, injury, predation on nests, and bad weather must also be important causes of death. Hence it is fundamentally difficult to detect selection on lifetime success unless simple and heritable traits account for most of the variation in fitness. Not only is this not the case in the song sparrow, but is is unlikely to be true in general (Gustafsson 1986). Although selection in stable environments tends to remove genetic variation in fitness (Fisher 1958), such environments are a myth. Reversing and episodic selection may frequently occur within the reproductive life span of an organism, although this is speculative at our present state of knowledge. We should therefore expect much of the variation in lifetime success to be associated with purely phenotypic traits.

What Is a Good Measure of Reproductive Success?

The answer to this question clearly depends on the demographic characteristics of the organism in question. For a species like the song sparrow, which is relatively short-lived and produces several small clutches of eggs each year, the following attributes are desirable:

1. At least an entire season's record of performance, preferably the first year after hatching; this should ideally be supplemented by measurements in one or more subsequent years.
2. At least a crude knowledge of the population trajectory, and the average breeding success before, during, and after the years in which detailed measurements of success are made.

If lifetime measurements can be obtained, the present study suggests that measurements like lifetime egg production, which are relatively easy to obtain, are reasonable measures of fitness. I found that lifetime egg production was closely correlated with the production of independent young, a characteristic that is much harder to measure. Egg production, however, was not closely correlated with survival of offspring from independence to breeding age, which was the greatest single factor influencing lifetime reproductive success.

For longer-lived species, it is clearly very difficult to obtain direct estimates of lifetime reproductive success. However, fine examples are already available (Clutton-Brock, Guinness, and Albon 1982). For short-lived species like hole-nesting passerines that produce few but larger clutches, it will be easier to measure lifetime success, and single-season measures of success may be adequate indexes of the fitness of most individuals.

10.5 Summary

In this chapter I present information on the lifetime reproductive success of four cohorts of song sparrows that hatched on Mandarte Island, British Columbia, during 1975–78. I attempt to explain variation in lifetime success using information on the overall demography of the population, and on the morphology and social status of individuals.

Individuals raised up to sixteen independent young during their lifetimes. The major sources of variation in lifetime reproductive success were the population trajectory during the history of each cohort; the survival of offspring from independence to breeding age; and the life span of individuals. A severe population crash had a major influence on lifetime success of many individuals. Several different measures of lifetime reproductive success were found to covary closely. I tested for selection on lifetime reproductive success using six morphological characters and two others, territory size and hatch date, but I found no directional or stabilizing selection for any character. Selection did, however, occur on morphology during shorter time intervals within the study. The social status of males during their first spring had a marked effect on their average fitness. Males that obtained a territory and a mate were the most successful over their lifetimes, and males that obtained neither a territory or a mate at this time were the least successful. Comparisons of siblings failed to identify a heritable component to lifetime reproductive success even when the major proximate sources of variation in lifetime success were removed statistically.

I discuss some restrictions on my results that stem from three critical assumptions, and I offer suggestions on how best to measure lifetime reproductive success. The main conclusions from my study are that many influences on reproductive success are due to year and cohort effects and that the major influences on lifetime success are phenotypic in nature.

10.6 Acknowledgments

I thank Peter Arcese, Sarah Groves, Rich Moses, Juan Merkt, Juanita Russell, Mary Taitt, and many other people for help with fieldwork. Peter Arcese, Steve Arnold, Kim Cheng, Tim Clutton-Brock, Linda and Larry Fedigan, Gilles Gauthier, Mart Gross, Ian McLean, Judy Myers, Bill Neill, Dolph Schluter, Reto Zach, and two anonymous reviewers gave many helpful comments on earlier versions of the manuscript. Dolph Schluter also helped greatly with computing and statistical advice, and I am particularly grateful to him and to Peter Arcese for their help and stimulating criticism. David Brown wrote the program to analyze components of variance in reproductive success, and Steve Albon kindly ran the analyses for me. I thank the Natural Sciences and Engineering Research Council of Canada for financial support and the Tsawout and Tseycum Indian bands for allowing me to work on their island.

11 Lifetime Reproductive Success of House Martins

David M. Bryant

REPRODUCTIVE SUCCESS OF BIRDS is affected by many factors. Food availability is often crucial, accounting for between-season as well as within-season differences (Lack 1966; Newton 1980). This and other factors in the wider regime are often shared by populations, yet these populations embrace individuals that differ greatly in their breeding success. Among house martins, *Delichon urbica*, for example, some birds fledge nine young per year whereas others are completely unsuccessful (Bryant 1979). The cause of such differences can be obscure when, as in the house martin, members of a colony derive no clear benefit from the siting of their nests and use a common feeding ground. Seasonal success, however, is just one component of fitness (Vehrencamp and Bradbury 1984); a full understanding of the factors contributing to fitness necessarily implicates the survival chances of both parents and offspring as well. Measuring lifetime reproductive success by recording the number of young raised during the lives of marked individuals is one way of accounting for all the main components of success and allows identification of attributes, behaviors, and circumstances that contribute to fitness (Clutton-Brock, Guinness, and Albon 1982). I undertook this study of the house martin to find the causes of variation in reproductive success for the population as a whole as well as for its constituent individuals throughout their lives. In this account I describe the broader relationships between house martins and their food supply and present a first analysis of variation in lifetime reproductive success and its components.

The house martin (Hirundinidae) breeds in the Palaearctic and winters in the Afrotropical and Oriental regions. In Britain it shares the aerial-insectivore habit with three diurnal species—the swallow, *Hirundo rustica*, the sand martin, *Riparia riparia*, and the swift, *Apus apus*. House martins begin to arrive in breeding areas in April and start nesting soon after. They are typically colonial, but small groups of nests and isolated pairs are frequent. All broods are reared monogamously. Laying within colonies is not tightly synchronized; hence first clutches are spread over a

Figure 11.1: House martin feeding the brood in its nest box at the study colony.

month or more. One or two broods are raised each year, followed by departures from August to October. Molt is protracted and largely confined to African winter quarters. Accounts of the breeding biology of house martins in Britain and elsewhere can be found in Lind (1960), Bryant (1975), and Rheinwald (1979).

11.1 Study Area and Methods

Observations were made at a single colony in central Scotland. Artificial nests were provided in excess and could be removed to inspect contents and capture adults (fig. 11.1). The farmstead site was in mixed farmland with a river nearby. Foraging birds usually formed small flocks and stayed around the nesting site, but they concentrated by the river during the arrival period and in adverse weather. Further details of the colony and the biology of its martins are given by Bryant (1979), Bryant and Westerterp (1980), and Bryant and Turner (1982).

This analysis of reproductive success is based on house martins ringed as nestlings from 1971 onward or as yearlings from 1972 and includes all cohorts known to have died out by 1983. Reproductive success is judged by several criteria, but the most important are the number of young reared each year and, for estimating lifetime reproductive success, the total young raised during the lifetime of individuals. House martins are unusual in that fledglings return to the nest to roost after their first flights and so can be observed until full independence. Many recruits to the breeding colony had been reared elsewhere and so were unringed on first capture. It is likely that nearly all of these birds, mostly females, were in their first summer (Bryant 1979), and they were therefore assigned to this age-class.

Although house martins offer certain advantages for the study of life histories—for example, their colonial habits give easy access to significant numbers of breeding birds and their nests are secure from predators—they also have some drawbacks. Principal among these is the high mortality (82%–88%) and dispersal rates of birds in their first year (Rheinwald 1975; Rheinwald, Gutscher, and Hormeyer 1976). Because little is known about the fate of these birds, this analysis is necessarily incomplete and is restricted to individuals that reached their first breeding year.

Three morphological characters were taken from adults at each handling. All measurements used here are means, and to minimize measurement errors they are based on samples of three or more for each bird; for long-lived individuals the samples are much larger. In nearly all cases body mass was measured during incubation so that changes due to stage of breeding cycle were minimized. Methods of measurement and interrelations between these normally distributed characters are given by Bryant and Westerterp (1982). The sexes show no consistent differences in size (wing length, keel length), body mass (about 19 g), or plumage. The assumption is made that these characters are heritable (see Smith and Zach 1979; Price and Grant 1984).

11.2 Food Supply

The north temperate region is characterized by an obvious seasonality. This is particularly marked among insects, the food of house martins, with alate populations low in the cold winter months and yet abundant, or perhaps superabundant, for much of the summer (fig. 11.2). The arrival

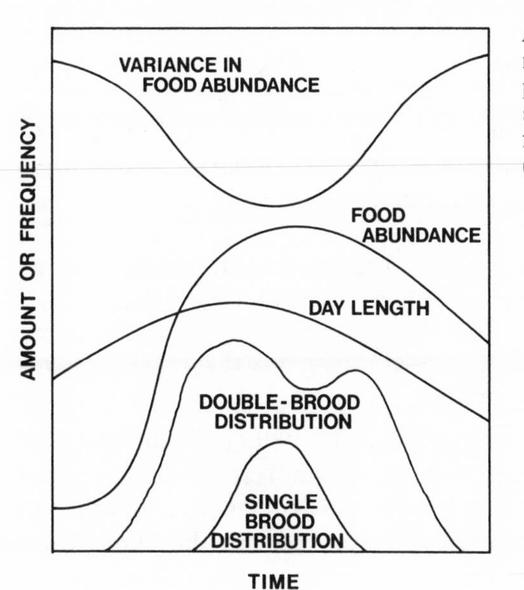

Figure 11.2: Breeding cycle of house martins in relation to abundance and predictability of the aerial-insect food supply. The time axis spans the interval from early spring (April) to late autumn (October).

and laying period of hirundines coincides with a rise in aerial insect numbers in spring (Bryant 1975). Nevertheless, food can be scarce during the laying period, generally linked to spells of cold, wet weather, though the chance of this happening becomes less as summer gets fully established (Turner 1982). Late June to early August provides stable conditions with only a small risk of food scarcity. Single-brooded pairs largely confine breeding activity to this most favorable "resource window." On the other hand, double-brooded pairs are likely to experience adverse conditions during the early stages of their first brood, the late stages of the second brood, or both. A consequence of the fluctuating food supply, both within and between seasons, is that much of the variation in seasonal reproductive success is linked to the timing and severity of poor weather (Bryant 1975, 1979).

11.3 Breeding Behavior

Arrival of birds is spread over a month or more, with early progress interrupted by locally bad weather or by unfavorable conditions along distant migratory routes. Males usually arrive before their mates, although there is extensive overlap for the colony as a whole. Pairing is discreet but accompanied by some song, mainly from males while in the nest. Between the first pairings and completion of laying, conflict between male colony members is obvious, usually involving intruders attempting to expel nest-box residents: accompanying tussles in and around nest boxes can be prolonged. It is reasonable to infer competition for nests or females, or both of these, from this behavior. Since males tend to arrive earlier than females (Bryant 1979), they are at first exposed to an adversely skewed sex ratio. During the days immediately before laying, mate guarding by males can be observed, and looser associations, even when at communal feeding sites away from the nest, are established from pairing onward. Courtship feeding does not occur at the study colony. Incubation and the brooding of hatchlings is done by both sexes, with the female taking a larger share, but feeding the brood is more equal. Second clutches are usually laid in the same nest, following quickly after fledging of the first brood. About 10% of pairs split after the first brood, leaving one individual paired with a previously unmated bird or with one that had rejected or lost a first partner, while the other stayed unpaired for the rest of the season. No examples of fidelity between years were recorded in this study, even though both partners lived over the "winter" in 17% of cases.

11.4 Age-Related Reproductive Success

Many reproductive parameters in house martins change with age. The main change was a significant increase in mean reproductive output between the first year and later years (table 11.1). A similar pattern is found when birds surviving two years are examined in the same way,

Table 11.1 Reproductive Success in Relation to Age for House Martins

Measure	Sex	Age in Years					
		1	*2*	*3*	*4+*	$\bar{x} \pm \sigma$	F^f
Number of reproductive attempts (y^{-1})[a]	♀	1.68 ± .47	1.74 ± .45	1.67 ± .49	2.00 ± 0	1.70 ± .46	n.s.
	♂	1.61 ± .49	1.77 ± .48	1.79 ± .41	1.77 ± .43	1.69 ± .48	*
Number of successful attempts (y^{-1})[b]	♀	1.44 ± .62	1.64 ± .53	1.33 ± .72	2.00 ± 0	1.49 ± .61	n.s.
	♂	1.37 ± .55	1.59 ± .66	1.58 ± .65	1.72 ± .46	1.49 ± .60	**
First clutch size	♀	3.79 ± .86	4.10 ± .76	4.13 ± 1.06	4.80 ± .45	3.92 ± .87	**
	♂	3.74 ± .87	4.11 ± .69	3.96 ± 1.08	4.50 ± .62	3.94 ± .87	**
Second clutch size	♀	3.03 ± .62	3.28 ± .53	3.30 ± .67	3.00 ± 1.00	3.11 ± .62	n.s.
	♂	3.05 ± .56	3.11 ± .63	3.16 ± .69	3.36 ± .63	3.12 ± .61	n.s.
Egg survival to fledging[c]	♀	76%	85%	74%	79%	78%	n.s.
	♂	80%	82%	72%	79%	79%	n.s.
Total eggs (y^{-1})[d]	♀	5.85 ± 1.96	6.48 ± 1.93	6.33 ± 2.09	7.80 ± .84	6.10 ± 1.97	*
	♂	5.56 ± 1.94	6.66 ± 1.82	6.46 ± 2.25	7.11 ± 1.64	6.10 ± 2.00	**
Total fledglings (y^{-1})	♀	4.54 ± 2.22	5.52 ± 1.82	5.00 ± 2.75	6.00 ± 1.41	4.85 ± 2.20	**
	♂	4.40 ± 2.05	5.61 ± 2.20	4.92 ± 2.36	5.56 ± 1.98	4.88 ± 2.17	**
Age of mate	♀	1.64 ± 1.00	2.20 ± 1.07	2.27 ± 1.63	4.40 ± 0.89	1.92 ± 1.14	**
	♂	1.22 ± .52	1.74 ± .74	1.86 ± .89	2.33 ± 1.65	1.56 ± .89	**
First egg date[e]	♀	40.58 ± 14.33	31.88 ± 13.78	33.8 ± 20.28	20.20 ± 4.66	37.33 ± 15.30	**
	♂	43.35 ± 15.01	31.35 ± 11.10	34.35 ± 20.96	26.44 ± 7.11	37.48 ± 15.75	**
Annual survival	♀	37%	26%	29%	20%	33%	n.s.
	♂	42%	56%	52%	37%	46%	n.s.
n	♀	117	42	15	5	179	
	♂	94	44	24	18	180	

[a]Number of clutches incubated, excluding clutches deserted before laying was completed.
[b]At least one nestling raised to independence.
[c]Number of independent young/number of incubated eggs.
[d]Includes only incubated clutches.
[e]Numbered so 1 May = 0
[f]Compares first-year birds with all others; one-way ANOVA.
$*p < .05$; $**p < .01$.

showing that such trends are apparent for individuals as well as for the population as a whole. These results suggest an age-related change in reproductive strategy. At the same time, the potential effects of differential mortality on population parameters cannot be ignored, especially since females with the highest annual success also experienced the greatest mortality (Bryant 1979). I therefore assume that both processes are involved, although their precise contributions remain to be evaluated. There were no significant trends in survival of eggs to independence, and yet in both males and females it peaked among two-year-olds (table 11.1).

Age-related trends in breeding can be misleading if individuals that fail to breed at all are omitted. Only single birds in their first year (one male, one female) were nonbreeding residents at the study colony for an entire season, and so "floaters" appear to be rare. Yet in most years one or two males would fail to pair over a period of weeks and then, I assume, would move on to seek and perhaps find a mate elsewhere. So the possibility remains that "floaters" are more frequent than could be confirmed for the study colony. If this is the case, mean success of males will be lower, and variance in success greater, than implied.

Survival rates differed between the sexes (table 11.1). Particularly marked was the low return rate for second-year and older females—birds that are themselves characterized by early laying dates and a high frequency of double brooding. In a previous analysis, it has been shown that females with such breeding patterns had a high mortality rate, probably because they depleted their reserves by persisting with brood feeding during adverse weather (Bryant 1979).

11.5 Seasonal Reproductive Success

The average breeding house martin lays (or fertilizes) 6.1 ± 2.0 eggs and rears 4.9 ± 2.2 young to independence per season. Success is higher in years with favorable feeding conditions, especially if they occur in May and June (Bryant 1975). In the most successful season an average of 6.8 young was raised per female (6.9 per male), whereas in the poorest year only 3.3 (3.1 per male) reached independence. The most successful individuals in any season tended to be older and to lay earlier (table 11.2). Among males, all body-size measures were positively correlated with two or more success parameters, implying pairing with potentially productive mates and subsequent cooperation in brooding rearing. Of these only body mass of males was correlated with seasonal production of young. Among females, keel length was correlated with a high frequency of second clutches and with annual egg production (Bryant and Westerterp 1982; table 11.2). This correlation with egg output improves if we exclude late-laying females (first egg after 22 June; $r = .34$, $p < .01$, $n = 64$) or first-summer birds ($r = .48$, $p < .01$, $n = 31$). The justification for this segregation is that some females may follow a single-brood strategy independent of current conditions (Bryant 1979) and therefore should not be compared directly with those that attempt two broods per year. No

Table 11.2 Pearson Product-Moment Correlations for Seasonal Reproductive Success and Other Success Parameters in Relation to Body Size for House Martins

Measure	Number of Reproductive Attempts (NRA)	Number of Successful Attempts (NSA)	First Clutch Size (CS1)	Egg Survival to Fledging (ESF)	Total Eggs (TOTE)	Total Fledglings (TOTF)	First Egg Date (FED)
Females (n = 179)							
NRA	—						
NSA	0.72***	—					
CS1	0.29***	0.32***	—				
ESF	0.08	0.60***	0.23***	—			
TOTE	0.61***	0.72***	0.69***	0.10**	—		
TOTF	0.61***	0.86***	0.54***	0.72***	0.77***	—	
FED	−0.50***	−0.53***	−0.55***	−0.18**	−0.64***	−0.39***	—
MASS[a]	0.06	0.01	0.08	−0.01	0.11	0.07	−0.08
WING	−0.02	0.02	0.04	−0.01	0.02	0.05	−0.01
KEEL	0.26**	0.09	0.06	−0.12	0.20*	0.03	−0.04
AGE	0.08	0.1	0.23***	0.04	0.18**	0.16*	−0.29***
Males (n = 181)							
NRA	—						
NSA	0.71***	—					
CS1	0.29***	0.38***	—				
ESF	−0.06	0.51***	0.27***	—			
TOTE	0.83***	0.74***	0.68***	0.11	—		
TOTF	0.55***	0.84***	0.58***	0.68***	0.75***	—	
FED	−0.53***	−0.15*	−0.55***	−0.23***	−0.67***	−0.58***	—
MASS[a]	0.06	0.07	0.18**	0.07	0.18**	0.18**	−0.15*
WING	0.04	−0.05	0.13*	−0.15*	0.10	−0.03	−0.10
KEEL	0.00	−0.00	0.27**	0.02	0.13	0.03	−0.20*
AGE	0.14*	0.19**	0.26***	−0.03	0.25***	0.18**	−0.35***

[a]None of the morphological correlations hold for first-year birds alone. MASS denotes body mass (g); WING, wing length (maximum chord, mm); KEEL, length of sternum (mm); AGE (years).

*$p < .05$; **$p < .01$; ***$p < .001$.

correlation was found between female body mass and number of young reared (table 11.2). The positive association identified in an earlier analysis among females two or more years old ($r = .57$, $p < .01$, $n = 25$; Bryant 1979) was not apparent in the present, larger sample ($r = .15$, $p < .10$, $n = 58$). For both sexes, therefore, large size was positively associated, albeit weakly and inconsistently among females, with one or more components of seasonal reproductive success.

11.6 Lifetime Reproductive Success

Many individuals die between fertilization and first breeding, most between independence and the following breeding season (Bryant 1975, 1979; Rheinwald, Gutscher, and Hormeyer 1976). Where and when this mortality strikes is unknown; a reasonable guess might identify as main factors postfledging deaths during spells of poor weather and losses during migration, especially for late departures and early arrivals. Irregular losses on wintering grounds also occur and probably fall most heavily on first-winter birds (Broekhuysen 1953; Skead and Skead 1970). Whether losses are otherwise distributed randomly, or nearly so (Ross and McLaren 1981), or nonrandomly (Klomp 1970; Perrins and Moss 1975) remains to be investigated.

Minimum and median lifetime reproductive success was the same in both sexes for all success parameters (table 11.3). The distribution of lifetime success differed, however (fig. 11.3), as did the maximum observed, which was higher in the sample of males. Standardized variance in lifetime reproductive success ($\sigma^2/\bar{x}^2$) was greater in males than in females (table 11.3), in contrast to seasonal success patterns, where variance was the same for both sexes (the normal pattern for a monogamous species).

Lifetime reproductive success may depend on chance factors. The cohort that hatched in 1972 and bred for the first time in 1973, for example, had the greatest lifetime success, averaging 9.7 young per female and 16.9

Table 11.3 Lifetime Reproductive Success of Male and Female House Martins That Reach Their First Breeding Year

	Females (n = 125)				Males (n = 103)			
Measure	*Minimum*	*Median*	*Maximum*	$\bar{x} \pm \sigma$	*Minimum*	*Median*	*Maximum*	$\bar{x} \pm \sigma$
Total eggs incubated (or fertilized)	0	7	38	8.5 ± 5.4	0	7	51	10.6 ± 0.9
Total young reared to independence	0	6	28	6.8 ± 4.6	0	6	42	8.5 ± 7.4
Total clutches started	0	2	9	2.4 ± 1.4	0	2	12	2.9 ± 2.2
Life span (years)	1	1	5	1.5 ± 0.8	1	1	6	1.8 ± 1.2
Variance in egg success $\sigma^2/\bar{x}^2$				0.40				0.68
Variance in young success $\sigma^2/\bar{x}^2$				0.46				0.76

Table 11.4 Pearson Product-Moment Correlations for Lifetime Reproductive Success and Related Parameters in the House Martin

	Females (n = 116)[a]			Males (n = 99)[a]		
Measure	*Total Young Reared*	*Total Eggs Incubated*	*Life Span (years)*	*Total Young Reared*	*Total Eggs Incubated*	*Life Span (years)*
Total young reared	—	0.94***	0.81***	—	0.97***	0.91***
Total eggs incubated	0.94***	—	0.86***	0.97***	—	0.94***
Life span (years)	0.81***	0.86***	—	0.91***	0.94***	—
Body mass (g)	0.06	0.11	0.09	0.26*	0.27**	0.24*
Keel (mm)	0.13	0.23*	0.17	0.25*	0.28*	0.29*
Wing (mm)	0.03	0.02	0.00	0.06	0.12	0.11
Body size (1)	0.19	0.31**	0.32**	0.34**	0.38**	0.38**
Body size (2)	0.19	0.30**	0.31**	0.33**	0.37**	0.37**

[a]For keel, body size (1) and (2), n = 59 females, n = 54 males.
[b]Body size (1) is body mass times keel; Bodysize (2) is standardized mass plus standardized keel.
*p < .05; **p < .01; ***p < .001.

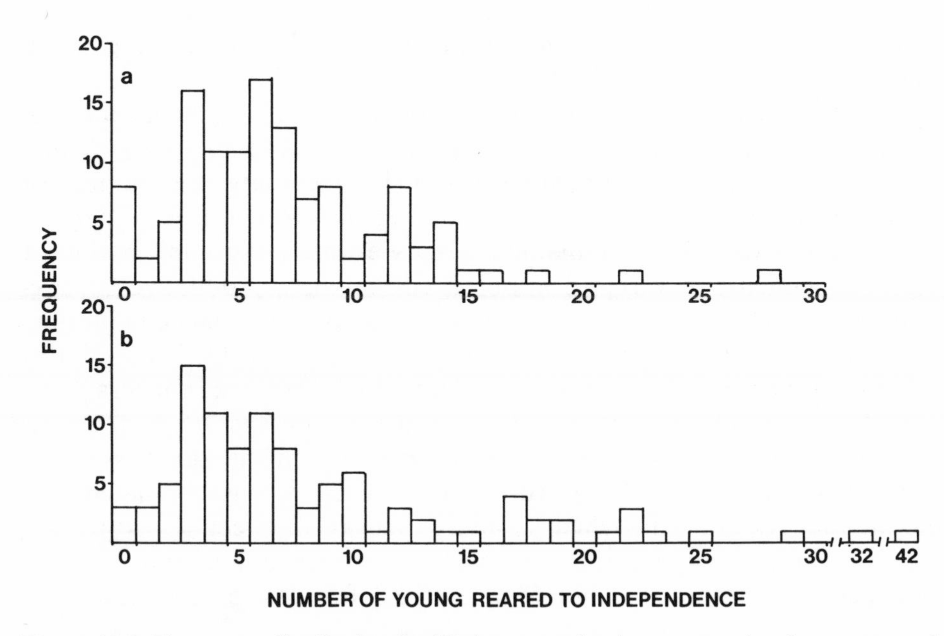

Figure 11.3: Frequency distribution for lifetime reproductive success (total young reared to independence) in house martins: (*a*) females; (*b*) males.

young per male. The poorest success was for the 1977 cohort, again occurring in the same year for both sexes, with 3.9 young per female and 4.4 per male. For the study period as a whole, significant differences in cohort success were found for males (one-way ANOVA, $p < .05$) but not for females. It is likely that some of these differences were associated with annual fluctuations in food supply; for example, the poor year of 1977 was notable for a persistent scarcity of aerial insects (Bryant and Westerterp 1980). At the same time, body mass and body size showed significant differences between cohorts (one-way ANOVA, $p < .01$ for mass and keel), and this is likely to account for some further intercohort variation in reproductive success.

Correlations between lifetime success parameters and three morphological characters are given in table 11.4. The only significant correlation among females was for keel length (and two related size measures) with lifetime egg production. In males body mass and keel length were both correlated with output of eggs and young and with life span (table 11.4). Least-squares regressions for lifetime success in males (LFS, total young reared) are LFS = $-26.65 + 1.82$ mass (g) ($n = 99$, $p < .01$) and LFS = $34.11 + 2.30$ keel (mm) ($n = 54$, $p < .05$). Repeating analyses of this type (Lande and Arnold 1983) for body mass and keel within each cohort revealed only a few significant correlations, though in some years samples were unsuitably small. These mainly conformed to the positive associations identified in the pooled samples. In the 1975 cohort, however, body mass was negatively correlated with lifetime success in females (eggs: $r = -.44$, $n = 18$, $p < .05$; young: $r = -.56$, $n = 18$, $p < .01$). In some years, therefore, directional selection may act counter to the generally prevailing trend.

Lifetime reproductive success, measured as the total young surviving to independence, may be considered as the product of three components: life span (i.e., number of breeding years) (L), fecundity, the number of eggs laid per year (or incubated) (F), and mean survival rate of eggs to independence (S). Brown's method lets us estimate the contribution of these components to variance in observed lifetime reproductive success (this volume, chap. 27). For both sexes, almost all the variance in lifetime success was due to breeders (98% females, 99% males) (tables 11.5–11.10), because proven nonbreeding was rare (see above). Among breeders, over half of the overall variance was associated with life span (51% females, 54% males). The residual variance for females was mainly due to S (22%) and F (17%) (tables 11.5–11.7). About 10% of variance was associated with the interactive term LF, which is interpreted as an effect of age-related increases in fecundity (tables 11.1 and 11.2). Variance in lifetime success of males was greater than for females (23.3 females, 55.2 males). The

Table 11.5 Mean and Variance of the Components of Lifetime Reproductive Success in Male House Martins

Component	Original		Standardized
	Mean	*Variance*	Variance
L	1.765	1.330	0.427
F	5.820	2.839	0.084
S	0.791	0.056	0.089
LF	10.800	76.405	0.724
FS	4.607	3.407	0.161
LS	1.412	1.009	0.518
LFS	8.668	54.965	0.833

Note: N = 102 breeding males.
L = life span; F = fecundity (number of eggs fertilized); S = survival of offspring to independence.

Table 11.6 Percentage Contribution of the Components of Lifetime Reproductive Success to Variation in LRS in Male House Martins

Component	L	F	S
L	51.29		
F	25.63	10.07	
S	0.18	−1.46	10.70
LFS	3.59		

N = 103; 1 nonbreeder
Proportion of breeders, 0.99
Overall variance (OV), 55.16
Percentage of OV due to nonbreeders, 1.32
Percentage of OV due to breeders, 98.68

Note: $V(LFS)/(LFS)^2$.
L = life span; F = fecundity (number of eggs fertilized); S = survival of offspring to independence.

interactive term LF was ranked second after life span and accounted for 25% of overall variance (tables 11.8–11.10). The importance of the LF term among males is taken to reflect a tendency for older males to pair with the most fecund females (tables 11.1 and 11.2). The possibility that promiscuous matings might distort the apparent success of males and its variance cannot be discounted. The occurrence of mate guarding, how-

Table 11.7 Percentage Contribution of the Components of Lifetime Reproductive Success to Total Variation in LRS in Male House Martins, Including Nonbreeders

Component	L	F	S
L	50.80		
F	25.29	9.93	
S	0.18	−1.44	10.55
LFS	3.54		

Note: V(LFS)/(LFS)^2.
L = life span; F = fecundity (number of eggs fertilized); S = survival of offspring to independence.

Table 11.8 Mean and Variance of the Components of Lifetime Reproductive Success in Female House Martins

	Original		Standardized
Component	*Mean*	*Variance*	Variance
L	1.452	0.575	0.273
F	5.991	3.178	0.089
S	0.774	0.068	0.113
LF	8.794	31.234	0.413
FS	4.723	4.146	0.193
LS	1.143	0.486	0.384
LFS	7.017	23.063	0.509

Note: N = 124 breeding females.
L = life span; F = fecundity (number of eggs laid); S = survival of offspring to independence.

Table 11.9 Percentage Contribution of the Components of Lifetime Reproductive Success to Variation in LRS in Female House Martins

Component	L	F	S
L	53.63		
F	10.14	17.41	
S	−0.24	−1.72	22.19
LFS	−1.41		

N = 125; 1 nonbreeder
Proportion of breeders, 0.99
Overall variance (OV), 23.27
Percentage of OV due to nonbreeders, 1.69
Percentage of OV due to breeders, 98.31

Note: V(LFS)/(LFS)^2.
L = life span; F = fecundity (number of eggs laid); S = survival of offspring to independence.

Table 11.10 Percentage Contribution of the Components of Lifetime Reproductive Success to Total Variation in LRS in Female House Martins, Including Nonbreeders

Component	L	F	S
L	53.20		
F	9.97	17.11	
S	−0.23	−1.69	21.82
LFS	−1.39		

Note: V(LFS)/(LFS)^2.
L = life span; F = fecundity (number of eggs laid); S = survival of offspring to independence.

ever, meant that opportunities for paired or unpaired males to gain access to fertile females were very few (mating occurs almost exclusively in the nest). I therefore suspect that the incidence of extrapair matings was low (see Fitch and Shugart 1984).

11.7 Mate Choice

Seasonal, and ultimately lifetime, reproductive success in those monogamous species that share nesting duties depends partly on mate quality. Although mates are retained for a season or less, the choice of partner will affect the success of the offspring phenotype and the nature of the genotype. It is therefore to be expected that both sexes will attempt to pair with a partner likely to maximize fitness. Although correlative evidence is unsuitable for demonstrating active mate choice (many birds must take partners by default, for example), it can show up associations that are likely to have involved choices by some individuals. Older females (which, being the earliest arrivals and generally the most successful breeders, would be expected to exercise choice) tended to pair with older males ($r_s = .39$, $p < .001$, $n = 166$, two-tailed test). Older males, as well as pairing with older females, tended to pair with early layers ($r_s = .55$, $p < .001$, $n = 162$, two-tailed test). This may come about because old females select early-established males (those males already present on their arrival). Large males also paired with females laying the largest first clutches (body mass with first clutch size: $r = .22$, $p < .01$, $n = 163$; keel: $r = .27$, $p < .01$, $n = 82$). They did not, however, show a tendency to pair with large females, even when the effect of age in either sex was partialed out.

11.8 Discussion

Seasonal Reproductive Success

The opportunity for reproduction available to migrants during the north temperate summer has both predictable and unpredictable elements. There is a predictable wealth of food suitable for house martins, many times the levels at low latitudes (Hails 1982). Prolonged food shortages in the warmest months are invariably rare or absent (Bryant 1975; Bryant and Westerterp 1980). Unpredictability emerges at the start and end of the breeding season. Double-brooded pairs run the risk that adverse conditions will disrupt breeding or affect migratory survival: only single-brooded pairs can confine activities to the stable midsummer spell (fig. 11.2). It has been proposed elsewhere that the trade-off between the obvious fitness benefits of double brooding and its attendant mortality costs (linked mainly to weather adversity) has led to a continuum of reproductive strategies among females (Bryant 1979). At one extreme some first-year females, often those small in size, arrive late and rear a single brood at the most favorable time, and at the other extreme older females arrive early and attempt two broods under more hazardous conditions. Both extreme strategies were found to have broadly similar fitness benefits (Bryant 1979). An apparent anomaly, that late layers had small clutches when conditions were at their best, was resolved by showing that this pattern could be considered adaptive for the typical individuals involved once changes in day length, foraging costs, and body size were considered together (Bryant and Westerterp 1983a,b). A central finding was that daily energy costs were generally higher in smaller birds of both sexes, not only on a unit mass basis, as might be expected, but also in total. Whether the same pattern holds in the different feeding and activity environment of winter quarters is not known, but I consider it unlikely. In contrast to females, males bore no mortality cost for an early start or for attempting two broods, and so all males would be expected to arrive early and compete for the potentially most successful females (Bryant 1979). An exception may be that small size makes success in intrasexual selection so improbable that even a slight mortality risk (too small to identify using present techniques) makes early arrival unprofitable for small yearling males.

Lifetime Reproductive Success of Females

Analysis of lifetime reproductive success in house martins has pointed to the potential significance of chance factors. Since many individuals breed for only one or two seasons, an unsuccessful year can have a dominant influence on lifetime success. In such an unpredictable environment, where both nesting success and yearling survival are likely to fluctuate, it may be advantageous for females to tend toward repeat breeding (Murphy 1968; Goodman 1984). The 32% of first-year females that opt for a single brood can perhaps be seen as one result of the irregular benefits that follow from this strategy. The relationship between morphology and the lifetime success of females is less clear: consistent directional selection on the three characters measured can probably be dismissed because of their lack of correlation with lifetime output of young. Presumably the benefits of large egg output (tables 11.2 and 11.4) were offset by the risks associated with rearing the resulting offspring to independence (Bryant 1979). Indeed, data for lifetime production of young by females in relation to body mass and keel were more easily compatible with stabilizing selective forces, although the fit of their quadratic regressions was not significant (Lande and Arnold 1983). It is likely that a better understanding of the functional significance of morphological characters, especially in the way they contribute to foraging success, energy storage, and survival (Bryant and Westerterp 1982), would allow size-related differences in the fitness of females to be investigated with more success.

Lifetime Reproductive Success of Males

Lifetime reproductive success was more variable in males than in females (ratio, males:females, 1.65) but was, as expected, lower than in examples from polygamous (Wade and Arnold 1980) and strongly territorial (Price 1984) birds. It can be inferred that intrasexual selection is only moderate, a view easily reconciled with the absence of plumage or size dimorphism between the sexes. This helps little, however, in distinguishing whether the apparent benefit of large body size for lifetime reproductive success in males was a consequence of intrasexual or intersexual selection, or if both were involved. Large males will presumably be dominant over small males in contests for nests and for the most productive females. Yet one group of females expected to be in a position to exercise choice—namely, the more productive older females—showed no significant tendency to pair with the larger males once the effect of male age was partialed out. Instead, older individuals of both sexes came together. A simple proximate mechanism by which this could occur is if female choice was involved and early arrivals (invariably older males) were favored above large males. If it is experienced birds that tend to survive the hardsips of an early migratory arrival, this would give females the benefit of a male with a proven survival history. At the same time there is enough "noise" in the timing of arrival and matings for unmated males to compete intrasexually for incoming females. Since most of the later arrivals are birds in their first year, any age-related criterion would then be irrelevant in female choice. Under these conditions large males will be at an advantage, though not sufficiently so to oust sometimes smaller males from previously established pairs.

That larger males enjoy a higher lifetime reproductive success implies directional selection for increasing body size (assuming the pattern found at the study colony pertains elsewhere). Yet there was no evidence for a continuing trend toward large size in the eleven years of study. It is possible that size variations were largely phenotypic in origin, perhaps reflecting feeding conditions while in the nest or in the weeks after fledging. Another prospect is that body-size variation is mainly of genetic origin (van Noordwijk, van Balen, and Scharloo 1980; Garnett 1981; Boag 1983). If this is so, either selection pressures on body size in males must be variable, or they must act at an early stage in the life cycle (i.e., between fertilization and first breeding), or both. The advantages of large size in males may be frequency-dependent, for example, giving the greatest benefit when large males are scarce. Another possibility is that the optimum size for overwinter survival is smaller than that favored for males

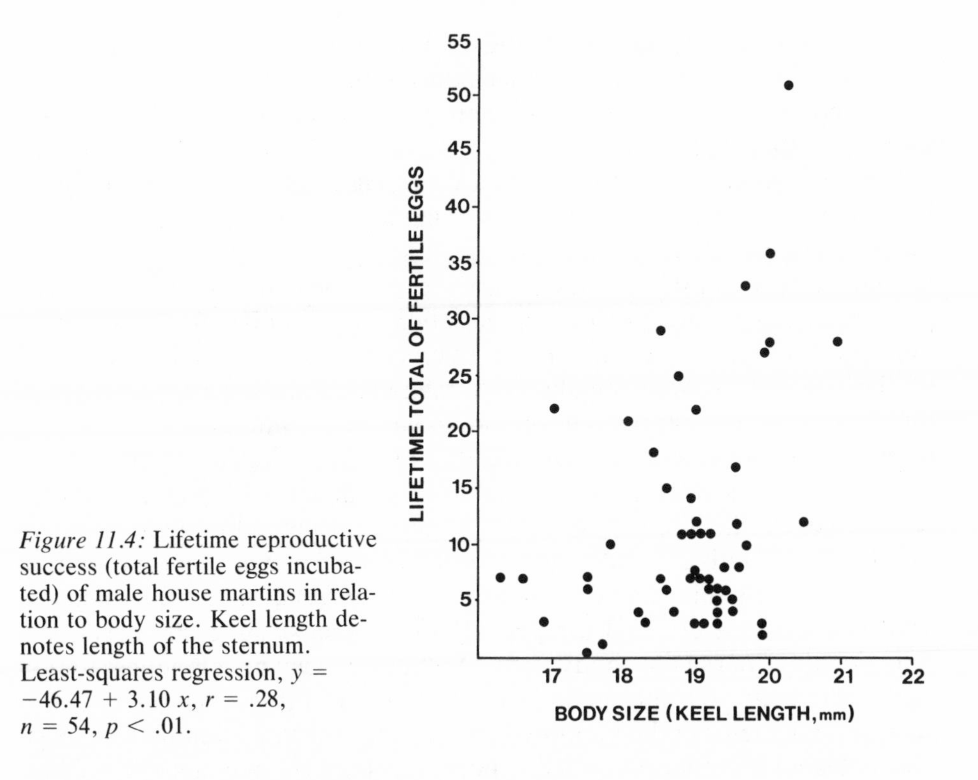

Figure 11.4: Lifetime reproductive success (total fertile eggs incubated) of male house martins in relation to body size. Keel length denotes length of the sternum. Least-squares regression, $y = -46.47 + 3.10\,x$, $r = .28$, $n = 54$, $p < .01$.

under sexual selection at the time of mating or under natural selection at any time during migration or in the temperate summer (see Price and Grant 1984). If selection against large size in "winter" is irregular, then over periods that lack such "winter" adversity, the breeding season benefits of large size for lifetime success can become apparent (fig. 11.4). Certainly the cluster of large but rather unsuccessful males (fig. 11.3) hints at least of a fluctuating benefit for large size. It may be useful to view large size as a survival handicap, the disadvantage of which lessens as birds gain experience with age. Hence large yearling males may find it difficult to complete the round trip to Africa, while the same journey carries less risk for survivors in later years. If this is the case, then females that accept large males as mates could employ the facultative handicap of large size as one mate choice criterion because it correlates with genetic quality, or more specifically with a propensity for survival (Maynard Smith 1976; Zahavi 1975). Potential benefits of this nature conferred on offspring would complement the more immediate advantages for females of gaining a mate with a high foraging efficiency (Bryant and Westerterp 1982, 1983a,b), potentially able to bring more food to the brood.

11.9 Summary

Lifetime reproductive success is reported for 125 female and 103 male house martins, studied over a period of eleven years. Their aerial-insect food supply is unpredictable, and this can affect seasonal reproductive success. Because few birds live more than one or two years, food scarcity may have a marked effect on the lifetime success of individuals that happen to experience food shortages during breeding.

Among birds that survived their first winter, the median lifetime number of offspring raised was six for both females and males. The most successful female and male reared twenty-eight and forty-two young to independence, respectively. In birds living for two years or more, success increased most sharply between the first and second years. Lifetime reproductive success was more variable in males than in females. The proximate source of most of this variability was similar for both sexes: life-span differences accounted for over half of the overall variance in lifetime success. Among females the remaining variance was mainly due to differences in fecundity and offspring survival, whereas in males an interactive effect of life span and number of eggs fertilized was important.

Morphological correlates of lifetime production of young included body mass and keel length in males: large males tended to gain access to the most productive females and to survive longer. Only certain combinations of characters showed similar positive correlations among females. That larger males enjoyed a higher lifetime success suggests directional selection for increasing body size. Yet there was no evidence for an increase in body size during the study. Counterselection probably operates against large males in the nonbreeding season, especially among yearlings. One possibility is that potentially successful females may accept pairings with large males both because greater size implies high genetic quality and because large males offer immediate benefits in terms of higher provisioning rates to the brood.

11.10 Acknowledgments

I am grateful to T. Clutton-Brock, P. Tatner, and anonymous referees for their comments on the manuscript. Very many people have given help during the study—colleagues, students, and landowners—and to all of these people I am much indebted.

12 Factors Influencing Reproductive Success in the Pied Flycatcher

Paul H. Harvey, Martyn J. Stenning, and Bruce Campbell

THE PIED FLYCATCHER (*Ficedula hypoleuca*) has the virtue of being the subject of more long-term population studies than any other bird species. Populations have been monitored for many years in most parts of the species' range, including the United Kingdom, Scandinavia, the USSR, and Germany. One reason for this popularity is the species' preference for nesting in boxes rather than in naturally occurring holes. Artificially provided nest boxes can be examined regularly to provide data on the breeding performance of the occupants. In this chapter we concentrate on one particularly intensive and long-term study that was carried out in the Forest of Dean, Gloucestershire, in the United Kingdom, but we supplement the results with those from other studies. We present two separate analyses. The first considers the factors influencing between-year differences in reproductive success and gives some insights into why some years are better than others for pied flycatchers. The second analysis examines the effects of breeding experience, age, and mate change on reproductive success.

12.1 The Pied Flycatcher

The species is a migratory passerine that breeds in northern Europe and in parts of Spain and North Africa (Heinzel, Fitter, and Parslow 1979). Adults weigh about 12 g and are usually sexually dichromatic, the females being brown-and-white while the adult males are black-and-white.

As their name implies, pied flycatchers feed on insects, which they often catch in flight. A dependence on insect larvae to feed their young probably influences their choice of breeding habitat. In the United Kingdom they are usually found in mature oak (*Quercus* spp.) woodland, but throughout their range they nest in a variety of mature and successional forests. They will even breed in nest boxes placed in coniferous forests, which support markedly less diverse communities of insect larvae (Lundberg et al. 1981).

Figure 12.1: Male pied flycatcher, *Ficedula hypoleuca*, from Morris 1891.

Unlike other passerines that nest in similar habitats, the pied flycatcher does not normally defend a feeding territory during the breeding season. The only defended area is that immediately surrounding the nest hole. Males display nesting cavities to prospective mates but after nest building and copulation, and usually when the female is starting to lay eggs (Haartman 1956), the male moves away in an attempt to attract a second mate at another nest hole. This pattern of behavior has been termed "polyterritorial polygyny" by Haartman (1949) and has now been observed in at least two other bird species (Lawn 1982; Temrin, Mallner, and Winden 1984). In the Forest of Dean study, only 12% of females that shared a mate were likely to be each other's nearest nesting neighbors, and in some cases they were separated by six or seven nest boxes occupied by other females (Harvey et al. 1984). In one Swedish study the distances between territories of the same male were generally similar to those in the Forest of Dean, though one individual had two territories that were separated by 3.5 km (Silverin 1980). As a result, the second female to pair with a male is probably unaware that her partner is mating bigamously (Alatalo et al. 1981). Even if a male is successful in "duping" a second female, he normally returns to the primary female and helps to raise her young. The secondary female feeds her young with little or no help from the male and fledges fewer offspring than do concurrently laying monogamous and primary females (Alatalo, Lundberg, and Stahlbrandt 1982; Stenning 1984).

The pied flycatcher not only is polygynous, it is also polyandrous in the sense that females copulate with other males when the primary male is away from his territory (Björklund and Westman 1983). On the basis of a heritability study, one population in Sweden was estimated to have about

24% of nestlings sired by a male other than the one attending the nest (Alatalo, Gustafsson, and Lundberg 1984).

12.2 The Forest of Dean Study

A History of the Forest of Dean study is given by Campbell (1968). Nest boxes were first placed in the Nagshead enclosure in the Forest of Dean (51°40' N, 2°40' W) in 1942, but they were monitored only casually until 1947. The data used here were collected by Campbell and his assistants between 1948 and 1964. During that period, about 250 nest boxes were available each year in a 24 ha area, many more than the number of birds of all species that occupied them. The number of breeding female pied flycatchers ranged between about fifty and one hundred.

12.3 Between-Year Variation in Reproductive Success

Results

Some years are better than others for pied flycatchers. In table 12.1 we summarize population characteristics for the seventeen years considered. Broods were divided into two sets: those started within fourteen days of the first egg laid in any brood that year ("early broods"), and those started after that date ("late broods"). Pied flycatchers fledge only one brood a year in the study area, so late broods were primarily of two types: either repeat clutches after failed first broods or delayed first-breeding attempts. Given the synchrony of breeding in each year (Stenning 1984), the vast majority of late broods were probably repeat clutches, though it should always be borne in mind that secondary broods of polgynously mated males will represent a higher proportion among late broods than early ones.

In addition to nestling data for early and late broods, table 12.1 also contains additional information: (1) the proportion of adult and fledged birds recaptured in the study area in later years; (2) estimates of the density of the winter moth (*Operophtera brumata*) in Wytham Wood (90 km from the study area), extracted from Varley, Gradwell, and Hassell (1973), because the larvae of this species form a major component of the adults' and nestlings' diets (Betts 1954); and (3) average estimates of oak-tree growth increment in the study area for each year as a further measure of the density of insect food. We used this last measure because Varley and Gradwell (1962) had demonstrated that this measure correlated negatively with defoliating caterpillar density in Wytham Wood.

Some variables could not be measured for the earlier and later years. For example, some males were not caught in the early years, and juvenile survival could not be assessed for the last two years because some returning birds may not breed until they are two years old (see below). Nevertheless, several patterns emerge from the available data.

Early-laying years are associated with larger clutches ($r = -.68$, $n =$

Table 12.1 Data on the Forest of Dean Pied Flycatcher Population and Associated Environmental Variables over the Seventeen Years of the Study

	1948	1949	1950	1951	1952	1953	1954	1955	1956	1957	1958	1959	1960	1961	1962	1963	1964
Number of first/early clutches[a]	57	66	82	87	97	83	70	59	51	46	66	64	56	60	56	55	53
Average date first egg laid in first clutches	37.6	35.2	39.9	48.4	43.2	44.4	44.8	51.2	47.0	40.7	40.2	43.1	41.2	37.7	41.5	49.4	43.2
Average clutch size in first clutches	7.37	7.46	7.13	6.89	6.90	6.66	7.14	6.38	6.90	7.20	7.11	6.73	6.82	6.58	7.00	6.51	6.91
Maximum number of hatched young averaged over first clutches	6.68	5.23	6.13	6.15	5.61	5.31	6.50	5.20	6.14	5.52	6.21	5.19	5.36	4.57	5.43	5.62	6.03
Average number of young fledged from first clutches	6.60	5.06	5.81	5.39	5.02	5.16	6.49	5.08	5.76	5.20	6.00	5.02	5.00	4.00	4.96	5.51	5.70
Number of failed first clutches	0	14	8	7	4	3	5	5	4	6	4	6	15	10	11	6	4
Average date first egg was laid in failed first clutches	—	37.4	41.1	47.1	44.3	42.7	44.8	51.2	46.3	43.0	38.3	42.3	42.2	36.1	41.4	50.3	45.5
Number of late/replacement clutches [a]	8	27	21	13	4	6	7	10	9	10	8	11	7	10	12	7	9
Average number of young fledged from late clutches	2.25	1.96	0.62	0.54	2.25	1.83	3.71	4.90	4.44	5.10	4.88	3.91	2.86	2.20	4.08	3.14	1.56
Percentage of breeding females recaptured in a later year	—	—	—	—	35.7	31.7	19.2	29.2	23.0	30.9	50.8	27.9	42.3	45.6	43.1	25.5	—
Percentage of breeding males recaptured in a later year	—	—	—	—	34.1	29.9	23.3	23.0	33.3	46.0	42.1	33.9	44.0	48.5	36.8	21.8	—
Percentage of fledged females found in study area in a later year	—	—	—	—	8.5	4.1	2.5	4.6	3.0	6.9	5.5	2.2	6.7	3.8	6.7	—	—
Percentage of fledged males found in study area in a later year	—	—	—	—	5.3	5.5	3.3	1.7	6.6	6.9	6.0	1.1	2.7	6.9	5.5	—	—
Density of winter moths in Wytham Wood[b]	—	—	2.05	2.07	1.74	1.26	1.20	1.89	1.98	2.44	2.28	1.76	1.33	0.88	1.13	1.61	2.12
Relative growth increment of mature oak trees in study area[c]	4.37	4.23	4.57	5.85	7.46	6.32	5.23	4.41	3.56	4.97	4.80	4.70	6.53	7.69	7.71	7.65	—

[a]See text.
[b]Relative measure from Varley, Gradwell, and Hassell 1973.
[c]From Stenning 1984.

17, $p < .01$), and with higher subsequent overwinter survival by adults of both sexes (males $r = -.87$, $n = 12$, $p < .001$; females $r = -.69$, $n = 12$, $p < .02$), and fledged juveniles ($r = -.48$, $n = 11$, n.s.). The only one of these relationships that is not formally significant is that between laying date and survival by juveniles, but this is not surprising since small numbers are involved each year (fewer than 10% of breeding birds were born in the area).

Although clutch size is well correlated with between-year differences in laying date, the maximum number of hatched young recorded in nests and the number fledging from early broods do not seem to be correlated with laying date ($r = .08$ and $r = .07$ respectively, $n = 17$). This indicates that the proportion of eggs failing to produce recorded hatchlings is higher in years when laying is earlier ($r = .45$, $n = 17$, $p < .10$). The average number of birds fledging from early broods is negatively correlated with oak-tree growth increment ($r = -.52$, $n = 16$, $p < .05$), which is itself highly negatively correlated with the estimates of winter moth density in Wytham Wood $r = -.64$, $n = 15$, $p < .01$).

The analysis so far has considered only early clutches. Far fewer young fledge from late than early clutches, an average of 2.95 compared with 5.40. However, there is considerable variation among years in the number of young fledging from late clutches (see table 12.1). The number of young fledging from early clutches is negatively correlated with the number fledging from late clutches ($r = -.57$, $n = 17$, $p < .02$).

Years when the proportion of eggs from early clutches that give rise to fledglings is high are those when adult survival to breed again is low ($r = -.56$, $n = 12$, $p < .10$ for females; $r = -.70$, $n = 12$, $p < .02$ for males).

Interpretation

Perrins (1970) has argued that the females of many bird species start laying as soon as food supplies are plentiful enough for egg production. The pied flycatcher seems to fit this pattern. Slagsvold (1975) has shown that pied flycatchers in the Forest of Dean tend to nest earlier in years with early springs, measured as the date of half-fall of the winter moth (i.e., the date by which half the winter moths under oaks in Wytham Wood pupated; Lack 1966). The advantages to pied flycatchers of early nesting years are that adults (and possibly juveniles) have a higher chance of surviving the subsequent winter, probably because they have more time to build up food reserves before migration. Furthermore, pairs that manage to raise an early first brood fledge more young, possibly because increased competition for limited food supplies later in the season (at which time fledglings from early broods are competing for food) limits the amount of food parents can bring to their offspring. This interpretation is supported by the finding that years in which there is high mortality among early broods are those in which late broods do relatively well. However, we might also expect this negative relationship between number fledging from early versus late clutches if food abundance peaked just once each year, leading to either relatively successful early broods or relatively successful late broods in any particular year.

The number of young fledged in early broods depends on available food supplies: the density of winter moth and of defoliating caterpillars in general (estimated through oak-tree increments) is well correlated with the number of young fledging.

Adult overwinter survival is high in years when few young have been fledged. Two possible explanations are density-dependent overwinter mortality and the costs of breeding (parents who have expended resources on rearing a brood are less likely to survive as a consequence).

One pattern that remains unexplained is the association of large clutch sizes with years in which breeding is early, since the number of young fledged from early broods does not seem to be related to laying date across years. There is some evidence that early breeding years are ones with *proportionately* higher nestling mortality.

12.4 Within-Season Variation in Reproductive Success

Some pied flycatchers have higher reproductive success than others within the same season. In this section we examine correlates of variation in individual reproductive success. In many bird species we know that older females, and sometimes the males also, have higher reproductive success than younger ones (e.g., Coulson 1966; Potts 1966; Mills 1973; Fisher 1969; Thomas 1980; Kluyver 1951; Perrins and Moss 1974; Bryant 1979; Rowley 1981; Harvey et al. 1979). In almost all cases it is not known whether age, previous breeding experience per se, or previous breeding experience with the same mate contributes to increased reproductive success when the effects of the other factors are held constant. A previous analysis of the Forest of Dean pied flycatcher data demonstrated that older birds are indeed more successful breeders (Harvey, Greenwood, and Campbell 1984). Although it could be shown that previous breeding experience with the same mate in successive years was not the cause of this pattern (which was detected irrespective of the fate of the previous mate), it was not determined whether age or previous breeding experience per se or both were involved. Below we distinguish between correlates of seasonal reproductive success associated with age and those associated with previous breeding experience. Parts of the analysis are presented in more detail elsewhere (Harvey, Stenning, and Campbell 1985).

Data

We have attempted to produce an unbiased sample of individual lifetime reproductive success descriptions drawn from a series of age cohorts. Between 1948 and 1958 inclusive, 6,001 eggs were laid. Of the 5,233 young that hatched, only 86 males and 100 females were subsequently found nesting in the study area. However, before 1952 many of the breeding males were not caught, so males hatched between 1948 and 1950 were not necessarily recorded during their first breeding season(s). If we included males that were ringed as nestlings during those years, then the sample would indicate that many males started breeding after they actually did. Accordingly, we have restricted our basic data set to the 68 males that were ringed as nestlings between 1951 and 1958 and were subsequently found breeding in the wood. We have examined the reproductive success of those 168 birds (100 females and 68 males). Data from birds born after 1958 were not used because the sample would have been biased against long-lived individuals (the maximum recorded life span was six years).

We have assumed that none of the birds in the sample bred outside the study area. This seems reasonable with the Forest of Dean study, because dispersal distances between successive breeding sites are very short in relation to the size of the study area (Harvey et al. 1984). In addition, there were twenty four opportunities for us to detect a gap in the apparent breeding records of individual birds used in the sample, but no cases were actually recorded, thus indicating that birds rarely left the wood to breed and then returned.

However, the assumption that we have the complete lifetime breeding records of all the birds in our sample is unrealistic. Some 9% of the birds in the sample apparently started breeding as late as their third, fourth, and fifth years. Although this is a small proportion of the birds, it is likely that some of them bred in the wood after previously breeding elsewhere. If indeed birds do delay breeding for several seasons, then a further possible source of bias is introduced: some individuals may survive several years without breeding and then die as unrecorded adults.

A final bias in the data results from about 10% of the males breeding polygynously each year (Stenning 1984). Occasionally the mates of secondary females were not caught, and in fact no males in the sample used were recorded as having more than one mate in any single year. This bias (which leads to estimating an artificially reduced variance in male reproductive success) is small because secondary females or those that attempted to raise their young unaided rarely fledged more than one or two young, compared with five or six for monogamous females.

With the provisos mentioned, we have data on lifetime reproductive success for the sample of 168 birds described above. However, we can also include data on lifetime reproductive success from some other birds—those born between 1959 and 1962 that bred in the wood for one or more seasons and then failed to breed subsequently. We can reasonably assume that those birds died. But this additional sample is biased toward individuals with short life spans because other birds, which were still breeding in 1964 when records ceased, could not be included in the sample since their reproductive life was not necessarily complete. When our data set is extended to include this sample of birds that is biased in favor of short-lived individuals, we shall refer to the use of the "extended data set" instead of the "basic data set."

Results

Over 90% of birds in the basic data set started breeding in the wood in either their first or their second year of life, and there were no records of birds' missing a year's breeding before dying. Our statistical analyses of age at breeding, life span, and breeding success are presented elsewhere (Harvey, Stenning, and Campbell 1985) and are summarized below.

We can find no significant sex differences in age at first breeding, in the average number or variance of breeding attempts, or in life spans. Once a bird starts breeding it has about a 30% chance of surviving to breed between each successive breeding season up to its fifth year of life; the data beyond that are two few for meaningful comparison. However, females that first breed in their second year have a lower chance of surviving to breed as three-year-olds (10%) than those that breed in both their first and their second years (46%).

Seasonal breeding success may be defined as the number of young produced that themselves survive to make a breeding attempt. We have data on three measures that contribute to seasonal breeding success: date of laying the first egg, clutch size, and number of young fledged. There are very few data on the number of fledglings surviving to breed, because fewer than 10% of fledged birds subsequently attempt to breed in the study area (see table 12.1).

Seasonal breeding success may improve with age and previous breeding experience. We have examined how previous breeding success is related to the two variables combined and to each when the effect of the other is held constant. Do birds of the same age but with different breeding histories have different patterns of reproduction? And do birds of different ages but with the same breeding histories have different patterns? Finally, we have looked for correlations between seasonal reproductive investment and the chances of surviving the subsequent winter: Do the birds that fledge most young in a breeding season have higher overwinter mortality?

The comparisons entailed making thirty-six statistical tests, eleven of which were significant at the 10% level, seven at the 5% level, and three at the 1% level. Sample sizes were small, and so in the descriptions that follow we list all the comparisons that provided significance at the 5% level or below, and we also indicate the directions of the other nonsignificant comparisons.

Does Reproductive Success Increase with Age and Breeding Experience?

We have sufficient data to compare results between breeding seasons for males and for females that breed as both two- and three-year-olds. Two-year-old males breeding for their second time fledge significantly more young than they did as one-year-olds, and the mates of three-year-old males produce significantly larger clutches than did the mates of the same males the previous year. Twelve comparisons were made for this analysis, and the older, more experienced birds had earlier laying dates in three of four comparisons, had larger clutch sizes in four of four comparisons, and fledged more young in three of four comparisons.

Does Reproductive Success Increase with Age When Breeding Experience Is Held Constant?

We have compared reproductive success among females breeding for the first and only time in their lives as one-, two-, and three-year-olds, and also between males breeding only as one-year-olds and those breeding only as two-year-olds. Older females have significantly earlier laying dates than younger females, and the mates of the older males have significantly earlier laying dates than those of the younger males. This latter result is not due to older males pairing with older females. The analysis involved making nine comparisons, and the older birds had earlier laying dates in three of three comparisons, had larger clutches in one of three comparisons, and fledged more young in three of three comparisons.

Does Reproductive Success Increase with Breeding Experience When Age Is Held Constant?

We have compared breeding success of both males and females breeding as two-year-olds that have not had breeding experience with those that have had breeding experience. There were also sufficient data to compare the reproductive success of breeding three-year-old females that had bred before with those that had not. Three-year-old females that had bred before produced significantly larger clutches than those that had not, and two-year-old males that had bred before mated with females that produced significantly larger clutches than the mates of two-year-old males with no previous breeding experience. The analysis involved making nine comparisons, and the birds with previous breeding attempts had earlier laying dates in three of three comparisons, had larger clutches in one of three comparisons, and fledged more young in two of three comparisons.

Is Breeding Success Associated with Subsequent Overwinter Survival?

We have compared the reproductive success for one-year-old breeding females, one-year-old breeding males, and two-year-old breeding males that survived to breed again with the reproductive success of those that did not. The mates of one-year-old males that survive to breed as two-year-olds start laying significantly earlier than the mates of one-year-old males that do not survive to breed again. We made nine comparisons, and the birds that survived to breed again had earlier laying dates in two of three comparisons, had smaller clutch sizes in three of three comparisons, and fledged fewer young in two of three comparisons.

Interpretation

Differences in laying date, clutch size, and fledging success among individual pied flycatchers within seasons are correlated with both age and previous breeding experience. However, following single birds through successive breeding seasons does not allow us to distinguish between the effects of age and those of previous breeding experience. Instead, birds of different ages that had not bred before have been used to show that breeding success tends to increase with age when previous breeding experience is held constant. This and the patterns described below do not apparently result because older birds tend to pair up with each other. We could detect no association between the ages of mates: older males are as likely to mate with young females as they are to mate with old females (Harvey et al. 1984).

Comparisons were also made between birds of the same age, some of which had bred before and some of which had not. Birds with breeding experience tended to have better breeding success. This was not a consequence of their having previous breeding experience with the same mate: the pattern is found irrespective of whether the experienced birds had the same mate or a new one (Harvey, Greenwood, and Campbell 1984). Similarly, mate change does not influence the date of egg laying in kittiwakes (*Rissa tridactyla*) if the ages of the two mates are taken into account (Thomas 1980).

Nevertheless, the pattern in the pied flycatcher data could still result from breeding experience per se. Birds that have bred before have had practice. Or it could be that birds that have bred are better in the sense of being of higher quality or fitter (that is *why* they were able to breed on a previous occasion). There are two pieces of evidence to support the latter interpretation of the pied flycatcher data. First, females that had bred before as one-year-olds and as two-year-olds were more likely to survive to breed again than those that bred only as two-year-olds. Second, males whose mates laid early in one season were more likely to survive to breed again than those whose mates laid late. Unfortunately, we do not have data on weights of individual birds.

A disturbing finding to emerge from this study is that in our basic data set about 40% of birds of both sexes did not attempt to breed as one-year-olds. If we assume that these birds did not breed outside the wood, this figure is likely to be a considerable underestimate, because birds that failed to breed during their first year and died subsequently without breeding were unrecorded. There were numerous empty nest boxes available each year, and once they have mated, pied flycatchers defend only the area immediately surrounding their nesting site (Haartman 1951, 1956). Neither nest-site limitation nor territoriality was apparently preventing these first-year birds from breeding.

Any unrecorded cases of polygyny would be likely to strengthen the patterns revealed by our analyses. Older males tend to have earlier laying dates, thus freeing them to attract secondary females early in the season when such females are available (Alatalo et al. 1981). Therefore our assessment of increased breeding success with age of males is likely to be an underestimate.

12.5 Discussion

We have presented two complementary analyses of the Forest of Dean pied flycatcher data. The first asks why some breeding seasons are better than others for reproductive success and concludes that differences relate mainly to food availability mediated by both climate and intraspecific competition. The second analysis examines the importance of age and breeding experience; failure to distinguish these effects has plagued the interpretation of several other studies (e.g., Davis 1976). In the case of the pied flycatcher, the answer seems to be that both older birds and those with previous breeding experience are more successful.

Furthermore, some birds seem to be better than others. For example, females that have bred in both their first and their second years have a higher chance of surviving to breed as three-year-olds than females of the same age that failed to breed as yearlings but did breed in their second year.

The reasons *why* some birds are more reproductively successful than others are not revealed by this study, and additional information would be necessary before some hypotheses could be tested. For example, in other population studies of birds, reproductive success has been shown to be correlated with body weight (e.g., Bryant 1979), and feeding efficiency can increase with age (e.g., Orians 1969a; Dunn 1972). We do not have data on feeding efficiency or body weight.

The positive association among females between number of breeding attempts and probability of surviving the subsequent winter highlights a further problem with the interpretation of observational studies. As with Smith's (1981b) study of song sparrows (*Melospiza melodia*), which also showed a positive association of reproductive performance and survival, the finding does not imply an absence of costs in terms of overwinter survival associated with reproduction. As we have already suggested, it may well be that female pied flycatchers (or song sparrows) with a good breeding record are fitter birds. Indeed, there probably are costs in terms of reduced overwinter survival that are associated with increased reproductive investment by male pied flycatchers (Stenning 1984), but it took manipulative field experiments to reveal them (Askenmo 1979). Sometimes reproductive costs can be inferred from straightforward observational studies (e.g., Bryant 1979), but manipulative field experiments are more likely to provide conclusive evidence (e.g., Nur 1984a,b).

12.6 Summary

We consider factors that influence reproductive success in pied flycatchers nesting in the Forest of Dean, United Kingdom. Our analyses are based on a long-term study of known individuals, and we use data from seventeen consecutive breeding seasons.

Early springs are associated with early laying dates and subsequent high probability of overwinter survival. Years in which most young fledge are those in which defoliating insect larvae are most abundant. Late broods are less successful in years when early broods are more successful.

Older birds and those with more breeding experience are more reproductively successful in any particular breeding season. An age cohort analysis demonstrates that age and previous breeding experience are independent correlates of breeding success. There are indications that "better" birds start breeding when they are younger and that this may account for an association between reproductive success and previous breeding experience.

13 Age and Reproduction in the Sparrowhawk

I. Newton

THIS CHAPTER CONCERNS REPRODUCTION and survival in the European sparrowhawk, *Accipiter nisus* (fig. 13.1). Data are given on variation in breeding success with age and on the number of offspring raised in a lifetime by different individuals. The sparrowhawk is a small woodland raptor which feeds on other birds. It breeds throughout the Palearctic region, in boreal, temperate, and Mediterranean zones. It is usually monogamous for each breeding season, and pairs breed solitarily, each well separated from the next. Most surviving

Figure 13.1: Female sparrowhawk at nest. Photo by R. J. C. Blewitt.

individuals remain faithful to their territories and mates in successive years, but some change their territories and mates from one year to the next. A pair raises no more than one brood each year, containing up to six young. As in other birds of prey, the female is larger than the male, but the sparrowhawk is extreme in this respect, with females (280 g) nearly twice the weight of males (150 g).

The data in this chapter were collected during 1971–84 in two areas of southern Scotland, where sparrowhawks were resident year-round. Each year I attempted to find all nests, record breeding performance, ring the young, and trap for individual identification as many breeders as possible. Females were easier to catch than males, so I gained more information for females, in some cases their complete life histories. Untrapped individuals of both sexes could often be classed as "yearling" or "adult," according to whether the dorsal plumage was brown or blue gray. For further details on study areas, methods, and results, consult the papers referred to later.

13.1 Life Span and Survival

Compared with some other birds of prey, sparrowhawks are short-lived. On average, females survive longer than males. From about 1,200 ringed birds recovered in Britain during 1908–82, the oldest male died at 7.6 years and the oldest female at 10.8. Both were from my study populations.

For these populations, the annual survival of each sex was calculated from birds ringed as nestlings (at eighteen to twenty-one days old) and then reported dead by members of the public. For males, survival through the first and each subsequent year was estimated at 31% and 67%, respectively, and for females at 49% and 71% (these differences were not amenable to statistical testing; Newton, Marquiss, and Rothery 1983). No evidence was found in either sex for appreciable variation in annual survival between the second and eighth years of life. Beyond this age, so few individuals were left alive that mortality must presumably have increased, or birds older than 10.8 years would have been found.

These estimates are based on all sectors of the population, breeders and nonbreeders. Separate estimates of 66% and 64% were obtained by two independent methods for breeding adult females caught alive each year in the nesting population (Newton, Marquiss, and Rothery 1983). They referred to females that survived to breed and, after first breeding, remained in the area (which was usual). They were so close to the estimates above for adult females based on general ring recoveries that there is little reason to suspect the latter. However, the larger sample of national ring recoveries (from the whole country) revealed no difference in first-year survival between the sexes, so the study area sample may have been unusual or biased in this respect.

On the basis of size, well-grown nestlings could be sexed without error. At fledging the sex ratio was effectively equal, as 651 broods included a total of 1,102 (51%) males and 1,061 (49%) females (Newton and

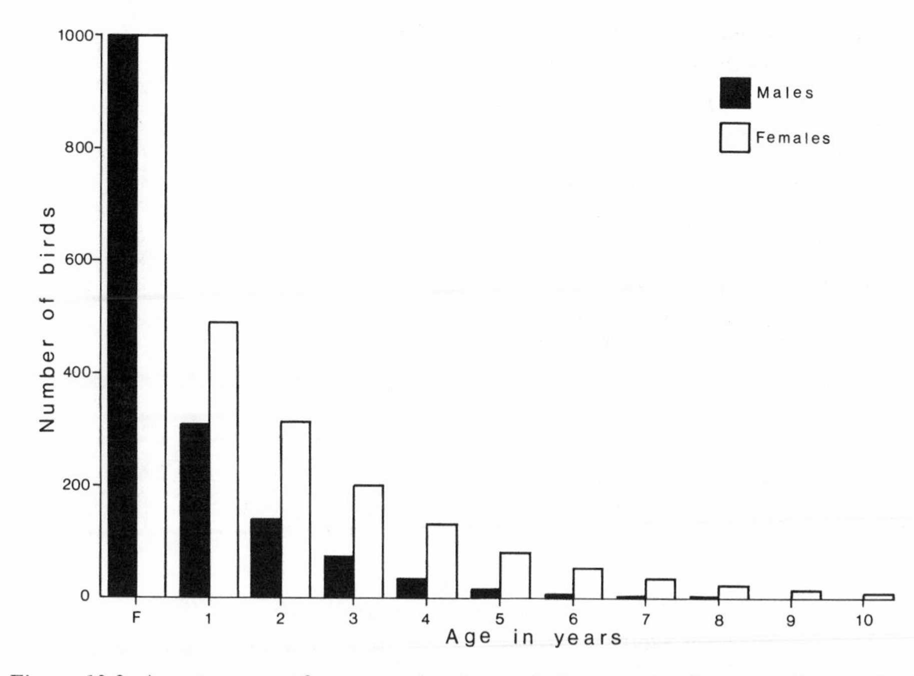

Figure 13.2: Age structure of a sparrowhawk population, starting from one thousand of each sex, based on mortality estimates obtained from ringing in southern Scotland. On the figures given, the number of females per male at different ages was as follows: year 1, 1.58; year 2, 2.20; year 3, 3.06; year 4, 4.25; year 5, 5.91; year 6, 8.35; year 7, 11.62; year 8, 15.43; year 9, 22.26; year 10, 30.34. In the population as a whole, taking account of the frequency of different age-classes, the overall sex ratio was 1 male to 2.36 females.

Marquiss 1979). The ratio was equal among young from broods of all sizes, from one to six. With the subsequent greater mortality of males, the sex ratio in the adult population became tipped in favor of females. According to the survival estimates above, at the start of each breeding season the total females in the population would have exceeded the total males by about 2:1 (fig. 13.2).

Despite surplus females, monogamy seemed usual. However, seven nests (0.6% of the total) found in 1971–84 contained up to ten eggs of two distinct types, the product of two females. On six occasions, both females were identified and found to be unrelated. None of these "double nests" produced young, for in each some eggs were pushed out and the rest deserted. I was unable to find how many males were involved, but because only one nest was used in each case, I suspected bigamy. Some males may have attended females on different nests, but I had no evidence of this. Fifteen radio-tagged males attended no more than one nest.

13.2 Overall Breeding Success

Of sparrowhawk pairs established on nesting territories each spring, some built no nest and did not breed that year. Others built a nest but did not produce eggs, while yet others produced eggs, some of which gave rise to young. Only those pairs that reached at least the nest-building stage could

Table 13.1 Summary of Overall Breeding Performance of Sparrowhawks, Southern Scotland, 1971–84

A. Overall Success in Terms of Individual Nesting Attempts	
Number of nests found	1,389
Number (%) in which eggs were laid	1,176 (85)[a]
Number (%) in which young were hatched	843 (72)
Number (%) in which young were reared	783 (93)
Mean clutch size (± s.d.)	4.58 ± 1.08
Mean brood size (± s.d.) at hatching	3.85 ± 1.24
Mean brood size (± s.d.) at fledging	3.43 ± 1.26
Mean number of young raised per clutch laid	2.3
Mean number of young raised per nest built	1.9
B. Overall Success in Terms of Total Eggs Laid by Population	
Total number of eggs laid	5,386
Number (%) of eggs that hatched	3,246 (60)
Number (%) of chicks that fledged	2,686 (83)

[a]Percentage of the previous figure.

Table 13.2 Causes of Failure in Breeding of Sparrowhawks, Southern Scotland, 1971–84

Cause	Number	Percentage of All Nests	Percentage of All Failures
Failure to lay eggs	213	15	35
Egg stage			
Total	333	24	55
Desertion	122	9	20
Predation	8	—[a]	1
Egg disappearance	15	1	2
Nest collapse	3	—	—
Two females at nest	7	—	1
Egg addling	13	1	2
Egg breakage	72	5	12
Addling and breakage	2	—	—
Female shot	23	2	4
Human egg robbing	5	—	1
Tree felling	10	1	2
Unknown	53	4	9
Nestling stage			
Total	60	4	10
Predation	21	2	3
Starvation	27	2	4
Nest collapse	1	—	—
Female shot	5	—	1
Tree felling	1	—	—
Unknown	15	1	2

[a]Less than 1%.

be counted accurately. The performance of 1,389 nests found during 1971–84 is summarized in tables 13.1 and 13.2; overall, the mean production was 1.9 young per nest. Knowing that 28% of fledged young survived long enough to breed (see later), we can calculate that 4% of eggs gave rise to potential breeders.

13.3 Age of First Breeding

Counting egg laying as a breeding attempt, seventy three females were first found breeding in their first year of life, seventy nine in their second year, fifty five in their third year, and eighteen in their fourth (Newton 1979). All older individuals that were trapped at the nest had bred on at least one previous occasion. Assuming that the mortality estimates given above could be applied to these individuals, the number of prebreeders present at different ages can be calculated (table 13.3). On this basis, 18% of all first-year females in the population bred, 49% of all second-year females, 83% of all third-year females, and practically all fourth-year females. This indirect approach was the only way to calculate such figures.

From the ratio of different female age-groups in the whole population, about 45% of all females were "prebreeders." The total number of "nonbreeders" would have been even greater, because some individuals that had bred occasionally had a "nonbreeding" year. Overall it seemed likely that less than half the total females in the population bred in any one year. For males, I had insufficient data to estimate reliably the number of either prebreeders or total nonbreeders, but from the ratio of first-year to second-year males in the nesting population each year (on average 1:1.7), it was clear that not all first-year males bred. Since the total population at the start of breeding contained far fewer males than females and polygyny was rare, a greater proportion of the male than the female population would have bred in any one year.

Table 13.3 Proportion of Breeders among Females of Different Age-Groups

Year of First Breeding	Fledglings Produced (% of Total)	Numbers Still Alive (% Breeding) in the Following Years after Birth						
		1	*2*		*3*		*4*	
1	149.0 (18.3)	*73 B* (18.3)	46.7 B	(49.2)	29.9 B	(82.8)	19.1 B	(100)
2	251.9 (30.9)	123.4 PB	*79 B*		50.6 B		32.4 B	
3	274.0 (33.6)	134.3 PB	85.9 PB		*55 B*		35.2 B	
4	140.1 (17.2)	68.7 PB	43.9 PB		28.1 PB		*18 B*	

Note: The numbers of females first found breeding in their first, second, third, or fourth year was known (italic in the table), and all other figures were calculated, assuming a 49% survival in the first year and a 64% survival in each succeeding year. These estimates were used as being most appropriate to the study populations.

B = breeding; PB = prebreeding.

Performance in Relation to Age of First Breeding

This section compares the performance of females that started breeding in their first, second, or third year. Too few started at four to warrant their inclusion.

I noted no significant differences in mean production at the first attempt between females that started in their first, second, or third year (table 13.4, panel A). Evidently the greater experience of older birds did not enable them to raise appreciably more young at their first attempt than younger birds. The mean lifetime production also showed no significant variation with age of first breeding, providing the bird lived to its first, second, or third year. The disadvantage of postponing first breeding until beyond the first year came from mortality, an aspect not accounted for in table 13.4, panel A.

If the analysis was restricted to females which lived three or more years, those that started in their first or second year produced significantly more young in their lifetimes than did those that started in their third year (table 13.4, panel B). But from the third year on, I found no significant differences in mean production related to age of first breeding. Thus these data also indicated an advantage in starting to breed early in life, and no penalty of poorer production in later life, compared with late starters.

In conclusion, the advantage of breeding at an early age derived mainly from the risk of mortality, which meant that birds failing to breed in any one year had only a 64% chance of surviving long enough to try the next year, or a 41% chance of surviving to try two years later. There was no great penalty in breeding at an early age, which resulted in lower lifetime production.

One weakness in the analysis presented was that production estimates could be compared only between different classes of birds that first bred at one, two, and three years of age. It could not indicate how those same individuals that first bred at two or three years would have

Table 13.4 Productivity of Females in Relation to Age of First Breeding

Year of First Breeding	Number of Individuals	Mean (± s.e.) Number of Young Raised[a]	Mean (± s.e.) Number of Young Raised in Lifetime
A. All Females			
1	49	2.10 ± 0.25	4.10 ± 0.57
2	49	2.59 ± 0.23	5.40 ± 0.71
3	38	2.21 ± 0.31	5.00 ± 0.56
B. Females That Survived Three or More Years			
1	13	5.31 ± 1.27	8.00 ± 1.58[b]
2	25	5.52 ± 0.97	8.40 ± 1.05[c]
3	38	5.00 ± 0.61	5.00 ± 0.61[d]

[a]At first attempt for all females; from third year on for those surviving three or more years.
[b,c,d]Significant differences between b and d, $t_{49} = 2.17$, $p < .05$; between c and d, $t_{61} = 2.99$, $p < .01$.

performed had they themselves bred earlier. The best age for one bird may not have been the best for another, and if birds started to breed at the age that was best for them, no proper assessment of the advantage of early breeding was possible.

13.4 Breeding in Relation to Age

This section considers breeding in relation to age for those individuals that produced eggs (updating Newton, Marquiss, and Moss 1981). Enough females were identified so that the performance of different age-groups could be examined separately, corrected for variation in performance associated with year and habitat. On the whole, clutches started earliest in the season were the most successful, so a ''good'' laying date was an early one.

Comparing different female age groups, laying dates and clutch sizes improved during the first few years of life and then deteriorated again (fig. 13.3). The average age of peak performance was calculated by quadratic regression at 3.9 years for laying date (birds laid earliest at that age), and 4.5 years for clutch size (fig. 13.3). On the other hand, the most crucial measure of production, the mean number of young raised per clutch, continued to increase at least to the seventh year of life. Beyond this age production seemed to deteriorate, but samples were too small to be sure. Improvements in production beyond the fourth year, in the face of reduced clutch size, were due to reduced losses between the egg and fledgling stages among older birds and to a greater proportion of successful nests. These were perhaps those aspects of breeding in which experience would count most. These various trends were equally apparent in different habitats. Although they were mean trends for the whole female population, they were shown by the majority of individuals for which I had long series of records.

In addition to a bird's own age, its mate's age also influenced breeding performance. For 230 pairs I knew whether both partners were yearlings or adults. Some selective mating occurred between yearling and adult birds, but not between adult birds of different ages (Newton, Marquiss, and Moss 1981). Table 13.5 therefore compares mean clutch size, percentage of successful nests, and mean young per nest between the four combinations: yearling male/yearling female, yearling male/adult female, adult male/yearling female, and adult male/adult female. Only in these three aspects of reproduction was significant variation between groups apparent, after correcting for variation associated with year and habitat.

Adult/adult pairs laid more eggs, on average, than did pairs containing yearlings. Adult/adult pairs also showed the highest percentage of successful breeding attempts, yearling/yearling pairs showed the lowest, and mixed-age pairs were intermediate. Mixed pairs in which the male was an adult did slightly better in both respects than mixed pairs in which the male was a yearling. Overall, the number of young produced per nesting attempt was 1.8 by yearling/yearling pairs, 2.3 by yearling male/adult female pairs,

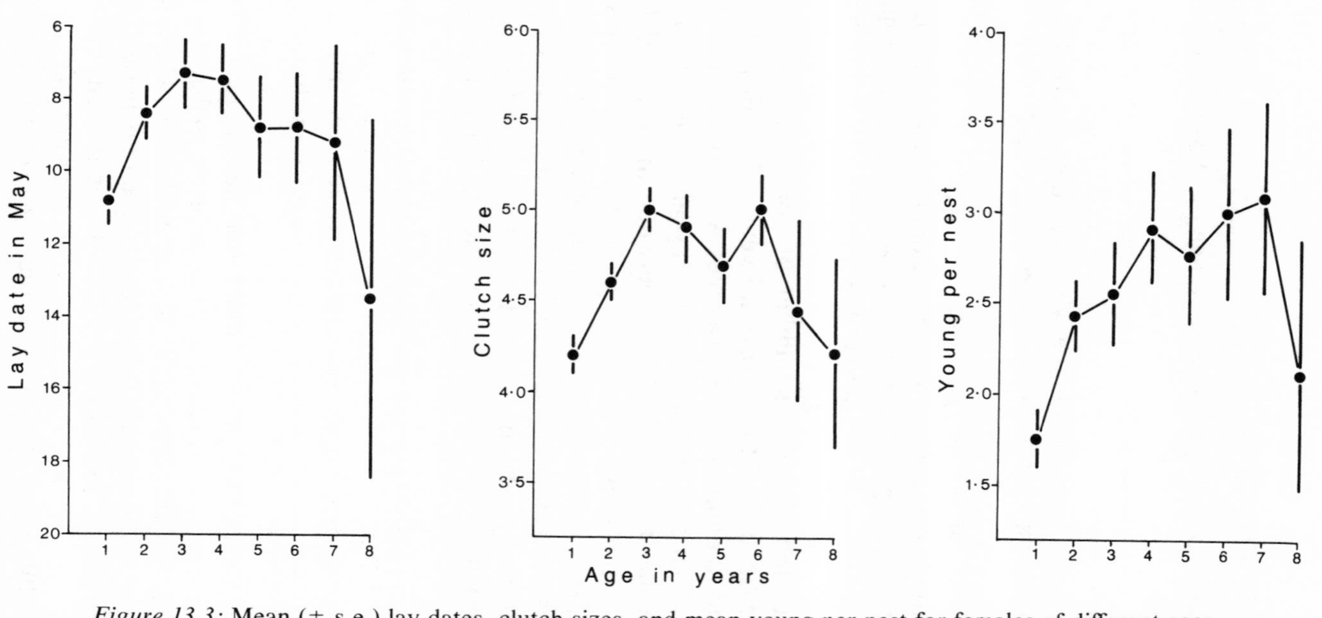

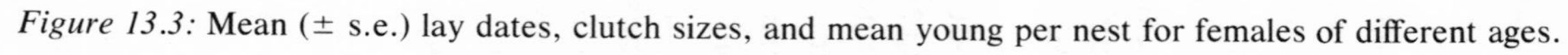

Figure 13.3: Mean (± s.e.) lay dates, clutch sizes, and mean young per nest for females of different ages.

Table 13.5 Breeding Performance in Pairs in Which Both Partners Could Be Classed as Yearling or Adult

	Clutch Size		Percentage of Nests That Produced Young		Young per Nest	
	n	*Mean ± s.e.*	*n*	*Mean ± s.e.*	*n*	*Mean ± s.e.*
Yearling male/yearling female	11	4.43 ± 0.29	13	53.1 ± 11.1	13	1.8 ± 0.30
Yearling male/adult female	23	3.99 ± 0.20	25	65.7 ± 8.0	25	2.3 ± 0.22
Adult male/yearling female	13	4.39 ± 0.27	14	80.3 ± 10.7	14	2.4 ± 0.26
Adult male/adult female	157	4.76 ± 0.08	178	82.9 ± 3.0	178	3.0 ± 0.08
Significance of variation between groups		$p < .01$		$p < .05$		$p < .05$

Note: Data corrected for variations associated with year and habitat (see Newton, Marquiss, and Moss 1981).

2.4 by adult male/yearling female pairs, and 3.0 by adult male/adult female pairs.

13.5 Production and Survival in Relation to Production the Previous Year

In the population at large, some 67% of clutches laid during 1971–84 produced young, with the remainder failing for one reason or another. For females that laid, I attempted to find whether high production in one year had any adverse effect on production or female survival the following year.

In fact, breeding success in any one year seemed to be independent of that the year before (table 13.6). Regression of brood size in the second year on brood size in the first showed a slight positive slope, suggesting a given production in the first year tended to be followed by better production in the next, rather than vice versa. This was consistent with the gradual improvement of breeding with age, as discussed above. Nor was evidence found that brood size influenced subsequent female survival.

Table 13.6 Number of Young Produced by Individual Females in One Year in Relation to Number Produced in the Previous Year

Brood Size in Following Year	Brood Size in One Year						
	0	1	2	3	4	5	6
0	13	5	10	16	16	4	0
1	4	4	2	2	3	3	0
2	2	1	6	4	5	1	1
3	9	3	4	8	14	6	0
4	9	2	6	13	11	7	0
5	5	2	2	3	11	7	0
6	0	0	0	0	0	0	0
Mean	2.29	1.94	2.00	2.20	2.57	3.07	(2.0)

Note: A regression analysis of brood size in the second year against brood size in the first year showed no significant trend (slope = 0.14, $r = .12$, $p > .1$).

Table 13.7 Proportion of Females That Were Retrapped, According to Number of Young Raised the Previous Year

	Number of Young Raised				
	0	*1–2*	*3–4*	*5–6*	*Overall*
Number trapped	82	66	174	51	373
Number (%) retrapped	44 (54)	42 (64)	108 (62)	29 (57)	223 (60)

Note: For this analysis, all age groups were pooled. Females that were retrapped in a later year were classed as survivors, whereas those not retrapped were classed as nonsurvivors. Individuals that were retrapped at a nest in more than one year were included more than once, since in each case the unit of comparison was one "bird-year." No significant variation in the proportion recovered was found between birds that raised no young and birds that raised one to six young ($\chi^2 = 1.3$, $p > .5$). In addition, a weighted regression analysis of proportion retrapped on brood size (in seven categories, 0–6) revealed no significant trend ($b = 0.2$, $p > .9$).

Overall, about 59% of females were retrapped in a later year, and this proportion did not vary according to the number of young raised in the preceding breeding attempt (table 13.7). In conclusion, therefore, I found no evidence for females that breeding performance in one year influenced either breeding performance or adult survival in the next.

13.6 Lifetime Production

For 142 females, I was able to calculate the total number of young reared to fledging during their lives, on the assumption that all breeding attempts occurred within the study area. For the majority of females this assumption was reasonable, but a minority may have made one or more attempts outside the areas and hence unknown to me. However, these were unlikely to have been numerous enough to affect the conclusions. The females for which I had lifetime reproductive estimates constituted only a small proportion of the females handled during the study but, in their various life spans they were representative of the female population as a

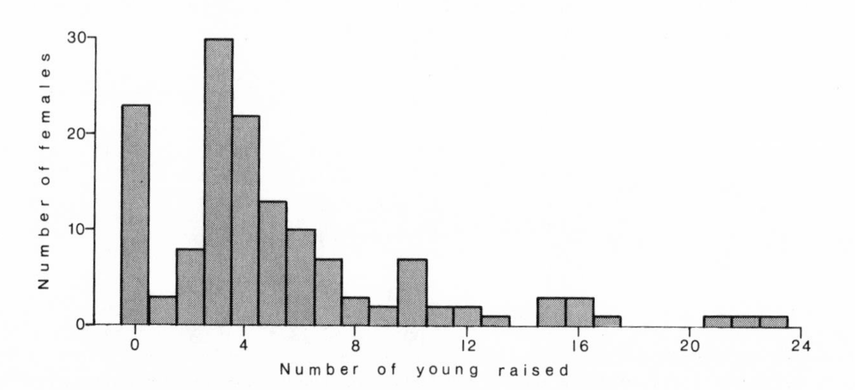

Figure 13.4: Lifetime production of 142 females. The sample of females was representative, in terms of longevity, of the whole female population.

whole (Newton 1985). For each individual, egg laying counted as a breeding attempt.

These females varied greatly in the number of young they produced. Some 16% of those that attempted to breed produced no young during their lives, and, for the 84% that did produce young, the number varied between one and twenty three in different individuals (fig. 13.4). With the distribution found (fig. 13.4), the median number of young produced per female was 3.4 and the mean was 5.1, but the estimates of variance were of limited value (table 13.8). On that distribution, 15% of breeding females produced 50% of all fledged young, and 42% of breeding females produced 75% of young.

From the data given on survival and age of first breeding (table 13.3), it could be calculated that 72% of all females that left the nest died before they could start breeding in their first to fourth year of life. A further 4.5% attempted to breed but produced no young, and 23.5% were productive.

Proximate Causes of Variations in Lifetime Production

The main proximate source of variation in lifetime production by females was longevity (fig. 13.5). Lifetime productions of up to five young were recorded from females that survived for only one breeding season, productions up to ten young were recorded for birds that survived up to

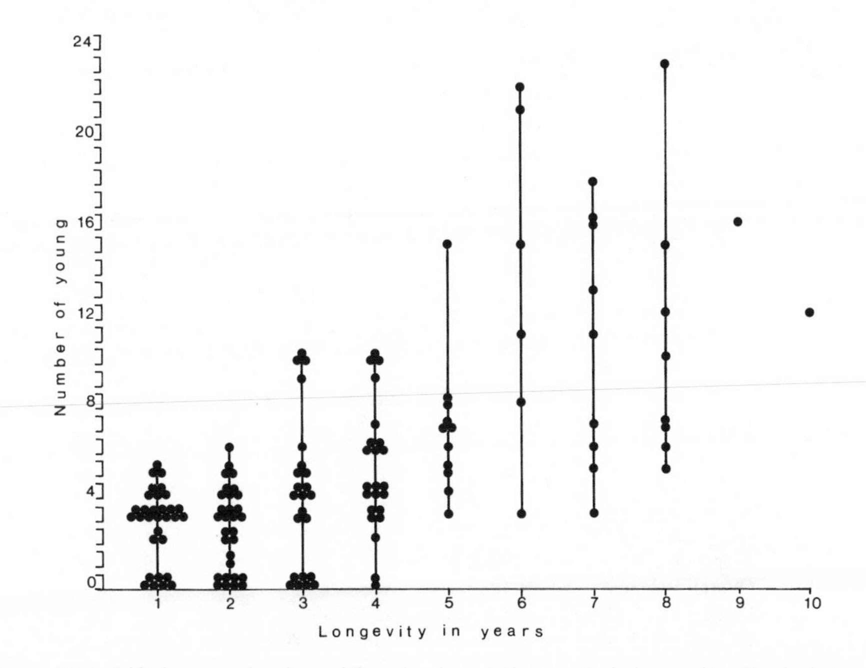

Figure 13.5: Lifetime production of females in relation to their longevity. Each dot represents the number of fledglings raised by an individual female. Significance of relationship: $b = 1.46$, $r = .67$ $p < .001$.

three or four years, and all productions greater than ten young were from those that survived more than four years. The three productions greater than twenty young were for birds that survived six to eight years. Within females of any one age-group, however, variation in productivity was great, and some individuals up to eight years old raised no more young

Table 13.8 Mean and Variance of the Components of Lifetime Reproductive Success in Female Sparrowhawks That Bred Successfully

Component	Original		Standardized Variance
	Mean	*Variance*	
L	3.65	5.07	0.38
P	0.74	0.05	0.10
F	4.25	1.04	0.06
S	0.62	0.05	0.13
LP	2.69	4.08	0.56
LF	14.70	72.40	0.30
PF	3.09	1.36	0.14
LS	2.09	1.85	0.36
PS	0.46	0.05	0.24
FS	2.65	1.41	0.20
LPF	10.58	55.96	0.42
LPS	1.50	1.24	0.44
LFS	8.53	29.80	0.32
PFS	1.92	1.11	0.29
LPFS	6.01	19.39	0.38

Note: $N = 120$ females that bred successfully. A successful breeder is one that raised one or more young to fledging.
L = life span; P = proportion of life span that is reproductive; F = annual fecundity (number of eggs laid); S = survival of offspring to fledging.

Table 13.9 Percentage Contribution of the Components of Lifetime Reproductive Success to Variance in LRS among Female Sparrowhawks That Bred Successfully

Component	L	P	F	S
L	99.28			
P	21.25	25.16		
F	−35.64	−4.10	15.07	
S	−37.94	4.00	5.10	33.27
LPFS	−25.54			

$N = 120$

Inclusion of 22 failed breeders[a]:
Proportion of successful breeders, 0.85
Overall variance (OV), 21.12
Percentage of OV due to failed breeders, 22.54
Percentage of OV due to successful breeders, 77.46

Note: $V(LFS)/(LFS)^2$.
[a]A failed breeder is one that laid eggs but raised no young.
L = life span; P = proportion of life span that is reproductive; F = annual fecundity (number of eggs laid); S = survival of offspring to fledging.

than did others that lived for only one year. The single individual that survived for ten years produced only twelve young.

The lifetime reproductive success (LRS) of an individual depended not only on its life span (L), but also on the proportion of that life span that was reproductive (P), in turn dependent mainly on age of first breeding, on fecundity (F) as assessed by the number of eggs laid per breeding attempt, and on the proportion of these eggs that gave rise to fledglings (S). The simple multiplicative model LRS = L × P × F × S, gave a means of estimating the contribution of each factor to variance in lifetime production (Brown, this volume, chap. 27). The analysis in tables 13.8–13.11 was done initially for 120 successful breeders only, defined as those that produced at least one young, then again including the 22 individuals that produced eggs but no young, and finally also including 365 females that died before they could produce eggs. This last figure was estimated from the earlier calculation that such females formed 72% of the total fledged. These three analyses thus gave estimates of the variance in lifetime

Table 13.10 Percentage Contribution of the Components of Lifetime Reproductive Success to Total Variance in LRS among All Breeding Female Sparrowhawks, Failed and Successful

Component	L	P	F	S
L	76.91			
P	16.46	19.49		
F	−27.61	−3.18	11.67	
S	−29.39	3.10	4.02	25.77
LPFS	−19.78			

Inclusion of all nonproductive birds:[a]
Proportion of successful breeders, 0.24
Overall variance (OV), 11.09
Percentage of OV due to females that died before breeding, 58.91
Percentage of OV due to breeders, 41.09

Note: V(LFS)/(LFS)^2.
[a]N = 22 failed breeders and 365 other females that died before they could breed.
L = life span; P = proportion of life span that is reproductive; F = annual fecundity (number of eggs laid); S = survival of offspring to fledging.

Table 13.11 Percentage Contribution of the Components of Lifetime Reproductive Success to Total Variance in LFS among All Female Sparrowhawks Fledged, Including Those That Died before They Could Breed

Component	L	P	F	S
L	40.80			
P	8.73	10.34		
F	−14.65	−1.69	6.19	
S	−15.59	1.64	2.13	13.67
LPFS	−10.50			

Note: V(LFS)/(LFS)^2.
L = life span; P = proportion of life span that is reproductive; F = annual fecundity (number of eggs laid); S = survival of offspring to fledging.

production for successful breeders only, all breeders including failed ones, and all fledged females, including those that died before they could breed.

Among 120 females that produced one or more young, the overall variance in lifetime production was 19.39. Including the failed breeders, the overall variance became 21.12, of which 77% was due to variation among successful breeders and 23% to the inclusion of failed breeders. Including in addition those birds that died before they could breed, the overall variance became 11.09, of which 41% was due to variation among successful breeders and 59% to the inclusion of all nonproductive birds, including those that died young. The total variance was lowered when the last group was included, because this last group was formed by a large number of birds with uniform nil production.

Of the total variance in the lifetime productions of successful breeders, 99% could be accounted for in terms of variation in L, 25% in terms of variation in P, 15% in terms of variation in F, and 33% in terms of variation in S. These features were themselves correlated, in some cases positively and in others negatively (tables 13.8–13.11). Particularly important contributions to the total variance were made by the −41% covariation between L and F, the −50% between L and S, and the −30% between L, P, and S.

These various estimates gave some indication of the relative importance of each of the four features in influencing lifetime reproductive success among female sparrowhawks that bred successfully. A substantial part of the variance in lifetime production was associated with covariation between different variables.

Ultimate Causes of Variation in Lifetime Production

I attempted to relate the individual variations in lifetime production of fledglings to features of individual anatomy, previous history, and habitat. The only useful anatomical feature I measured was wing length, with the feathers flattened and straightened on the rule. This value varied among females between 226 mm and 247 mm, but within individuals it was fairly consistent (± 3 mm) beyond the first year of life. Longevity was significantly related to wing length ($n = 113$, $b = 0.09$, $p < .05$), as was lifetime egg production ($n = 113$, $b = 0.37$, $p < .02$), and lifetime young production ($n = 113$, $b = 0.25$, $p < .02$), but in each case it explained only a small percentage of the variance. Age of first breeding showed no relationship with wing length ($n = 113$, $b = -0.002$). Assuming that wing length reflected body size, larger females tended to live longer and produce more eggs and young than smaller ones. There was no relation between wing length and territory grade (unpublished information), so this did not occur because the larger females occupied the better territories.

Birds that had been raised in the study areas as nestlings and later retrapped there as breeders were classed as "residents." Breeders that were unringed at first capture had mostly been raised elsewhere and were

Table 13.12 Lifetime Production of Resident and Immigrant Females

Status	Number of Females That Produced the Following Numbers of Young		
	0	*1–10*	*11+*
Bred once in study area			
Residents (raised in study area)	2	17	0
Immigrants (raised elsewhere)	18	27	0
Bred more than once in study area			
Residents (raised in study area)	2	13	4
Immigrants (raised elsewhere)	2	38	8

Note: For "once only" birds, $\chi^2 = 4.12$, $p = .02$, for others, $\chi^2_2 = 2.98$, $p < .3$.

classed as "immigrants." On average, the immigrants would have moved farther between birthplace and breeding place than the residents. Once they had started to breed, relatively more of the residents than of the immigrants produced young (table 13.12). This was especially marked in birds that bred only once in the study areas, and it became less marked in birds that bred more than once. It was consistent with another finding, that short-distance dispersers performed better in individual breeding attempts than did long-distance ones (Newton and Marquiss 1983).

Habitat was divided into two types: lowland with mixed farmland and scattered woods, and upland with open sheepwalk and plantation forest. No association was apparent between natal habitat and subsequent lifetime production or between lifetime production and breeding habitat. However, nonbreeding may have been relatively more frequent in upland areas, since a greater proportion of the nests built there were not laid in (Newton, Marquiss, and Moss 1979).

Summarizing, significant relationships were apparent between lifetime production on the one hand and wing length and dispersal distance on the other, but not between lifetime production and natal or breeding habitat.

13.7 Recruitment to Breeding Population

Ideally, lifetime production should have been assessed by the number of offspring that entered the breeding population. Of the 717 fledglings produced by females in the sample, only 36 (both sexes) have so far been found breeding in the study areas, though others may have settled elsewhere or may breed in the areas in the future. However, the number of recruits derived from individual females was significantly correlated with the number of young they had reared to fledging (excluding females that reared no young to fledging, $n = 119$, $r = .44$, $p < .001$). This indicated that the number of young reared to fledging gave a realistic index of contribution to future breeding populations.

Since the regression of number of young entering the breeding population on number of young fledged passed through a point very close to zero ($a = -0.11$), fledglings from all types of females, whether poor or good producers, apparently had equal chances of breeding. In other words, individual young from good producers were no more likely to become breeders than were young from poor producers, once these young had reached the fledging stage.

13.8 Discussion

With maximum life spans of little over seven years for males and ten for females, sparrowhawks were not much longer lived than the small songbirds that formed their prey (Lack 1954). As in the songbirds, survival among sparrowhawks was lower in the first year than subsequently and then remained fairly constant until near the maximum life span.

There were at least two likely proximate causes of the lower annual survival in males. Smaller body size alone would have made males more susceptible than females to temporary food shortages (having relatively less body reserves) and predation (having a greater range of enemies). On food shortage, weights and sex ratios of trapped birds from different months suggested that young males suffered greater mortality than females in the first two months of independent life, mainly from starvation (Newton, Marquiss, and Village 1983). The absolute food requirements of males were less than those of females, but they took smaller prey and their metabolic requirements per unit of body weight were greater. Thus only three to four days without food was sufficient to kill a male, whereas a larger female could last six to seven days (unpublished information). I several times found males killed and eaten by larger females, but no cases of the reverse. Both sexes were occasionally taken by tawny owls, *Strix aluco*, and goshawks, *Accipiter gentilis*, but cases were too few to be sure whether one sex was killed more often than the other.

Whereas mean annual survival remained fairly constant after the first year, individuals continued to improve in breeding success well beyond this age, at least to the seventh year in females. Since almost every aspect of breeding was dependent on food supply (Newton, Marquiss, and Village 1983), the improvement in performance with age in both sexes may have been due partly to increased experience at foraging, and in the female perhaps also to increased skill in parental care. The situation was complicated because many individuals moved to slightly better territories as they aged. In fact, the mean grade of territory occupied by each female age-group increased progressively up to the oldest birds of seven to nine years. This may have partly accounted for the improvement in overall production up to the seventh year. On the other hand, the deterioration in clutch size beyond the fourth year could only be attributed to a decline in abilities associated with aging. In addition to laying later and smaller clutches than middle-aged females, old females produced smaller eggs (unpublished data). The only outward sign of senility in some old birds was the presence of swollen "arthritic" joints in their feet, a feature not seen in younger birds.

Breeding in relation to age has been studied in several other bird species. In some, clutch size and breeding success improved for several years of increasing age and then leveled off (e.g., various seabirds: Coulson 1966; Coulson and Horobin 1976; Mills 1973; Ollason and Dunnet 1978; Richdale 1957; snow goose, *Anser caerulescens*: Finney and Cooke 1978; white stork, *Ciconia ciconia*: Schüz 1957). In other species clutch size reached a peak in middle age and then declined again (e.g., great tit, *Parus major*: Perrins 1979; arctic tern, *Sterna paradisaea*: Coulson and Horobin 1976). Hence the trends found in sparrowhawks had parallels in several other species.

It was expected in the sparrowhawk that age of the male as well as of the female would affect breeding, because the male provides the food for a large part of the breeding cycle, from before the start of egg laying until well after hatching. During this stage the female stays on the nest almost continuously, fed by her mate. The hunting success of males would also have been expected to change during their lives, according to their experience and abilities.

At the time nesting began, food reached its lowest point of the year. To judge from their weights, some males were prevented from breeding by inability to obtain enough food. At this time of year if they were to breed, males had to feed a female in addition to themselves. This was a more acute problem for young males, presumably in part because of their lesser hunting experience and in part because, through competition with older males, they usually occupied habitat with fewer prey. General food shortage in spring may also have been the factor that prevented the development of large-scale polygyny in the presence of surplus females. If males had difficulty feeding one female at this time, it is unlikely that many could have fed more than one.

Food shortage in spring also prevented some females from breeding. Before males began to supply their food, well-fed females could afford to spend more time guarding nesting places than could poorly fed ones, which were obliged to spend more time hunting and thus lost out in competition for nesting places occupied by competent males. This too ensured that in any one year only the better-condition females bred. The relative shortage of males was a further factor preventing many females from breeding, so again the disparate sex ratio was involved, with a greater proportion of males than females breeding in any one year and males breeding at a younger average age than females. In southern Scotland the combined shortages of food and males meant that well under half of all females bred in any one year. The nonbreeders were mainly young individuals but included some older ones that had bred in previous years. Many of the pairs that formed broke up before eggs were laid, in some cases before a nest was started and in others at various stages of nest construction, apparently because the male was unable to maintain food deliveries. In conclusion, therefore, food shortage in spring was a major constraint that prevented many young birds of both sexes from breeding, and shortage of mates was a further constraint for females.

The comparison of lifetime success among females that started breeding at one, two, or three years of age gave no hint that birds might refrain from breeding early in life in order to promote greater production overall (Williams 1966b). Among females that lived three or more years, mean lifetime productions were significantly greater in those that started at one or two years than in those that started at three, mainly because many of those that missed the first one or two years were dead before they could start at three. This lent further support to the "constraint" hypothesis of delayed breeding as opposed to the "restraint" hypothesis, that birds hold back to save themselves for later (Curio 1983). With such a high risk of dying before the next season and no obvious penalty in breeding, it would pay a sparrowhawk of any age to breed if it possibly could.

The females that attempted to breed (28% of those fledged) produced an average of 5.1 young each. If the females that died before they could breed are included, the mean number of young produced per female fledged was 1.4. Since the sex ratio at fledging was equal, the latter figure would also have applied to males, but, owing to an earlier mean age of first breeding in males, the distribution of reproduction throughout the male life span would have differed from that of females.

A mean lifetime production of 1.4 young per pair fledged fell considerably short of the 2.0 per pair needed to maintain population stability in the long term. In the seven years 1974–80, when most data were obtained, breeding numbers fell from 137 to 98 pairs, a decline of 28%. With a replacement of 1.4 recruits per pair that died, a 69% annual survival (the mean of males and females), and no immigration, the population would have reached 76 pairs in 1980, a decline of 42%. This implied a net immigration, which was sufficient to slow the decline but not to prevent it. An increase in lifetime production, needed to sustain the population, could have been achieved among breeders by an earlier start to breeding, a reduction in the frequency of nonbreeding years in later life, improved annual survival, or better breeding success at individual attempts. For reasons given earlier, the figures on lifetime production obtained for certain females may have been too low, biasing the estimate of the mean value. If the females that attempted to breed had raised an average of one more brood each, this would have been more than sufficient to sustain the population without net immigration.

13.9 Summary

1. In the sparrowhawk, females weigh almost twice as much as males. In southern Scotland the sex ratio at fledging was equal, but owing to subsequent greater mortality in males, females predominated among adults, possibly to the extent of 2:1. The study population may have been unusual in this respect. Despite the surplus of females, monogomy was usual.

2. In both sexes, annual mortality was greater in the first year of life than subsequently. Ringed males lived up to 7.6 years and females to 10.8.

3. Females first bred in their first, second, third, or fourth year of life. Owing to the risk of mortality, it was advantageous for females to start breeding as early in life as possible. Shortage of males and shortage of food were two factors that prevented some females from breeding at a younger age.

4. Among breeding females, the mean production of young improved steadily from the first to at least the seventh year of life. On the other hand, clutch size increased to about the fourth year of life, then declined again.

5. The lifetime production of females varied greatly. About 72% of fledgling females died before they could reach breeding age, 4.5% attempted to breed but produced no young, and the remaining 23.5% produced between 1 and 23 young each. The mean lifetime production was 5.1 young per female which attempted to breed and 1.4 young per female that fledged. This was not sufficient to maintain population stability in the long term, and the population declined.

6. Proximate factors that influenced lifetime production of young by breeding females included longevity, age of first breeding, egg production at individual attempts, and the proportion of eggs that gave rise to fledglings. These four factors could "explain" respectively 99%, 25%, 15%, and 33% of the total variance in lifetime production among successful breeders; to some extent these factors were intercorrelated, some positively and others negatively. Ultimate factors related to lifetime production included wing length (reflecting body size), which gave a positive relationship, and dispersal distance between natal and breeding site, which gave a negative relationship.

13.10 Acknowledgments

I am grateful to M. Marquiss, who contributed to the fieldwork on which this chapter is based, to D. Brown for help with statistics, and to J. P. Dempster, T. Clutton-Brock, and two anonymous referees for helpful comments on the manuscript.

14 Breeding Success in Bewick's Swans

D. K. Scott

In monogamous birds, analysis of the factors affecting lifetime breeding success is complicated because an individual's success may depend as much on characteristics of its mate as on its own characteristics. For example, in kittiwakes the number of young fledged by different pairs is influenced by the laying date of the female, which is affected by the age of the male (Thomas and Coulson, this volume, chap. 16). Similarly, in sparrowhawks the age of the male affects the clutch size and nest success of the female (Newton, Marquiss, and Moss 1981; Newton, this volume, chap. 13). These effects must also apply to the long-term success of pairs, but it has not yet been possible to separate the contributions of male and female characteristics to long-term success.

Histories of individual Bewick's swans (*Cygnus columbianus bewickii*) and their mates have been followed by the Wildfowl Trust since 1964, and it is now possible to examine male and female characteristics affecting the success of pairs. In this chapter, I investigate success in relation to adult body weight and size of male and female partners separately as well as the overall dominance of the pair.

14.1 Methods

Records of the life histories of individual Bewick's swans in the United Kingdom have been kept by the Wildfowl Trust at Slimbridge (Gloucestershire) since 1964 and at Welney (Cambridgeshire) since 1972 (Evans 1978; Scott 1978). Over 4,000 individual swans have been identified by differences in the pattern of black and yellow on the bill (Scott 1966; Rees 1981) (fig. 14.1), and over 1,500 of these have been caught and ringed with large plastic rings, legible from 200 m in good light (Ogilvie 1972; Evans 1979). Each year, from October to March, individuals returned to the wintering grounds at Slimbridge and Welney after spending the summer on the breeding grounds in the Soviet arctic, and their dates of arrival, paired status, and breeding success were recorded.

Studies of the conspecific whistling swan (*Cygnus columbianus columbianus*) have shown that nestlings marked with their parents travel with them to the wintering grounds (Sladen 1975), and in Bewick's swans we know that cygnets then remain closely associated with their parents throughout their first winter (Scott 1966). It is thus possible to measure the annual breeding success of a pair by counting the number of cygnets raised to four months.

Lifetime breeding success has been calculated by summing annual success for all individuals whose winter histories have been followed from the first or second winter through to death. Death was determined by recovery of rings from dead birds in 65% of cases, and in the remaining 35% of cases death was assumed when an individual's mate returned without it. This assumption was made because no case of divorce has yet been observed among 500 breeding pairs, and only 20 cases have been noted among 1,000 pairs that never bred successfully.

"Complete" life histories were available for 43 individuals (21 males

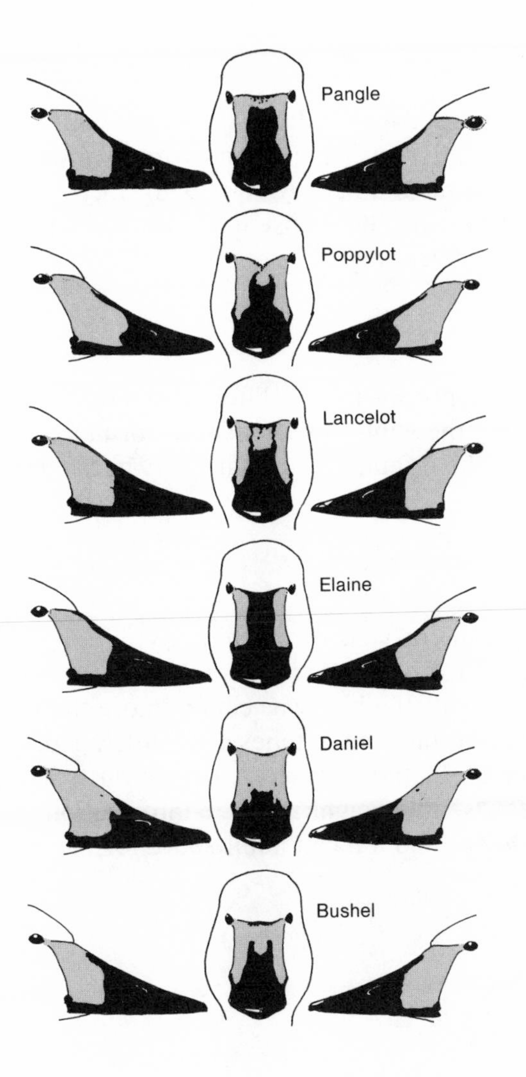

Figure 14.1: Illustration of individual differences in bill pattern in Bewick's swans.

and 22 females). A further 15 individuals that missed only one winter between first identification and death were included, but individuals that missed more than one winter were omitted. Forty of the birds were ringed as cygnets (aged four to seven months) or yearlings (aged sixteen to nineteen months), and 18 were first identified by bill pattern as yearlings. Five of the latter were ringed as adults, and 13 were not caught at all during their lives.

Sex

All birds caught were sexed by cloacal examination, aged, weighed, and in some cases measured (wing chord and skull length in mm). Individuals not caught were sexed by estimating size: males are on average 13% heavier than females (Evans and Kear 1978). Evidence from 150 birds sexed by estimation and subsequently caught and examined showed that sex had been estimated with 89% accuracy. For paired birds (n = 100) the accuracy was higher (95%) than for single birds (78%, n = 50).

Age

Age was determined from plumage color. Adult swans are white, cygnets (first winter birds) are gray, and yearlings (second-winter birds) have traces of gray feathering on the head and neck (Scott 1978). Beyond this, age was determined only by following the subsequent histories of individuals first identified as cygnets or yearlings.

Dominance

Dominance was calculated as the proportion of different opponents beaten by the individual or pair during the winter (Scott 1980). For paired birds it was not possible to separate the dominance of partners, because the birds were seldom apart.

Weights

Weights were standardized for time of year of catching by dividing each bird's weight by the mean weight for its age/sex-class for each catch to provide a measure of "relative" weight. If a bird was caught more than once in a winter, its relative weights in different catches were averaged. All references to weight below refer to relative measures. Of the birds in this study, 50% had dominance records, and weight and size information, in more than one year during adulthood, and a mean adult score for each variable was calculated for these birds.

Survival

Survival rates were investigated in two samples. The first comprised individuals ringed as cygnets between 1968 and 1974, and second comprised birds first identified as adults before 1968 (minimum age two years on first identification). In the first sample, data included many observations of individuals at sites other than Slimbridge or Welney, both in Britain and on the Continent, where observers read rings on living swans and returned rings from swans found dead. Because of this it was necessary to incorporate a correction for plastic ring loss in analyzing this sample. Among birds returning to Slimbridge, ring loss was relatively constant with time, at 6.1% per year among males and 4.2% among females. These values represent the maximum rate of loss, since rings that were worn thin or cracked when a bird was recaptured were taken off and counted as lost when they might otherwise have remained on the bird for several more years.

The second sample of adults first identified before 1968 was restricted to birds with histories of five or more years, which had a high rate of return to Slimbridge. Only five males and three females were not ringed. However, since identification was mainly by bill pattern, no correction was made for plastic ring loss. In calculating percentage survival from both samples, birds that disappeared were assumed dead, and consequently mortality may be overestimated. Survival is therefore minimum survival.

Breeding Success

To investigate the components of success, data from all dead birds (death determined only by recovery of birds and return of mate) were used. Life span was taken to start at two years, when the birds attain adult plumage, weight, and size and are capable of breeding. Four measures of breeding success used below are defined as follows:

Total success = total number of cygnets raised to four months during life or known history.
Mean brood size = mean brood size in years of breeding.
Breeding rate = number of broods raised in life/number of years of life (or known history).
Productivity = total number of cygnets raised in life/number of years of life (or known history).

The records were examined to discover how many years' information were necessary to give a stable estimate of variance in total success. All birds with ten or more years' information were investigated (43 males and 28 females), and standardized variance in success was calculated for cumulative numbers of years. Standardized variance ($\sigma^2/\bar{x}^2$) declined rapidly as the number of years' data included increased, stabilizing after four years' data had been included (fig. 14.2*a*). This indicated that

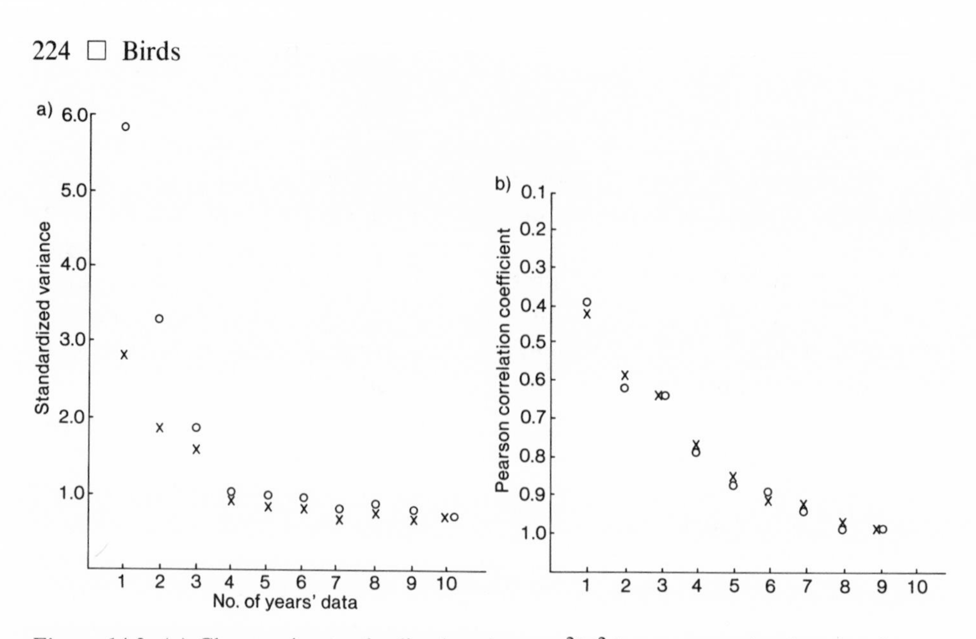

Figure 14.2: (*a*) Changes in standardized variance $\sigma^2/\bar{x}^2$ in success using increased numbers of years' data; (*b*) Correlations between success based on ten years' data and success based on different numbers of years' data, one to nine. (*x's*, males; *open circles*, females.)

estimates of variance in success based on fewer than four years' data would be unreliable but that data covering the entire life span were not necessary. Two samples of individuals were used to examine variance in success: all those with complete life histories (28 males and 33 females), and all those with histories of five or more years as adults (235 males and 205 females).

To determine a reliable estimate of long-term productivity, mean success per year for individuals with ten or more years' data was correlated with mean success per year calculated for cumulative numbers of years. The correlation coefficients increased rapidly with one to four years' data included and then increased more slowly (fig. 14.2*b*). On the basis of this, birds with five or more years' history as adults were used in the analysis of factors affecting long-term success.

Since success changes with age (see below), it was necessary to cover a restricted age period when examining factors affecting success. For this reason, I investigated the success of pairs over a five-year period starting at age five. This starting point was chosen since it both maximized sample size (few pairs had histories beyond age ten) and included peak breeding age (nine years). In sixteen cases one partner was of known age and the other of minimum age, and in thirty-six cases both were minimum-age birds (i.e., at least two years on first identification). For ten pairs where the ages of partners differed by one year, the younger bird's age was taken as the starting point. For the remaining six pairs where partners differed by more than one year, half the difference in age was added to the older bird at age five, and this point in time was taken as the start. Pairs where partners differed in age by more than four years were omitted.

In the analysis of factors affecting success of pairs, total success refers to the total number of cygnets raised in the five-year period; number of broods raised was that in the five-year period, and mean brood size was calculated only from broods raised within that period.

14.2 Results

Survival

The oldest wild Bewick's swan was at least twenty-five years old, and thirty other individuals with histories of fifteen or more years are still alive at this time. The oldest captive Bewick's swan died at thirty-six years. Both sexes showed over 60% minimum survival between their first and second winters (tables 14.1 & 14.2). Thereafter minimum survival rose above 80% in both sexes. No significant sex differences in survival could be detected, though from age four onward females appeared to survive slightly less well than males. Since 190 of the 198 birds involved in these analyses were ringed, and since ringed birds are not infrequently resighted elsewhere in Europe in years when they do not visit Slimbridge or Welney, the results should not be affected by any tendency for males and females to differ in loyalty to a particular wintering ground. In any case, the proportion of swan-years when individuals that were ringed as cygnets

Table 14.1 Life Table Information for Bewick's Swans: Known-Age Birds Caught as Cygnets between 1969 and 1974

Age (years)	Number Seen[a]	Number Known Dead	Number Disappeared[a]	Minimum Percentage Surviving from One Year to the Next
Males (n = 70)				
1	45	3	22	64.3
2	40	5	25	88.9
3	35	5	30	87.5
4	32	5	33	91.4
5	29	5	36	90.6
6	26	6	38	89.7
7	21	6	43	80.8
8	19	6	45	90.5
9	14	6	50	74.7
10	11	7	51	78.6
Females (n = 69)				
1	47	2	20	68.1
2	43	2	24	91.5
3	39	4	26	90.7
4	35	5	29	89.7
5	27	7	35	77.1
6	24	7	38	88.9
7	19	8	42	79.2
8	16	9	44	84.2
9	14	9	46	87.5
10	13	9	47	92.9

[a]Corrected for loss of plastic leg rings; figures rounded to whole numbers.

Table 14.2 Life Table Information for Bewick's Swans: Minimum-Age Birds First Identified before 1968, with Histories of Five or More Years

Age (years)	Number Seen	Number Known Dead	Number Disappeared	Minimum Percentage Surviving from One Year to the Next
Males (n = 32 at age seven)				
8	29	1	2	90.6
9	28	2	2	96.6
10	27	3	2	96.4
11	23	5	4	85.2
12	21	6	5	91.3
13	19	7	6	90.5
14	15	8	9	78.9
15	14	8	10	93.3
16	12	9	11	85.7
17	11	9	12	91.7
18	7	9	16	63.6
Females (n = 27 at age seven)				
8	23	1	3	85.2
9	22	2	3	95.7
10	20	3	4	90.9
11	19	4	4	95.0
12	14	6	7	73.7
13	12	8	7	85.7
14	8	10	9	66.7
15	6	10	11	75.0
16	5	10	12	83.3
17	3	11	13	60.0
18	2	11	14	66.7

were resighted elsewhere compared with at Slimbridge did not differ between the sexes.

Age at First Pairing

Median ages at first pairing did not differ significantly between the sexes. They were 2.4 years in males (n = 45, range 1–9 years) and 2.35 years in females (n = 41, range 1–6 years). These values may be underestimates, since birds with long histories before pairing were more likely to have missed one or more years and been excluded from the analysis.

Age at First Successful Breeding

There was no significant difference in age at first successful breeding between males and females. Median values were 4.64 years among males (n = 19, range 2–10 years) and 5.50 years among females (n = 18, range 2–9 years), all of which had been seen each year before rearing cygnets. One male and one female each reached age ten without breeding successfully. As with age at first pairing, these figures may be underestimates.

Changes in Breeding Success with Age

Success varied with age. When breeding success was calculated across all known-age individuals at each age, the mean number of cygnets raised to four months increased with age until ten years, then appeared to decline and to rise again toward the end of the life span in both sexes (fig. 14.3*a*). This was due to changes in the proportion of birds of different ages that raised broods (fig. 14.3*b*), not to changes in mean brood size with age (fig. 14.3*c*).

The increase in proportion of pairs raising broods was significant between ages three and four ($\chi^2 = 7.46$, d.f. = 1, $p < .01$), ages four and five ($\chi^2 = 6.16$, d.f. = 1, $p < .02$), and ages five and six ($\chi^2 = 5.33$, d.f. = 1, $p < .05$), and also between ages six and nine ($\chi^2 = 4.11$, d.f. = 1, $p < .05$), with data from the two sexes combined. The decrease between ages nine and 12 was not quite significant ($\chi^2 = 2.84$, d.f. = 1, $p < .10$), and the subsequent increase between ages twelve and fifteen was also nonsignificant ($\chi^2 = 0.07$, d.f. = 1, $p > .10$).

The apparent changes in breeding success during the life span could be due to differences in longevity between successful and unsuccessful breeders. To overcome this problem, each bird's average breeding rate was determined, and deviations from this were calculated for each age. Figure 14.3*d* shows the mean of these deviations and reveals a similar curve.

Nor were these results a consequence of annual variation in success. The sample comprised individuals whose histories began in different years from 1964 onward, so each age-class (including those for ages sixteen, seventeen, and eighteen) contained information from several different years. There were no runs of successive good or poor years that might have influenced the results.

Variance in Breeding Success

Standardized variance in seasonal breeding success greatly exceeded variance in lifetime success (fig. 14.4*a*). Brood sizes ranged from zero to five (two broods of six and one of seven were recorded in over one thousand broods). The total number of cygnets raised to four months varied from zero to forty-two in both sexes among birds with histories of five or more years. Paired birds did not successfully raise broods every year, and many pairs did not bring young for periods of several years.

Among birds with complete life histories, the mean lifetime success of males was 0.79 cygnets raised to four months (range 0–7 cygnets), variance = 3.73 (n = 28) and that of females was 0.85 cygnets raised (range 0–10 cygnets), variance 4.38 (n = 33). Since these birds died young, individuals with histories of five or more years were also investigated. Mean success of males was 5.1 cygnets raised (range 0–42), variance = 33.3, and that of females was 5.4 cygnets raised (range 0–42), variance = 35.6. Variance

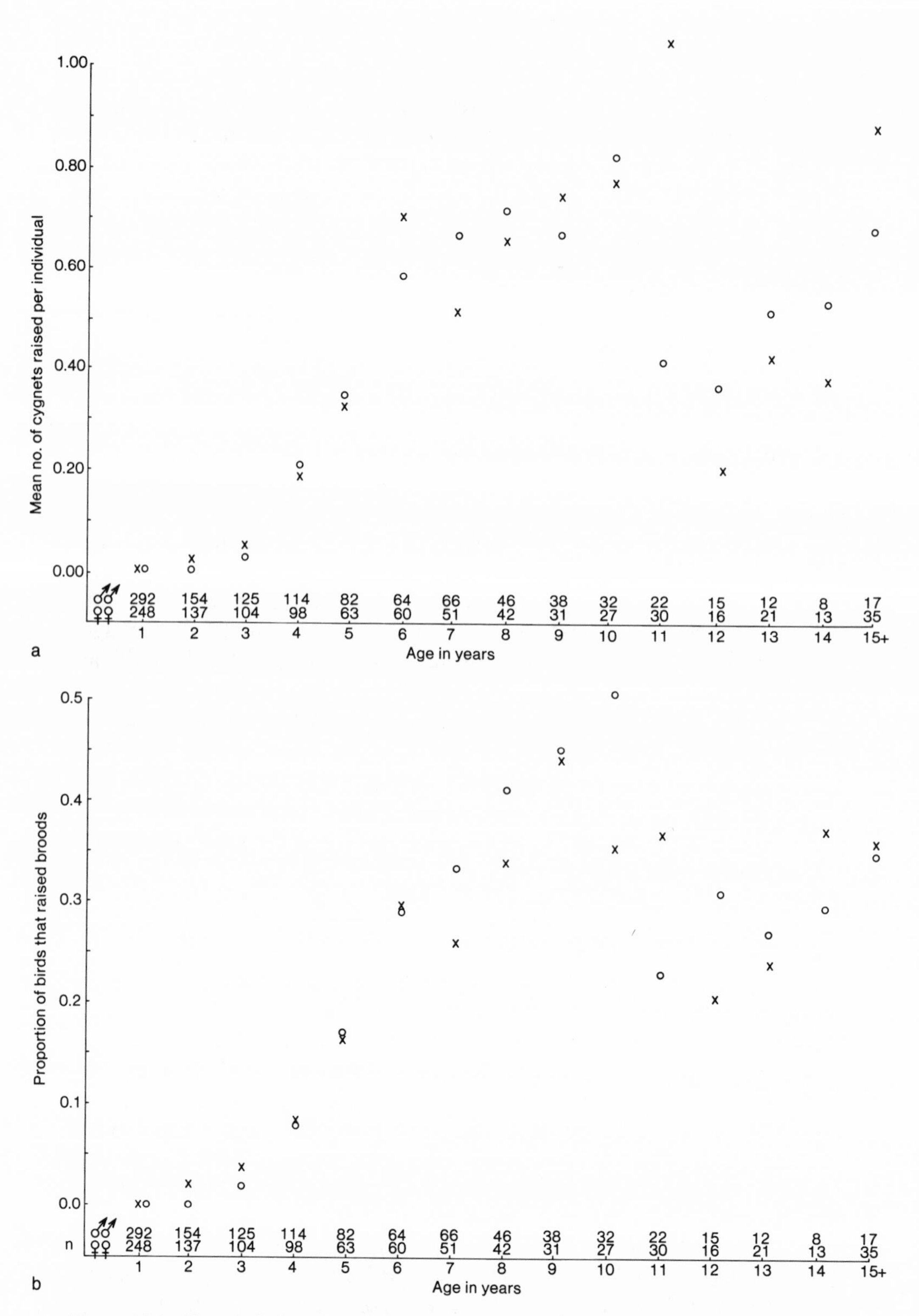

Figure 14.3: Changes in breeding success with age: (*a*) mean number of cygnets raised at each age; (*b*) proportion of pairs that raised broods; (*c*) mean brood size for pairs that bred; and (*d*) mean of deviations from average success in raising broods (= deviations from number of broods raised/number of years present for each individual). (*x's*, males; *open circles*, females.)

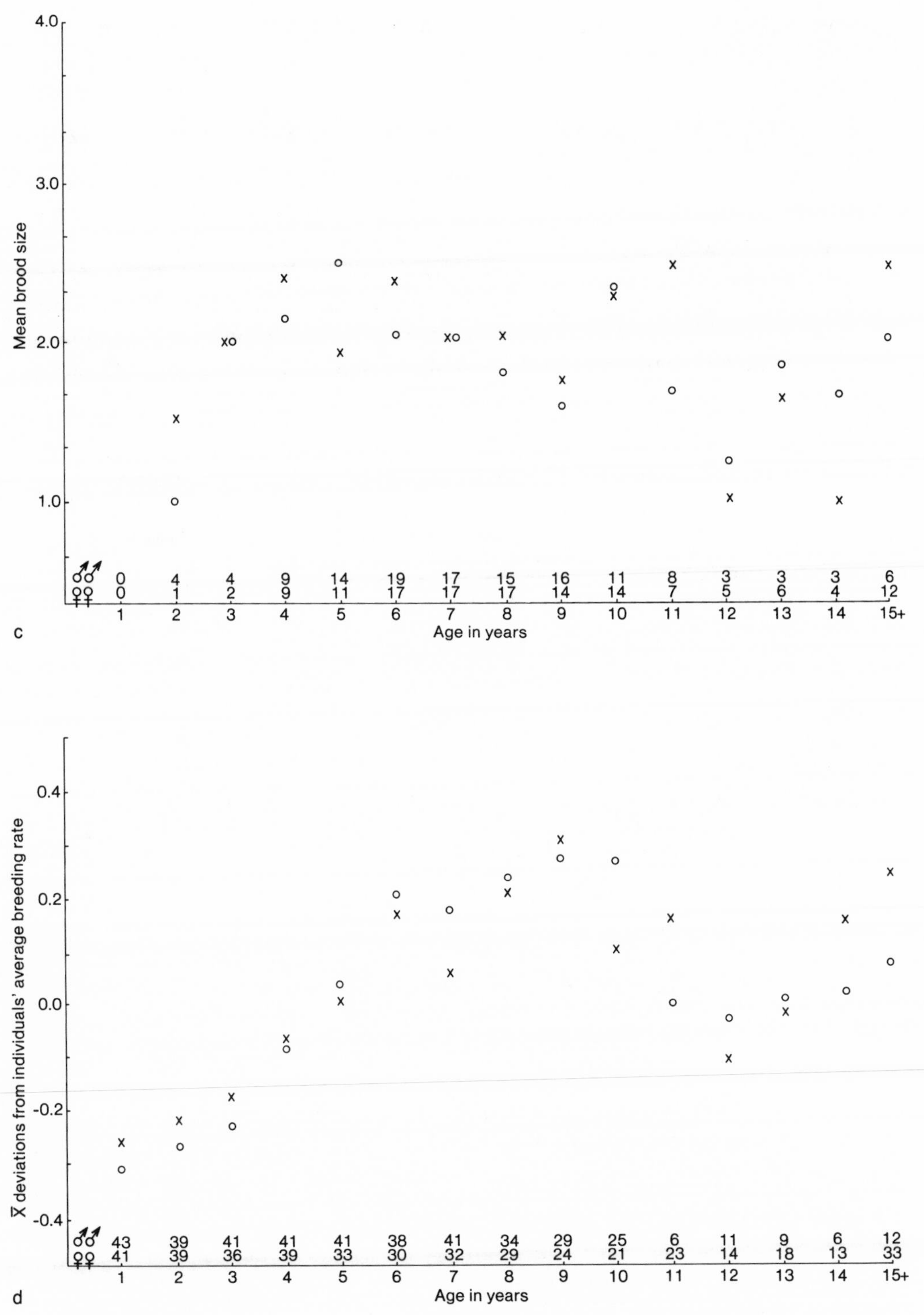

Figure 14.3 (continued)

was not significantly greater among males than among females in this sample (fig. 14.4, table 14.3).

Relative Importance of Life Span, Breeding Rate, and Brood Size

Analysis of minimum-age individuals known or assumed dead showed that breeding rate contributed more to the variance in total success than mean brood size or life span (table 14.4). Among males, life span was more important than brood size, while in females the reverse was true. This might be a consequence of higher adult mortality among females, resulting in shorter average life spans. A difference between male and female life spans and resulting male-biased adult sex ratio would also

Table 14.3 Mean Success and Variance among Male and Female Bewick's Swans with Complete Life Histories and Long Histories (Five or More Years)

	Mean Number of Cygnets Raised to Four Months		Variance in Number of Cygnets Raised		Standardized Variance		
Individuals	*Males*	*Females*	*Males*	*Females*	*Males*	*Females*	Im/If
With complete life histories[a]	0.786	0.849	3.73	4.38	6.02	6.08	.995
With long histories[b]	5.149	5.361	33.32	35.64	1.26	1.24	1.01

[a] N = 28 males, 30 females.
[b] N = 235 males, 205 females.

Table 14.4 Percentage Contribution of the Components of Long-Term Reproductive Success to Variance in LRS in Minimum-Age Male and Female Bewick's Swans

Component	L	F	BS
Males			
L	35.6		
F	8.4	38.4	
BS	14.7	1.6	24.8
N = 36 breeding males			
Proportion of breeders, 0.41			
Overall variance (OV), 19.00			
Percentage of OV due to nonbreeders, 55.60			
Percentage of OV due to breeders, 44.40			
Females			
L	24.7		
F	−29.7	72.5	
BS	6.6	9.5	24.0
N = 41 breeding females			
Proportion of breeders, 0.57			
Overall variance (OV), 22.50			
Percentage of OV due to nonbreeders, 41.70			
Percentage of OV due to breeders, 58.30			

L = life span; F = fecundity (breeding rate); BS = mean brood size.

explain the higher proportion of breeders among females than among males.

With the incorporation of mortality between four months and two years, the overall variance was similar in each sex (table 14.4). Among birds with complete life histories, most of which died young, over half the overall variance was due to nonbreeders. Among minimum-age birds, nonbreeders contributed most to variance in overall success in males, while breeders contributed most in females.

Factors Affecting Breeding Success: Presence and Life Span of Mate

The probability of breeding in any year is low for birds unpaired in the previous winter. Only 14 of 439 birds that returned with cygnets did not have a mate in the preceding winter, and on only five occasions did birds return with cygnets and a different mate from that in the previous winter.

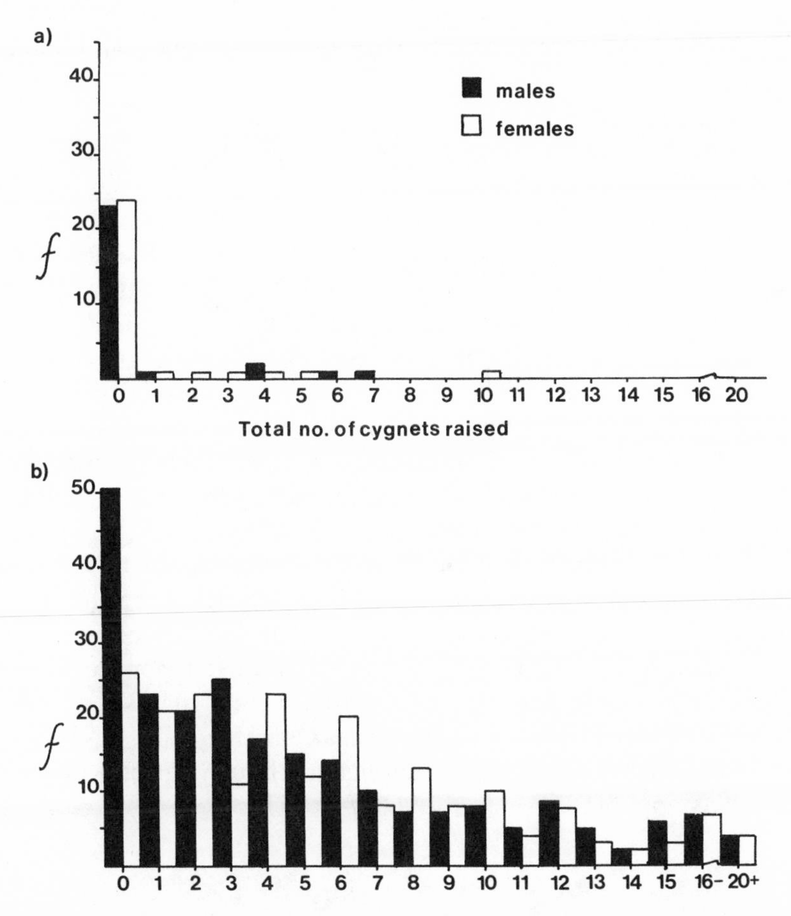

Figure 14.4: Frequency distribution of total number of cygnets raised by males and females for (*a*) birds with complete life histories and (*b*) birds with histories of five or more years as adults.

Breeding success increased with the duration of the pair-bond. The mean number of cygnets raised per year was significantly correlated with bond duration ($n = 10$, $r_s = .964$, $p < .001$ for all bonds, complete and incomplete). However, since this relationship might be a consequence of the increase in success with age, the data were reanalyzed, calculating success for each bird at each age as the deviation from the average probability of raising young and then computing the mean of these deviations for each bond. The results showed that there was still a significant positive correlation between success and duration of the pair-bond, both using success of different pair-bonds ($n = 43$, $r_s = .508$, $p < .001$) and using mean success per year for cumulative numbers of years of the same pair-bonds ($n = 8$ years, $r_s = .970$, $p < .001$). The increase was greatest between the first and the fifth year of the bond, after which no significant increase could be detected, though the samples were small.

Most birds had only one mate during their lives. Only among individuals living to age thirteen did more than 50% of birds have two or more mates. As individuals aged, the probability of losing a mate accumulated. From age three, some birds returned with second mates. Some individuals bred more successfully with their first mates and some with their second mates.

Mate loss temporarily reduced a bird's breeding rate. This was partly because some birds took more than one year to remate: although 56% of males and 63% of females remated in one year, 35% of males and 33% of females took two years and the remainder took between three and six years. In addition, individuals seldom reared cygnets in the first year with a new mate. Mean latency to successful breeding after return with a new mate was 1.9 years for ten birds aged between five and ten years, compared with 0.70 years, the mean latency to successful breeding among ten randomly chosen established pairs of the same ages ($n_1 = 10$, $n_2 = 10$, $U = 17$, $p < .05$, Mann-Whitney U test). Males losing their mates from age nine were slower to remate than those losing their mates before age nine ($\chi^2 = 5.404$, d.f. $= 1$, $p < .02$). The mean latency to breeding with a new mate after age nine was 1.23 years.

Among birds living to minimum-age ten, individuals whose first mates survived throughout their lives produced more cygnets per year than those with more than one mate (table 14.5). Analysis of variance showed significant differences in number of cygnets raised per year between individual females who had one, two, three, or four mates, and the sums of squares were partitioned to show a linear effect ($F_{1,99} = 8.95$, $p < .01$). Among males the effect was not significant, though a trend in the same direction was apparent ($F_{1,104} = 2.966$, $p < .1$).

Male and Female Characteristics and the Breeding Success of Pairs

Within pairs aged five to ten years, the best predictor of total success was male size (skull length) (table 14.6). Male weight, the dominance of the pair, and the size and weight of the female were also significantly but less strongly associated. These same characteristics were also related to the number of broods raised and mean brood size (table 14.6). In each case, male characteristics were more closely associated with success than were female characteristics.

Since male and female size and weight were all intercorrelated (table 14.7), it was difficult to assess the effect of each variable separately. However, multiple-regression analysis showed that the only significant predictor of success was male size. Female size did not contribute

Table 14.5 Reproductive Success of Bewick's Swans with Different Numbers of Mates (Individuals Surviving to Minimum Age of Ten Years)

Mean Number of Cygnets Raised per Year ± s.e.	Number of Mates			
	1	*2*	*3*	*4*
Males				
Mean	0.818	0.610	0.680	0.508
s.e.	0.105	0.079	0.318	0.215
n	50	44	6	6
Females				
Mean	0.918	0.631	0.550	0.328
s.e.	0.092	0.074	0.133	0.140
n	56	32	10	5

Table 14.6 Correlation of Components of Reproductive Success in Bewick's Swan's with Pair Dominance and Measures of Body Size

	Pair Dominance	Weight		Skull Length		Wing Length	
		Male	*Female*	*Male*	*Female*	*Male*	*Female*
Total success	0.335*	0.466***	0.327	0.527***	0.376*	0.114	0.189
Number of broods raised	0.328*	0.473***	0.297	0.572**	0.316	0.088	0.186
Mean brood size	0.396**	0.481***	0.410**	0.456*	0.449*	0.198	0.335
n	45	45	42	26	27	32	31

*$p < .05$; **$p < .01$; ***$p < .001$.

Table 14.7 Correlations between Pair Dominance and Measures of Body Size of Male and Female Bewick's Swans

Measure	Pair Dominance	Male Weight	Male Skull Length	Female Weight	Female Skull Length
Male weight	0.434**	1.000			
n	39				
Male skull length	0.546**	0.609***	1.000		
n	25	26			
Female weight	0.243	0.403*	0.188	1.000	
n	35	35	22		
Female skull length	0.109	0.439*	0.490*	0.351	1.000
n	26	24	20	27	

*$p < .05$; **$p < .01$; ***$p < .001$.

significantly to the variance in total success, number of broods raised, or brood size when the effect of male size was held constant. Nor did male or female weight contribute significantly to the variance in these measures of success when the effect of male size was taken into account. Dominance was closely related to male size but did not explain a significant proportion of the variance in success when male size was held constant.

Although it is possible that characteristics of females as yet unmeasured will prove as important as male size in determining success, one prediction can be made: If male characteristics are more important than female characteristics, females that remate should show a greater change in breeding rate than males that remate. Among eighty-six breeding males and sixty-four females, the mean change in number of cygnets raised per year between partners was 0.676 for males and 0.918 for females (χ^2 = 3.28, d.f. = 1, $p < .05$, median test, one-tailed p value).

14.3 Discussion

Bewick's swans are typical large monogamous birds with high adult survival and delayed age at first breeding (Lack 1968). Though no significant sex difference in mortality was detectable in the analysis of survival, several findings are consistent with a slightly higher mortality among adult females than among adult males. First, the sex ratio among unpaired adults caught at Slimbridge was significantly male biased (Evans 1979). Second, presumably as a result of the preponderance of single males, most of the variance in success among minimum-age birds was due to nonbreeders among males but to breeders among females. Third, among minimum-age birds, females did appear to suffer higher mortality than males in nine of eleven years.

As in other Anserinae, breeding success increased with age during the first part of the life span (snow geese: Rockwell, Findlay, and Cooke 1983). The results showed that after the initial significant increase, there was a peak followed by a nonsignificant decline and apparent rise toward the end of the life span. While a peak and decline in performance with age have been found in other species (Davis 1975; Brooke 1978), the apparent rise in proportion of birds raising broods from age fourteen is surprising. One possibility is that this is merely a result of the small sample sizes involved. This seems unlikely, however, since a similar pattern with peak, trough, and terminal rise was found in a separate and larger data set comprising minimum-age birds (Scott, unpubl. MS). Another possibility, that the entire pattern of success throughout the second part of the life span results from patterns of mate loss and remating, was discounted by analyzing changes in breeding rate with age among birds with only one mate (and taking only first mates for birds with more than one mate). There was a similar gradual increase in breeding rate up to age nine and a subsequent decline in both sexes, with the same terminal rise.

An individual's breeding success gradually increased with the duration of its pair-bond, the increase being greatest between the first and the fifth year of the bond. Most individuals had only one mate during their lives, but a proportion of birds lost their first mates and subsequently remated. Loss of mate caused a temporary reduction in breeding rate, since some birds failed to remate within one year and there was also a greater latency to breeding with a new mate than with a familiar mate. Individuals with more than one mate tended to be less successful than those whose mates survived throughout their lives. While it is possible that this is because poor-quality birds lost their mates sooner than good birds, there was no evidence that birds that lost their mates were smaller or lighter than those that did not. It seems likely that the difference is due both to the temporary reduction in success associated with mate loss and remating and to the gradual increase in success with duration of the pair-bond. These results may explain why divorce is extremely uncommon among Bewick's swans even though pairs frequently fail to rear young successfully. A substantial reduction in breeding success in the year after a change of mate has been shown in barnacle geese (Owen, Black, and Liber, n.d.).

Male characteristics, particularly size, predicted the long-term breeding success of pairs more accurately than female characteristics. Observations have shown that dominance relationships between pairs are established by disputes involving gradually escalating displays that culminate in physical fights. The males finally grab each other by the base of the neck with their bills and beat each other with their wings. In such aggressive encounters, it is fights between males that determine dominance relationships between pairs and therefore access to resources (Scott 1980). Occasionally the female partners fight each other while the males are fighting, but it is only the outcomes of fights between males that determine the direction of subsequent relationships between pairs. In winter, dominant pairs displace subordinate pairs at feeding sites and spend a higher proportion of time feeding. Male fighting ability is therefore probably crucial in determining the pair's access both to food and to breeding territories.

14.4 Summary

Bewick's swans are large monogamous birds with high adult survival and delayed age at first breeding. Success in raising broods increased up to nine years of age. Variance in long-term success did not differ significantly between the sexes. Breeding success increased with the duration of the pair-bond. Although most birds had only one mate during their lives, a proportion lost their first mates and subsequently remated. Loss of mate caused a temporary reduction in breeding rate. Individuals with more than one mate were less successful than those whose mates survived throughout their lives.

The long-term success of pairs appeared to be more strongly influenced by male characteristics, particularly size, than by female characteristics. Since dominance relationships, and hence access to food in winter and, presumably, breeding territories in summer, are determined by male and not female fighting ability, this may explain the importance of male size in determining the success of pairs.

14.5 Acknowledgments

I would like to thank everyone who has helped to collect data on individually identified Bewick's swans at Slimbridge, Welney, and elsewhere at all stages since the study began. I am particularly grateful to Eileen Rees for all her help and to Steve Albon for assistance with analysis of the data. This research would not have been possible without the exceptional support of the Wildfowl Trust and its staff or the original foresight of Sir Peter Scott in 1964. I thank the Conder Trust and World Wildlife Fund (USA) for funds. I am very grateful to Steve Albon, Eileen Rees, the referees, and the editor for comments on the manuscript.

15 Reproductive Success in a Lesser Snow Goose Population

F. Cooke and R. F. Rockwell

In this chapter we document the components of fitness within breeding seasons in the lesser snow goose (*Anser caerulescens caerulescens*) and provide mean values and variances associated with them in order to assess the opportunity for selection. We discuss the temporal variability of this opportunity for selection and attempt to assess the proximate causes of the variation at each stage.

The data span seventeen years of continuous study of the snow goose population that breeds at La Pérouse Bay in northern Manitoba, Canada, and the colony comprises two thousand to seven thousand breeding pairs (see fig. 15.1). Two thousand to three thousand nests are found and

Figure 15.1: Mixed pair of lesser snow geese: blue male with banded white female.

monitored each year. Currently this represents 30%–50% of all the nests on the colony, and the large sample sizes allow precision of the various estimates. Four thousand to eight thousand geese, both adults and goslings, are individually marked each year, but because of the size of the population, we are rarely able to monitor the breeding success of the same individuals in several successive seasons.

15.1 Field Methods

We arrive in the field before the geese, and as they arrive and begin to nest we find as many nests as possible on the day when the first egg is laid. This allows us to discover exactly how many eggs are laid in the nest. Four hundred to eight hundred nests of this type are found each year. They provide the most complete information of fecundity. When the laying period is over and the geese are incubating (nesting in snow geese is highly synchronized, and second clutches are unknown), a further fifteen hundred to two thousand nests are found. These augment the information about later aspects of reproductive performance.

During the hatching period, we visit each nest daily and record information on hatching goslings (weight, color phase, sex), identity of parents, and causes of egg and gosling mortality. Each gosling is given a numbered Monel Metal web tag that identifies it uniquely. The goslings remain in the nest for approximately twenty-four hours and then leave with their parents for the feeding areas—salt marshes peripheral to the nesting colony. Approximately five weeks later 20%–40% of the flightless adults and their almost fledged offspring are rounded up and ringed, with Canadian Wildlife Service bands and individually identifiable plastic leg bands, and then are weighed, measured, sexed, and aged (gosling, yearling, or adult). Birds ringed in previous years are recaptured and form the major basis for our calculations of survival. Almost all the birds caught at this time are successful breeding adults and their young. Most nonbreeders and failed breeders leave the area during the hatch period and fly northward to other molting areas. The family groups begin their fall migration around mid-September, reaching the Gulf Coast wintering grounds of Texas and Louisiana in December. They are heavily hunted during fall migration and in the winter. The return migration begins in late February. Nutrients for nesting and egg production are acquired mainly in the northern prairies in April, and birds arrive on the nesting grounds in May or early June.

The snow goose is monogamous, though extrapair copulations occasionally occur. The breeding pair stay together throughout the year, and goslings remain with their parents through the first year of life. First pairing of geese occurs generally in the second winter or early spring when the birds are a year and a half old. Most two-year-old females that return to La Pérouse Bay are paired, but only a fraction of them breed at this age, and some delay until four years of age. They remain paired during the life of the pair.

15.2 Components of Fitness

Figure 15.2 is our summary of the measurable components of fitness that delineate the life cycle of a snow goose and our estimates of the values of the fitness components and of the transition probabilities between them.

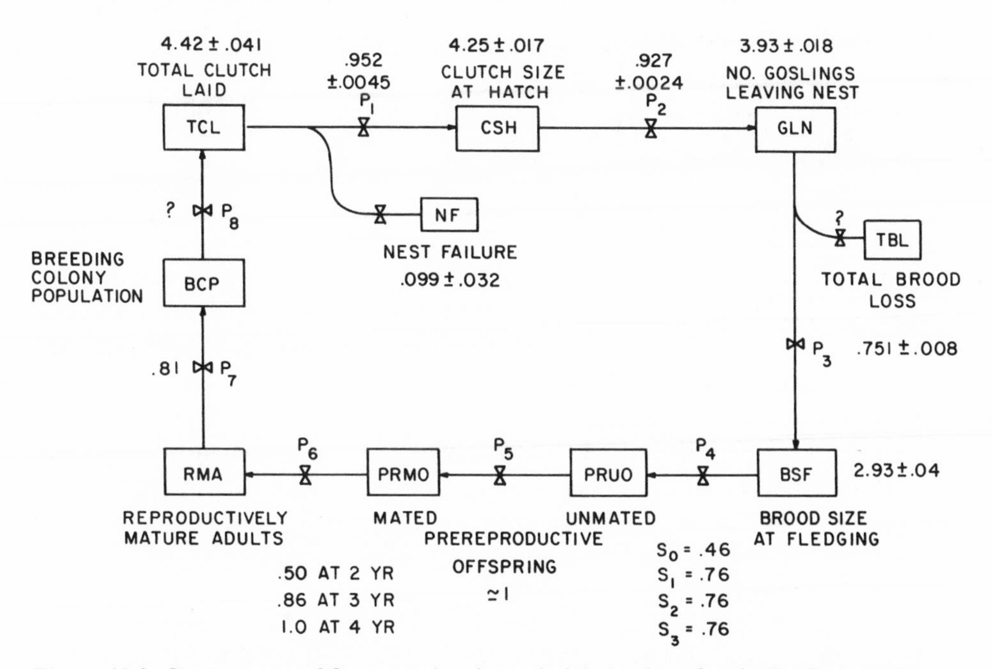

Figure 15.2: Components of fitness and estimated global values for the La Pérouse Bay lesser snow goose population:

Components of fitness: **TCL** (total clutch laid), the total number of eggs laid in a nest. Includes only nests found at the one egg stage that reach the beginning of incubation. Excludes clutches of more than seven eggs, since they were presumably laid by more than one female. **CSH** (clutch size at hatch), the number of eggs present in a given nest at hatch. Nests are included if at least one egg hatched. Nest failure is a nest in which CSH = 0. Only nests that reached the beginning of incubation were used for an assessment of the frequency of nest failure. **GLN** (goslings leaving nest), the total number of goslings leaving a nest. Nests are included only if CSH > 0. **BSF** (brood size at fledging), the number of goslings from a nest that survive to the time of banding. This is assessed by noting the composition of families whose goslings were web tagged at hatch. **TBL** (total brood loss), those families where at least one gosling left the nest but none reached fledging. **PRUO** (prereproductive unmated offspring), the number of fledged female goslings reaching an age where reproduction becomes possible. Since first reproduction may occur at ages two to four, the value is defined in terms of survival probabilities during the first four years of life. **PRMO**, the number of PRUO that successfully form pairs. **RMA**, the number of reproductively mature females; it incorporates the variable age of first breeding.

Transition probabilities: **P1**, egg survival (CSH/TCL); differences between TCL and CSH reflect partial clutch loss (mainly predation) during the laying and incubation periods. **P2**, hatching success (GLN/CSH); differences between CSH and GLN reflect unhatched eggs or goslings that die or disappear before the rest leave the nest. **P3**, fledging success (BSF/GLN); differences between BSF and GLN reflect partial brood loss during the posthatch period. **P4**, survival from fledging to breeding age; the values here are in the form of a vector of survival probabilities S_0, S_1, S_2, S_3, for the first four years of life. **P5**, Probability of acquiring a mate, given that the bird has survived to pairing age. **P6**, Probability of first breeding at ages 2, 3, or 4. **P7**, Annual adult survival rate. **P8**, Breeding propensity, a measure of opting out of a formerly breeding pair.

Those components traditionally associated with fecundity are defined in terms of the average number of offspring produced by a breeding pair of geese. Viability measures are defined as the populations of females that reach the various stages from among those that fledge successfully. The number of reproductively mature females for a particular cohort in a particular breeding season will depend on survival in the prereproductive years, acquisition of a mate, and age of first breeding. When one has calculated the number of reproductively mature females for a particular cohort, the breeding colony population (BCP) over all cohorts can be found by estimating an annual adult survival rate.

15.3 Fitness Values: Fecundity

We initially ignore known differences among seasons, ages, and color phases for these values, simply to provide a reference point against which specific deviations can be assessed. Pooling data from years 1973 to 1981 and using only data from banded geese, we have calculated means and variances for total clutch laid (TCL), clutch size at hatch (CSH), goslings leaving nest (GLN), brood size at fledging (BSF), and transition probabilities P1 to P3. (See legend to fig. 15.2 for further explanation of variables.) These are presented in table 15.1 The mean fecundity values are defined in terms of total number of offspring produced. These values refer only to successful nests.

Total nest failure (NF) owing to predation or nest abandonment is another important component of fitness. Table 15.2 shows the frequency of nest failures from 1973 to 1983. Nest failure during incubation varies from 1% in 1973 to 18% in 1982 and averages about 10%.

Total brood loss (TBL) is one of the few measures of fecundity that is difficult to carry out. Only a sample of the families that hatch are caught in the banding drives. Therefore absence of a brood in the sample does not necessarily mean that brood loss has occurred.

Despite an inability to calculate brood loss, it is possible, using the values in tables 15.1 and 15.2, to generate a value for the mean number of fledged goslings per nesting pair. This value is TCL $\times$ P1 (1 $-$ NF) $\times$ P2 $\times$ P3 (1 $-$ TBL). We can generate a series of values by postulating a series of hypothetical TBL values. If total brood loss is negligible, the expected number of fledged goslings per nest is 2.64. With a 20% TBL, the number would be 2.11. In all likelihood the mean of fledged goslings per nest lies between these values.

There are several sources of variance in these fecundity measures, both intrinsic and extrinsic. Rockwell, Findlay, and Cooke (1983) and Davies and Cooke (1983) have both shown significant seasonal variation in clutch size. In delayed nesting seasons clutch sizes are low, but in the two extremely early seasons 1977 and 1980 values were also low. In these years Davies and Cooke attributed the low values to drought conditions on the Dakota and Manitoba prairies, where the female geese acquire nutrients for the nesting period.

Table 15.1 Fecundity Measures and Transition Probabilities for Lesser Snow Geese at La Pérouse Bay, 1973–81

	TCL	P1	CSH	P2	GLN	P3	BSF
Mean	4.424	0.9518	4.248	0.9272	3.928	0.7509	2.932
s.e.	0.041	0.0045	0.017	0.0024	0.018	0.0079	0.040
n	726	684	4126	3933	3979	818	821

TCL = total clutch laid; CSH = clutch size at hatch; GLN = goslings leaving nest; BSF = brood size at fledging.

Table 15.2 Nest Failure Rate for Lesser Snow Geese at La Pérouse Bay, 1973–83

Year	Failure	Total Nests	Failure Rate
1973	2	164	.012
1974	57	342	.167
1975	16	298	.054
1976	11	204	.054
1977	6	127	.047
1978	13	221	.059
1979	3	21	.143
1980	27	297	.091
1981	35	322	.109
1982	70	377	.186
1983	17	222	.077
Total (± s.e.)	257	2,595	.099 ± .0165

Clutch size also varies with age (Finney and Cooke 1978; Rockwell, Findlay, and Cooke 1983). Two-, three-, and even four-year-old birds have lower clutch sizes than older birds. Average values, pooled over the years 1974–80 for two-year-old, three-year-old, four-year-old, and older birds, are 3.43, 3.87, 4.09, and 4.31, respectively. There is no evidence of clutch size decline among old birds, such as that reported by Newton (this volume, chap. 13). Second-time breeders have a lower clutch size than those that have bred at least three times (Cooke, Bousfield, and Sadura 1981).

Findlay, Rockwell, and Cooke (1985) found no evidence for a cohort effect on clutch size. This suggests that environmental factors operating during the fledging period do not have a lasting influence on this component of fitness.

Intraseasonally there is a positive correlation between clutch size and laying date. In common with many bird species, birds that lay earlier in the season have larger clutches. This is true not only of the population as a whole but also within an age-class. In general, birds laying early in the season produce more recruits into the breeding population (Cooke, Findlay, and Rockwell 1984), but those laying in the middle part of the season fledge more young (Findlay and Cooke 1982b).

There is a genetic component to both clutch-size and hatch-date variance (Findlay and Cooke 1982a, 1983, 1987). This is reflected in both repeatability (comparison within birds among seasons) and heritability (comparison of mothers and their daughters) measurements. This shows that some of the variance in clutch size and laying date reflects underlying genetic variance among birds and indicates that there is indeed opportunity for natural selection to occur in these characters.

From the beginning of incubation until fledging, the number of offspring per pair drops on average from 4.42 to 2.11–2.64, depending upon how severe we assume total brood loss is. We can compare the loss of fecundity due to P1, P2, P3, nest failure, and total brood loss and identify the causes of this loss. It is sometimes difficult to be sure whether nest failure is due to predation or to abandonment followed by scavenging of the eggs. In years when severe weather conditions occur just before hatch, females in poor body condition may abandon their clutches or in extreme cases die on the nest. Predation by caribou, gulls, jaegers, cranes, or bears is more common than abandonment and varies with the frequency of the predators.

In general, partial clutch loss is of minor importance and may be partly an artifact of human disturbance combined with predation. Partial clutch loss accounts in an average season for fewer lost eggs than does nest failure. From our figures, nest failure accounts for an average loss of 438 eggs from a colony of one thousand nests, whereas partial clutch loss results in a loss of 192 eggs.

The loss of fecundity during the hatching period is also minor. It is mainly a result of eggs that fail to hatch and goslings that die before leaving the nest. Brood loss is more important, with partial brood loss leading to a 25% reduction in brood size. Young goslings are vulnerable to predators such as herring gulls and ravens. Starvation may contribute to this loss, since families spend many hours each day grazing on the brood-rearing salt marshes. Disease may also be important (D. Rainnie, pers. comm.).

As mentioned above, we have little information on the cause of total brood loss. Insofar as it seems to occur mainly among early-hatching nests, it may be due primarily to the relative vulnerability of the early-hatching goslings to avian predators in the brood-rearing areas. As more goslings hatch, predator swamping occurs, and the goslings are less vulnerable. Observations show that goslings in the early-hatching nests are often harassed by groups of herring gulls, and in the course of the disruption the total brood is lost.

15.4 Fitness Values: Viability

All survival estimates are calculated from the banding period (just before fledging), not from hatching date. In calculating survival we used only birds that were ringed as goslings and are therefore of known age. This allowed us to calculate age-specific phenomena. Our calculations are based on birds recaptured at La Pérouse Bay in years subsequent to their original capture as goslings (recaptures) or on birds that are recovered dead (recoveries). Recaptured birds can be assumed to be successful nesters. Recoveries are of known age but may be breeders or nonbreeders.

Our analysis of recaptures is restricted to females, since they show a high degree of natal and breeding philopatry, whereas males seldom return to their natal colony (Cooke, MacInnes, and Prevett 1975; Rockwell and Cooke 1977). In the following calculations we have assumed that by the age of four all surviving snow geese have entered the breeding population. This is consistent with the age distribution of the colony. We will regard the age of four as a convenient point at which to separate prereproductive and adult components of viability.

Prereproductive Viability

For a female to enter the pool of breeding birds, she must survive to reproductive age, pair, and establish a nest in the colony. These three aspects of prereproductive selection are classified in figure 15.2 as P4, P5, and P6.

P4

P4 is a vector of survival probabilities in the first four years of life, S_0, S_1, S_2, and S_3. These values can be calculated directly using standard life-history ratio methods. Despite recent criticisms of the method (Lakhani 1985), the values obtained are very close to, though slightly higher than, those obtained by our more indirect methods based on recaptured birds. In our indirect method we calculate the fraction of females that returned and bred successfully as four-year-olds (l_4). This is possible because we know the number of females banded (B) in a particular cohort, the fraction of the females in the colony that are banded in the recapture year (M/N) and the number of females of the particular cohort recaptured (R); l_4 is simply RN/MB. Any l_x value can be calculated in this way. In

Table 15.3 Survival Estimates (l_4) for 1969–74 Cohorts of Lesser Snow Geese at La Pérouse Bay

Cohort	Recapture Year	Number of Females of Cohort Banded (B)	Colony Size in Recapture Year (N)	Total Adult Females Captured in Recapture Year (M)	Number of Cohort Birds Recaptured in Recapture Year (R)	Estimated Number of Females from the Cohort on Colony in Recapture Year (L)	l_4
1969	1973	285	2,738	802	17	58.04	.2036
1970	1974	1,409	2,800	708	61	241.24	.1712
1971	1975	440	3,338	631	25	132.25	.3005
1972	1976	1,183	3,830	798	49	235.18	.1988
1973	1977	1,281	3,704	1,008	66	242.52	.1893
1974	1978	802	5,785	774	16	119.58	.1491
Mean							.2021 ± 0.021

Table 15.4 Estimated Annual Survival Rates (S_0–S_3) for 1969–74 Cohorts of Lesser Snow Geese at La Pérouse Bay

Cohort	l_4	S_0	S_1, S_2, S_3
1969	.2036	.4580	.7633
1970	.1712	.4386	.7310
1971	.3005	.5048	.8413
1972	.1988	.4553	.7588
1973	.1893	.4497	.7495
1974	.1491	.4237	.7061
Mean (± s.e.)	.2021 ± .021	.4571	.7619

table 15.3, l_4 values are calculated for six cohorts and show that 15%–30%, with an average of 20.2% of all the fledged females from those cohorts, are successfully breeding in the colony at four years of age. Since we know from Boyd, Smith, and Cooch (1982) that first-year survival of snow geese is 60% of that of older birds, with appropriate algebra, and assuming equal S_1, S_2, and S_3 values, we can arrive at S_0 values of from 42% to 50% and S_1–S_3 values of from 70% to 84% for the six cohorts examined, with averages of 46% and 76% respectively (table 15.4). These values are slight underestimates, since female emigration would be included as mortality and does occur rarely at La Pérouse Bay (Geramita and Cooke 1982). This can account for the slightly higher values obtained by the ratio method (details of which are in preparation by Richards and Cooke). Thus prereproductive mortality particularly in the first year contributes in a major way to variance in reproductive success.

P5

We assume that most, if not all, suitably aged snow geese acquire a mate in this monogamous species. We have no way of assessing this, since pair formation occurs away from the breeding area and unmated birds would not necessarily return to the breeding colony. The sex ratio is approximately 1:1 at the fledging stage, and there is no evidence of overall differential survival of the two sexes, although males have lower S_0 values (unpublished data). Thus we would not expect a large number of birds to forgo breeding because they could not find a mate.

P6

The probability that a bird that has survived and acquired a mate breeds for the first time in years 2, 3, or 4 can be calculated using recapture data in a way similar to that described for P4. We know the number of successful breeders of each class (b_x). However, we can also estimate the number alive at each age (1_x) from the P4 and P7 values. From this we can calculate the proportion of those alive of each age class that did breed $b_x/1_x$, as shown in figure 15.3. Using data for 1974 through 1981, which have many known-age birds, the expected numbers alive can be obtained by first calculating the surviving number, assuming average survival values, from P4 and P7 and multiplying these by the number of females in the original banded cohort. These numbers are then corrected to allow for sampling intensity (M/N). We should point out that although our method samples only successful breeders, because of this correction factor it estimates the number of birds that attempted to breed. The assumption is, however, that each segment of the population has an equal probability of being successful. If, as seems likely, younger birds are less successful than older ones, our method underestimates the proportion of birds nesting for the first time as two- and three-year-olds.

Table 15.5 shows the proportion of two- or three-year-olds that breed relative to the frequency of birds four years old or older. The frequency of

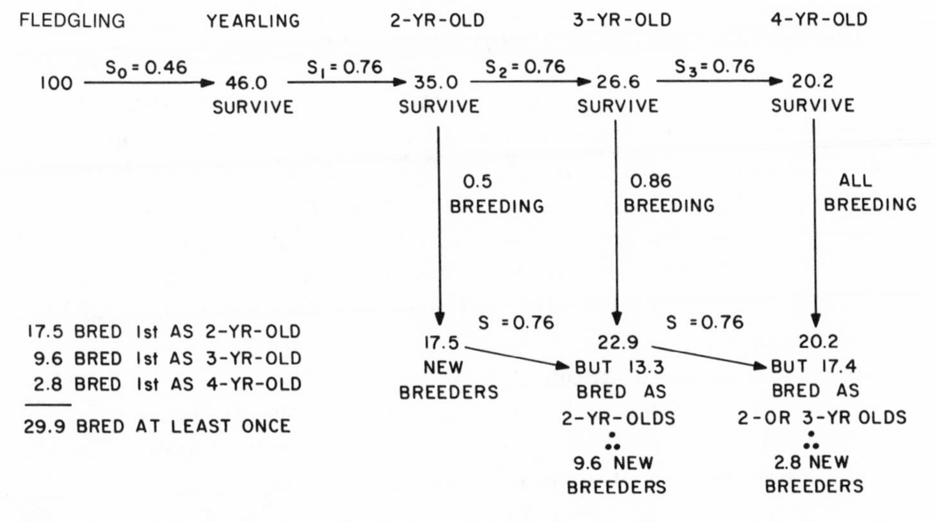

Figure 15.3: Method of calculating the breeding frequency of two- and three-year-old birds (see text for details).

Table 15.5 Estimated Percentage of Two- and Three-Year-Old Lesser Snow Geese That Bred, La Pérouse Bay

	Percentage of Age-Class Breeding	
Year	*Two-Year-Olds*	*Three-Year-Olds*
1974	48	81
1975	64	61
1976	33	117
1977	28	62
1978	64	82
1979	80	89
1980	28	91
1981	43	70
Mean (± s.e.)	50 ± 6.8	86 ± 6.5

Note: See text for details.

two-year-olds breeding varies from 28% to 80%, with a mean of 50%. The frequency of three-year-olds breeding varies from 61% to 100%, with a mean of 86%.

Despite the crudeness of the calculations necessary to estimate the P6 values, it seems clear that there is considerable variation in the proportion of two-year-old breeders in different years. A proximate explanation for this has been proposed (Davies and Cooke 1983). In 1977 and 1980, years of prairie drought when conditions before nesting were particularly severe and the total number of breeding birds was lower than expected, the relative proportion of breeding two-year-olds was low. This is thought to be due to the poor competitive ability of young birds when food is in short supply. An inadequate nutrient supply at this time could lead to nonbreeding (Ankney and MacInnes 1978).

Recruitment into the Breeding Populations

From the P4 and P6 values generated above, we can estimate the proportion of fledged female goslings that enter the breeding population. Birds enter either as two-, three-, or four-year-olds, and we know the probability of surviving to these ages and the probability of breeding, assuming survival (see fig. 15.4). We assume that the probability of survival is the same for breeding and nonbreeding birds, but we have no data on this point. From the figure it appears that 29.9% of fledged goslings eventually nest (not necessarily successfully) at La Pérouse Bay.

From this calculation, one can see that most fledged goslings never breed, and in common with most species, a large proportion of the variance in fitness arises from variance in prereproductive survival. We can also conclude that, of those birds that do eventually breed, 17.5/29.9, or 59%, do so for the first time as two-year-olds, 32% as three-year-olds, and 9% as four-year-olds.

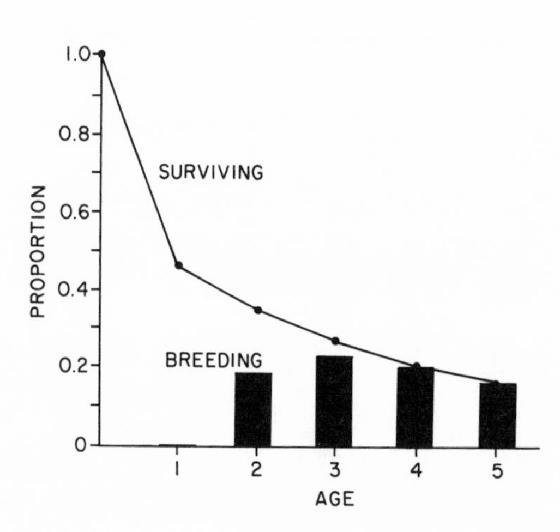

Figure 15.4: Predicted fate of one-hundred fledged female goslings during first four years of life.

Table 15.6 Estimate of Adult Survival Probabilities of Female Lesser Snow Geese at La Pérouse Bay

	Census Period							
Method	*1976*	*1977*	*1978*	*1979*	*1980*	*1981*	Mean	s.e.
Tanner	.84	.76	.93	.74	.73	.84	.81	.027
Regression	.83	.80	.88	.73	.75	.80	.80	.027

Adult Viability

P7

Methods of estimating adult (age > 4 years) viability based on recaptures of birds ringed as goslings are described in detail in Rockwell et al. (1985). We have calculated annual adult survival for the years 1976–81 by two methods. The Tanner estimate is essentially a relative age-specific survivorship value pooled over all ages from four upwards (Tanner 1978). The regression estimate is derived by expressing log l_x as a linear function of x and solving for the slope (b) with a standard least-squares approximation. These estimates are documented in table 15.6. They range from 73% to 93% and average 81%. Values are close to those obtained for all adult birds (i.e., nonyearlings) calculated by Boyd, Smith and Cooch (1982) using band recovery data and those obtained by the ratio method (unpublished data). The method, based on capture/recapture, scores any permanent emigration as death, which suggests that permanent emigration is rare among this older segment of geese. Our best estimate of annual adult survival is therefore 81%.

From band recovery data and estimates of reporting rates, it is clear that hunting by man is a major source of mortality in both prereproductive and adult lesser snow geese, and since this is a relatively recent cause of death, the selective pressures acting on the population may be very different now from what they were before Europeans appeared on the North American continent. Starvation and disease may also contribute to mortality, particularly in prereproductive birds, but their relative importance is not well understood.

P8

P8, breeding propensity, is the tendency for birds that have bred at least once to fail to breed in a subsequent season. If birds fail to breed while still alive and having bred in a previous season, we would be unlikely to detect such an occurrence. We feel that these cases are rare at La Pérouse Bay, but we cannot put a value on P8.

Longevity

Some of the birds ringed as adults in our first major banding season (1970) are still alive and breeding in the colony. They are at least eighteen years of age. Clapp, Klimkiewicz, and Kennard (1982) record a twenty-eight-year-old snow goose. We have some indications of a higher annual mortality among older birds, but this has yet to be analyzed in detail. Assuming the P4 and P7 values reported earlier and no change in P7 with old age, approximately 10% of fledged goslings should reach the age of seven years; 1% should reach the age of eighteen years. Assuming a stable colony, roughly 10% of the breeding birds in the colony would be expected to be more than twelve years old.

Breeding Colony Population

Breeding colony population can be estimated by two methods. The first is by direct count supplemented by estimates from transects. In the early years this method was relatively accurate, but as the colony expanded more areas had to be estimated, and our values appear to have been too low. The second approach uses the Jolly-Seber (J-S) method (Seber 1982) and depends on capture/recapture techniques. This method gives good estimates except in 1977 and 1980, when large numbers of birds did not breed. In this situation, J-S values are too high. Table 15.7 shows our two sets of values for the breeding population. The visual estimates are made during the incubation period. The Jolly-Seber estimates are at the end of the fledging period. Both methods show that the colony has expanded during the study period. Some of the variation associated with the numbers can be accounted for by temporary emigration (nonbreeding of adults) in 1977 and 1980 or by temporary immigration in 1978, as documented by Geramita and Cooke (1982). In 1978, breeding in the more northerly colonies was severely restricted. From table 15.7 one can see that the colony has grown almost threefold in twelve years. This is consistent with an annual growth rate of approximately 9%.

Table 15.7 Estimates of Number of Breeding Female Lesser Snow Geese at La Pérouse Bay, 1970–82

	Colony Counts (Number of Breeding Females)	
Year	*Visual Estimate*	*Jolly-Seber Estimate* (± *s.e.*)[a]
1970	2,500	2,290 ± 811
1971	1,500	2,951 ± 410
1972	2,500	2,672 ± 215
1973	3,000	2,738 ± 185
1974	3,000	2,800 ± 182
1975	3,000	3,338 ± 239
1976	3,300	3,830 ± 259
1977	2,500	3,704 ± 225
1978	4,500	5,785 ± 440
1979	4,400	5,570 ± 296
1980	3,500	4,356 ± 182
1981	5,200	6,987 ± 394
1982	7,200	6,533 ± 424

[a]Courtesy of R. F. Healey.

The various fecundity and viability values described in this chapter can be used to calculate an annual growth rate for the colony, given a stable age distribution. Because of the uncertainties involved in estimates of total brood loss and breeding propensity, it is not possible to give precise values, but an annual growth rate of between 5% and 15% is predicted. Thus it appears that the fecundity values reported here more than compensate for the mortality and are consistent with the observed colony growth of 9%. It is gratifying that observed and predicted colony growth rates are similar.

15.5 Discussion

In snow geese there is more opportunity for selection for viability than for fecundity. Fewer than 30% of all fledged goslings enter the breeding population, and most of these provide offspring for the next generation. Most birds that enter the breeding population breed successfully, fewer than 10% failing completely during the incubation period and an unknown though small proportion failing during the fledging period. Nest-failure rate varies considerably from season to season, probably owing to variation in the abundance of predators. A recent increase in nest failure coincides with an increase in the number of caribou, a major nest predator. Partial clutch loss and loss during hatch owing to unhatched eggs or abandoned goslings are minor and contribute little to the variance in reproductive success. Gosling loss is more important, with partial brood loss reducing brood size by 25%. Large broods lose relatively more goslings than small ones, but nevertheless those families with the largest clutches in general produce the most fledged goslings (Rockwell, Findlay, and Cooke, n.d.). Most variance in fecundity is associated with variation in clutch size, which varies from 1 to 7 eggs, with a mean of 4.42. Most of this variance is associated with females' ability to acquire sufficient nutrients while in the spring staging areas in the Dakotas in southern Manitoba. Young birds produce fewer eggs, probably because they are at a competitive disadvantage during feeding. In years of prairie drought when food is scarce, mean clutch sizes decline. There is a heritable component to clutch-size variability that would allow clutch size to respond to natural selection. Birds that lay the largest clutches fledge the most young. Variation in the number of fledged young is therefore mainly a function of females' ability to acquire nutrients before nesting.

When age and seasonal effects are eliminated, individual geese tend to lay clutches of the same size each year, so lifetime reproductive success is a function of annual clutch size and longevity. Most of the variation in lifetime fecundity is due to the length of the breeding life span. Since life span is uncorrelated with clutch size (Rockwell, Findlay, and Cooke, n.d.), annual clutch size is a good measure of the relative fecundity of a particular bird.

Age of first breeding may also be related to the ability to acquire nutrients before nesting. In good years, when food is readily available, most two-year-old birds nest, but in poor years few do so. This suggests that although some birds do not enter the breeding population until they are three or four years old, they would do so at a younger age if they were able to acquire sufficient nutrients. Food supply in the spring staging areas and the competitive ability to acquire it seem to be the key to reproductive success in snow geese.

Survival, particularly in the first year of life, is the major source of variance in reproductive success. Yet the major cause of mortality is the North American hunter, an instrument of selection for only the past hundred generations of snow geese. He (or she) may be a relatively undiscriminating selective agent in terms of phenotypic characteristics that were important during most of the bird's evolutionary history. Thus one is led to the paradox that although variance in reproductive success is greatest for components of fitness affecting survival, selection may be acting most strongly on those components affecting fecundity, particularly those that improve the efficiency with which female geese can acquire nutrients before egg laying.

15.6 Summary

1. The life cycle of the lesser snow goose can be subdivided into components of fitness whose means and variances can be measured.

2. The overall mean number of eggs laid per breeding pair is 4.42 and, for fledged goslings, 2.11–2.64.

3. Approximately 30% of the fledged female goslings eventually enter the breeding population.

4. Approximately 50% of two-year-olds and 86% of three-year-olds breed. Probably all birds attempt to breed by the age of four. There is considerable seasonal variation in the proportion of two-year-olds that breed, suggesting that environmental constraints limit the opportunities for breeding early in life.

5. Survival in the first four years of life is estimated as 46%, 76%, 76%, and 76%. Annual adult survival is estimated at 81%.

6. The fecundity and viability values presented in this chapter are consistent with an expanding population. Since 1968 the colony has increased at an average annual growth rate of approximately 9%.

7. Annual and lifetime fecundity is largely constrained by the ability to acquire nutrients in the spring staging areas before nesting.

16 Reproductive Success of Kittiwake Gulls, *Rissa tridactyla*

Callum S. Thomas and J. C. Coulson

THE MOST APPROPRIATE MEASURE of an individual's reproductive success is the number of offspring it produces during its lifetime that survive to breed. In practice, measures of this type are often impossible to obtain for long-lived, free-living species such as seabirds. This is particularly true in the kittiwake, because after fledging the young disperse. Although about half of those that survive return to the natal colony to breed, the rest are recruited into other colonies and are difficult to find. As a result, the best practicable estimate of reproductive success in the kittiwake is the number of chicks fledged.

For the past thirty years, detailed information has been collected on the reproductive performance of kittiwakes breeding in a colony in which the entire breeding group has been individually marked (figs. 16.1 and

Figure 16.1: The study colony. Kittiwakes nest on the window ledges of this riverside warehouse at North Shields, Tyne and Wear, England. Photo by J. C. Coulson.

Figure 16.2: An adult kittiwake with three chicks at the North Shields colony. Photo by J. C. Coulson.

16.2). Breeding birds do not change colonies (Coulson and Wooller 1976), so we can assume that individuals lost from the breeding population have died and that new birds nesting in the colony are breeding for the first time. As a result, an accurate measure of the lifetime reproductive success is available for this species.

This chapter examines some of the factors associated with variations in reproductive success and survival, paying particular attention to differences between males and females.

16.1 Methods

The study colony is situated on the window ledges of a warehouse at North Shields, Tyne and Wear, England. In 1954 a program of individually marking each of the breeding adults was started, and details were recorded annually of all breeding attempts made in the colony up to the present day. These included details of nest location, the members of each pair, the date of egg laying, the number of eggs in the clutch, the number of chicks hatched, and the number fledged. In addition, we calculated two proportionate measures of reproductive success from these data: percentage hatching success (the proportion of eggs laid that hatched) and percentage fledging success (the proportion of chicks hatched that survived to fledging).

The breeding experience of all birds has been calculated (in years) from the year of first breeding. Some birds could be aged from their plumage (Coulson 1959), and others had been ringed as chicks. Most birds started to breed when three to eight years old (Wooller and Coulson 1977), and so breeding experience underestimates the age of a bird by about five years on average.

Sex was determined from the behavior of the birds and their body size (Coulson et al. 1983) and confirmed from the sex of their partners. A few marked adults were found dead, and the sex of these was checked by dissection.

Values given in the text are presented as means ± 1 standard error.

16.2 Age, Reproductive Success, and Survival

On average, male kittiwakes breed for the first time at an earlier age than females (males: 4.7 ± 0.1 years; females: 5.1 ± 0.2 years; Wooller and Coulson 1977), and significantly more males than females first breed at three or four years old or less ($\chi_1^2 = 4.35$, $p < .05$). The number of chicks fledged per pair per year increases with experience, reaching a plateau in birds that have previously bred for four to nine years (fig. 16.3). In the oldest males, there is evidence of a decline of about 15% in the number of young fledged per year (5–10 years: 1.43 ± 0.04; 11–19 years: 1.22 ± 0.08; $t = 2.35$, d.f. = 605, $p < .05$); however, there is no significant difference between the number of chicks fledged by females with a breeding experience of five to ten years (1.36 ± 0.04) and those aged eleven to nineteen (1.25 ± 0.07; $t = 1.36$, d.f. = 736).

Age changes in the number of chicks fledged arise from variation in the number of eggs laid (fig. 16.3) and in hatching success (fig. 16.4), but not from variation in fledging success. There is no evidence of any age-related variation in the ability of birds to raise chicks from broods of one and two to fledging (fig. 16.4), but the fledging success from broods of three increased with the breeding experience of the parents, from approximately 73% among birds breeding for the first or second time to 92% among those with a breeding age of eleven to nineteen years (Thomas

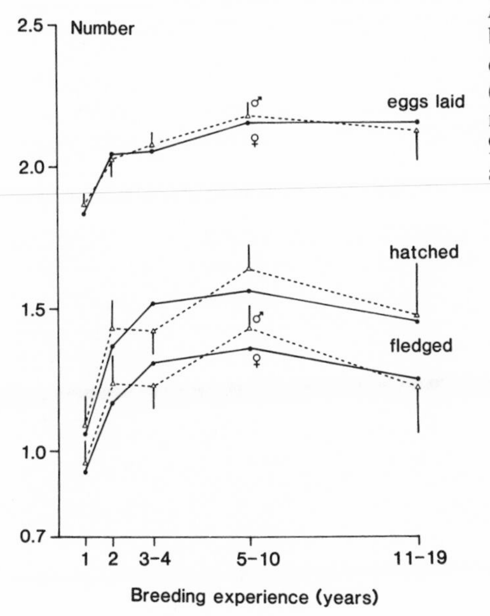

Figure 16.3: The relationship between breeding experience and reproductive success in male (*open triangles*) and female (*solid circles*) kittiwakes. Sample sizes of means all greater than one-hundred pairs; 95% confidence level similar for each sex and shown only for males.

1983). No differences between the sexes were detected in the relationships above, partly because kittiwakes tend to breed with partners of similar breeding experience (Coulson and Thomas 1980).

Once they recruit into the breeding population, most kittiwakes continue to breed each year until they die. However a proportion of birds show intermittent breeding. This is higher among females than in males, and it is highest following the first breeding attempt (19% of males, 29% of females) and declines in older birds (2% of males, 5% of females) (Wooller and Coulson 1977).

The survival rate of kittiwakes during the first year after fledging is approximately 0.79 (Coulson and White 1959), and it is assumed that, as for many other birds, the survival rate in the second year approaches that of adults. There are no data from which the survival rates of immature kittiwakes can be calculated separately for each sex. Among breeding adults at North Shields, the mean annual survival rate of females is 0.86 ± 0.008 and that of males is lower (0.81 ± 0.010) (Coulson and Wooller 1976), a difference that is highly significant ($p < .01$) and is found in all age groups (fig. 16.5). The annual survival rate among breeding birds declines with increasing age such that the mortality rate in females with at least eight years breeding experience is twice that of first breeders. A similar though less marked decline is found among males.

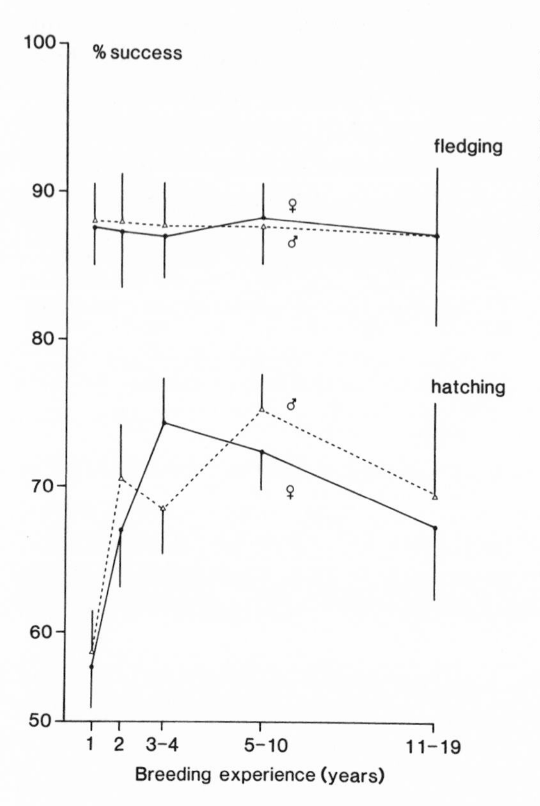

Figure 16.4: Hatching success (± 95% confidence level) and fledging success (± 95% confidence level) of chicks from one- and two-chick broods in relation to breeding experience of male (*open triangles*) and female (*solid circles*) kittiwakes. After Coulson and Thomas 1985b).

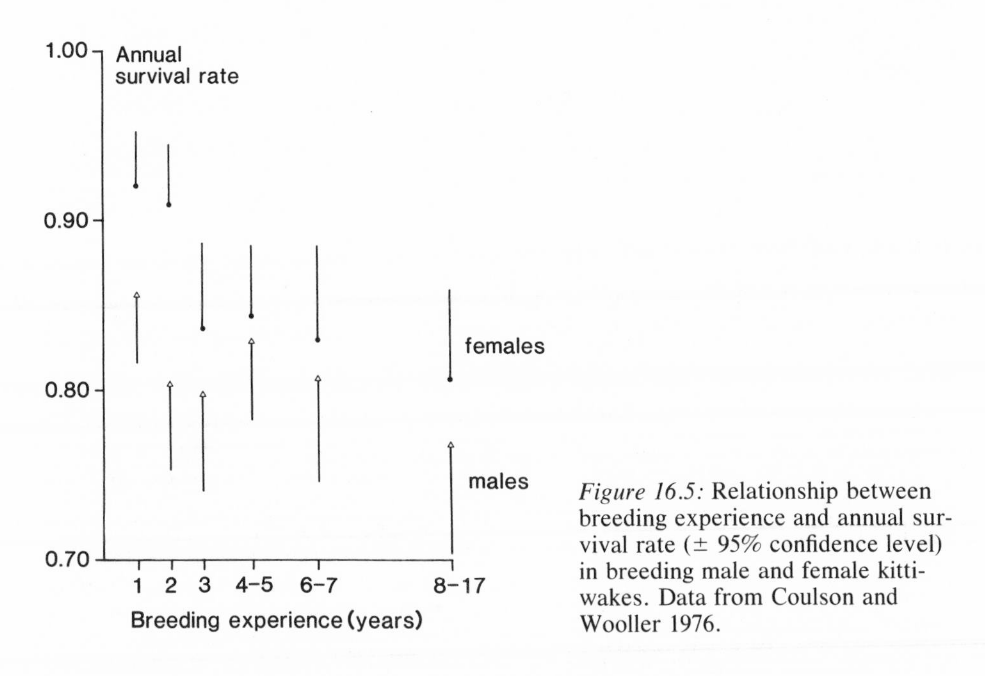

Figure 16.5: Relationship between breeding experience and annual survival rate (± 95% confidence level) in breeding male and female kittiwakes. Data from Coulson and Wooller 1976.

16.3 Variation in Breeding Success

Of 1972 pairs that laid eggs at North Shields between 1954 and 1983, approximately 27% failed to fledge any chicks in a year, 30% fledged one chick, 40% fledged two chicks, and the remainder fledged three chicks. On average, 1.19 young were fledged per pair each year in this colony. Since this is a monogamous species, the same values of breeding success apply to both sexes.

Lifetime reproductive success is highly variable in the kittiwake (fig. 16.6). Over three-quarters of the birds that reach independence fledge fewer than five chicks during their lifetime, and about 62% die having fledged none. At the other extreme, one female in this colony fledged twenty-eight chicks while two males each fledged twenty-six chicks. The variance of lifetime reproductive success (calculated as $\sigma^2/\bar{x}^2$ after Wade and Arnold 1980) in a sample of 250 males and 220 females recruited into the breeding population before 1970 (thereby ensuring a representative distribution of life spans) is not significantly different between the sexes (males: 0.83; females: 0.69). Although the annual survival rate of male kittiwakes is lower than that of females (and this may lead one to expect that males would produce fewer chicks during their lifetime), males start to breed at an earlier age than females and the incidence of intermittent breeding is higher among females, with the result that males and females, on average, breed on the same number of occasions.

In animals generally, current thinking suggests that there is an inverse relationship between reproductive effort and further expectation of life, arising from a balance between the costs of breeding and the potential

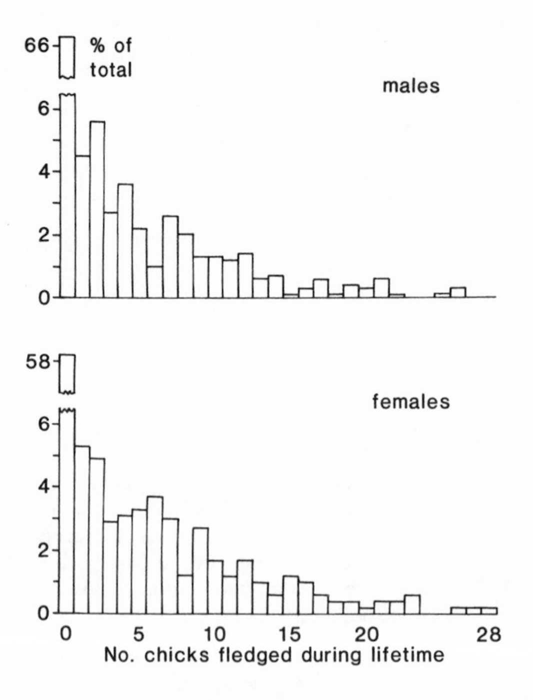

Figure 16.6: Distribution of lifetime reproductive success in male and female kittiwakes. The histograms show the proportion of all individuals fledged that produced particular numbers of fledged young during their lifetime. Numbers dying before breeding were calculated from known mortality rates and age of first breeding (see text).

output of young in future years (Gadgil and Bossert 1970; Wiley 1974a; Williams 1966b). Such a correlation occurs in the kittiwake, where reproductive success increases and survival declines in older birds. However, an investigation of the relationship between breeding success and life span (table 16.1) reveals that long-lived individuals on average breed more successfully when younger than those that survive fewer years (a trend found in thirty out of thirty-six pairs of means compared in table 16.1). In other words, despite their greater involvement and effort in

Table 16.1 Mean Number of Chicks Fledged per Pair for Male and Female Kittiwakes of Differing Breeding Experience According to Their Breeding Life Span

Breeding Experience (years)	Breeding Life Span (years)			
	1–2	*3–5*	*6–9*	*10–19*
Males				
1	1.04 (113)[a]	1.07 (87)	1.31 (51)	1.39 (28)
2	1.15 (46)	1.27 (78)	1.45 (49)	1.22 (23)
3–4		1.20 (138)	1.45 (97)	1.47 (55)
5–10		1.35 (29)	1.46 (160)	1.64 (154)
Females				
1	0.99 (93)	1.10 (70)	1.21 (52)	1.29 (28)
2	1.02 (45)	1.24 (58)	1.39 (41)	1.27 (26)
3–4		1.36 (112)	1.32 (93)	1.40 (58)
5–10		1.32 (44)	1.26 (182)	1.63 (176)

Note: Birds that live longer breed more successfully than birds of similar breeding experience that have a shorter life span.

[a]Number of pairs in parentheses.

breeding when young, the cost was not sufficient to reduce their life span below that of birds with lower reproductive investment. Clearly, these differences indicate variation in the quality of individual kittiwakes. The increased proportion of better-quality birds in older age-classes contributes, in part, to the increase in reproductive success with increased breeding experience reported above, but age-related change in reproductive performance is still evident among long-lived individuals (table 16.2).

Almost two-thirds of kittiwakes that fledge fail to leave any young, primarily because of prebreeding mortality, although 5% of birds that nest fail to fledge any chicks before dying (fig. 16.6). The variance of lifetime reproductive success in a sample of male and female kittiwakes was partitioned (using the techniques developed by D. Brown, this volume, chap. 27) according to its three components—fecundity (clutch size),

Table 16.2 Relationship between Breeding Experience and Number of Chicks Fledged per Pair among Male and Female Kittiwakes That Survived to a Breeding Age of at Least Seven Years

Breeding Experience (years)	Number of Chicks Fledged per Pair					
	Males			*Females*		
	$\bar{x}$	*s.e.*	*n*	$\bar{x}$	*s.e.*	*n*
1	1.26	0.09	84	1.20	0.08	98
2	1.41	0.10	71	1.24	0.09	87
3–4	1.39	0.07	160	1.34	0.06	185
5–10	1.46	0.04	404	1.35	0.04	482
11–19	1.24	0.08	100	1.26	0.07	157

Table 16.3 Percentage Contribution of the Components of Lifetime Reproductive Success to Variance in LRS in Male and Female Kittiwakes

Component	L	F	S
Males			
L	33.35		
F	7.23	1.42	
S	4.28	−0.76	13.12
LFS	2.66		
N = 310 breeding males			
Proportion of breeders, 0.36			
Percentage of variance due to nonbreeders, 38.7			
Percentage of variance due to breeders, 61.3			
Females			
L	37.16		
F	9.20	1.60	
S	0.72	−1.01	12.79
LFS	1.91		
N = 271 breeding females			
Proportion of breeders, 0.43			
Percentage of variance due to nonbreeders, 37.6			
Percentage of variance due to breeders, 62.4			

L = breeding life span; F = fecundity (clutch size); S = survival of offspring to fledging.

offspring survival, and life span—also taking into account the nonbreeding constituent of the population (which is determined by prebreeding mortality). The results of this analysis have been summarised in table 16.3. Approximately 38% of the overall variance in lifetime reproductive success in this study can be attributed to prebreeding mortality. A further 33% arises from variation in breeding life span, and the rest is mainly due to variation in mean annual clutch size. There are no apparent sex differences in the relative importance of each source of variation in lifetime success in the kittiwake.

16.4 Causes of Variance in Breeding Success

There is a high consistency in the reproductive performance of the same individual kittiwakes in successive years (Coulson and Thomas 1985b). Birds that fledge more chicks in one year fledge significantly more the next time they breed (table 16.4). This trend, which is statistically significant, is particularly apparent in those individuals that fledge three chicks. As a whole, only 4% of pairs fledge three young, but of those that did so in one year, 15% did so again the following year. The relationship is still evident and significant when birds that fledged three chicks the previous year are excluded (table 16.4).

The consistency in the number of chicks fledged, which is apparent over a number of years (tables 16.5 and 16.6), arises from consistencies in the number of eggs laid in successive years as well as in hatching and fledging success (Coulson and Thomas 1985b).

Variance in annual reproductive success in this data set was partitioned, using analysis of covariance, into what can be attributed to consistent differences between birds and what occurs within individuals (see Coulson and Thomas 1985b). This was further partitioned to assess the amount of variance that arises from the independent effects of the proximate factors measured in this study (including date of egg laying, breeding experience, year of breeding, and whether an individual retained the same mate from the previous year). Approximately 33% of the

Table 16.4 Reproductive Success of Male and Female Kittiwakes in the Present Year According to Their Success the Last Time They Bred

Number of Chicks Fledged in Last Attempt	Number of Chicks Fledged This Year					
	Males			*Females*		
	$\bar{x}$	s.e.	*n*	$\bar{x}$	s.e.	*n*
0	1.17	0.05	271	1.18	0.05	281
1	1.31	0.04	390	1.25	0.04	386
2	1.37	0.04	542	1.34	0.04	571
3	1.58	0.15	43	1.67	0.14	43

Chi-square test of linear trend in proportions (Maxwell 1961):
All birds: males, $\chi^2_1 = 13.1$, $p < .001$; females, $\chi^2_1 = 12.8$, $p < .001$.
Birds that fledged 0, 1, or 2 chicks last year (see text): Males, $\chi^2_1 = 9.3$, $p < .01$; females, $\chi^2_1 = 7.1$, $p < .01$.

Table 16.5 Reproductive Success of Male and Female Kittiwakes: Chicks Fledged during the First Two and Following Two Breeding Attempts by Birds Breeding for between Four and Eight Years

Number of Chicks Fledged in First Two Years	Number of Chicks Fledged in Following Two years					
	Males			*Females*		
	$\bar{x}$	s.e.	*n*	$\bar{x}$	s.e.	*n*
0–1	2.47	0.22	34	2.42	0.16	36
2	2.46	0.17	48	2.67	0.19	43
3	3.04	0.22	28	2.76	0.23	29
4–5	2.71	0.22	35	3.00	0.25	28
	(r = .14, n = 145)*			(r = .15, n = 136)*		

Note: Individuals that fledged more chicks when young also fledged significantly more when older.
*$p < .05$.

Table 16.6 Reproductive Success of Male and Female Kittiwakes: Chicks Fledged during the First Four and Following Five Breeding Attempts by Birds That Bred in at Least Nine Years

Number of Chicks Fledged in First Four Years	Number of Chicks Fledged in Following Five Years					
	Males			*Females*		
	$\bar{x}$	s.e.	*n*	$\bar{x}$	s.e.	*n*
0–1	6.67	0.33	3	5.33	0.88	3
2–3	6.79	0.71	14	6.80	0.83	10
4–5	7.31	0.70	13	7.56	0.45	27
6–9	8.17	0.60	12	7.60	0.79	10
	(r = .28, n =42)*			(r = .26, n = 50)*		

Note: Individuals that fledged more chicks when young also fledged significantly more when older.
*$p < .05$.

variance in clutch size, 26% of that in hatching success, 23% of that in fledging success, and 23% of that in the numbers of chicks fledged per pair was attributed to consistent differences among birds, that is, individual variation. Less than 9% of the variance in each parameter of reproductive success was explained by the effects of the proximate factors studied. It appears, therefore, that there is considerable variation in the "quality" of individuals in the population and that this is the single most important factor determining reproductive success in individual kittiwakes.

16.5 Discussion

Although the annual reproductive success of individual kittiwakes breeding at North Shields has varied over a thirty-one-year period (Coulson and Thomas 1985a) and changes in relation to breeding experience, pair status (Coulson and Thomas 1980), the location of the nest site (Coulson 1968), egg size (Thomas 1983), and date of egg laying (Coulson and Thomas 1985b), a major proportion of the overall variation in breeding success arises from consistent individual differences in reproductive success. There is a high degree of similarity between the reproductive performance of the same individuals in consecutive and subsequent years, a tendency also reported in other avian studies (Koskimies 1957; Perrins and Jones 1974; Findlay and Cooke 1982a; Batt and Prince 1979; van Noordwijk, van Balen, and Scharloo 1981a). Consistent differences of this nature can arise from a number of causes. They may have a genetic basis, as demonstrated by Perrins and Jones (1974), Batt and Prince (1979), van Noordwijk, van Balen, and Scharloo (1981a), may reflect differences in body size (Bryant and Westerterp 1982), may result from physiological differences between individuals (for example, arising from the ability of an individual to find food [Perrins 1970]) or from debilitating factors such as disease or injury (all of which may be classified under the nebulous term "overall body condition"). In addition, consistent differences may arise from learned behavior. The precise causes of this consistency of breeding success in the kittiwake have yet to be identified.

Lifetime reproductive success is a function of annual reproduction and survival rates, both of which are age dependent in the kittiwake. Several aspects of reproductive performance of the kittiwake increase with age, a tendency also reported in other seabird species (Mills 1973; Ryder 1975; Coulson and Horobin 1976; Haymes and Blokpoel 1980; Pugesek 1981). The annual survival rate declines in older birds, as predicted by Botkin and Miller (1974), as shown for eiders, *Somateria mollissima* (Coulson 1984), and as suggested in the fulmar, *Fulmarus glacialis* (Dunnet and Ollason 1978), but in contrast to the results obtained in other studies of seabirds (Potts 1969; Richdale and Wareham 1973; Coulson and Horobin 1976; and others), where significant deviations from a constant adult survival rate were not detected (possibly because some of these studies lack large enough samples of old birds).

An increase in reproductive success with age may result either from increased efficiency arising from experience or from increased effort. The decline in adult survival among older birds could be caused by increased effort being put into each successive breeding attempt, or it could be due to the cumulative costs of previous breeding attempts. As a result of a study on the california gull, *Larus californicus*, Pugesek (1981) questioned the importance of increasing experience upon reproductive success among older birds and doubted that the learning process could continue over a long enough period ("upward of nine years") to explain it. However, the study of Greig, Coulson, and Monaghan (1983) shows that improved feeding continues for several years in the herring gull, *Larus argentatus*, and they have additional evidence of further improvement in birds well over four years of age. As an alternative, Pugesek has suggested that older birds may be putting more effort into breeding than younger individuals, basing this conclusion upon the assumption that fewer breeding opportunities are available to older birds. In fact, this interpretation requires that age-specific mortality of adults occurs in this species, but this remains to be confirmed. In bird species with constant adult mortality rates, the number of future breeding opportunities does not change. It remains to be demonstrated whether the theoretical models of Gadgil and Bossert (1970), Wiley (1974a), and others apply to gulls. Indeed, the finding that kittiwakes with the highest annual reproduction rates are among the longest-lived and that birds therefore differ in their "quality" itself brings into question the whole idea that there is a simple relationship between success and future survival. High-quality birds are able to maintain a high annual reproductive output without, apparently, reducing their life span. Further, high-quality birds, because of their higher survival rates, will form an increasing proportion of the cohort as it becomes older. This in itself is an alternative to Pugesek's interpretation.

The concept of variation in quality between individuals was recognized in differences between the biology of birds nesting in different parts of the North Shields colony (Coulson 1968). Males nesting in the center of the colony (the area of the building that was occupied when the population was half its maximum size) were found to have a higher annual reproductive rate and lived longer on average (Coulson 1968) than those nesting on the edge. Although there was no significant difference in wing length between these two groups of birds, those nesting in the center were found to be slightly, though significantly, heavier in their year of recruitment than those on the edge (Coulson 1968). Birds born in the center of the colony do not necessarily return to this part to breed (Wooller and Coulson 1977), suggesting that these particular differences in quality may not be inherited but may be environmental in origin. Since the male kittiwake normally acquires the nest site, this sex difference in body-weight variation appears to indicate that competition for nest sites is a major factor segregating males according to their quality. Differences between the quality of individuals may also explain why those shags, *Phalacrocorax aristotelis*, that had a high reproductive rate in a particular year were also more likely to survive the following winter than those with a lower reproductive rate (Potts 1969).

Consistent individual differences in annual reproductive success, coupled with an association between high annual reproductive rate and high annual survival rate, leads to a situation in which lifetime reproductive success in the kittiwake is highly variable. By definition, annual reproductive success in a monogamous species is the same for both members of the pair, but lifetime reproductive success varies according to the survival rates of the two sexes or, in a species such as the kittiwake, which exhibits intermittent breeding, according to the number of breeding attempts made during an individual's lifetime. Breeding life span is, in fact, the major factor that determines lifetime reproductive success in the kittiwake. Differences in survival rates between the sexes arise from physiological and behavioral differences between males and females. There is no evidence from ringing recoveries to suggest a differential pattern of winter dispersal in male and female kittiwakes, and at the colony most aspects of breeding behavior are shared equally by the sexes, the only exception being nest-site acquisition. Differences in survival of the sexes have been found only during the period from colony reoccupation to nest building (Coulson and Wooller 1976). In the kittiwake the survival rate of males is lower than that of females but the incidence of intermittent breeding and age of first breeding are higher in females, with the result that lifetime reproductive success in this species, though highly variable, is similar in both sexes.

16.6 Acknowledgments

This project has been made possible by the continued goodwill and protection of the colony provided by the staff and management of Swan Hunter Ship Repairer, Smith's Dock Company, and Jim Marine. We are indebted to a number of people who collected data during part of this study, including E. White, A. F. Hodges, G. Brazendale, R. D. Wooller, J. W. Chardine, and J. Porter.

17 Variation in Breeding Success in Fulmars

Janet C. Ollason and G. M. Dunnet

THE ATLANTIC FULMAR, *Fulmarus glacialis* (L.), is a pelagic seabird related to the albatrosses and shearwaters (see frontispiece). Over the past two hundred years its geographical range has increased dramatically from the high Arctic southward, so that fulmars are commonly found breeding around the coasts of Britain, northern France, and Scandinavia (Fisher 1952, 1966) and, more recently, Newfoundland and Labrador (Nettleship 1974; Montevecchi et al. 1978). In Britain this spread in range has been accompanied by an increase in breeding numbers of 7% per annum (Cramp, Bourne, and Saunders 1974), maintained over the past one hundred years (Fisher 1966). Explanations for the rapid increase include a response to food made available by whaling and trawling (Fisher 1952) and the appearance of a new genotype in Iceland (Wynne-Edwards 1962) that allowed the exploitation of the macroplankton in the warmer offshore boreal waters (Salomonsen 1965).

Though in the Atlantic the fulmar is polymorphic in color, in Britain virtually all birds are light phase. With the exception of prefledglings, fulmars of all ages look similar, but males are on average 20% heavier than females in the breeding season (Ollason and Dunnet 1978) and have longer and deeper bills (Wynne-Edwards 1952).

17.1 The Life History

The fulmar breeds on coast cliffs and slopes, laying a single white egg in May in an earthy or rocky hollow. The life history is represented in figure 17.1. Chicks fledge in September and spend the next three or four years almost entirely at sea in the Atlantic, Norwegian Sea, and European waters (Macdonald 1977a). During the next four years they begin to appear at breeding areas (Fisher 1952, chap. 20), probably prospecting for nest sites and forming pair-bonds. Of fifty birds ringed as nestlings that were recaptured as breeders, the modal age at assumed first breeding is eight years (range six to nineteen), with females possibly older than males (Dunnet, Ollason, and Anderson 1983, plus unpublished data). Indirect

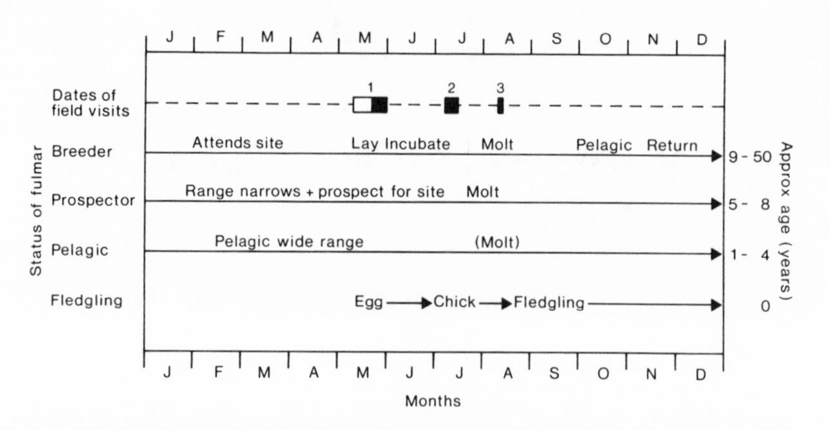

Figure 17.1: Life history and annual cycle of the fulmar and dates of field visits. Visit 1 covered all of laying in four years; in most years it occurred toward the end of laying (*shaded*). Visit 2 occurred during hatching. Visit 3 occurred shortly before fledging.

calculations indicate that prebreeders have a high survival rate of about 95% per annum. Only 10% of new breeders nest close to their natal area (Dunnet, Ollason, and Anderson 1983), and we know nothing about the distribution of the remaining 90%. Immediately before laying, both males and females leave the breeding grounds (the prelaying exodus) and go to sea for about nine and twenty days respectively (Macdonald 1977b), presumably to build up reserves for breeding. Within a day or so after the egg is laid the male begins incubating, usually for about a week (Dunnet, Anderson, and Cormack 1963). Thereafter males and females incubate alternately in brooding the chick for the first two weeks (Mougin 1967). During this period (July/August) they may forage hundreds of kilometers from the breeding area (Dunnet and Ollason 1982). Most egg losses occur early during incubation (Mougin 1967), and fulmars do not lay replacement clutches.

Fulmars form monogamous pairs and are highly faithful to both mate and nest site (Ollason and Dunnet 1978). The same mate and site are retained in 91% (632/696) of cases. Changes of mate (regardless of site) occur on 5% (33/696) of occasions and changes of site (regardless of mate) on 6% (43/696) of occasions. Breeding adults have a high survival rate, which does not differ significantly between males and females (Dunnet and Ollason 1978). An attempt (Dunnet 1982) to detect differences in survival in cohorts with average breeding histories ranging from 2.5 to 18.5 years failed to find any sign of increasing mortality with age. These data, extended to include a further year, were plotted by Buckland (1982) as an age-specific survival curve, and again no decrease in mean survival of older birds was found, though the 95% confidence band inevitably widens for older birds (fig. 17.2). Buckland estimates a constant mean annual survival rate of 0.968, s.e. 0.0042, and an expectation of further life of breeding adults of over thirty years. After breeding, adults molt and disappear from their breeding sites for some months. They return from October onward (Macdonald 1980) and attend their nest sites during the winter and early spring, going to sea in rough weather (Coulson and Horobin 1972).

17.2 Methods

This detailed study of the breeding biology of fulmars began in 1950 on the island of Eynhallow in Orkney (Carrick and Dunnet 1954). Annual visits have continued uninterrupted to the present. Normally there are three field visits of two to five days each during laying, hatching, and fledging (fig. 17.1). However, most of the egg-laying period (mid-May to early June) was covered in six years. Unringed breeders are caught and individually ringed after their egg has hatched. On each visit ringed birds are recorded and their nest sites and contents noted. Nest sites are photographed and labeled in August after the adults have departed.

Fulmars reached Orkney about 1900 and probably began breeding on Eynhallow in the third decade of this century (Robertson 1934; Fisher 1952). When this study began in 1950, the breeding population would have contained birds with a wide range of ages and experiences. In the first five years 114 breeders (54% of 1954 breeding population) were caught and ringed. We assume that from 1955 onward the unringed breeders being caught were mostly inexperienced—that is, near the beginning of their breeding lives. Since fulmars are long-lived, including data from birds first caught after their first breeding year is unlikely to produce serious biases, though it may increase variances. On average about 60% of the breeders on Eynhallow are ringed.

Fulmars can breed in consecutive years whether or not they are successful in the previous year (Carrick and Dunnet 1954). We therefore assume that once a fulmar has bred it will attempt to do so every year

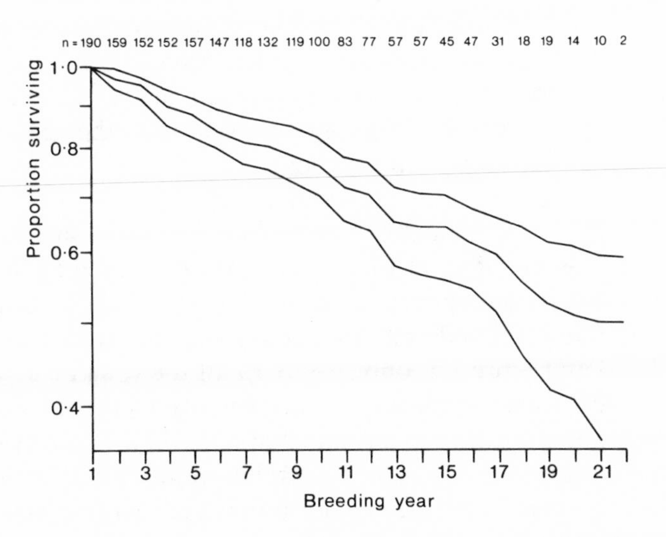

Figure 17.2: Age-specific survival curve of breeding fulmars, with 95% Monte Carlo confidence band. After Buckland 1982, fig. 2.

thereafter. Thus a gap in the record of an individual either results from our failure to detect it or represents a year when breeding was attempted but failed quickly, possibly even before an egg was laid. Since not every surviving ringed adult was recorded every year, there are gaps in the records of many individuals. Gaps of four years or longer occur with a frequency of less than .05, hence birds not seen for at least four years after their last record are assumed to have disappeared from the breeding population and to have completed their breeding lives.

Most birds with completed breeding lives were first caught after 1954, and so we can estimate the beginning of their breeding histories—that is, their entire breeding life span is recorded. A frequency histogram of lengths of these life spans shows significantly more birds than expected (from survival estimates) with life spans only one year long, which may result from greater mobility of new breeders. Hence analyses requiring entire life spans use birds with life spans of two or more years.

Even though this study is thirty-four years long, the maximum entire breeding life span available to us is still only twenty-four years (1955 to 1978), owing to our procedures for determining the beginning and end of breeding life. Thus our sample of birds whose entire life spans are known is restricted to histories no longer than twenty-four years, whereas the expectation of further life for a new breeder is thirty years. To cover longer breeding life, we repeated some calculations using all birds in the study, with incomplete histories up to thirty-two years. The results were similar, hence we decided that the more rigorously defined sample, of birds with entire life spans, would be satisfactory, and only results from this sample are presented.

Males are distinguished from females with a linear discriminant function based on bill length and bill depth (Dunnet and Anderson 1961). Using this and information about pairings and repairings, 95% of the fulmars measured can be sexed with 95% confidence. Throughout the thirty-four years of the study we have accumulated over eight hundred individually ringed adults, of which fifty are of known age. Breeding occurred at over a thousand individually marked nest sites. The number of active sites per year (corrected for egg losses early in the season; Ollason and Dunnet 1980) has increased by about 5% per annum and is currently about three hundred (Ollason and Dunnet 1983).

In years when one member of an established pair was not recorded while its mate was breeding at their usual site, the breeding success was attributed to both parents provided there were no changes to the pair or their site in the year immediately preceding and following the gap. Such gaps were filled and represent only 1% to 3% of the total bird-years in the different data sets used for analyses. Unringed breeders, of unknown age, are usually caught after the egg has hatched, and therefore this initial year is excluded from analyses of success in relation to early breeding experience because the breeding success in the first year will be biased.

Table 17.1 Breeding Success in First Known Breeding Year of Fulmars of Known Age

Age at First Breeding	n	Total Eggs Hatched	Proportion Hatched	Total Chicks Fledged	Proportion Fledged	Proportion of Total Fledged
Males (mode = 8 years)						
Younger than or equal to modal age	11	3	0.27	1	0.33	0.09
Older than modal age	12	4	0.33	4	1.00	0.33
Females (mode = 12 years)						
Younger than modal age	10	4	0.40	1	0.25	0.10
			**		*	***
Older than or equal to modal age	12	12	1.00	10	0.83	0.83

*$p < .02$, **$p < .003$, Fisher exact test; ***$p < .001$, $\chi^2_1 = 8.983$, two-tailed tests.

17.3 Breeding Success Throughout Life

Age at First Breeding and First-Year Success

Age at first breeding was determined from birds previously ringed as nestlings. The modal age at which twenty-three males first bred was eight years, and for twenty-two females it was twelve years. However, this difference may not be as great as it appears because males take the long first-incubation stint, and since breeding attempts by inexperienced birds often fail early in the incubation period, we may catch males earlier in their breeding lives.

In estimating breeding success in the first recorded breeding year, when often only one parent is known, the fact that 86% of males were first caught off eggs, compared with 55% of females, increases the estimated breeding success for females. Therefore in the first year the breeding success of each sex must be analyzed separately.

There is considerable variation in age at first recorded breeding in males (six to seventeen years) and females (seven to nineteen years). Are birds that begin breeding relatively early in their lives as successful in their first year as those that delay? In twenty-three males and twenty-two females the relatively young birds were less successful at hatching eggs, at fledging chicks, and in overall success (proportion of eggs producing fledglings) than the older birds, and these differences are significant in females (table 17.1).

We have too few birds of known age with completed histories to determine whether the lifetime production of younger first-time breeders differs from that of older first-time breeders.

Experience and Success

Data are available for thirty-nine males and thirty-two females with breeding histories of at least twenty-one years. This length of history is chosen because it is reasonably long and yet the sample sizes are adequate.

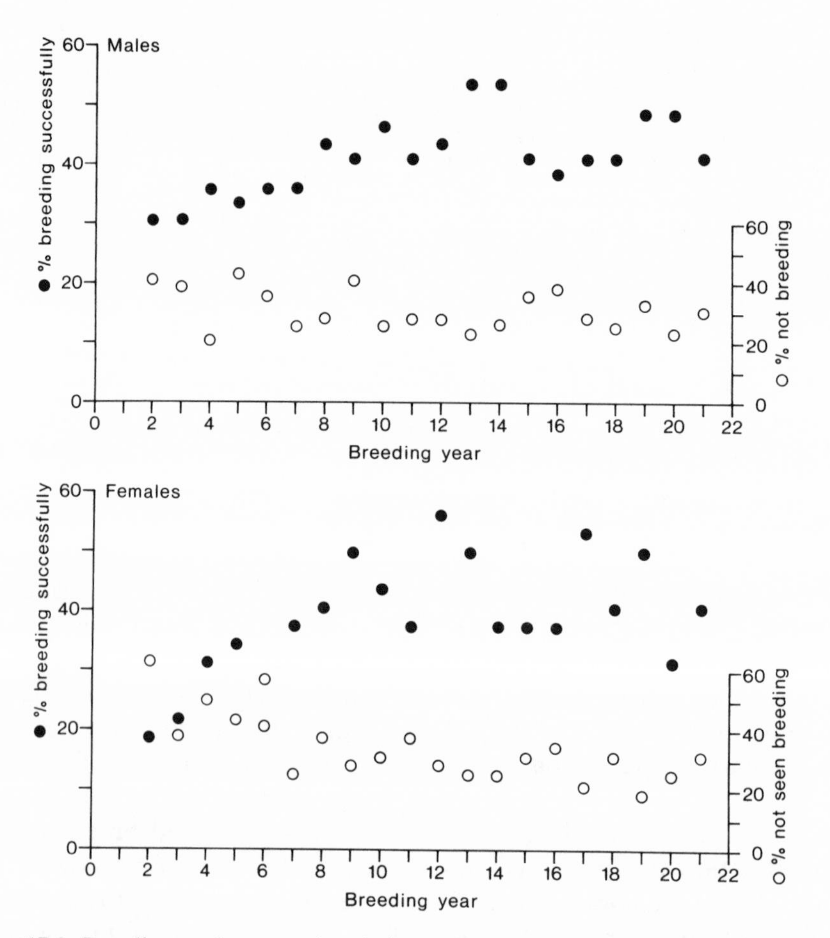

Figure 17.3: Breeding performance in relation to breeding year of (*a*) thirty-nine males and (*b*) thirty-two females with at least twenty-one years of breeding history: percentage breeding successfully, that is, fledging a chick (*solid circles*), and percentage not seen breeding (*open circles*).

Both males and females improve their breeding success until about the tenth breeding year, and thereafter success remains at approximately 45% with no indications of any decline in later years (fig. 17.3). In females the improvement in success is steeper than in males, largely because we are more likely to miss females than males owing to early failure, and therefore females have a higher-than-average proportion of gaps in their records until their seventh breeding year. Thereafter the percentage of gaps varies around 30%, similar to males.

Breeding Success toward the End of Breeding Life

We have twenty-three fulmars with at least thirty years of breeding history, and even these birds do not show any obvious decline in breeding performance in their later years. However, it is possible that the older age categories may contain both birds that are close to the end of their breeding life span and others that will continue to breed successfully. If so, any tendency for breeding success to decline with age might be obscured.

Does a decline in breeding performance occur among birds approaching the end of their lives regardless of their absolute age? To answer this question, two categories of birds were defined: "disappearing" breeders with completed breeding histories (as defined in sec. 17.2); and "continuing" breeders that continued breeding for four years beyond the years being analyzed (fig. 17.4).

The first and last years of breeding histories were eliminated from the analyses because all birds would be present and breeding, and hence fecundity is bound to be 1.0. The remaining years were divided into a "beginning" period consisting of eight years after the first year, covering the improvement in success identified above; an "end" period of four years, which in disappearing birds was four years before but excluding their last year and in continuing birds was four years before their last four years of continued breeding; and a "middle" period of years between the beginning and the end (fig. 17.4).

In the disappearing breeders, paired *t*-tests indicate a significant decrease in fecundity (eggs per year) and overall breeding success (fledglings per year) from the middle to the end period (table 17.2). Chicks per egg and fledglings per chick were also reduced, though these differences were not significant. There are no such decreases among continuing breeders. Comparisons between disappearing and continuing breeders show that the beginning periods are very similar. In the middle period the disappearing breeders tend to have higher fecundity ($.10 < p > .05$) and higher overall success ($.10 < p > .05$) than continuing breeders, whereas the end periods of disappearing and continuing breeders are similar. Thus the decrease in performance of disappearing breeders toward the end of

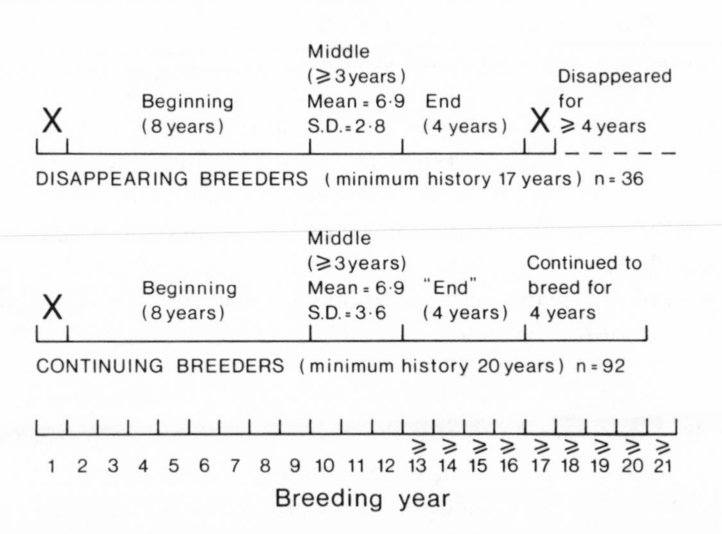

Figure 17.4: Definitions of continuing and disappearing breeders, and beginning, middle, and end periods of breeding history. Breeding years excluded from analyses are represented by large crosses.

Table 17.2 Comparison of Middle and End Periods of the Breeding History of Disappearing and Continuing Breeders

	Middle (≥ 3 years)			*p* of paired *t*-tests	End (4 years)		Decline
	Mean	*s.e.*	*n*	(two-tailed)	*Mean*	*s.e.*	(%)
Disappearing breeders (*n* = 36)							
Eggs per year	0.7590	0.040	36	0.039	0.6458	0.052	11.42
Chicks per egg	0.7464	0.041	33	n.s.	0.6540	0.062	9.24
Fledglings per chick	0.8894	0.026	29	n.s.	0.7902	0.062	9.92
Fledglings per year	0.4913	0.038	36	0.016	0.3681	0.051	12.32
Continuing breeders (*n* = 92)							
Eggs per year	0.6672	0.031	92	n.s.	0.6630	0.033	0.42
Chicks per egg	0.6992	0.035	81	n.s.	0.6163	0.039	8.29
Fledglings per chick	0.8757	0.033	63	n.s.	0.8437	0.026	3.20
Fledglings per year	0.4088	0.028	92	n.s.	0.3723	0.033	3.65

Note: For definitions, see figure 17.4 and text.

their breeding histories appears to be due to their greater fecundity and success during the middle period: their performance in the end phase is not absolutely lower than that of the continuing breeders.

17.4 Variation in Long-Term Breeding Success

Effects of Individuals and Calendar Years

Though breeding performance alters throughout the life of a fulmar, not all individuals exhibit the same pattern. In addition, breeding success (fledglings per egg) on Eynhallow has varied between years from 12% in 1973 to 38% in 1969, averaging 27%. Thus, to analyze differences in success between individuals, we must examine their performance in the same set of years. In addition, since an individual's success changes as it gains experience, the individuals compared should be at the same stage of life history in the same calendar year. Cohorts were chosen that spanned a long series of years and also had sample sizes as large as possible. Breeding success was determined for each individual in each year.

Log-likelihood ratio tests (Zar 1980) examined the variation in success between years (individuals combined) and between individuals (years combined) for the cohort of the eight males and fourteen females first ringed in 1955, using their breeding data from 1956 to 1967 (table 17.3), and the cohort of the six males and eight females first ringed in 1967, using their breeding data from 1968 to 1979 (table 17.4).

Of the four tests on variation in success between years, one is statistically significant; of the four tests on individuals, two are statistically significant. Thus variation between individuals may account for more of the variation in breeding success than does variation between years.

Effect of Breeding Life Span

Does extended breeding experience lead to improvements in all aspects of breeding performance? We have eighty-two males and seventy-six females with entire breeding life spans ranging from two to twenty-four years. For each individual the measures of breeding success (table 17.5) were calculated and the relationship with breeding life span was determined. For both sexes, there are significant positive correlations between length of breeding life span and lifetime success (total fledglings) and fledging success (fledglings per chick hatched). In females, but not in males, there is a significant negative correlation between breeding life span and fecundity (eggs per year of breeding life span).

For birds with at least five-year breeding life spans, early breeding performance (in the second and third breeding years) was compared with the duration of their breeding life span. In females only, there are significant negative correlations between the following parameters of early breeding and life span: fecundity ($r = -.2764$, $n = 53$, $p < .05$); hatching success (chicks per egg, $r = -.3497$, $n = 42$, $p < .03$); and overall success (fledglings per year, $r = -.3068$, $n = 53$, $p < .03$). Since these analyses are based on entire life spans, these findings suggest that females that are more fecund and more successful in their early breeding years live shorter lives.

Thus, in general, although fulmars that live longer do produce more offspring, females that put great effort into breeding in their first few years are likely to shorten their breeding lives and produce fewer offspring than those that do not do so.

Components of Breeding Success

Does breeding life span, fecundity, or survival of egg and chick to fledging have the greatest effect on lifetime fledgling production? The relative contributions of these multiplicative components to variance in lifetime success were estimated, using females whose entire breeding life

Table 17.3 Comparison of Variation in Breeding Success in Male and Female Fulmars Due to Year and Individual: 1955 Cohort

	Percentage Breeding Success (range)	d.f.	G
Males ($n = 8$)			
Breeding from 1956 to 1967			
Year (individuals combined)	27 (13–50)	11	7.609
Source of variation			
Individual (years combined)	27 (8–58)	7	12.697
Females ($n = 14$)			
Breeding from 1956 to 1967			
Year (individuals combined)	37 (14–64)	11	24.429**
Source of variation			
Individual (years combined)	37 (8–67)	13	14.976

Note: Log-likelihood ratio test: $**p < .02$.

Table 17.4 Comparison of Variation in Breeding Success in Male and Female Fulmars Due to Year and Individual: 1967 Cohort

	Percentage Breeding Success (range)	d.f.	G
Males ($n = 6$)			
Breeding from 1968 to 1979			
Year (individuals combined)	29 (0–50)	11	17.688
Source of variation			
Individual (years combined)	29 (8–58)	5	11.978*
Females ($n = 8$)			
Breeding from 1968 to 1979			
Year (individuals combined)	45 (25–75)	11	7.761
Source of variation			
Individual (years combined)	45 (8–92)	7	36.717***

Note: Log-likelihood ratio test: $^{*}p < .05$; $^{***}p < .001$.

Table 17.5 Correlation between Various Components of Breeding Success and Breeding Life Span in Male and Female Fulmars

Component	Males[a]	Females[b]
Lifetime success (number of fledglings)	.7838***	.7340***
Fecundity (eggs per breeding year)	−.1452	−.3034**
Hatching success (chicks per egg)	.0914	.0795
Fledging success (fledglings per chick)	.2180*	.2564*

Note: Pearson correlation coefficients.
[a]$N = 82$; life spans 2 to 24 years, mean = 9.20.
[b]$N = 76$; life spans 2 to 23 years, mean = 8.63.
$^{*}p < .05$; $^{**}p < .01$; $^{***}p < .001$; two-tailed tests.

spans were known (tables 17.6 and 17.7). Breeding life span contributed 60% of the variance, offspring survival 48%, and fecundity 12%. Covariation in life span and fecundity (significant negative correlation, table 17.5) contributed −15% to the variance; covariation in life span and survival contributed −6%; and covariation in fecundity and survival contributed 3%. Separating survival into survival of egg to hatching and survival of chick to fledging, the former contributed 42% of the variance due to survival, the latter 5%, and covariation 53%. The complete analysis, repeated for males, gave broadly similar results.

Thus breeding life span is the most important component of lifetime success, and this was used in a regression analysis to predict the number of fledglings that would be produced by a fulmar with a breeding life span of thirty years (the expectation of further life of a new breeder). The regression equation for females is: Number of fledglings = 0.3582 × breeding life span + 0.1453. The standard error of the regression coefficient is 0.0385, that is, 95% confidence intervals ± 0.0755. The equation predicts that a female with a breeding life span of thirty years will produce

10.89 ± 2.26 fledglings. Taking the age at first breeding as eight years, and survival of 95% per annum, 7.22 ± 1.50 fledglings will survive to breed, and 2.64 ± 0.55 will survive (adult survival 96.8% per annum) to their thirtieth breeding year, which is slightly more than the required two birds. However we must bear in mind that the regression equation is based on data from birds with entire breeding life spans of two to twenty-three years: in a complete population there will be some birds that breed for only one year and probably produce no fledglings, and some that breed for more than twenty-three years.

17.5 Other Causes of Variation in Breeding Success

The previous section described the relative importance of the different components of reproductive success. This section considers the effects of extrinsic factors on success, including environmental conditions, physical constraints, and quality of the mate.

Date of Breeding

Fulmars on Eynhallow lay over a three-week period commencing in mid-May and have a remarkably consistent mean laying date from year to year (Ollason and Dunnet 1980). The precise date on which an individual

Table 17.6 Mean and Variance of the Components of Lifetime Reproductive Success in Female Fulmars

Component	Original		Standardized Variance
	Mean	*Variance*	
L	2.6316	26.5291	0.3561
F	0.7550	0.0395	0.0692
S	0.4999	0.0716	0.2866
LF	6.2105	14.3551	0.3380
FS	0.3738	0.0529	0.3715
LS	4.5058	11.2873	0.6062
LFS	3.2368	6.3166	0.5952

Note: N = 76 females with entire breeding life spans of 2 to 23 years.
L = breeding life span; F = fecundity (number of eggs laid); S = survival of offspring to fledging.

Table 17.7 Percentage Contribution of the Components of Lifetime Reproductive Success to Variation in LRS in Female Fulmars

Component	L	F	S
L	59.8		
F	−14.7	11.6	
S	−6.1	2.6	48.2
LFS	−4.5		

Note: V(LFS)/(LFS)^2; N = 76 females with entire breeding life spans of 2 to 23 years.
L = breeding life span; F = fecundity (number of eggs laid); S = survival of offspring to fledging.

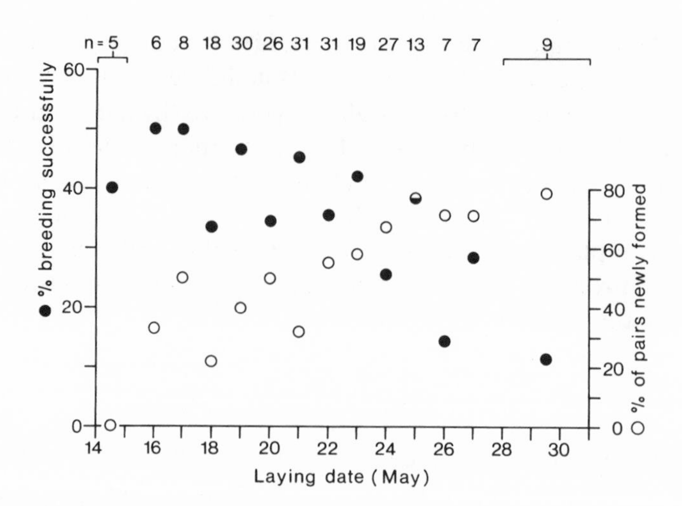

Figure 17.5: Numbers of pairs (n) laying on each date, and the percentage of these that are successful (*closed circles*) and breeding together for the first time (*open circles*).

lays is likely to affect breeding success because of changes in weather and food availability as the season progresses.

Complete data on egg laying are available for 1960, 1961, 1962, and 1978. Pairs that lay earlier in the season are more successful, and there is a significant negative correlation between breeding success and laying date, $r = -.7628$, d.f. = 12, $p < .01$. However, established pairs and pairs with a male that has bred before breed significantly earlier than newly formed pairs or pairs with a male with no experience (Ollason and Dunnet 1978). In addition, the proportion of newly formed pairs laying on any particular date increases as the season progresses (fig. 17.5), and this could explain the decline in breeding success. Thus newly formed pairs and established pairs were analyzed separately: the decline in success with season remains significant in newly formed pairs, ($r = -.6047$, d.f. = 11, $p < .05$), but not in established pairs ($r = -.4183$, d.f. = 10). The timing of breeding therefore has a greater effect upon the success of newly formed pairs than on that of established pairs, with those that lay late being relatively less successful.

Egg Size

The size of the egg may be a function of the previous breeding experience of the female, since experience influences the timing of breeding, and food may be more available early in the season.

Egg size was measured by its length, breadth, and weight within a day of laying. Egg volumes were determined using the formula $0.85\ (\pi\ \mathrm{LB}^2)/6$, where L is length and B is breadth (Romanoff and Romanoff 1949). There is a close correlation between egg volume calculated with this formula and fresh egg weight ($r = .9197$, d.f. = 87, $p < .001$), suggesting that weight gives a reasonable estimate of volume.

Eggs were measured in 1975, 1976, and 1978. Significant positive correlations were found between length of breeding experience of females and their egg breadth (r = .2724, d.f. = 99, p = .006), egg volume (r = .2256, d.f. = 99, p = .023), but not egg length. Thus more experienced females tend to lay broader and larger eggs. Egg volume varies significantly more between females than within females between years (F ratio = 2.682; d.f. = 15, 18; $p < .05$). There is also significant constancy of laying date between years: in a sample of twenty-two females, laying date in 1960 correlates significantly with laying date in 1961, $r_s = .5251$, $p < .012$. Thus the constancy of laying date within females may explain the lower variation in egg volume within females than between females.

To test whether larger or broader eggs are more successful than smaller or narrower ones, we compared the success of eggs laid by the same female in different years. In fifty-six females that laid at least twice, the success of their larger eggs was not significantly different from that of their smaller ones, nor were broad eggs better than narrow.

The Pair

A pair may consist of two birds without any known previous breeding experience; one bird without known experience and one with, or two birds with experience that may or may not have bred together before. Pairs in which the birds have bred together for at least one year before are termed "established" pairs. The remaining pairs, "newly formed" pairs, have been divided into those in which neither bird has any known experience (new inexperienced) and those in which at least one bird has breeding experience (new experienced).

Comparing the breeding success of new inexperienced pairs with that of new experienced pairs shows that the latter fledge a significantly greater percentage of their chicks (table 17.8) and that established pairs are significantly more successful than new experienced pairs at hatching eggs and in overall fledging success. An experienced mate is therefore a significant advantage in terms of breeding success.

If new inexperienced and new experienced pairs are combined and compared with established pairs, established pairs are significantly more successful at hatching ($\chi^2_1 = 17.495$, $p < .001$), fledging ($\chi^2_1 = 11.706$, $p < .001$), and overall ($\chi^2_1 = 29.492$, $p < .001$). Birds with a familiar mate do significantly better than those with a new mate.

Does breeding failure increase the probability of changing mates or sites? In the year immediately following failure, 14% (32/228) of birds change mates or sites, compared with only 7% (32/468) after success (χ^2_1 = 8.669, $p < .01$). Considering changes of site (regardles of mate): after failure, 11% (34/228) change sites, compared with only 4% (19/468) after success ($\chi^2_1 = 9.972$, $p < .01$). When the mate is retained and the pair change sites together, success shows some improvement: of forty-nine established pairs that changed sites, success before the change was 37% (18/49), compared with 51% (25/49) after, though this difference is not significant.

Table 17.8 Breeding Success of Newly Formed Pairs and Established Pairs of Fulmars

Measure	Newly Formed Pairs: *Both Birds without Experience*		Newly Formed Pairs: *At Least One Bird with Experience*		Established Pairs
Hatching success (chicks/eggs)	130/204		448/702	$\chi^2_1 = 15.211$	726/997
%	63.7		63.8	$p < .001$	72.8
Fledging success (fledglings/chicks)	92/130	$\chi^2_1 = 6.343$	365/448		627/726
%	70.8	$p < .02$	81.5		86.4
Overall success (fledglings/eggs)	92/204		365/702	$\chi^2_1 = 19.677$	627/997
%	45.1		52.0	$p < .001$	62.9

Note: Two-tailed tests.

Considering all changes of mate (regardless of site): after failure, 7% (15/228) change mates, not significantly different from the 4% (18/468) that change after success. Changing mates does not improve success: of forty-one birds that changed mates, success before the change was 56% (23/41), not significantly different from 46% (19/41) after.

Thus change of site is rare, and allowing for mortality of mate, change of mate—divorce—is even rarer. Change of site is significantly more frequent in the year following breeding failure than in that following success, and where the mate is retained, breeding success tends to improve.

17.6 Discussion

In many long-lived birds such as the fulmar, there is considerable range in the ages of first breeding of different individuals. In some, including the Adélie penguin, *Pygoscelis adeliae*, females are on average younger than males when they begin breeding and have poorer survival (Spurr 1975; Ainley and DeMaster 1980). Conversely, in kittiwakes, *Rissa tridactyla*, males begin breeding sooner than females (Wooller and Coulson 1977). As in the fulmar, males of the Laysan albatross, *Diomedea immutabilis*, are younger than females at first breeding, but their survival rates do not differ significantly (Fisher 1975). In both sexes of this species, individuals that begin breeding earlier than usual live shorter lives, possibly owing to strain in the early breeding years. In the fulmar, females that begin breeding sooner than the average age at first breeding are significantly less successful in their first year than those that first breed at older ages; and females that are more fecund in their early breeding years have shorter breeding life spans.

Across mongamous bird species, there appears to be no consistent pattern in whether males or females breed first, making it difficult to explain deferred maturity in terms of feeding ability and nutritional requirements for laying and incubation (Carrick 1972). In addition there is considerable variation in age at first breeding in different species, from an average of two years in the common diving petrel, *Pelecanoides urinatrix*, (Richdale 1965) to thirteen years in the gray-headed albatross, *Diomedea chrysostoma* (Croxall 1981, citing Prince et al., unpublished), often exceeding 50% of the mean adult life expectancy. No comparative study of the significance of the length of prebreeding life has been undertaken.

Perrins (1970) suggests that in general the earliest-hatched young survive best, provided the female has sufficient time to obtain resources to produce the clutch. Thus in many species there is a negative relation between success and laying date, and between length of breeding experience and laying date (e.g., Nelson 1966; Serventy 1967; Mills 1973; Fisher 1975; Lloyd 1979; Harris 1980; Mills and Shaw 1980; Thomas 1980). This is also true in fulmars, though the negative relation between success and laying date holds only for new breeders, as in the gannet *Sula bassana* (Nelson 1966).

Breeding success commonly improves with experience (e.g., Nelson 1966; Coulson and Horobin 1976; Davis 1976) and duration of the pair-bond (Coulson 1972; Mills 1973; Davis 1976; Brooke 1978), as in fulmars (Ollason and Dunnet 1978). A decline in various components of breeding (clutch size, egg volume, laying date) in older birds has been found by in some studies (e.g., Richdale 1957; Coulson and Horobin 1976; Mills and Shaw 1980), but a decline in production of young has seldom been demonstrated.

In older kittiwakes Thomas (1983) discovered some decline in fledgling production independent of egg volume. In fulmars, we show a significant decline in fecundity and overall success in the years near the end of breeding life, which can occur at different absolute ages and is associated with better-than-average success early in life. It has not so far been possible to demonstrate any age-related decline in adult survival rate.

A number of studies show that divorce and re-pairing is more likely after breeding failure than after success (e.g., Coulson 1966; Brooke 1978; Newton and Marquiss 1982). In the fulmar, change of site (which sometimes involves change of mate) is more likely after breeding failure. When an established pair moves to a new site there is some increase in success.

Curio (1983) suggests that the poor breeding performance of young birds may occur either because poor breeders have poor survival and short lives (i.e., are "poor-quality" birds) and better breeders have better survival and longer lives (i.e., are "good-quality" birds) or because of reproductive constraint or restraint. In fulmars the former is unlikely because of the weak correlation between breeding life span and fledging success. There is large variation in success between individuals, but both good and bad breeders can have short or long breeding life spans.

In the fulmar, variation in breeding success, laying date and egg size is greater between individuals than that within individuals (similar to results of Nelson 1966; Spurr 1975; Brooke 1978; and Thomas 1980). Thus birds tend to be individually consistent in their performance.

17.7 Summary

1. Fulmars, *Fulmarus glacialis* (L.), are pelagic, iteroparous, monogamous seabirds. Breeding adults have a low mortality rate (3.25%, s.e. = 0.43%) and expectation of further life of over thirty years, not significantly different in males and females.

2. The modal age at first breedings is eight years, ranging from six to nineteen years. Males begin breeding sooner than females.

3. Fulmars that first breed at younger-than-average ages tend to be less successful in their first year than those that are older. This difference is significant in females.

4. On average, about one-third of breeders are not identified in the breeding area each year. Although some may have been missed, there is almost certainly some intermittent breeding that is more frequent early in breeding life, especially in females.

5. Breeding success improves with experience over approximately the first nine breeding years. Thereafter it remains relatively constant until at least the twenty-first breeding year.

6. Comparing fulmars that have completed their breeding lives with those that continue breeding shows that the former tend to be more successful in their middle years, with a significant decline toward the end of their breeding life.

7. In females, high fecundity or success in the early or middle breeding years is negatively correlated with length of breeding life span.

8. The most important factor affecting lifetime fledgling production is length of breeding life span. Thus those fulmars that live longest produce most offspring.

9. There is more variation between individuals than within individuals in breeding success, laying date, and egg size.

10. Established pairs are more successful than new pairs that have one bird with previous breeding experience, and these are more successful than new pairs in which neither bird has previous breeding experience.

11. Established pairs lay earlier than new pairs. New pairs that lay early are more successful than new pairs that lay late.

12. Change of nest site is more frequent after breeding failure than after success. Established pairs that change site are more successful at their new site than at their old one.

17.8 Acknowledgments

We are very grateful to A. Anderson of the University of Aberdeen for his long-term assistance, and to the numerous others who have helped at some time over the years. We appreciate the comments of I. J. Patterson of the University of Aberdeen and T. H. Clutton-Brock of the University of Cambridge on earlier versions of the manuscript. The University of Aberdeen Computer Centre provided facilities for the analysis of data.

18 Individual Variation in Reproductive Success in Male Black Grouse, *Tetrao tetrix* L.

J. P. Kruijt and G. J. de Vos

In black grouse pair-bonds do not exist, and males compete with each other for territorial ground and for copulations with females. Territory owners spend much time displaying in spring, and each is strongly attached to one or several small display sites within the territory. Males may display solitarily, but often the display sites of several males are aggregated on places called arenas or leks. Especially on arena centers, where display sites of different males are clustered most closely together, intermale competition is severe, and as a consequence, central males on arenas often have much smaller territories than peripheral ones or males with solitary display sites.

In spring, females usually visit several displaying males before choosing a mating partner and copulating with him on his territory. Males vary strongly in mating success, and this is caused by both intermale competition and female choice. In this chapter we synthesize results of a study on the influence of these factors on mating success of males in a population in the Netherlands between 1965 and 1981 (Kruijt and Hogan 1967; Kruijt, de Vos, and Bossema 1972; de Vos 1979, 1983).

18.1 Study Population and Methods

In most study years at least 80% of all males were individually marked. The estimated number of males alive in the mating period varied from about fifty (in 1965) to ten (in 1981). The number of females cannot be estimated as accurately because a much smaller percentage was marked. In most years the adult sex ratio probably approximated one, but for unknown reasons, during six years (1968–73) females undoubtedly outnumbered males.

The population range consists of about 1,400 ha of heather and moorland surrounded by meadows and arable land. It contained seven traditional display grounds (locations where display sites of one or more territory owners are situated), 1.0 to 1.4 km apart. Five were on meadows, the other two on heather and moorland.

Observations on one or more display grounds during the early morning were carried out almost daily in the mating period. Most data were gathered on arena A, which was situated on meadows. This was the only display ground closely observed in all study years and also the only one where an arena was present all the time.

18.2 Life History of Males and Females

Males caught and marked in their first year of life stayed alive 0–8 mating periods, averaging 4.1. Males defended territories throughout the year, but the percentage that had a territory was highest in the mating period and varied with age of the males. About 30% of the first-year males (juveniles) and 85% of the other males (adults) defended a territory in the mating period. As a rule, juveniles stopped territory defense during molt in summer, whereas adults remained territorial.

Males mostly defended a territory on a single display ground in the course of their life. Territory shifts were common but usually occurred between positions on the same display ground. On arenas (i.e., display grounds with two or more territorial males) newly established males mostly defended a marginally located territory first and subsequently shifted their territory toward the arena center. As a result, older males were more often in possession of a centrally located territory than younger ones. However, males older than five years seldom had a central territory: the ability of males to maintain themselves in positions with intense competition decreased after their fifth year (de Vos 1983).

Females that were caught and marked in their first year stayed alive 0–7 mating periods, averaging 2.5. On average, about 35% of the marked females known to be alive in a mating period were not observed to copulate, but their copulations may simply not have been observed. Both juvenile and adult females participated in copulating, and the percentage of individuals observed to copulate in both groups was approximately the same. Most females (about 80%) mate only once per season; the remaining ones mate two or three times, usually with the same male. After mating females may continue to visit display grounds for foraging.

18.3 Variation in Copulation Success among Males

Copulation frequency varied widely between males. Males without a territory seldom copulated. For instance, on arena A on average about 40% of all males observed possessed no territory, but this was the case for only 5.7% of the males observed to copulate. Males without a territory performed only 2.4% of all copulations. Though territorial males often trespassed on other males' territories when these were visited by females, they copulated only sporadically during such excursions, and copulations were almost exclusively reserved for males that were on their own territories.

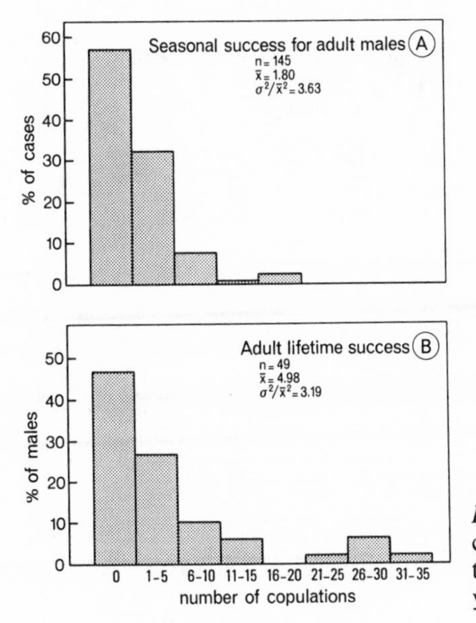

Figure 18.1: Variation in seasonal and lifetime copulation success for adult males with territories on arena A. Data from the years 1965–81.

Copulation success also varied among territorial males. Table 18.1 and figure 18.1*a* show that in all mating seasons except 1969, many of the males with territories on arena A were not observed to copulate (58% on average), whereas others sometimes performed a large number of copulations (maximally twenty). Lifetime copulation success, shown in Figure 18.1*b*, also varied highly. The standardized variance ($\sigma^2/\bar{x}^2$) in seasonal and lifetime copulation success was about the same.

Lifetime copulation success of a territorial male depends on two factors: number of years with a territory (L), and mean number of copulations per year (F). Tables 18.2 and 18.3 show that variation in F was most important in causing variance in adult lifetime copulation success among territorial males. Of the overall standardized variance, 52.4% was due to variation in F and only 8.9% to variation in L. The remaining 38.7% was due to simultaneous variation in L and F (simultaneous independent variation, 23.6%; covariation, 15.1%).

The chance of copulating was influenced by sex ratio in the period when males defended a territory. For instance, of males with a territory on arena A, 60% copulated on average for the years 1968–73, when females outnumbered males, and 35% copulated on average for other years; the mean number of copulations per male per mating period was 3.20 and 1.15, respectively. The influence of sex ratio is partly responsible for the variation in seasonal copulation success among males shown in Figure 18.1*a*, but it cannot explain why several males had such extreme success compared with others with a territory on arena A in the same year (see table 18.1). Undoubtedly, these large differences reflect differences between males in their ability to achieve copulation success (see below).

Table 18.1 Distribution of Copulations over Individual Black Grouse Males (Coded in Parentheses) with Territories on Arena A

Year	Number of Copulations (Male)	Number of Males	Percentage with Copulations	$\bar{x}$	$\sigma^2/\bar{x}^2$
1965	5(019), 4(030), 1(006,014,032), 0(009,029,035,X0)	9	56	1.33	1.97
1966	4(019), 3(014), 2(031), 0(005,008,009,013, 016,021,028,029,032,035,039,045)	15	20	0.60	4.68
1967	8(009), 4(031), 3(029), 2(002,008), 1(006), 0(013,014,016,033,035,039,041)	13	46	1.54	2.37
1968	20(009), 5(054), 1(024), 0(008,013,016,047,059)	8	38	3.25	4.62
1969	16(054), 4(059)	2	100	10.00	0.72
1970	17(059), 8(054), 1(064,067,073), 0(058,X1)	7	71	4.00	2.54
1971	14(059), 2(064,079,080,085), 1(070), 0(058,060,077)	9	67	2.56	2.95
1972	6(064), 5(085), 3(080), 1(060), 0(084,088)	6	67	2.50	1.07
1973	10(085), 4(064), 1(060,080), 0(083,088,089,X2)	8	50	2.00	3.07
1974	9(085), 2(088,089), 1(064,092), 0(060,X3,X4,X5)	9	56	1.67	2.97
1975	9(088), 6(089,103), 5(101), 0(092,093,097,098,X6)	9	44	2.89	1.54
1976	6(101), 4(115), 1(088,089,103), 0(092,097,098, 099,100,106,113,119)	13	38	1.00	3.50
1977	7(115), 2(119), 1(101), 0(089,092,097,098,099, 100,106,109,111,121)	13	23	0.77	6.52
1978	9(115), 5(108), 1(119,121), 0(092,097,098,099, 100,101,109,123,126)	13	31	1.23	4.86
1979	5(115), 2(108), 1(100,101,119), 0(092,097,098, 121,123)	10	50	1.00	2.44
1980	3(108), 0(097,100,101,114,115,117,119,121,123)	10	10	0.30	10.00
1981	4(126), 0(115,119,121,123)	5	20	0.80	5.00

Note: $\bar{x}$ = mean number of copulations per male; $\sigma^2/\bar{x}^2$ = standardized variance in number of copulations performed by different males.

Table 18.2 Mean and Variance of the Components of Lifetime Copulation Success in Adult Black Grouse Males with Territories on Arena A

Component	Original		Standardized Variance
	Mean	*Variance*	
L	2.612	3.076	0.451
F	1.512	6.065	2.652
LF	4.980	78.979	5.061

Note: Data from the years 1965–81.
L = number of years male held a territory while adult; F = mean number of copulations per year.

Table 18.3 Percentage Contributions of the Components of Lifetime Copulation Success to Its Variance for Adult Black Grouse Males with Territories on Arena A

Component	L	F
L	8.91	
F	38.70	52.40

Note: Data from the years 1965–81.
L = number of years male held a territory while adult; F = mean number of copulations per year.

18.4 Copulation Success of Territorial Males in Relation to Positional Status and Age

Among arena males, those with a central territory were most successful in copulating, as is illustrated in figure 18.2. Central males and others are distinguished by whether a territory is at least partly located on the arena center and completely surrounded by territories of other males. This distinction is not completely satisfactory, since a male may be strongly attached to a display site on the arena center but would still not be considered a central male if no other males established territories at the periphery of his territory. Undoubtedly, the degree of clustering of males' display sites with those of other males would show an even better correlation with copulation success. However, this criterion, because it requires measuring time spent displaying in the neighborhood of other males, is much more laborious to obtain.

Figure 18.2 shows that seasonal copulation success of males with a territory on an arena is correlated both with positional status (Mann-Whitney U test, two-tailed, $p = .012$) and age (Kruskal-Wallis test, d.f. = 6, $p < .02$). The data show that first- and second-year males as well as males beyond the age of five years are less successful in copulating than third- to fifth-year males. Up to their fifth year, older males were more often in possession of a centrally located territory than younger ones, but sixth-year and older males seldom had central territories. This relationship between age and positional status is about the same as that between age and copulation success, and this raises the question whether the age effect on copulation success might be entirely due to a position effect, or vice versa. Table 18.4 shows that probably both effects are important: within the group of males with a central territory the younger and oldest males have lower success (Kruskal-Wallis test, d.f. = 2, $p < .05$); and within the group of third- to fifth-year males those with a central position have more success than others (Mann-Whitney U test, two-tailed, $p = .0120$).

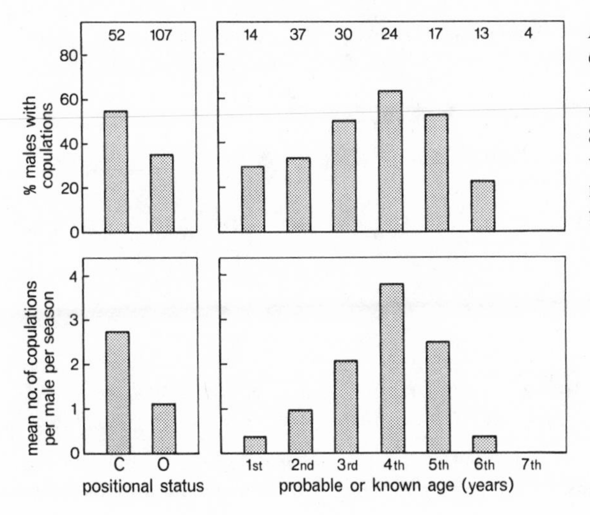

Figure 18.2: Copulation success of males with territories on arena A in relation to age and positional status. Data from the years 1965–81. Positional status: C, males with central territories, O, other males with territories. For definition of central territory, see text.

With regard to lifetime copulation success, figure 18.3 illustrates the importance of possessing a territory in the optimal age period (fig. 18.3*a*), of number of mating periods with a territory in these years (fig. 18.3*b*), and of central position of the territory (fig. 18.3*c*) (Mann-Whitney U test, one-tailed, a: $p < .0003$; b: $p = .0233$, c: $p < .05$). However, a central

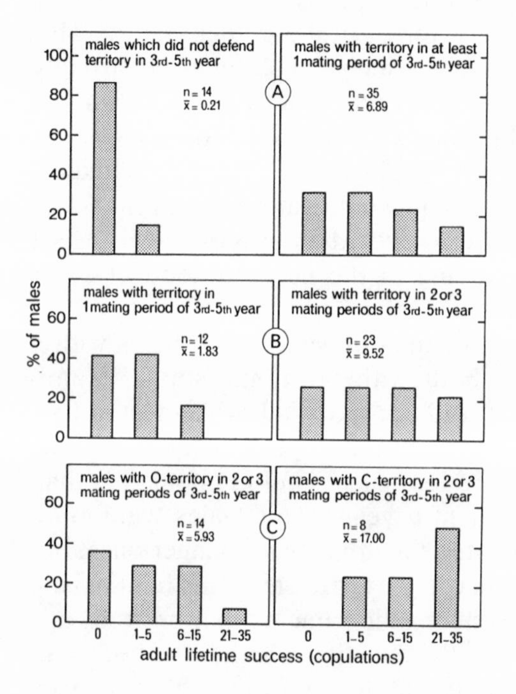

Figure 18.3: Lifetime copulation success of adult males with territories on arena A in relation to age period in which they defended the territory and to location of the territory. Data from the years 1965–81. For definition of C (central) and O (other) territory, see text.

Table 18.4 Number of Copulations Performed by Black Grouse Males with Territories on Arena A in Relation to Their Age and Positional Status

Probable or Known Age and Position	Number of Copulations per Male per Mating Season: *0*	*1*	*2*	*3*	*4*	*5*	*6*	*7*	*8*	*9*	*10*	*14*	*16*	*17*	*20*	Number of Males	Percentage with Copulations	$\bar{x}$	$\sigma^2/\bar{x}^2$
First or second year																			
Central	8	1	2	0	1	0	0	0	0	0	0	0	0	0	0	12	33	0.75	2.95
Other	27	4	4	1	0	2	1	0	0	0	0	0	0	0	0	39	31	0.79	3.76
Both	35	5	6	1	1	2	1	0	0	0	0	0	0	0	0	51	31	0.78	3.53
Third, fourth, or fifth year																			
Central	9	6	3	0	1	2	1	1	1	2	1	1	0	1	1	30	70	4.20	1.66
Other	23	8	1	0	3	1	2	0	1	1	0	0	1	0	0	41	44	1.76	3.50
Both	32	14	4	0	4	3	3	1	2	3	1	1	1	1	1	71	55	2.79	2.55
Sixth or Seventh year																			
Central	3	0	0	1	0	0	0	0	0	0	0	0	0	0	0	4	25	0.75	4.00
Other	11	2	0	0	0	0	0	0	0	0	0	0	0	0	0	13	15	0.15	5.96
Both	14	2	0	1	0	0	0	0	0	0	0	0	0	0	0	17	18	0.29	6.89

Note: Data from the years 1965-81; $\bar{x}$ = mean number of copulations per male; $\sigma^2/\bar{x}^2$ = standardized variance in number of copulations performed by males. For definition of central status, see text.

territory and optimal age are no guarantee that a male will achieve high copulation success. Table 18.4 shows that per mating season about 30% of the third- to fifth-year central males obtained no copulations at all, and another 30% obtained at most two copulations. Figure 18.3 shows that males with a central territory in two or three mating periods of their third to fifth year always had *some* success during their lifetime, but 25% still had little success (one to five copulations). This indicates that properties of males other than age and possession of a central territory influence copulation success.

18.5 The Role of Intermale Competition and Female Choice in Causing Variation in Copulation Success among Males

An important aspect of sexual selection in black grouse is that males probably cannot constrain the choice of females. Forced copulations have never been observed, and males do not guard females for prolonged periods. Females clearly have the opportunity to copulate with any male, and we never saw a male refuse to copulate with a soliciting female. Occasionally males attempt to interrupt the copulation of another male. However, in our population this happened seldom, and we obtained no evidence that it had an important influence on copulation success of interrupting or interrupted males. Furthermore, even if it had, such a result would not imply that female choice is constrained, because if a female's copulation is interrupted, she still has the option to copulate with the same male later.

The inability of males to coerce females does not mean that intermale competition is unimportant, since females may use the outcome of intermale competition as a criterion of their choice. For example, males compete strongly for territorial ownership, and the whole population distribution range is divided into territories. Establishment of territories is often preceded by vigorous fighting. Despite this, territorial males often tolerate other males on their territory for prolonged periods as long as intruding males behave inconspicuously, even when females are visiting. Display by an intruding male, however, usually elicits a chase by the territory owner, so that nonterritorial males are unable to court females for a prolonged period. This outcome of intermale competition affects the choice of females: females usually require prolonged periods of courtship before they are willing to copulate, and consequently nonterritorial males have little success in copulating.

18.6 Which Cues Are Used by Females?

The density of displaying males is highest at the center of an arena, and females might use the latter characteristic as a cue in their choice between territorial males. Kruijt, de Vos, and Bossema (1972) found that females tended to land preferentially near a cluster of six stuffed males as compared with a cluster of three males exposed simultaneously at some

Figure 18.4: Rookooing black grouse male.

distance near an arena. However, clustering of males is not the only criterion females use, and solitarily displaying males can be very successful in competition for copulations with males displaying on an arena.

What cues might females use in addition to clustering of males? There is no indication that morphology is important. Though juvenile males often have smaller wattles than adult males, actively courting adult males do not differ obviously in this respect. Nor was there any consistent relationship between copulation success and body weight measured in the previous winter.

Is there any evidence that females use behavioral properties of males as a cue? It has been suggested that in lekking grouse species, females preferentially mate with dominant males (Scott 1942; Wiley 1973). In black grouse each male is dominant over all others on his own territory in terms of priority of access to ground or ability to expel all other males, so that central arena males cannot be considered dominant over marginal males, and there is no reason to believe that dominance can be used as a cue in mate choice by females. Dominance could possibly be defined in a more subtle way—for example in terms of behavioral differences between males during boundary disputes. If such differences were detectable by females and affected their choice, male "dominance" might still be related to female choice (cf. Bradbury and Gibson 1983). However, this has yet to be demonstrated. Males differ in courtship tactics, and this might also influence their chance of being chosen by females (Kruijt and Hogan 1967; Kruijt, de Vos, and Bossema 1972). For example, males confronted with a female about to leave their territory differ in their tendency to drive her back, which increases the duration of her visit. However, so far there is no direct evidence that differences in courtship tactics contribute to differences in copulation success.

During the last few years of our study period (1979–81) we have examined variability in rookooing performance of males. One indication that rookooing (see fig. 18.4) may act as an important signal for females is that rookooing frequency is clearly correlated with the presence of females. However, the percentage of time spent rookooing did not appear to correlate with copulation success of males. We therefore examined the possibility that females discriminate between males on the basis of rookooing structure. Rookooing consists of strings of phrases, each lasting about 2.5 to 2.9 sec. Each phrase may or may not be preceded by one or several incomplete phrases (intention utterances), each of about 0.25 sec (Hjorth 1967, 1970). Males differed in the ratio between number of intention utterances and number of phrases produced (intention/phrase ratio), but though successful males usually produced few intention utterances, less successful males sometimes surpassed them in this respect. The average time it takes for each phrase to be produced appears to be a more promising measure. First, when females were present, males produced phrases which were about 0.1 sec shorter than when females were absent (fig. 18.5). Second, males differed in the duration of phrases when they were in the presence of females (one-way analysis of variance, $p < .001$ for all three study years) and those of males successful in copulating were significantly shorter (fig. 18.5). The differences in phrase duration between males already existed before the copulation period had started, so that it is unlikely they are consequences rather than causes of female preferences.

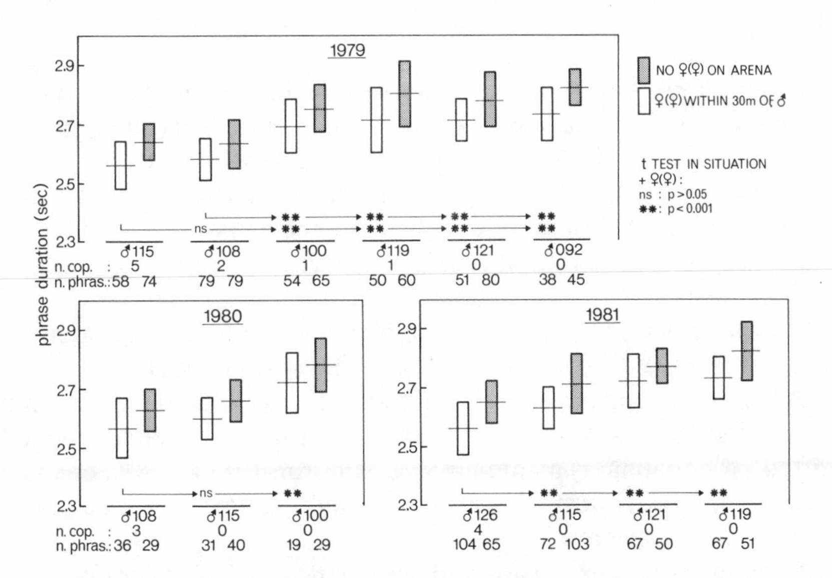

Figure 18.5: Variation in rookooing phrase duration and copulation success for males with territories on arena A. Horizontal lines and vertical bars indicate mean ± standard deviation.

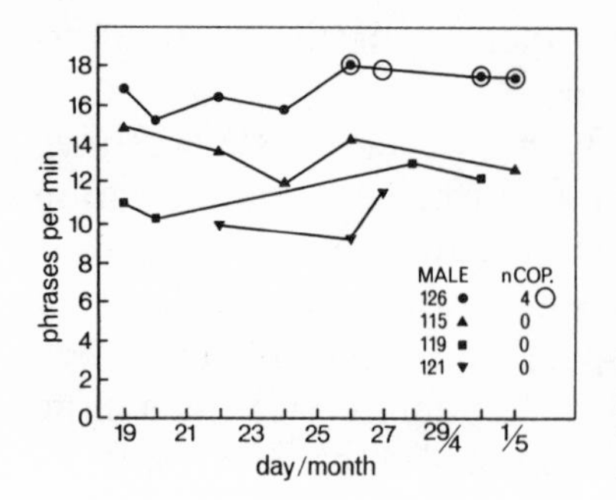

Figure 18.6: Intermale differences in rate of rookooing phrase production in presence of females. Data from 1981 for males with territories on arena A in situations where females were within 30 m.

Females might use the rate at which complete phrases are produced as a cue for discrimination. Phrase production rate depends on the intention/phrase ratio, but also on the duration of phrases. As figure 18.6 illustrates for one year, rate of phrase production was highest for the most successful male.

We conclude that of all possible cues mentioned, rookooing performance and maintaining a central status on arenas appear to be the most likely ones females use in choosing between males.

18.7 Discussion

As we have seen, seasonal as well as lifetime copulation success among adult black grouse males territorial on an arena varies highly. Variation in seasonal success does not necessarily reflect variation in lifetime success (Clutton-Brock 1983), but in our case they appeared to be similar. It is worth noting here that our measures may not accurately reflect variation among all adult males in the population. For instance, we did not systematically measure copulation success of solitary males and males without a territory. The latter males had little chance of copulating either on arenas or elsewhere, because territorial males interfered with their courtship. Several solitary males, however, were observed copulating. A number of these obviously were not very successful, but one, observed during eight years, appeared to be extremely attractive to females and had a higher lifetime copulation success than the most successful male on arena A in our study period (fifty-four vs. thirty-five copulations). It can be concluded from these observations that copulation success among males without a territory on an arena is also highly variable, but we do not know whether it is more or less variable than among arena males.

Age has a strong effect on copulation success of arena males: third- to fifth-year individuals were more successful. Similar age effects have been reported for polygynous mammals (e.g., Clutton-Brock 1983), but as far as we know our study provides the first well-documented example for a polygynous bird species.

In addition to age, position of the territory is an important determinant of copulation success of arena males: those with a central territory were most successful. This effect is most clearly expressed in third- to fifth-year males, that is, males that are in the optimal age period.

Because males cannot coerce females to copulate with them, copulation success of black grouse males depends on female choice. Possession of a central territory in the optimal age period is no guarantee that a male will achieve high copulation success. Our study indicates that an additional aspect of attractiveness for females is rookooing performance: females choose males that produce short phrases in rapid succession.

Possession of a territory is necessary for a male to be successful in copulating. In general, territory maintenance requires good physical condition of males, and central arena males are put to the test in particular (de Vos 1983). There is also evidence that the physical condition of males deteriorates after their fifth year (de Vos 1983). So by choosing third- to fifth-year arena males with centrally located territories, females select males in good physical condition. This may also be true for choice of males on the basis of their rookooing performance: the ability of males to produce short phrases in rapid succession might well depend on their physical condition.

How should males maximize their lifetime copulation success? It is clearly important that they optimize their attractiveness to females, because this is the most important determinant of lifetime copulation success. Male options to increase attractiveness to females include decisions about time and site of territory establishment and shifts in the location of the territory. Consequences of these decisions with respect to physical condition must also be taken into account because they influence life span as well as attractiveness to females. For instance, de Vos (1983) presented evidence that occupying a territory increases mortality and thus probably affects physical condition negatively. This cost is inevitable because nonterritorial males are not attractive to females. However, a male should decide to establish a territory only at a time when he can cope with the increased demands on his physical condition. Options of males to increase their attractiveness to females need not necessarily involve costs with respect to physical condition and survival. For instance, over the years arena males attempt to shift their territories gradually to the center. This results in increased defense costs. However, de Vos (1983) showed that mortality rate is somewhat higher for marginal males than for central arena males, possibly because clustering yields ecological advantages (de Vos 1979). For sharp-tailed grouse, Moyles and Boag (1981) also found that males occupying marginal territories disappeared in greater proportions, and probably died, than those with central territories. Shifting the territory toward the center may thus on balance affect survival positively as well as improving chances of copulating.

Apart from opting for high attractiveness, males should establish their territories on sites with a high female-encounter rate. In our population, males establish territories on sites where they have obtained the opportunity to court females (Kruijt, de Vos, and Bossema 1972; de Vos 1983). The chance of nonterritorial males to obtain courtship experience on a particular site depends both on the number of females visiting that site and on the number of already established males. The mechanism underlying territory establishment promotes the choice of hot spots (Bradbury and Gibson 1983) with good chances for high female-encounter rates.

18.8 Acknowledgments

We are grateful to I. Bossema, O. H. Bruinsma, and M. D. Smit for cooperation in research, and to T. H. Clutton-Brock for criticism of the manuscript.

19 The Effect of Breeding-Unit Size on Fitness Components in Groove-billed Anis

Sandra L. Vehrencamp, Rolf R. Koford, and Bonnie S. Bowen

FOR COOPERATIVELY BREEDING ANIMALS, the relationship between fitness and breeding unit size is the issue of greatest interest to evolutionary biologists (Brown 1983). By discovering which components of fitness increase and decrease as a function of unit size and identifying the ecological and social variables that affect these components, one can evaluate the benefit and cost trade-offs that favor the evolution of cooperation. However, there are at least three practical complications in the analysis of fitness as a function of breeding-unit size, particularly in iteroparous species. First, it is often difficult to know which offspring in a communal brood belong to which mother. This is especially true for egg-laying and burrow-living cooperative breeders (Rood 1975; Bertram 1979; Noonan 1981; Strassmann 1981b). Invasive observational procedures, time-consuming continuous monitoring, and probabilistic genetic analyses can be used, but each technique has its drawbacks (Metcalf and Whitt 1977a; McCracken and Bradbury 1977). Second, breeding-unit size is often correlated with ecological variables, so that observed correlations between unit size and fitness could be a result of ecological parameters (Brown et al. 1982). Third, the unit size in which an individual breeds may vary over its lifetime, making difficult a meaningful measure of lifetime fitness as a function of breeding unit size.

In this chapter we illustrate the ways we have circumvented these difficulties in analyzing the effect of breeding-unit size on fitness in communally nesting groove-billed anis (*Crotophaga sulcirostris*, Cuculidae). All three ani species (*Crotophaga* spp.) are iteroparous cooperative breeders in which several females lay eggs in a single nest (Davis 1940, 1942; Skutch 1959; Köster 1971). Preliminary studies of the groove-billed ani (Vehrencamp 1978) indicated that both annual reproductive success and annual survivorship were affected by the size of the breeding unit, making an estimate of lifetime fitness highly desirable for comparisons of individuals breeding in different unit sizes. However, birds did not breed in the same-sized group year after year. Ecological variables such as amount of vegetative cover on the territory were strikingly correlated with unit size and confounded the analysis. Finally, although all females in a cooperative unit usually contributed eggs to the commual clutch, biases in egg ownership often occurred as a function of social role. While it was relatively easy to determine which eggs belonged to each female, it was not possible to determine which nestling came from which egg without attracting predators to the nest. Therefore, to evaluate the effect of unit size on fitness, we separately analyzed two components of fitness: mean annual reproductive success, computed here by dividing the total number

Figure 19.1: Groove-billed ani.

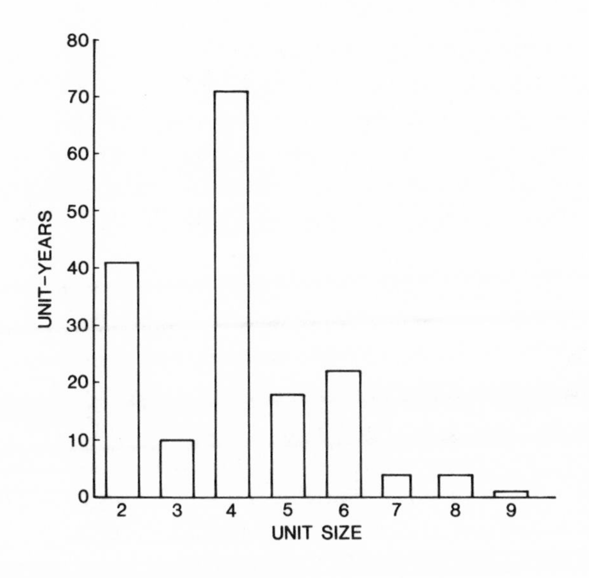

Figure 19.2: Frequency histogram of breeding unit size in Guanacaste population, 1978–82.

of offspring raised by the number of females or pairs in the unit, and annual adult survivorship. For each component we first examined the simple (bivariate) relationship between unit size and fitness and then performed a multivariate analysis to control for the possible confounding effects of ecological variables (Dixon 1981; Nie et al. 1981). To evaluate the effect of social roles on fitness, we determined the average bias in egg ownership as a function of role and the effect of role on survivorship; we then employed multivariate analyses to determine the traits correlated with roles for each sex. The two components of fitness could then be calculated as a function of sex, role, and unit size and combined multiplicatively to estimate lifetime reproductive success for a hypothetical bird that spends its entire life in a particular unit size and a particular role.

19.1 Breeding Biology of Groove-billed Anis

Groove-billed anis are medium-sized black birds (males = 83 ± 5 g, females = 74 ± 5 g) with long tails and crested bills (fig. 19.1). They are common throughout the lowland Neotropics and inhabit most nonforested habitats such as swamps, secondary growth, agricultural fields, parks, and pastures. Anis are weak fliers and forage terrestrially for insects.

During our five-year study (1978–82), breeding units ranged in size from two to nine birds. The frequency distribution of unit sizes is shown in fig. 19.2. Most units consisted of an even number of birds, with an equal number of males and females. Unit members sorted out into monogamous pairs; once mated, pairs remained together for the duration of the season and frequently longer. Breeding units therefore comprised one to four pairs, with two-pair units being by far the most common in our study area. Although we were not able to obtain electrophoretic data to confirm monogamous mating, we observed many copulations between mated pairs and only four between unmated birds; all four occurred at times that were not likely to result in fertilizations. Males guarded their females continuously during the egg-laying period and relaxed their defense only when laying had ceased. We therefore assumed that the egg ownership of a male was equivalent to that of his mate. Unmated extra males probably did not contribute genes to the communal clutch, as a result of mate guarding. Extra females usually bonded bigamously with one male in the unit and laid eggs along with the other females.

Each breeding unit defended a communal nesting and feeding territory during the six-month breeding (rainy) season in Guanacaste, Costa Rica. Territoriality often broke down in the dry season, and neighboring units frequently merged to form roving flocks. However, breeding units usually retained their identity and reformed on the same breeding territories in subsequent years. At the onset of the breeding season, each unit cooperatively constructed a nest, in which all the females laid eggs. The number of eggs incubated averaged about four per female, regardless of unit size. All members contributed to incubation, nestling feeding, nest defense, and territory defense, but the amount of effort expended depended on sex and breeding role. Nest predation was very high in our population. Most breeding units renested within a month if a nest was lost. Because the breeding season was relatively long, several nesting attempts were usually recorded for each unit. If a nest was successful (one or more nestlings raised to independence), a second brood was often attempted. In our study area, 51% of all groups produced no young for the season, 41% raised one brood, and 8% raised two broods (n = 170 units).

The young were independent of adult feeding about one month after hatching. They remained on the parental territory, and if the adults attempted a second brood, the young helped with nestling feeding and nest defense. All surviving young stayed with the parental unit until midway into the dry season. At this point they began to disperse, oldest birds first. By the beginning of the next breeding season, most yearlings had become established as breeders either in the parental unit (8% of all surviving independent young) or in a new unit (13%) or had disappeared from the study area (75%). Young that hatched very late in the previous breeding season often remained on the parental territory as nonbreeders (4%) but eventually dispersed (Bowen, Koford, and Vehrencamp, in prep). Therefore, no evidence existed for delayed reproduction in our population.

19.2 Factors Affecting Annual Reproductive Success

We defined annual reproductive success as the number of young raised and still alive at the end of the breeding season. In our case it was an average value for the females (or pairs) in a breeding unit. Normally one would break down this component of fitness into at least two multiplicatively combined subcomponents, such as the number of young born or hatched and the probability of their survival to some specified age. Because of our inability to match nestlings with eggs in the anis, and because many nestings may be attempted during the long breeding season, with highly variable success, we have broken down annual reproductive success into two somewhat different, but still multiplicatively combined, subcomponents: the probability that a breeding unit will successfully raise one or more young in that season, and the average number of young raised per female in units that were successful. We analyzed these subcomponents separately for two reasons. First, since 51% of our units failed to produce any young for the season, mean annual reproductive success was not distributed in a manner that met the assumptions of parametric statistical analysis. Second, we had reason to believe that the two subcomponents of annual success might be differentially affected by unit size and ecological variables such as vegetation cover. Details of these analyses are presented in Koford, Bowen, and Vehrencamp (1986).

To analyze whether breeding units of certain sizes were more likely to be successful, we considered the proportion of successful unit-years for units with one, two, three, and four females. The proportions were .45, .56, .45, and .60, respectively, but these were not significantly different ($\chi^2 = 1.88$, $n = 170$ units, $p > .50$). An ordered χ^2 contingency test that takes into account a possible trend with increasing unit size (Everitt 1977) also was not significant. We then performed a stepwise logistic regression analysis on successful versus unsuccessful units that included the following independent variables: year, unit size, number of females, territory size, known nest sites, suitable nest sites, number of trees, tree area, brush area, total cover area, and percentage of tree cover; these variables are defined in table 19.1. Unit size and number of females both were significantly positively correlated with territory size, tree area, and percentage of tree cover. Only year and number of trees entered the stepwise multivariate model (table 19.1). Territory size, known nest sites, tree area, and total cover area had significant univariate χ^2 values but presumably did not enter the model because of their correlation with number of trees. Territory area, known nest sites, tree area, total cover area, and number of trees were all positively associated with successful units. Neither number of females nor unit size entered the model. This suggested that territory quality (specifically the amount of tree cover), but not unit size, was a critical determinant of successful reproduction. Since the single most important cause of failure was nest predation, we concluded that no nest defense advantage arises from communal nesting versus solitary pair breeding.

To analyze the reproductive success of successful units, we performed both an ANOVA and a linear regression analysis on the reproductive success data from successful units with one, two, three, and four females. For units of increasing size, the means were 2.25, 1.98, 1.29, and 1.0 young per female, respectively. These means were not significantly different in the ANOVA ($F = 1.98$, $n = 85$, $p = .12$), which treats the unit sizes as unordered categories and ignores trends with increasing group size. The regression analysis revealed a significant negative relationship between number of females in the unit and reproductive success, but number of females explained only a small proportion of the variance in reproductive success ($r^2 = .119$, $F = 12.6$, $p < .01$). The negative effect of more females resulted from the fact that larger units lost more nestlings from overcrowding, competition for food in the nest, and increased conspicuousness to predators. We then performed a stepwise multiple linear regression on mean reproductive success against the same dependent variables as above (table 19.2). Number of females was the first variable to enter the model. Territory size and number of known nest sites also entered the model, and both were positively associated with high

Table 19.1 Correlates of Unsuccessful Units versus Units Producing One or More Young for the Season in Groove-billed Anis

Variable	Unsuccessful Units (n = 45)	Successful Units (n = 49)	χ^2 Test	Logistic Regression
Year[a]	—	—	.0001*	.0001*
Unit size[b]	3.98 ± 1.32	4.01 ± 1.41	.832	—
Females[c]	2.02 ± 0.66	2.04 ± 0.74	.896	—
Territory size[d]	29,317 ± 19,492	40,430 ± 25,205	.009*	—
Known nest sites[e]	2.32 ± 1.41	3.14 ± 2.39	.031*	—
Suitable sites[f]	4.00 ± 2.54	5.12 ± 3.05	.053	—
Number of trees[g]	25.8 ± 21.8	40.0 ± 27.1	.006*	.008*
Tree area[h]	1,531 ± 1,345	2,386 ± 1,282	.006*	—
Brush area[i]	2,942 ± 4,685	5,381 ± 9,629	.115	—
Total cover area[j]	4,572 ± 4,897	7,767 ± 9,638	.039*	—
Percentage of tree cover[k]	6.2 ± 4.8	7.0 ± 4.4	.388	—

Note: Univariate χ^2 test and logistic regression analysis; means ± s.d. and p values are given in the table.

[a]Each year from 1978 to 1982 considered as a categorical variable.
[b]Number of adult birds in the breeding unit.
[c]Number of breeding females in the unit.
[d]Area of the territory in m^2.
[e]Number of known nesting sites per territory.
[f]Number of subjectively judged suitable nest sites per territory.
[g]Number of trees per territory.
[h]Area of tree crown cover in m^2.
[i]The area of brush cover more than 1 m high in m^2.
[j]Tree crown cover plus brush cover.
[k]Tree crown area divided by territory area.
*$p < .05$.

Table 19.2 Multiple Regression Analysis of Annual Offspring Production per Female

Variable	Simple r	Cumulative Multiple r^2	Beta	F	p
Females	−.40**	.158	−.42	11.66	<.005
Known nest sites	.31**	.282	.31	6.09	<.025
Territory size	.33**	.342	.25	4.16	<.05
Unit size	−.39**	—	—	—	n.s.
Percentage of tree cover	−.19	—	—	—	n.s.
Number of trees	−.10	—	—	—	n.s.
Total cover area	−.07	—	—	—	n.s.
Brush area	−.07	—	—	—	n.s.
Tree area	.15	—	—	—	n.s.
Suitable sites	.11	—	—	—	n.s.

Note: n = 49 unit years.
**$p < .01$.

reproductive success. Known nest sites entered because it was an indicator of the number of high-quality nest sites on the territory. Large territory size appeared to improve reproductive success because it increased the probability of raising a second brood, increased the amount of food available to the young, and increased the distance between neighboring nests, thereby reducing predation on the young. Number of females was still significant and negatively associated with high reproductive success even when the territory variables were present, indicating that the unit size relationship was not a spurious effect of ecological variables. In fact, larger units had significantly larger territories, so the negative effect of number of females on reproductive success was even stronger after territory size had been partialed out.

When reproductive success for successful units was multiplied by the probability of success, the resulting mean annual reproductive success per female was 1.01, 1.11, 0.58, and 0.60 young per female for units of one, two, three, and four females, respectively. Thus one- and two-pair units were about equally successful, and larger units were approximately half as successful.

19.3 Factors Affecting Proportion of Eggs Owned in Communal Nests

Within communal units of two or more pairs, there are consistent within-sex differences in egg-laying behavior, egg ownership, and amount of parental care provided (Köster 1971; Vehrencamp 1977). We defined distinct breeding roles for males and females based on these differences. Because breeding roles affected the number of eggs each female and her mate placed in the final incubated communal clutch, they altered the relative annual reproductive success of the pairs in a communal unit. Breeding roles also affected the level of parental investment and hence could lead to differences in adult survivorship within a unit as well. In this section we examine the correlates and possible causes of role acquisition in males and females, outline the probable basis of mate selection, and compute the average proportion of eggs owned in a communal clutch as a function of these roles. The effect of breeding roles on survivorship is analyzed in the section on annual adult survival. Details on the analyses presented here can be found in Vehrencamp, Bowen, and Koford (1986).

Male breeding roles were defined on the basis of conspicuous differences in incubation effort. One male in each group performed roughly 50% of all diurnal incubation and 100% of the nighttime incubation. We called this male the nocturnal incubator, and all other breeding males in communal units were called non-nocturnal incubators. The non-nocturnal incubators and the females shared the remaining 50% of the diurnal incubation. In single-breeding pairs, the male performed the same duties as the nocturnal incubator in a communal unit, and the female performed the remaining 50% of the diurnal incubation. Male incubator roles were extremely stable from nest to nest and in subsequent years if male

Table 19.3 Correlates of Male Roles: Paired t-Tests and Discriminant Function Analysis on Nocturnal Incubators versus Non-nocturnal Incubators

Variable	NI[a]	Non-NI[b]	Number of Pairs	Paired t-Test	Discriminant Function
Age (years)	2.79 ± 1.44	1.97 ± 1.03	43	.001*	.007*
Bill height	18.82 ± 0.64	18.64 ± 0.72	39	.150	.082
Wing length	139.2 ± 3.1	138.0 ± 3.8	39	.084	.122
Weight	83.0 ± 3.3	83.0 ± 5.4	39	.977	—

Note: Means ± s.d. and p values are given in the table.
*$p < .05$.
[a]Nocturnal incubators.
[b]Non-nocturnal incubators.

Table 19.4 Number of Eggs Laid, Tossed, and Incubated Plus or Minus the Standard Deviation for One-, Two-, and Three-Female Units as a Function of Laying Order

Unit Size	Laying Order	Laid	Tossed	Incubated
One-female units (n = 13)[a]	Only	4.08 ± 0.64	0	4.08 ± 0.64
Two-female units (n = 51)	First	5.98 ± 1.33	1.94 ± 1.24	4.04 ± 1.08
	Second	4.75 ± 0.85	0.08 ± 0.34	4.67 ± 0.93
Paired t-test analysis		$t = 5.81$ $p < .001$	$t = 11.24$ $p < .001$	$t = 3.64$ $p < .001$
Three-female units (n = 9)	First	6.22 ± 0.97	2.78 ± 1.64	3.44 ± 1.13
	Second	5.56 ± 1.42	1.78 ± 1.30	3.78 ± 1.20
	Third	4.56 ± 0.53	0	4.56 ± 0.53
Regression		$F = 78.7$ $p < .001$	$F = 11.9$ $p < .005$	$F = 5.72$ $p < .025$

[a]Number of nests.

Table 19.5 Correlates of Relative Clutch Size for Females: Univariate F-Test and Discriminant Function Analysis on Females with the Larger versus the Smaller Number of Incubated Eggs Averaged over All Nests of the Season

Variable	Smaller Clutch (n = 26)	Larger Clutch (n = 25)	Univariate F-Test	Discriminant Function
Age (years)	2.42 ± 1.21	3.08 ± 1.47	.087	.046*
Bill height	17.17 ± 0.52	17.44 ± 0.63	.091	.090
Wing length	133.1 ± 3.6	135.1 ± 3.3	.046*	.069
Egg size	807.9 ± 48.1	804.5 ± 51.7	.807	—
Laying order	1.31 ± 0.47	1.60 ± 0.50	.036*	.031*

Note: Means ± s.d. and p values are given in the table.
*$p < .05$.

membership was unchanged and was not affected by changes in female membership or roles.

To investigate the possible factors causing males to become either nocturnal incubators or non-nocturnal incubators, we compared the two types of males with respect to age, bill height, wing length, and weight. These results are shown in table 19.3. Using univariate (t-test) analyses, we found that age was the only significant variable: nocturnal incubators averaged approximately one year older than non-nocturnal incubators. Nocturnal incubators tended to be larger in all the body-size variables, but none was significant. In a discriminant function analysis of nocturnal versus non-nocturnal incubators using these same variables, age was still the only significant predictor of incubation role, but bill height increased in its predictive value after we controlled for age. The nocturnal incubator was also the most active bird in nest defense and territorial defense, and he and his mate had priority of access to the nest. These observations and the age and size differences outlined above suggested that the nocturnal incubator was behaviorally dominant over other males in the unit.

Females differed most conspicuously in timing of the onset of laying. We could assess laying order several days before laying on the basis of conspicuous differences in gravidness, and several weeks before laying on the basis of weight differences: first-laying females were significantly heavier (85.3 ± 6.8 g, $n = 6$) than last-laying females (72.7 ± 1.56 g, $n = 4$) (t-test, $p < .01$). The order in which females began laying affected the number of eggs they successfully placed in the nest. Last-laying females tossed the eggs of first layers out of the nest, leading to a consistent bias in the ownership of incubated eggs that favored the last layer (table 19.4). First layers differed from last layers with respect to several variables; they laid more eggs in total, laid larger eggs, and had smaller bills and wings than last layers. Age, however, did not differ for first and last layers, and no role of dominance in determining laying order was obvious.

Female laying order was not entirely stable from nest to nest. Laying order occasionally changed, but always in a highly specific context: rapid renestings. When a nest was lost or abandoned during the early egg-laying stage, the unit usually rapidly constructed a new nest and resumed laying in less than a week. In these renests, laying order frequently was switched, and often few or no eggs were tossed. We suspected that many cases of abandonment, in which all eggs were tossed out of the nest, were deliberate attempts by first layers to force a renest and improve their relative egg ownership. This strategy, along with the laying of several extra eggs, occasionally resulted in first layers owning the largest fraction of incubated eggs in the nest. To investigate whether there were any phenotypic correlates of relative fitness in females, we analyzed the traits that differentiated females with the larger incubated clutch from females with the smaller incubated clutch. The results of this analysis are shown in table 19.5. Age was the most significant variable, with older females associated with the larger relative clutches. Females with the larger clutches also had larger wings and bills.

Nocturnal incubators tended to pair with the last-laying female (68% of fifty groups) as opposed to the first layer (32%). However, the association between the nocturnal incubator and the female laying the larger relative clutch was much stronger (80% of nocturnal incubators were mated to the female with the larger clutch). In units where the nocturnal incubator was mated to the first layer, the first layer tended to produce the larger clutch either by laying many extra eggs or by forcing a renest. In almost all cases in which the nocturnal incubator was mated to the female obtaining the smaller relative clutch, the breeding unit disbanded as a result of the emigration of the nocturnal incubator, his mate, or both. These data suggested that the birds were "aware" of their relative reproductive success within communal units. Since pair-bonding took place at the beginning of the rainy season, well before the onset of breeding, incubation roles and laying success must have been anticipated. We believe that male roles were determined by dominance and were known to the birds at the time of group formation. Female laying success appeared to be much less predictable, and males may have used indirect cues such as female age and perhaps plumage traits to identify the most successful female. The dominant nocturnal incubator was then able to sequester the female of his choice, and he was correct in his assessment 80% of the time. In the typical stable two-pair unit, the nocturnal incubator and his mate owned an average 54% of the eggs in the communal clutch, whereas the non-nocturnal incubator and mate owned 46%. The difference was small but was consistent and significantly different (paired t-test, n = 42 pairs, $p < .001$).

19.4 Factors Affecting Annual Adult Survival

Unit size, role, sex, and ecological variables may also affect fitness through their influence on adult survival probabilities. Unit size and ecological variables were likely to affect groove-billed ani survival only during the breeding season; during the dry season birds frequently abandoned their breeding territories and lived in units of highly variable size. The differential reproductive effort arising from between-sex and within-sex roles might have had a residual effect on dry-season survival, but we found no evidence of this. We therefore analyzed only breeding-season survival in detail as a function of these independent variables. Differences in survival probabilities are notably difficult to discern, since large samples of individuals are required and one must be careful to distinguish betwen individuals that have probably died and individuals that have probably dispersed. We recorded several categories of disappearances: definite known death (body found), disappearance associated with a nest predation, disappearance of a single known mated bird, disappearance of an unmated, nonbreeding, or uncertain status bird, disappearance of a bird following the disappearance of its mate, simultaneous disappearance of a mated pair, and disappearance of an entire group. Only the first three categories were probable deaths: the others were more likely to be

Table 19.6 Correlates of Breeding-Season Survival: Logistic Regression Analysis

Variable	Males	Females
Unit size	.581	.016*
Age	.058	.250
Status	.048*	.136
Territory size	.899	.993
Tree area	.197	.817
Percentage of tree cover	.164	.785

Note: Probability values from χ^2 test are given in the table.
*Significant variable that entered the multivariate model.

dispersing birds. In the following analyses, cases of likely dispersal were omitted. Because the breeding activities for the two sexes are somewhat different, males and females have been analyzed separately. For both sexes, a stepwise logistic regression analysis was performed for surviving versus dead individuals, and the independent variables included were unit size, status, age, territory size, tree area, and percentage of tree cover.

For males, the only significant determinant of survival was role: nocturnal incubators (including single-pair males) had a lower probability of surviving than non-nocturnal incubators (81% versus 91%, $\chi^2 = 4.35$, $n = 194$, $p < .05$). Single-pair males were indistinguishable from nocturnal incubators in larger units. Neither age, unit size, nor ecological variables even approached significance in the multivariate analysis (table 19.6). The difference between nocturnal and non-nocturnal incubators was entirely accounted for by the category of disappearances associated with nest predation. Nest predation was known to frequently occur at night. On numerous occasions we discovered a nest and its contents completely gone, the nocturnal incubator gone, and adult tail feathers on the ground under the nest. Nocturnal predators such as raccoons, coatimundis, opossums, bats (*Vampyrum spectrum*), and snakes (e.g., *Trimorphodon*) were known to take adult anis.

For females, the only significant variable associated with survival was unit size: survival increased as unit size increased (ordered χ^2 test on lumped unit sizes of 2 + 3, 4 + 5, 6 + 7, and 8 + 9, $\chi^2 = 5.83$, 1 d.f., $n = 132$, $p = .016$) (fig. 19.3). No ecological variables were remotely associated with survival in the multivariate analysis (table 19.6). Role was entered in several different ways for females, for example, as laying order and as mate affiliation, but though there were slight differences, none was significant. The greatest unit-size difference was between single-pair females and females in multiple-pair units. There were several instances of single-pair female disappearances associated with evidence of nest predation, including adult tail feathers on the ground. No cases of this existed for females in communal units. The difference between the incubation efforts of single-pair females and communal females was very large: about 50% of the day for single-pair females, 17% for females in two-pair groups, and 10% for females in three-pair units. Mortality clearly increased with the amount of time spent on the nest, an explanation for the unit-size-

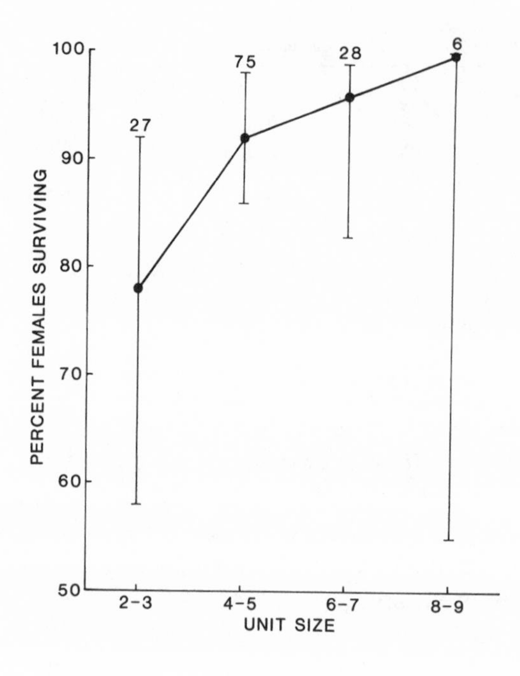

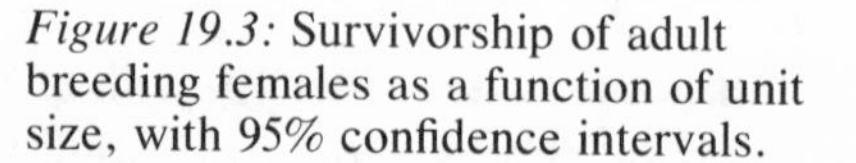
Figure 19.3: Survivorship of adult breeding females as a function of unit size, with 95% confidence intervals.

related effects on female survival that is consistent with the results for males. Thus the primary benefit of communal nesting in the anis for all but the nocturnal incubator is the reduction of risk during incubation that arises from sharing this duty.

19.5 Lifetime Reproductive Success

Lifetime reproductive success can be calculated for birds that have successfully entered the breeding population by multiplying the average annual total breeding unit production of young by the average fraction of eggs owned in the communal clutch and by the mean life expectancy for a given sex, role, and unit size. Nonbreeding-season survivorship was 85% for all birds. The results are plotted separately for females (fig. 19.4) and males (fig. 19.5). This calculation assumes that the bird remains in the same role and unit size all its life. For both sexes, role changed with age, so lifetime reproductive success for most individuals probably lay between the values for the two roles. Communal breeding in two-pair units was the most successful reproductive strategy for both sexes, and this unit size occurred with the highest frequency in our study area. Communal breeding, as opposed to solitary-pair breeding, was much more advantageous for females than for males.

The cost and benefit trade-offs of communal nesting are somewhat different for the two sexes. Both sexes suffer the costs of communal nesting, which include reduced efficiency in raising young through the nestling and fledgling stage and reduced probability of rearing a second brood. Females may in addition suffer an extra egg production cost in competing to obtain the largest number of incubated eggs. There is no

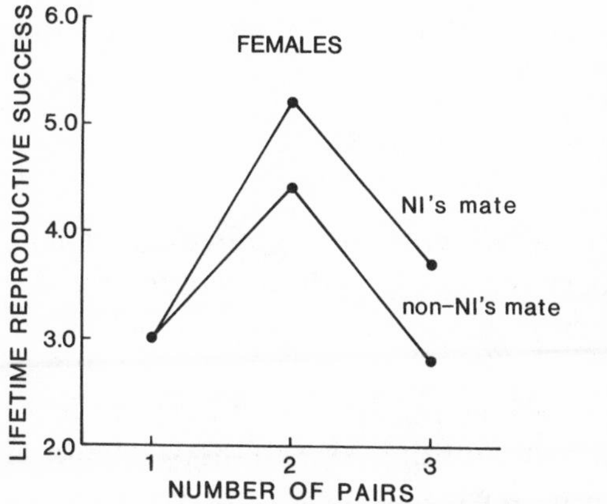

Figure 19.4: Estimated lifetime reproductive success for females as a function of unit size (number of breeding pairs) and status (mate of nocturnal incubator and mate of non-nocturnal incubator).

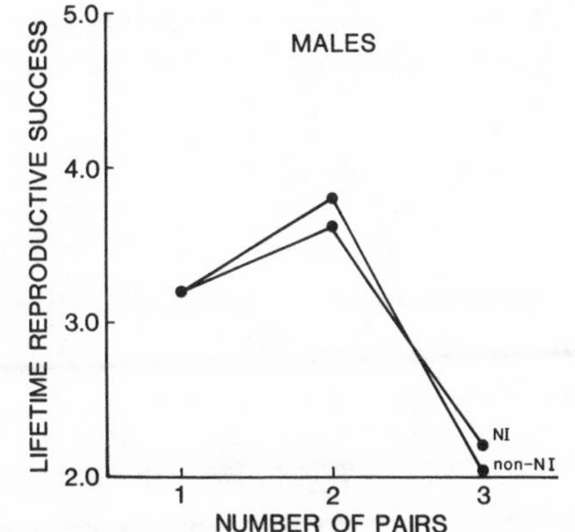

Figure 19.5: Estimated lifetime reproductive success for males as a function of unit size (number of breeding pairs) and status (nocturnal incubators and non-nocturnal incubators).

consistent trend for larger units to compensate for these losses with improved probability of nest success. While larger units do appear to benefit from cooperative nest defense against predators during the egg stage of nesting, they suffer the cost of increased conspicuousness to predators during the nestling feeding stage. Two-pair groups have the highest probability of success, and this offsets the inherent reproductive inefficiency of communal nesting to make them equal to single pairs in terms of annual reproductive success. The primary benefit of communal nesting for females arises as a result of their increased survivorship in larger units, which is attributed to their reduced diurnal incubation duties when parental care is shared among more individuals. This benefit makes breeding in two-pair units extremely successful and breeding in three-pair units approximately equal to single-pair breeding. Mates of nocturnal incubators have higher lifetime reproductive success than mates of non-nocturnal incubators because of their ability to place more eggs in the nest. Males do not show such differences as a function of role because of the trade-off between egg ownership and survival probability. Non-nocturnal incubators in two-pair units own fewer incubated eggs but enjoy a slight survival advantage over solitary-pair males, and nocturnal incubators benefit from the increased egg ownership of their mates but experience higher mortality. Neither of the benefits to males in larger units outweighs the reproductive inefficiency costs of three-pair units; hence this breeding strategy is very disadvantageous for males. Three-pair units can improve their success by forming on high-quality territories (e.g., large territories with high vegetation cover). High-cover areas have become relatively rare in the study area during the past ten years as a result of human activity, so that the opportunities, the benefits, and the observed frequencies of nesting in three-pair units have declined. Ecological factors obviously affect the critical components of fitness and alter the costs and benefits as well as the optimal breeding-unit size in this and undoubtedly other cooperatively breeding species.

19.6 Summary

To evaluate the effect of breeding-unit size on fitness in groove-billed anis, we analyzed each component of fitness separately with respect to unit size and other confounding variables such as age, year, and territory characteristics.

The probability of successfully raising one or more young during a season was not significantly associated with unit size, although units of two pairs had the highest success rate. Year and number of trees on the territory were the only significant variables that distinguished successful from unsuccessful units. Territories with more trees not only had more nest-site options, but also had breeding sites that were better concealed from predators.

The number of young raised per female in successful units was slightly negatively affected by the number of females in the unit. Territory size was positively associated with higher success. This appeared to be because units with large territories were more likely to raise a second brood and had a larger number of good nesting and feeding sites. When probability of success was multiplied by average productivity per pair in successful units to give annual reproductive success, units of one pair and two pairs were found to have the highest annual reproductive success and larger units somewhat lower success.

Adult survival probabilities were significantly affected by role in males and by unit size in females. These differences for the two sexes were consistent, in that the risk of mortality was parallel to the amount of incubation performed. Survival was higher for females and non-nocturnal incubator males in larger units because of reduced incubation duties when parental care was shared among more birds.

Breeding role affected the proportion of eggs obtained in the communal nest. The nocturnal incubator and his mate in a typical two-pair group owned 54% of the eggs, while the non-nocturnal incubator and his mate owned 46%. Both the nocturnal incubator and his mate tended to be older than the other birds in the group.

The components of fitness were multiplicatively combined to estimate lifetime reproductive success for a hypothetical bird that spends its entire life in the same unit size and role. Breeding in two-pair units yielded the highest lifetime reproductive success for both sexes, but breeding in very large groups was more advantageous for females than for males.

20 Components of Lifetime Reproductive Success in the Florida Scrub Jay

John W. Fitzpatrick and Glen E. Woolfenden

FLORIDA SCRUB JAYS (*Aphelocoma c. coerulescens*) (fig. 20.1) live in territorial family groups that exhibit cooperative breeding. Our study of a wild population of these jays began in 1969. Since then we have continuously monitored reproduction, dispersal, and long-term genealogies of about thirty scrub jay family groups, occupying about 400 ha of oak scrub. We measure survival by censusing the entire marked population each month and by periodically searching surrounding habitat for dispersers. About 90% of the jays in our study population are now of known age and parentage. In this report we present our first analyses of several patterns in lifetime reproduction among

Figure 20.1: Adult-plumaged Florida scrub jay (*Aphelocoma c. coerulescens*) holding a ripe acorn (*Quercus inopina*), probably in preparation for caching.

Florida scrub jays, based on the records of one hundred jays whose complete lifetime breeding histories we have documented.

20.1 Methods

Our study is carried out in an extensive tract of native Florida oak scrub permanently protected at the Archbold Biological Station, in south-central Florida, USA. The core of the study tract consists of about 400 ha of scrub in which essentially all resident jays have been individually color ringed and in which the nestlings of all annual nest attempts are individually ringed before they fledge. Dispersal by young jays does not begin until at least one year after fledging; therefore we can assess reproductive success extremely accurately, measured in both fledgling production and yearling production. Most dispersal is over short distances, and we find many of the few long-distance dispersers in the surrounding habitat (Woolfenden and Fitzpatrick 1984). Therefore we are confident that our measures of offspring survival and production of new breeders also are extremely accurate.

Between 1970 and 1984 we documented the complete breeding histories of exactly one hundred Florida scrub jays. These individuals became breeders after our study began and are now dead. We define breeding life span as the number of breeding seasons in which an individual produced at least one clutch of eggs. Therefore the minimum possible breeding life span in this study is one year; individuals who, for whatever reason, failed to produce a clutch of eggs are considered not to have become breeders. In our sample of one hundred breeding lifetimes, the mean breeding life span (3.92 ± 3.02 years) is substantially shorter than the predicted value for a general breeding population with an annual mortality of .19 (about 4.8 years). This difference indicates that our total sample is biased toward breeding jays that died relatively early; members of numerous age-class cohorts still remain alive in our population, and these longer-lived breeders are not included in the sample. Such a bias does not directly affect certain correlations of interest to us, but it does affect our estimates of mean and variance in lifetime fitness within the population as a whole. For the latter analyses, therefore, we use a subset of the total sample, consisting only of those breeders who *began* breeding in 1977 or before (but after 1969, when the study began). Four jays in this category remain alive at this time. For these four jays we assign hypothetical remaining life spans and reproductive outputs equivalent to the expected average values, already calculated for older breeders (as in Clutton-Brock, Guiness, and Albon 1982). The restricted sample of breeding lifetimes totals sixty-seven jays, whose average breeding life span was 4.79 years. This sample produced a total of sixty-two individual jays known to have become breeders, a figure that is reasonably close to the sixty-seven replacements expected within a stable population.

20.2 The Social System

The social organization of Florida scrub jays is an interplay between nonbreeders and breeders, who compete for territories within a chronically crowded environment (Woolfenden and Fitzpatrick 1984). Permanently bonded, monogamous pairs defend relatively large, permanent territories, which completely fill the relict patches of open oak scrub scattered through central peninsular Florida. Once paired, breeding jays reside permanently in the same patch of scrub, living as long as ten to fifteen years. Successful breeders accumulate nonbreeding offspring, often of several year-classes, who participate in territory defense, sentinel behavior, predator detection and mobbing, and the care of dependent young. On average, these helpers improve the survivorship and reproductive output of the breeders they assist. Male and female helpers show important differences in behavior around the nest (Stallcup and Woolfenden 1978), with males generally being more active food providers and females tending to perch for longer periods near the nest, observing but not actively feeding. In dispersing, females generally leave the territory temporarily, wandering greater distances through the surrounding neighborhood, resulting in earlier and longer-distance permanent dispersal than for males (Woolfenden and Fitzpatrick 1978, 1984, 1986). Inheritance of territory by helpers, almost always males, appears to represent an evolutionarily intermediate condition between social systems exhibiting early dispersal of offspring (e.g., Atwood 1980) and those in which at least some offspring routinely breed within their natal territories in multiple-pair groups (see Fitzpatrick and Woolfenden 1986, for review and model).

20.3 Breeding Success in Relation to Sex

With rare exceptions, breeding Florida scrub jays remain with the same mate until one of the pair dies (Woolfenden and Fitzpatrick 1984). Between age one year, when the sex ratio is equal, and first breeding at two years or older, behavioral asymmetries develop between the sexes that cause a higher annual death rate among females (Woolfenden and Fitzpatrick 1986). However, males tend to have a longer prebreeding period than females, and the total probability of death before breeding is equal between the sexes. Once paired, male and female breeders have essentially identical demographic profiles.

Death rates of male and female breeders are equal. From May 1969 through November 1984 we recorded the deaths of 185 breeders: 94 males and 91 females. Figure 20.2 illustrates the similar survival of male and female breeders among the 131 jays whose year of first breeding we established with certainty (including many that remain alive at this time). Annual mortality fluctuates from 7% to 45% (Fitzpatrick and Woolfenden 1986), but the average of 19% is identical between males and females, and the sexes show similar death rates each year. When one member of the pair dies, the widowed jay almost always remains in the territory and pairs with a dispersing jay from outside the group; close inbreeding, especially within members of an established group, is strictly avoided.

Permanent monogamy and identical survivorship between the sexes as breeders necessarily result in statistically similar patterns of variance in lifetime reproductive success between males and females. Demographic similarities between breeding males and females in this population eliminate any opportunity for differences to arise in the components of variance in male and female lifetime reproductive success. This similarity is

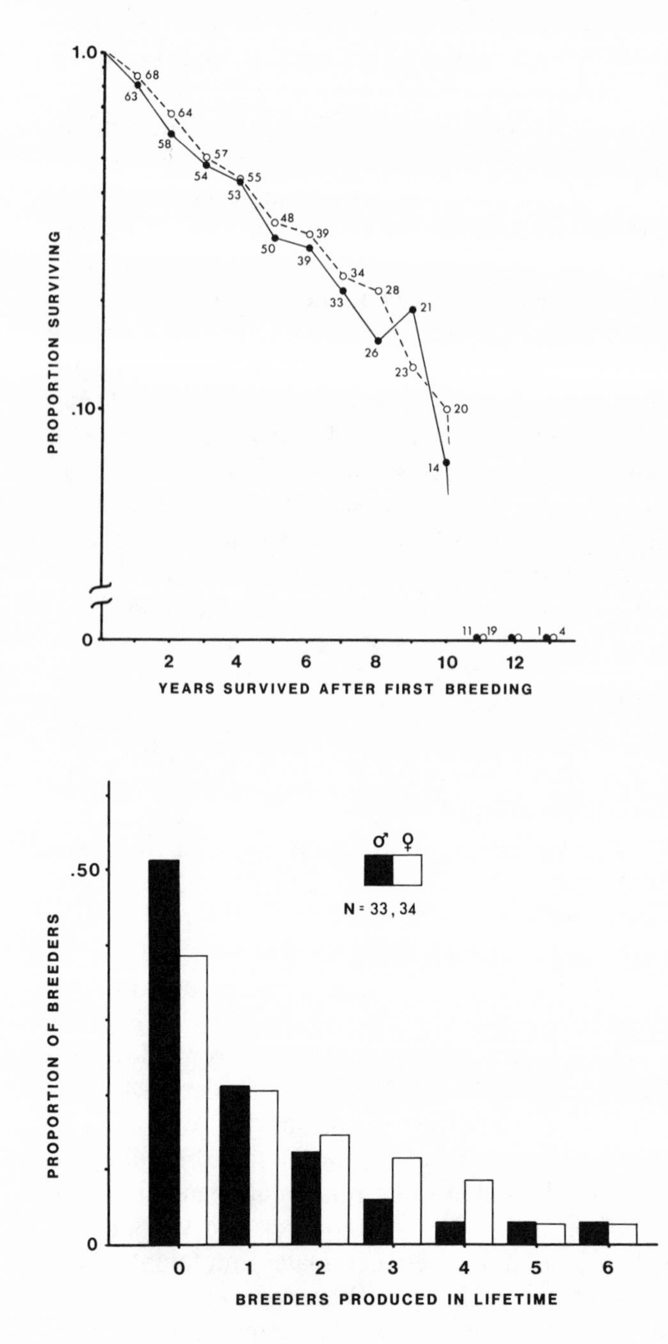

Figure 20.2: Survivorship of males (*closed circles*) and females (*open circles*) as breeders. Each year's proportions represent the ratio between the number of breeding jays surviving to that year and the sample (shown for each sex) of jays that became breeders early enough in our study to have survived at least to that year.

Figure 20.3: Relative frequency distribution of the number of breeding offspring produced in their lifetimes by thirty-three male and thirty-four female breeders that commenced breeding before 1978 (see text for discussion of this restricted sample). These sixty-seven breeders produced sixty-two known breeding offspring and possibly a small, unknown number of long-distance dispersers that we never encountered. The figure suggests that eighty-eight offspring were produced: twenty-six breeding offspring are tallied once each under their mothers and their fathers, because both parents were known-history breeders.

illustrated in figure 20.3, where the slight difference in average lifetime output is not statistically significant (see below). Because of this important (and simplifying) similarity, we pool the sexes in most of the analyses to follow.

20.4 Breeding Success in Relation to Age

Age at First Breeding

Age at first breeding in the Florida scrub jay varies from one to seven years, with most individuals becoming breeders at ages two to four (table 20.1). On average, males become breeders slightly later than females. Individuals that remain longer as helpers appear to show slightly reduced survival once they become breeders (table 20.2). This results in an inverse correlation between age at first reproduction and life span as a breeder, as suggested in table 20.2 (Spearman correlation coefficient, $r = -.28$, $p < .01$, sexes pooled). This effect is not related to senescence among breeders who are chronologically older, since mortality is age independent through the first nine years of breeding (fig. 20.2).

Table 20.1 Ages at First Breeding of 130 Florida Scrub Jays, 1971–84

	Age in Years							
	1	*2*	*3*	*4*	*5*	*6*	*7*	*Mean Age at First Breeding*
Males (n = 71)								
Number	1	30	25	8	4	2	1	2.91 ± 1.14
Proportion of first breeders	.014	.423	.352	.113	.056	.028	.014	—
Females (n = 59)								
Number	2	33	23	0	1	0	0	2.41 ± 0.65
Proportion of first breeders	.034	.559	.390	.000	.017	.000	.000	

Table 20.2 Life Spans as Breeders and Lifetime Production of Offspring among Eighty Florida Scrub Jay Breeders with Known Ages at First Breeding, 1971–84

	Age at First Breeding	
	1–2 Years	*3–7 Years*
Males		
n	19	24
Life span as breeder (years)	4.44 ± 2.89	2.96 ± 2.82
Fledglings	9.26 ± 8.74	4.29 ± 4.02
Breeding offspring[a]	1.17 ± 1.63 (.56)	0.58 ± 1.14 (.42)
Females		
n	26	11
Life span as breeder (years)	4.48 ± 2.96	3.18 ± 2.48
Fledglings	8.58 ± 6.33	5.82 ± 5.88
Breeding offspring[a]	1.24 ± 1.56 (.56)	1.18 ± 1.60 (.46)

[a]Proportion of breeders producing at least one breeding offspring shown in parentheses.

Regarding reproductive output, our earlier analysis (Woolfenden and Fitzpatrick 1984) demonstrated a slight trend toward *increased* annual reproductive success among older first breeders. This advantage appears to cancel some of the impact of reduced survival. As a result, the net production *of breeding offspring* by later-breeding jays is not significantly lower than that of early breeders (table 20.2), although a trend in this direction is evident among males (Student's $t = 1.37$, $p = .20$). Small sample sizes of lifetime histories in the four categories listed in table 20.2 prohibit meaningful further analyses of these results. In subsequent analyses, we assume for simplicity that age at first reproduction indeed does not significantly affect lifetime reproductive success.

Breeding Experience

Table 20.3 summarizes our data on reproduction in relation to experience, for the entire sample of 131 breeders whose first breeding we established with certainty. We measure annual reproductive success in terms of three classes of offspring: fledglings, yearlings, and breeders. We do not include clutch-size comparisons, because virtually all clutches contain either three or four eggs.

Two major trends in reproductive success through individual life-

Table 20.3 Production and Survival of Offspring, and Numbers of Helpers, in Relation to Breeding Experience (Sexes Pooled)

	Year of Breeding								
	1	*2*		*3*	*4*	*5*	*6*		*7–11*
n (individual jays)	131	101		68	54	42	25		50
Offspring production									
Fledglings ($\bar{x}$)	1.56	1.71	***	2.16	2.30	2.29	2.28	n.s.	2.24
s.d. (σ)	1.55	1.37		1.40	1.50	1.52	1.51		1.92
$\sigma^2/\bar{x}^2$[a]	.99	.64		.42	.43	.44	.44		.73
Yearlings ($\bar{x}$)	.39	.44	***	.71	.78	.48	.92	n.s.	.60
s.d. (σ)	.71	.73		.81	.96	.89	1.04		1.05
$\sigma^2/\bar{x}^2$	3.31	2.75		1.30	1.51	3.44	1.28		3.06
Breeders ($\bar{x}$)	.27	.25	*	.46	.44	.21	.48	†	.26
s.d. (σ)	.55	.56		.70	.69	.56	.77		.50
$\sigma^2/\bar{x}^2$	4.15	5.02		2.32	2.46	7.11	2.57		3.70
Offspring survival									
Fledgling–yearling	.25	.26	*	.33	.34	.21	.40	†	.27
Yearling–breeder	.69	.57	n.s.	.65	.56	.44	.52	n.s.	.43
Helpers per pair	.33	.48	***	.54	.67	.71	.69		.69

Note: Significance values for differences between means of pooled groups 1–2, 3–6, and 7–11: *$p < .05$; ***$p < .001$; †not significant (= n.s.), but $p < .1$ (Student's t).

[a]Estimates of relative variance in tables and text are calculated by dividing variance (σ^2) by the square of the mean ($\bar{x}$), following Wade and Arnold (1980); these estimates represent the square of the coefficient of variation (CV).

times are apparent (table 20.3). First, breeders in their first and second years of nesting show lower average reproduction than do older breeders. Significant differences exist between young breeders (years 1 and 2) and middle-year breeders (years 3–6) for average annual production of fledglings (1.63 versus 2.24), yearlings (0.41 versus 0.68), and new breeders (0.26 versus 0.41). Variation in reproductive success ($\sigma^2/\bar{x}^2$) shows a corresponding decrease during the most successful breeding period (years 3–6). For all three classes of offspring, variation typically is highest among the younger breeders. The apparent increase in variability at year 5 of breeding (table 20.3) occurred by chance, because unusually large numbers of jays happen to have become new breeders exactly five years before the two breeding seasons (1976, 1979) in which juvenile survivorship was atypically low. Indeed, the 1979 season was marked by an apparent epidemic (Woolfenden and Fitzpatrick 1984).

A second trend suggested in table 20.3 is that the oldest breeders may be less successful at producing breeding offspring. Because of small sample sizes, we lumped breeding years 7–11 (to date no jay has bred more than eleven years). Fledgling production is similar throughout the middle and later years, although the variance increases slightly in the older group (2.24 ± 1.92 for older breeders versus 2.24 ± 1.47 for younger). Average survival of fledglings to age one year is somewhat lower for the older breeders (.27 versus .32), as is survival from yearling to breeding status (.43 versus .55). These differences result in a drop in average annual production of new breeders by this sample of older jays (.26 versus .41 per breeding year), which leads us to suspect that a prime period of breeding life is reached between three and six years after first breeding.

The Role of Helpers

Average production of breeding offspring increases during the first few years of the breeding life span (table 20.3). However, changes in experience at breeding are confounded with the effects of helpers on the reproductive success of the breeders. Helpers usually are the young from earlier years' breeding by one or both parents, and beginning breeders are less likely to have helpers than are long-established breeders. First breeders acquire helpers in several ways, usually by pairing with an experienced widow who retains offspring from previous breeding seasons. The probability of having helpers rises from .33 for first-time breeders to .67 for fourth-year breeders, at which point it levels off at an average probability of about .70. Probability of having at least one helper declines somewhat among the very oldest breeders (.53 for breeding years 9–11).

To test the hypothesis that experience per se increases reproductive success, we compared the average production of breeding offspring in relation to breeding experience, for breeders with and without helpers (table 20.4). The results are equivocal. For both samples, production of breeding offspring is higher during the middle years, paralleling the pattern already shown (table 20.3), but the differences between age-classes are

Table 20.4 Production of Offspring in Relation to Breeding Experience, with and without Helpers (Sexes Pooled)

	Fledgling Offspring			Breeding Offspring		
	1–2[a]	*3–6*	*7–11*	*1–2*[a]	*3–6*	*7–11*
With helpers						
n	91	118	35	91	118	35
$\bar{x}$	2.13	2.28	2.53	0.29†	0.42	0.17*
s.d.	1.51	1.57	2.00	0.52	0.70	0.38
Without helpers						
n	142	70	16	142	70	16
$\bar{x}$	1.31***	2.16	1.63	0.22†	0.36	0.25
s.d.	1.36	1.25	1.54	0.55	0.61	0.58

[a]Year of breeding.

*Significantly below middle years, $p < .05$; ***significantly below average production of middle-year breeders, $p < .001$; †below middle years, but $.1 > p > .05$.

significant only for production of fledglings (and yearlings, not shown), not for production of breeders (table 20.4).

Table 20.4 also shows the expected tendency toward increased production of breeding offspring by breeders with helpers, although again the differences are not significant. These analyses suggest that both experience and the effects of helpers independently contribute to reproductive success. It appears that these two features combine to cause a peak in success among breeders in their middle years of reproduction.

20.5 Individual Variance in Breeding Success

Patterns of Variance

Figure 20.3 illustrates the frequency distribution of breeding offspring produced by the restricted sample of sixty-seven breeders who first bred in 1977 or before. Variance in lifetime output of breeders was approximately equal between males and females, a result of the demographic symmetries discussed above. Nearly half of the indvidual jays *that become breeders* within the population fail to produce breeding offspring of their own. About 20% of the breeders produce 65% of the replacement breeders. The largest number of breeders produced by any known-age individual was six, a record achieved by one jay of each sex (fig. 20.3). We know of one unknown-age breeder that produced at least twelve breeding offspring (Fitzpatrick, Woolfenden, and McGowan, n.d.).

Table 20.5 summarizes variation in annual and lifetime reproductive success of Florida scrub jays. Estimates of variation among individuals increase as reproductive output is measured in progressively older classes of offspring. As mentioned above, variation in clutch size is extremely low.

Total lifetime production of fledglings is a good, but imperfect, predictor of ultimate biological fitness among individuals (fig. 20.4). The implications of this result to field measurements of fitness among birds in general are profound: merely tallying fledgling production among individ-

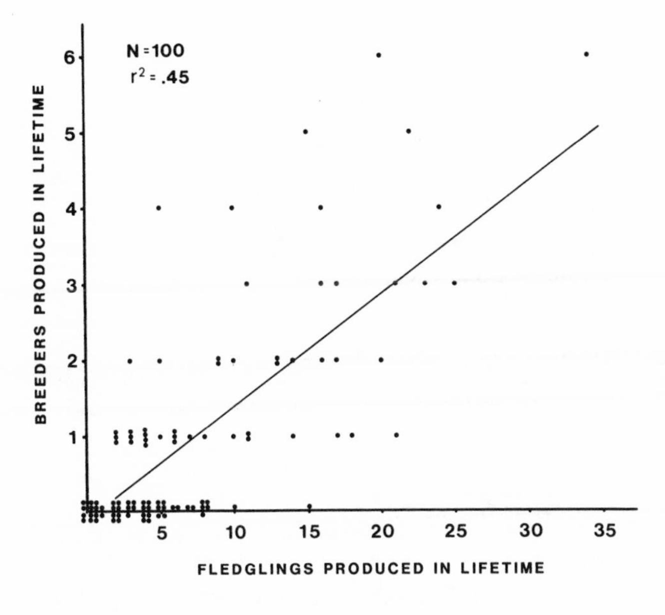

Figure 20.4: Lifetime production of breeding offspring plotted against lifetime production of fledglings, for one hundred jays with known breeding histories. Linear regression (line shown) is highly significant ($p <$ .0001), but nearly half the variation remains unexplained ($1 - r^2 = .44$).

Table 20.5 Individual Variation in Reproductive Success, among Years and among Lifetimes, Measured for Four Classes of Offspring

	Within-Year Reproduction (n = 13 years)				Lifetime Reproduction (n = 67 individuals)		
	Mean Productivity		*Annual Variation* ($\sigma^2/\bar{x}^2$)				
Class of Offspring	$\bar{x}$ of $\bar{x}$'s	*Range*	*Average*	*Range*	$\bar{x}$	*s.d.*	$\sigma^2/\bar{x}^2$
Clutch size	3.20	3.18–3.75	0.04 ± 0.01	0.02–0.05	—	—	—
Fledglings	2.01	0.89–2.93	0.70 ± 0.60	0.12–2.10	8.97	7.66	0.729
Yearlings	0.69	0.03–1.29	4.36 ± 8.46	0.50–32.10	2.57	2.79	1.176
Breeders	0.38	0.03–0.92	5.61 ± 8.18	0.81–32.11	1.31	1.62	1.540

Note: Reproductive success measured in offspring per breeding pair for within-year comparisons, total offspring production per individual for lifetime comparisons.

uals—by far the most common currency still in use—provides a disappointingly weak index of actual biological fitness within natural populations. Variables relating to offspring survival must be added to the picture, as elaborated below.

Within-year variance in reproductive success is higher than variance in lifetime production (table 20.5), because a substantial proportion of the breeding pairs normally fail to produce offspring in any one year, while the successful pairs produce up to four or more fledglings (mean successful brood size = 2.7 ± 0.87). Exceptions occur during the occasional highly productive breeding seasons, when among-individuals variance reaches its lowest levels (fig. 20.5). Highly successful years for offspring production result from unusually low rates of predation on nests and fledglings (Woolfenden and Fitzpatrick 1984), and this reduces the variance by lowering the proportion of pairs that fail entirely. Florida scrub jay reproduction is characterized by the irregular occurrence of excellent years "across the board" scattered among more frequent years of low success and high individual variance.

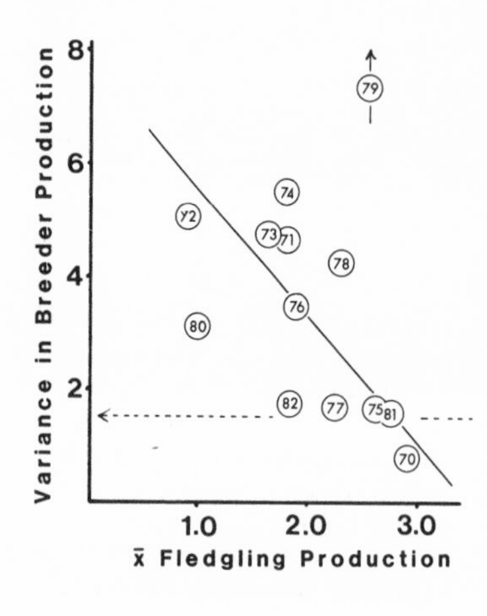

Figure 20.5: Relationship between variance among individuals in the production of breeding offspring from a given breeding season ($\sigma^2/\bar{x}^2$) and the mean fledgling production (fledglings per pair) during that season. Two-digit numbers represent the thirteen breeding seasons, 1970–82. Excluding the 1979 epidemic year, a significant negative correlation exists (line shown). Dashed line shows variance in lifetime breeder production among sixty-seven jays.

Components of Reproductive Success

Table 20.6 shows a matrix of Pearson product-moment correlations among six variables relating to lifetime fitness in Florida scrub jays. Three components of lifetime reproduction are identified: average production of fledglings per year (FPY, or fecundity), average survival of those fledglings to breeding status (OFS, or offspring survival), and the number of breeding seasons in which reproduction was attempted (LAB, or life span as a breeder). Lifetime reproduction is measured in three classes of offspring: fledglings, yearlings, and breeders. Of the three components, life span as a breeder shows the strongest simple correlation to lifetime reproduction. This correlation is strongest when lifetime reproduction is measured in fledglings (Spearman rank correlation, r = .883). The correlation grows weaker with later classes of offspring (Spearman r = .735 for yearlings, .590 for new breeders; $p < .0001$ for all three correlations). The correlations shown in table 20.6 are calculated from our restricted sample of sixty-seven breeders, but a similar correlation matrix, with virtually identical values, characterizes the full sample of one hundred breeders.

The diminishing correlations between life span as a breeder and lifetime fitness measured with older classes of offspring result from effects relating to prebreeding survival of offspring. We define offspring survival as the proportion of fledglings that eventually became breeders. Offspring survival also is strongly correlated with lifetime production of breeding offspring (table 20.6; Spearman r = .860). However, offspring survival is not significantly correlated with life span as a breeder (table 20.6; Spearman r = .206, p = .055). The apparent reduction in survival potential among offspring of the oldest breeders (table 20.3) probably accounts for the weakness of this last correlation.

Table 20.7 partitions the total variance in lifetime reproductive success into its components. This analysis was performed using the restricted sample of sixty-seven jays whose breeding commenced before 1978. Of these, only thirty-seven produced at least one offspring that eventually became a breeder. The average lifetime reproductive success *among these thirty-seven jays* was 2.4 breeding offspring.

Percentage contributions of all three variables total well over 100 (table 20.8) because of *negative* correlations between offspring survival and breeding life span. These correlations arise because the only birds whose offspring showed very high survival (near 1.00) are those that produced only a small number of fledglings. Such limited lifetime fledgling production results from both short breeding life spans and below-average annual production of fledglings.

Table 20.6 Pearson Product-Moment Correlations between Important Measures of Lifetime Fitness in Florida Scrub Jays

	Lifetime Reproduction			Components to Reproduction		
Measure	*LRB*	*LRY*	*LRF*	*FPY*	*OFS*	*LAB*
Lifetime reproductive success						
Breeders (LRB)	—	.840	.727	.481	.471	.599
Yearlings (LRY)		—	.888	.522	.218 (n.s.)	.773
Fledglings (LRF)			—	.578	−.057 (n.s.)	.877
Fledglings per year (FPY)				—	.039 (n.s.)	.256[a]
Offspring survival (OFS)					—	−.098 (n.s.)
Lifespan as breeder (LAB)						—

Note: All correlations highly significant ($p < .0001$) unless otherwise noted; $n = 67$ for all coefficients except those involving OFS ($n = 62$).
[a]FPY and LAB significant at $p < .05$.

Table 20.7 Mean and Variance of the Components of Lifetime Reproductive Success in Florida Scrub Jays (Sexes Pooled)

	Original		Standardized
Component	*Mean*	*Variance*	Variance
L	5.081	10.108	0.392
F	1.869	0.498	0.143
S	0.155	0.042	1.730
LF	9.694	56.314	0.625
FS	0.726	0.693	1.114
LS	0.296	0.138	1.639
LFS	1.419	2.707	1.245

Note: $n = 67$ jays whose breeding commenced before 1978.
L = reproductive life span; F = fecundity; S = survival of offspring to breeding status.

Table 20.8 Percentage Contribution of the Components of Lifetime Reproductive Success to Variation in LRS in Florida Scrub Jays (Sexes Pooled)

Component	L	F	S
L	31.46		
F	7.27	11.45	
S	−80.95	18.77	138.97
LFS	10.57		

Note: $V(LFS)/(LFS)^2$; $n = 67$ jays that attempted breeding.
L = reproductive life span; F = fecundity; S = survival of offspring to breeding status.

Offspring survival as we measure it here has numerous components. Fledglings experience extremely low survival to age one year, when they still have not reached breeding age (juvenile survival = .34; Woolfenden and Fitzpatrick 1984). Annual survival of older nonbreeders is much higher (Fitzpatrick and Woolfenden 1986), but not all yearlings become breeders at age two (table 20.1). Older helpers may attempt to become breeders for several years before either succeeding or dying in the process.

One-half of the overall variance in lifetime reproductive success among breeding Florida scrub jays is contributed by jays that fail to produce any breeding offspring at all (53.7%). Mean breeding life span in this sample was 3.54 ± 1.8 years, compared with 6.22 ± 3.5 years for the sample of thirty-seven that produced at least one breeding offspring (t = 3.84, $p < .0001$). Mean fecundity in the unsuccessful breeders was also lower than that of the successful breeders (1.3 ± .8 versus 2.1 ± .7, $p <$ .0001), but the difference was not as great as in life span or offspring survival. Therefore, even among unsuccessful breeders the overwhelming importance of breeding life span and offspring survival are apparent.

Brood Size and Breeding Success

The preceding analysis suggests that breeding Florida scrub jays should best apportion their annual reproductive investments into maximizing the survival and reproductive opportunities of a few offspring while minimizing the risk that their own future years' breeding opportunities will be jeopardized. As described in classical K-selection models (e.g., Horn and Rubenstein 1984; Boyce 1984), increasing the annual production of juvenile offspring appears to be of relatively less importance. This conclusion led us to examine directly the relationship between the number of fledglings produced in a single breeding season by a given pair of jays (n = 261 broods in 250 "batches," 1970–82) and the average number of *breeders* that resulted from each "batch" size. (The term "batch" is distinguished from "brood" because of the rare production, by the same pair, of two true broods of fledglings in the same breeding season.) Figure 20.6 illustrates the results: additional fledglings beyond three do not, on average, contribute to the net production of additional breeders from a given season's breeding efforts.

The mean number of breeders produced from annual "batches" of four or more fledglings is not significantly higher than the mean from batches of one to three fledglings (.59 ± .80 for large batches versus .46 ± .66 for small ones, Student's t = 1.16, $p < .20$). This, of course, could occur only because the overall survival *to breeding status* of fledglings in larger batches (.119) is lower than that of fledglings in smaller batches (.190; $\chi^2 = 3.1$, $p = .10$). The reasons for this pattern are not yet entirely clear, but at least two factors probably contribute. First, parents with unusually large numbers of fledglings cannot provide the same degree of vigilance to each offspring during the critical early months after fledging. Second, and probably more important, offspring of larger broods may be

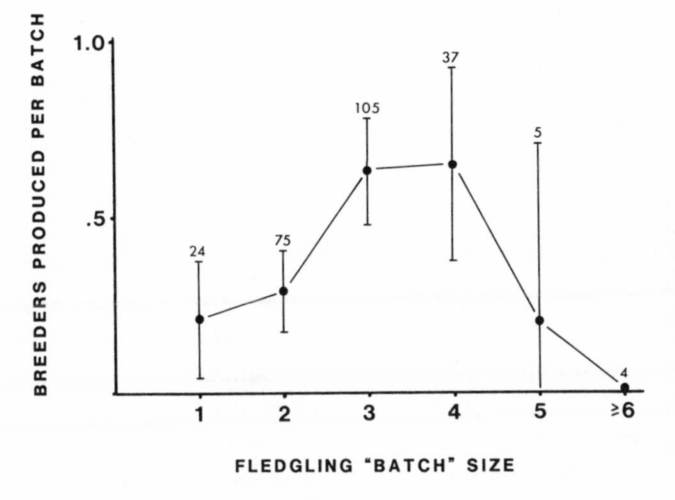

Figure 20.6: Relationship between number of fledglings produced per pair in a given year (= "batch" size; see text) and the number of breeders produced by those batches. Mean values ± 2 s.e., and sample sizes, are shown for each batch size.

forced to wait an extra year or two before becoming breeders, because dominant sibs experience statistical priority in filling breeding vacancies or inheriting breeding space (Fitzpatrick and Woolfenden 1986). Waiting longer to become a breeder also increases the probability of dying before breeding.

20.6 Discussion

Ecological Factors Influencing Reproductive Success

In the ecological regime of the Florida scrub jay, variance in reproductive success is extremely high, both within years and among years; jays that can succeed at breeding over several years greatly increase their probability of encountering one or more years of high average production. (Approximately 80% of all new breeding jays in the population were produced during five of the thirteen breeding seasons between 1970 and 1982.) Variances in clutch size and in willingness to attempt replacement clutches following nest predation are extremely low, and starvation of nestlings and fledglings is rare (Woolfenden and Fitzpatrick 1984). The chief source of variation in the overall production of offspring is predation.

The best defense against predation on nestling and fledgling jays is vigilance, and Florida scrub jays have a keenly developed sentinel system for spotting potential predators. Breeding with at least one helper significantly lowers the breeders' own mortality and reduces the probability of nest predation. Helpers do not, however, affect fledgling survival beyond independence (Woolfenden and Fitzpatrick 1984). Experience at breeding, especially in pairs that have bred together over several years, also is correlated with greater nest success and offspring survival. These factors all contribute to the maintenance of permanent monogamy and the tolerance of helpers in the breeding system of Florida scrub jays.

Our evidence regarding the biotic factors causing wide fluctuations in annual reproduction remains circumstantial. Variance in annual reproductive success (see above) can be traced directly to annual variance in predation pressure (Woolfenden and Fitzpatrick 1984; see list of predators there). The chief predators on both nest contents and juveniles include several species of snakes, mammals, and birds, and we remain uncertain why their densities, activity patterns, or preferences for young jays vary from year to year.

Factors that limit annual production of potential breeders do not appear to include fluctuations in food supply. Indeed, we have some evidence that Florida scrub jays often could feed more dependent offspring than they typically raise (Stallcup and Woolfenden 1978; Woolfenden and Fitzpatrick 1984). Instead, annual production is determined largely by the rate at which offspring are eliminated after they are produced. Occasional years of low predation represent important "windfall" years for most breeding pairs. The more frequent years of high predation favor a conservative strategy, in which parental investment is extended to maximize the survival of just a few offspring each year.

Extended Parental Care

The conclusion of our work to date has been that because of ecological restriction to a rare, patchy, and sharply defined habitat, the Florida scrub jay exists in a permanently "supersaturated" environment. Opportunities for dispersal by potential breeders into marginal habitats are severely limited (Fitzpatrick and Woolfenden 1986). These ecological constraints (Emlen 1982; Emlen and Vehrencamp 1983) result in the evolution of delayed dispersal and helping behavior (Woolfenden and Fitzpatrick 1984), principally because the alternatives facing nonbreeders in such a crowded environment are worse.

Most monogamous songbirds care for their offspring throughout the nestling period and for a few weeks after fledging, as the young birds develop physical independence. After this the offsping depart, and some are even expelled. Often the parents aggressively defend their breeding territory from recently fledged young as they attempt to produce second and third broods within a breeding season. Presumably such a system remains viable because the gain in total breeder production offsets any increased losses of fledged young resulting from their expulsion.

Among Florida scrub jays, expulsion from the territory would place offspring at extreme risk, because the habitat already is saturated with aggressively territorial jays of other families. Survival in marginal habitat is low (Fitzpatrick and Woolfenden 1986). Even *within* the home territory, the probability that an offspring will be taken by a predator before breeding is high. In such a regime the successful placement of even a few offspring into the breeding population represents unusually successful reproduction. If it can sufficiently improve the survival and reproductive potential of independent offspring, extended parental care (Ligon 1981) should be favored in such circumstances.

Parental care should be apportioned preferentially to those offspring with the highest reproductive value, especially if the extra care increases their likelihood of reproducing. In the Florida scrub jay the reproductive value of offspring at age one year or older is substantially higher than even an entire nestful of younger ones (Fitzpatrick and Woolfenden 1986), a pattern that arises specifically because of high mortality among eggs, nestlings, and young juveniles (see above). Parental care can take many forms, and we have argued elsewhere that it can even include gaining and defending space within which present or future offspring may eventually breed. Such a system may be most stable in an open habitat such as the Florida oak scrub, in which both predators and conspecific intruders can be spotted and dealt with relatively easily.

Intense competition for breeding space favors iteroparity and extended parental investment. The ultimate reproductive success of individuals depends not on the number of offspring they produce, but on how many become established as breeders. Success therefore may depend on a group investment that can extend over several years and encompass several breeding seasons. The most successful Florida scrub jays are those that succeed in breeding over the longest possible span of years and in providing the protection and even the very breeding space that ensures their offspring not only survival but also a place in succeeding generations of breeders.

20.7 Summary

We measured lifetime reproductive success for one hundred Florida scrub jays that became breeders and died between 1970 and 1982. The sample includes sixty-seven jays that first bred more than seven years ago and therefore constitute an unbiased sample of life spans. True reproductive success is most accurately measured as total production of breeding offspring, which does correlate with total production of fledglings (r^2 = .56) and yearlings (r^2 = .71). Variance in lifetime reproductive success among breeders is equal between the sexes, reflecting the permanent monogamy and equal mortality between males and females. Nearly half the breeders fail to produce any breeding offspring, while only 20% of the breeders produce 65% of the recruits. This results primarily because reproductive success increases with experience for several years after jays become breeders and because a large fraction of breeders die young. Within-year and among-year variances in reproductive success across the population are extremely high, generally higher than variance among lifetimes. Lifetime success therefore is highly correlated with life span as a breeder, since long-lived breeders have the highest probabilities of reproducing during the occasional highly productive years. Offspring survival is even more important than breeding life span as a determinant of lifetime reproductive success. Fecundity contributes little to total variance in lifetime success. In this intensely crowded population, parents are selected to promote the eventual success of a few offspring to become breeders rather than to produce the maximum possible number of offspring. Reduced clutch size, single broods, group living, sentinel behavior, communal defense of large territories, and cooperative breeding represent attributes of parental care and offspring behavior that reduce mortality and increase the probability that offspring will eventually gain the space necessary to breed.

20.8 Acknowledgments

We continue to express our gratitude to the Archbold Biological Station and its staff for their cooperation and their support of our long-term research. We are grateful to numerous friends and colleagues who have helped with various phases of fieldwork and analyses relating to this report, especially Molly Fitzpatrick, Wayne Hoffman, Bobbie Kittleson, Fred Lohrer, Kevin McGowan, Debra Moskovits, and Jan Woolfenden. Wayne Hoffman and Doug Stotz provided valuable insights during the preparation of the manuscript, and Tim Clutton-Brock and an anonymous reviewer greatly improved the chapter by commenting on an earlier draft. Fieldwork has been supported by the American Philosophical Society, a Frank M. Chapman Fellowship to Woolfenden, the Conover Fund of the Field Museum of Natural History, and the National Geographic Society.

4 Mammals

In all six mammals represented in this section, females live and breed in social groups and males are polygynous. In red deer, *Cervus elaphus* (chap. 21), and northern elephant seals, *Mirounga angustirostris* (chap. 22), males defined harems of females during a brief mating season. In lions, *Panthera leo* (chap. 23), vervet monkeys, *Cercopithicus aethiops* (chap. 24), and savannah baboons, *Papio cynocephalus* (chap. 25), breeding groups consist of several mature individuals of each sex, and though individual males may guard individual females over a part of their estrous cycle, no long-term, exclusive mating bonds are maintained between individual males and females within the group. Finally, in the Kipsigis tribesmen studied by Borgerhoff Mulder (chap. 26), men initially marry a single wife, bartering cattle for extra wives over their life span.

The problems of comparing the results of different studies resemble those of the previous sections. The studies of red deer and the Kipsigis measure breeding success in terms of offspring reaching sexual maturity, while the other four studies measure the number of offspring alive at or soon after weaning age. Nonbreeders include all individuals born in red deer and vervet monkeys but only animals weaned in elephant seals and lions. The extent to which life spans are truncated by the duration of observations or biased by overrepresentation of animals dying young differs between studies, as do the methods used to correct these biases. Finally, in vervet monkeys, lions, and baboons, estimates of the breeding success of males (and in lions, of females too) are based on assumptions likely to minimize individual differences in breeding success.

Estimates of the total opportunity for selection in breeding females fall within the distribution of previous studies, with the exception of estimates for elephant seals, which are unusually high, possibly because there is a high rate of immigration into the population Le Boeuf studied (see chap. 22). In all six species, individual differences in offspring survival are a major source of variation in lifetime success among females. As in birds, both fecundity and rearing success commonly increase after the first breeding attempt(s) (see especially chaps. 22, 23, and 25) while most measures of reproductive performance decline in the later years of the life span (see chaps. 21, 23, 25, and 26, and see chap. 22 for a contrast).

Climatic and demographic factors have an important influence on female breeding success directly through effects on food availability (see chaps. 23 and 24). In addition, environmental variables often influence juvenile growth, thus affecting breeding success in adulthood and in some cases generating substantial differences in breeding success between successive cohorts (see chaps. 21, 22, and 26). As in group-living birds, the social environment also affects female success. Both in red deer and in lions, the size of the group they live in has an important effect on the breeding success of females, though for entirely different reasons (chaps. 21 and 23).

Phenotypic factors are important too. In red deer, the body size of females appears to affect the birth weight of their offspring, influencing their early growth, survival, and eventual breeding success (see chap. 21). These effects are probably caused partly by the physical consequences of early development and partly by behavioral ones, for a female's social rank is established early in life and is related to her breeding success and body size (see also chap. 22). Similar relationships between female rank and breeding success have been found in social primates (Dunbar 1980, 1985; Silk 1983; Hrdy and Williams 1983; Fedigan 1983), though they may be more complex than in red deer (see chaps. 24 and 25).

In all four mammals where it was possible to compare the total opportunity for selection in the two sexes, this was greater in males than females, and in both harem-breeding species I_m/I_f ratios are higher than in any of the monogamous birds. In contrast to red deer, where individual stags rarely hold more than twenty to thirty hinds, dominant male elephant seals can guard over a hundred females, and the total opportunity for selection is particularly high (see chap. 22). In both species, differences in mating success are the predominant source of variation in male success.

In harem-forming mammals, the effective breeding life span of males is short because males that have passed their physical prime cannot compete successfuly. It is not clear why selection has failed to produce a male phenotype that deteriorates less rapidly with age, for there is little evidence that the mortality of prime males is substantially higher than that of mature females until the breeding life span is effectively over. The effects of age on male success appear to be less pronounced in social primates that live in multimale groups (see chaps. 24 and 25), perhaps because alliances between males reduce the effects of senescence (see Packer 1979a). Finally, in some human societies the association between wealth and age has led to the unusual situation where mating success continues to increase throughout much of the life span, and the period of effective reproduction is longer in males than in females (see chap. 26).

This is presumably associated with particularly strong selection for longevity.

Male breeding success is related to fighting ability and body size in red deer and elephant seals (chaps. 21 and 22). Although body size also affects female succeess, selection for large size is more intense in males on account of the greater variance in male success. Since early development is related to adult size (see chap. 21) and is strongly affected by maternal investment, females might be expected both to invest more heavily in individual sons than daughters and to vary the sex ratio of their progeny in relation to their capacity for investment (Trivers and Willard 1973; Clutton-Brock and Albon 1982). Previous studies have shown that the costs of rearing males exceed the costs of rearing females in both species (Clutton-Brock, Albon, and Guinness 1981; Reiter, Stinson, and Le Boeuf 1978), while in red deer the proportion of male calves born varies with the mother's social rank (Clutton-Brock, Albon, and Guinness 1984, 1986).

Finally, it is clear that female choice can exert an important influence on male success both in harem-breeding mammals (Cox and Le Boeuf 1977; Bell 1983) and in the complex, multimale societies of cercopithecines (see chaps. 24 and 25). In particular, as both studies of nonhuman primates in this book show, females commonly prefer immigrant males as mating partners.

21 Reproductive Success in Male and Female Red Deer

T. H. Clutton-Brock, S. D. Albon, and F. E. Guinness

In this chapter we compare the extent and causes of variation in breeding success in red deer (*Cervus elaphus*) stags and hinds on the island of Rhum, Scotland (fig. 21.1). Our analysis is divided into four parts: the first compares the effects of age on survival and breeding success in the two sexes; the second describes the distribution of lifetime breeding success and its components; the third describes the behavioral, morphological, and ecological factors that affect breeding success in females; and the fourth describes the factors affecting success in males.

The breeding system of red deer is a classical example of temporary harem polygyny (Darling 1937; Clutton-Brock, Guinness, and Albon 1982). Females aggregate in unstable groups of two to thirty or more that

Figure 21.1: Red deer stag and hinds.

vary in membership from hour to hour. Daughters adopt home ranges overlapping those of their mothers and associate with their matrilineal kin throughout their adult lives, whereas sons disperse from their mother's groups between the ages of two and three years to join stag groups that occupy areas adjacent to the main hind populations. At the end of September, stag groups fragment as individuals move to traditional rutting sites, mostly based in areas of *Agrostis/Festuca* greens preferred by hinds, where they collect and defend harems. Fights between rutting stags are regular events, and the size of a stag's harem is closely related to his success in fights (Clutton-Brock et al. 1979).

Breeding is seasonal. Over 75% of hinds conceive between 5 October and 25 October, and calves are born the following June (Guinness, Albon, and Clutton-Brock 1978). On average, 17% of calves die during the first four months of life (mostly in their first two weeks). Summer calf mortality varies between years from less than 10% to over 27% and is apparently unrelated to population density (Clutton-Brock, Major, and Guinness 1985). A second peak in calf mortality occurs at the end of winter between February and April (Guinness, Clutton-Brock, and Albon 1978) when, depending on weather and population density, between 5% and 45% of calves entering the winter die. Yearling mortality, which is also density dependent, varies between years from zero to 30%. After animals have reached two years of age, annual mortality is low until the deer approach the end of their life span at ten to twelve years.

21.1 Methods and Samples

Data were collected in the North Block of Rhum between 1968 and 1984. All deer regularly using an area of about 12 km^2 could be recognized as individuals, and regular observation throughout the year allowed us to monitor their survival and reproductive success (Clutton-Brock, Guinness, and Albon 1982). In 1972 the annual cull was terminated, and the numbers of hinds one year old or older using the area rose from 57 to 166 in 1983. The number of resident stags one year old or older rose from 116 to 155 in 1978, then declined to 86 by 1983 (Clutton-Brock, Major and Guinness 1985).

The majority of analyses described in this chapter are based on thirty-five hinds from the 1966–70 cohorts and thirty-three stags from the 1966–72 cohorts. These samples represented all animals born in the study area during these years that survived to breeding age (three years for hinds, five for stags). All the males had died by 1984, but some hinds were still alive. However, the probability of successful breeding after the age of fifteen years is slight (Clutton-Brock, Guinness, and Albon 1982). The proportion of animals belonging to the same cohorts that failed to reach breeding age was estimated from age-specific survival schedules from cohorts born between 1971 and 1975. A minority of analyses focus on the 1972 cohort of twelve hinds and fifteen stags for which our data were unusually complete.

All the distributions of values of each component of reproductive success were checked for skewness. Among hinds only calf summer survival was skewed. Since calf summer survival explained little of the variance in lifetime reproductive success (see below), the factors influencing it were not investigated in detail, and we did not transform the original data. The residuals calculated in the multiple regression analysis of other components were also checked for skewness, but it was not necessary to transform the data. The distribution of lifetime reproductive success for stags was skewed, but transformation had no substantial effect on the results.

Age and Life Span

The ages of most animals included in the analysis were known because we had followed them since birth. Ages of a small minority of the oldest animals were assessed from tooth wear or analysis of cementum rings after death. A hind's reproductive life span (RLS) was the number of years it lived after the age of two. A stag's RLS was the number of years it lived after the age of four. We could usually identify an animal's date of death to within a week from regular censuses of the study population, and most carcasses were found. Emigration could be distinguished from death by a gradual shift in the animal's ranging pattern toward the edge of the study area.

Fecundity (FEC)

Fecundity is measured by the proportion of years during its reproductive life span in which a hind produced a calf. Birth weights and birth dates were collected in the course of daily observation during the breeding season (Clutton-Brock, Guinness, and Albon 1982).

Summer calf survival (SCS) was the proportion of a parent's offspring that survived until 1 October in their year of birth.

Winter calf survival (WCS) was the proportion of a parent's offspring alive on 1 October that survived until May of their second year.

Yearling survival (YS) was the proportion of a parent's offspring surviving their first winter that survived their second year of life. In some analyses the three components of calf survival were multiplied together to provide a single measure of offspring survival (OS).

Lifetime reproductive success (LRS) was the number of offspring that survived to the age of two years: 97.4% of all hinds and 95.1% of all stags reaching two years survived to breeding age (three years for hinds, six for stags). For hinds, LRS was the product of fecundity, summer calf survival, winter calf survival, and yearling survival. For stags, fecundity was replaced by mating success (see below).

Mating Success of Stags (MS)

The date of conception of each calf born in the study area was estimated by backdating from its birth date by 236 days for males and 234 days for females. For five days before and five days after the estimated conception date (the standard deviation of gestation length) we identified the stag in whose harem the mother was seen, awarding each stag a fraction of the calf corresponding to the fraction of the eleven-day period in which the mother was in his harem. Estimates of mating success based on this method were closely correlated with other measures of breeding success, including the number of observed copulations, and likely sources of consistent error have been checked (Clutton-Brock, Guinness, and Albon 1982). In particular, there is no evidence that hinds in estrus are consistently attracted to the harems of particular stags (Gibson and Guinness 1980). Although stags less than five years old sometimes abduct hinds from harems, they have not been observed to copulate successfully. A stag's mating success was the estimated number of hinds fertilized per year and was calculated as the product of the number of days during the rut when he held hinds (DH); the mean number of potentially fertile hinds (over two years old) in his harem (HS); and the estimated number of matings per hind per day (MH) (based on the method of estimating mating success described above).

Matriline Size (Hinds)

Matriline size is the average number of an individual's known matrilineal female relatives one year old or older calculated over her lifetime. Members of the same matrilineal group shared a common home range (Clutton-Brock, Guinness, and Albon 1982).

Range Area

Range area is measured by the total number of diffrent hectare grid squares in which the hind was seen over her lifetime.

Dominance Rank

An individual's dominance rank is a measure of its ability to displace or threaten other individuals. This increased with age. To produce an age-independent measure of rank, we calculated the number of different animals the same age or older that each individual displaced or threatened and divided this by the number of animals the same age or younger that displaced her (Clutton-Brock, Albon, and Guinness 1986). The same ratio was used to produce an age-independent measure of fighting success in stags. In red deer, different dominance indexes usually produce very similar rankings (ibid.).

21.2 Results

Age, Breeding Success, and Survival

Age-related changes in breeding success showed different patterns in the two sexes (fig. 21.2*a*). The number of offspring raised to two years old by females per year increased until age four and then remained approximately constant until hinds were over age twelve. Thereafter it fell as a result of declining fecundity and increasing neonatal mortality of calves (Clutton-Brock 1983). Reduction in these components of breeding success in old hinds was partly offset by an increase in overwinter survival among calves (Clutton-Brock 1984).

Stags began to breed later than hinds—usually in their sixth or seventh year of life. After the age of eleven, breeding success declined

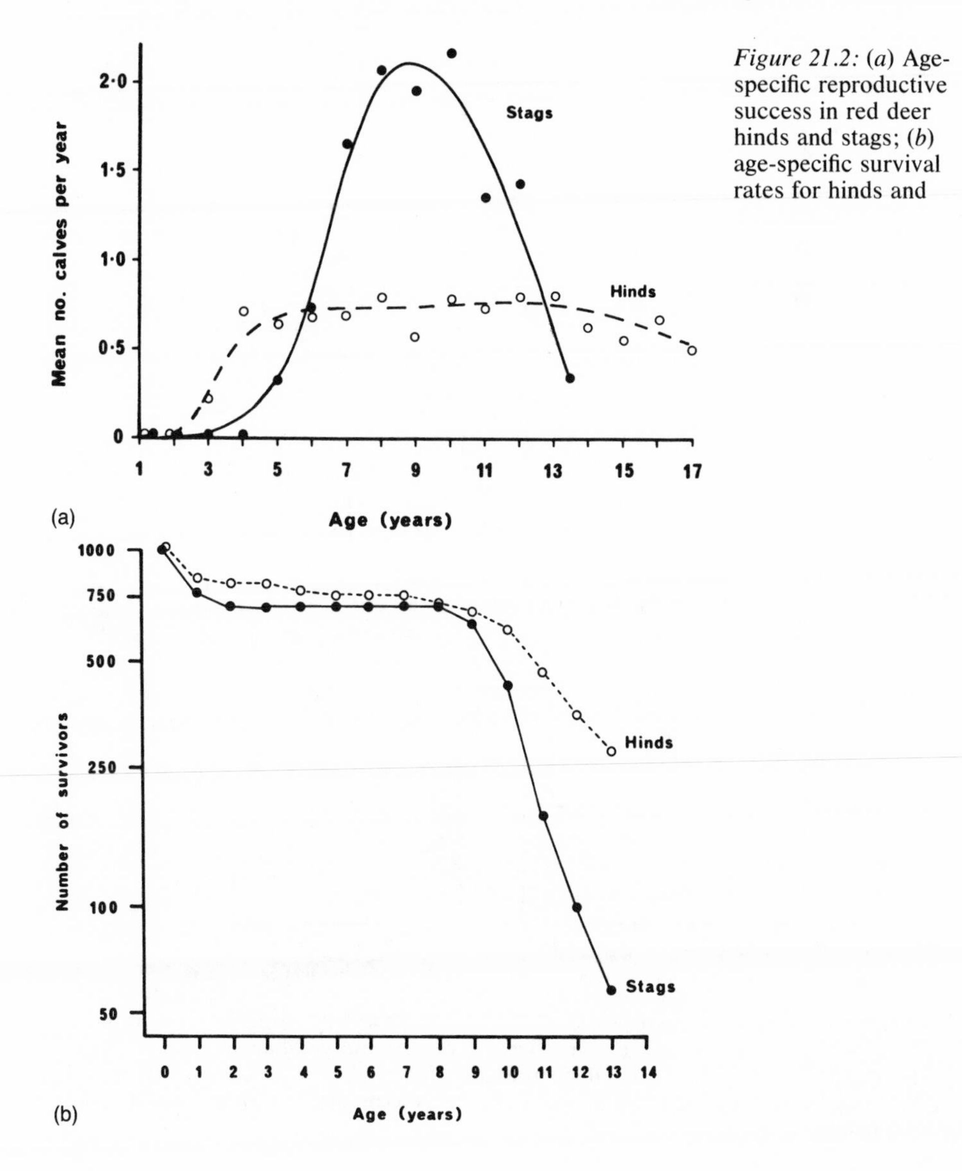

Figure 21.2: (*a*) Age-specific reproductive success in red deer hinds and stags; (*b*) age-specific survival rates for hinds and

rapidly, and few stags over twelve were able to win control of harems (Clutton-Brock et al. 1979). Among animals of both sexes surviving to two years old, survival rates were high until over the age of eight (fig. 21.2*b*), when mortality began to increase.

Variation in Breeding Success

Hinds

During the first period of the study (1971–75), 20% of hinds born died before reaching breeding age at three years old. Among the thirty-five animals belonging to the 1966–70 cohorts that reached breeding age, all calved at least once. LRS varied from 0 to 9, with a mean of 5.03, a

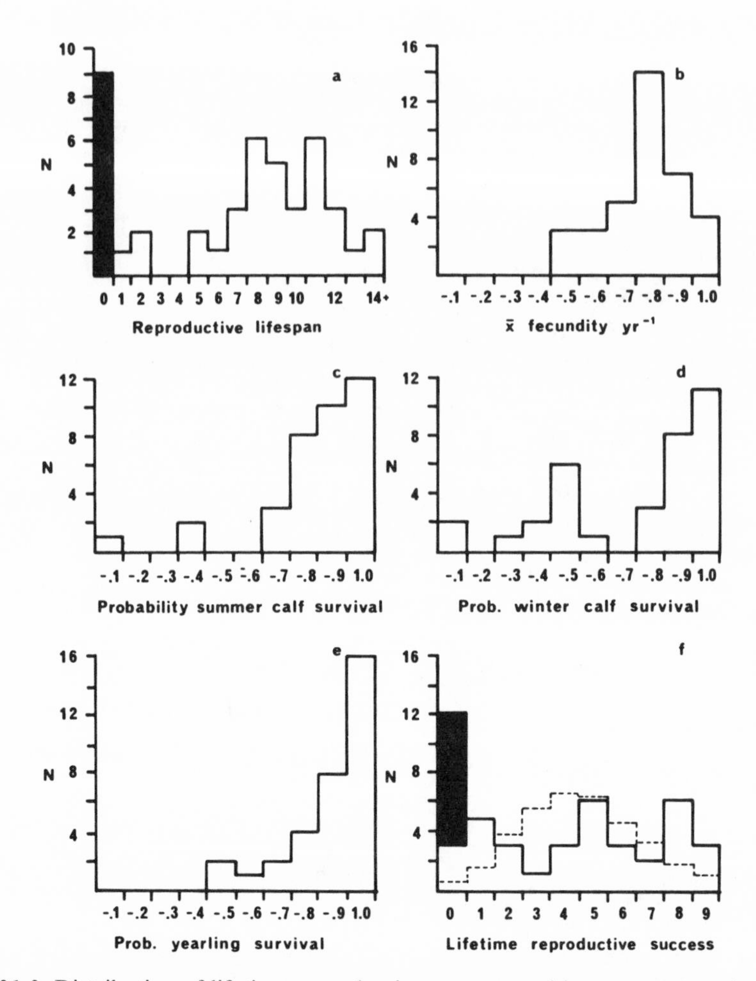

Figure 21.3: Distribution of lifetime reproductive success and its components (reproductive life span, fecundity, calf survival through the winter, and yearling survival) for thirty-three hinds born between 1966 and 1970. Black bars (*a*, *f*) indicate animals that failed to reach breeding age; the dashed line (*f*) shows the Poisson distribution for the sample of breeding hinds.

variance of 9.09, and a standardized variance ($\sigma^2/\bar{x}^2$) of 0.36. Inclusion of the estimated number of nonbreeders increased the variance to 11.40, of which 36.9% was contributed by nonbreeders and 63.1% by breeders (calculated by the method described in chap. 27). Standardized variance ($\sigma^2/\bar{x}^2$) in the total sample was 0.71. This figure is more than three times as large as standardized variance in the number of calves born over the life span.

All components of breeding success varied widely between individuals that reached breeding age: reproductive life span ranged from one to sixteen years, fecundity from 0.45 to 1.0 calves born per year, calf survival through the summer and winter from 0 to 100%, and yearling survival from 40% to 100% (see fig. 21.3). The components of breeding success were not closely intercorrelated (table 21.1), with the exception that calf survival through the summer increased with the hind's life span.

Partition of variance in LRS using Brown's methodology (this volume, chap. 27), shows that variation in offspring survival made the largest contribution to variance in LRS among breeding hinds (57.4%) (table 21.2).Covariances were all relatively small.

Breakdown of offspring survival into summer calf survival, winter calf survival, and yearling survival showed that differences in winter calf survival were the most important of the three and were responsible for more than a quarter of the total variance in LRS (25.8%). They were, in addition, the key factor limiting population density (Clutton-Brock, Major

Table 21.1 Correlations between the Five Main Components of Breeding Success in Red Deer Hinds

Component	RLS	FEC	SCS	WCS
Reproductive life span (RLS)				
Fecundity (FEC)	−.247			
Summer calf survival (SCS)	.601***	−.121		
Winter calf survival (WCS)	.268	.156	.378*	
Yearling survival (YS)	−.023	.303	.003	.167

Note: Pearson's correlation coefficients calculated across thirty-three hinds for which estimates of lifetime reproductive success (LRS) were available. Transforming the data for summer calf survival see sec. 21.2) changed the correlation with RLS to $r = .510$, $p < .01$.
*p <0.5; **p < .01; ***p < .001.

Table 21.2 Percentage Contribution of Three Components of Lifetime Reproductive Success to Variation in LRS in Thirty-three Red Deer Hinds Reaching Breeding Age

Component	RLS	FEC	OS
Reproductive life span (RLS)	*26.5*		
Fecundity (FEC)	−3.1	*7.7*	
Offspring survival (OS)	9.9	1.8	*57.4*
Three-way contribution, −0.2			

and Guinness 1985). Yearling survival contributed 11% of the total variance in LRS, and summer calf survival contributed slightly less than 10%.

Stags

An estimated 31% of all male calves born between 1966 and 1972 failed to reach breeding age. Among stags reaching breeding age, LRS varied from 0 to 32, with a mean of 5.41 and a variance of 41.9. Standardized variance for breeders was 1.43. Including the 31% of stags born between 1966 and 1970 that failed to reach breeding age reduced the variance to 35.0, of which 18.4% was due to nonbreeders and 81.6% to

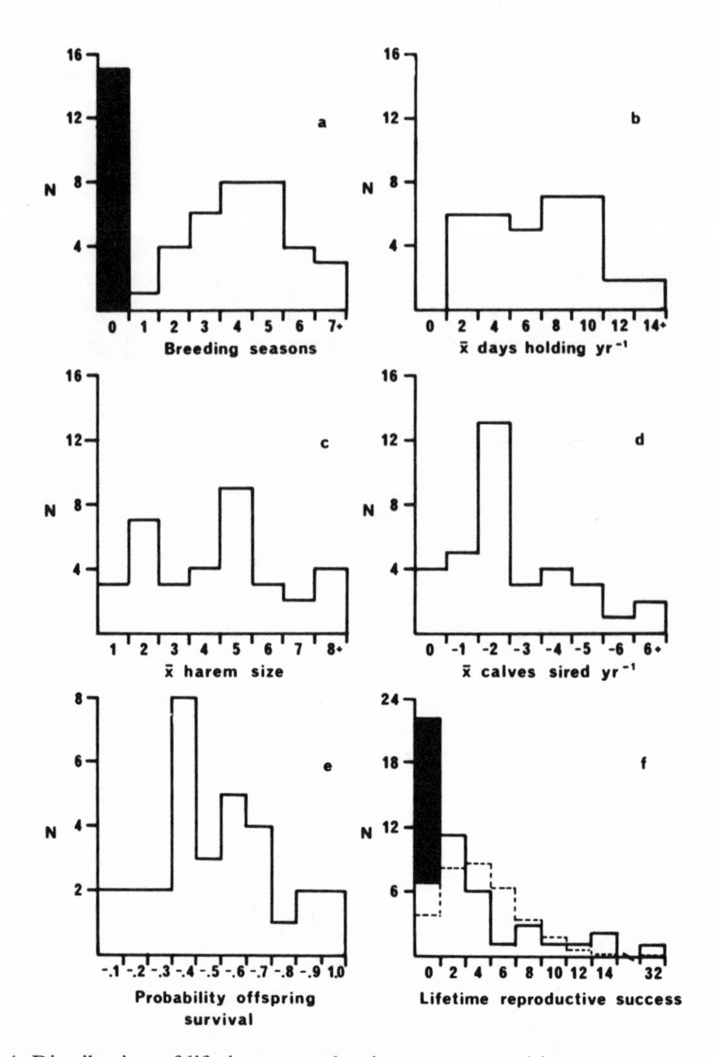

Figure 21.4: Distribution of lifetime reproductive success and its components (number of breeding seasons, days of harem holding per year, harem size, calves sired per year, and probability of offspring survival) for thirty-five stags born between 1966 and 1972. Black bars (*a*, *f*) indicate animals fialing to reach breeding age; the dashed line (*f*) shows the Poisson distribution for the sample of breeding stags.

breeders. Standardized variance in this sample was 2.51, giving an I_m/I_f ratio (Wade and Arnold 1980) of 3.53. Because much of the variance in the breeding success of females is due to variation in calf survival, I_m/I_f calculated in terms of calves born was substantially higher (9.85).

Among the stags that reached five years, reproductive life span ranged from 0 to 8, the average number of days per season when they were seen holding hinds from 2 to 15, mean harem size from 2 to 15, the average number of calves sired per year from 0 to 7, and the percentage of their offspring that survived to two years from 12 to 100% (see fig. 21.4). In contrast to hinds, the different components of breeding success were mostly intercorrelated (table 21.4). Partitioning variance in lifetime breeding success into its components showed that mating success accounted for

Table 21.3 Percentage Contribution of Five Components of Reproductive Success to Variation in LRS in Hinds Reaching Breeding Age

Component	RLS	FEC	SCS	WCS	YS
Reproductive life span (RLS)	*20.3*				
Fecundity (FEC)	−2.0	*7.7*			
Summer calf survival (SCS)	4.4	0	*8.7*		
Winter calf survival (WCS)	5.4	−0.8	2.2	*25.8*	
Yearling survival (YS)	0.4	7.5	0.2	2.5	*11.0*
Three-, four-, and five-way contribution, 7.1					

Note: The sample used in this analysis was the same as that for table 21.2 except that three hinds that failed to raise any offspring through their first winter were excluded.

Table 21.4 Correlations between the Four Main Components of Breeding Success in Red Deer Stags

Component	RLS	DH	HS	MS
Reproductive life span (RLS)				
Mean days harem held (DH)	.446**			
Mean harem size (HS)	.239	.613***		
Mating success (MS)	.322⁺	.793***	.873***	
Offspring survival (OS)	.075	.428**	.285⁺	.338*

Note: Pearson's correlation coefficients calculated across thirty-five stags for which estimates of lifetime reproductive success (LRS) were available.
⁺$p < .10$; *$p < .05$; **$p < .01$; ***$p < .001$.

Table 21.5 Percentage Contribution of Three Components of Lifetime Reproductive Success to Variation in LRS in Red Deer Stags

Component	RLS	MS	OS
Reproductive life span (RLS)	*6.86*		
Mating success (MS)	25.82	*31.73*	
Offspring survival (OS)[a]	−0.90	6.76	*19.56*
Three-way contribution, 10.17			

Note: Calculated from a sample of thirty-one stags for which estimates of lifetime reproductive success were available.
[a]Percentage of offspring surviving to age two.

Table 21.6 Percentage Contribution to Variation in Mating Success of Three Components of Mating Success in Red Deer Stags

Factor	DH	HS	MH
Mean days harem held (DH)	*21.0*		
Mean harem size (HS)	31.0	*25.2*	
Matings/hind/day (MH)	−12.8	−5.1	*25.3*
Three-way contributions, 15.4			

Note: Sample as for table 21.5.

32% of the variance in lifetime success, while the covariance between mating success and reproductive life span accounted for a further 26% (table 21.5). Offspring survival contributed 20%.

The average number of days per season on which stags held harems was closely correlated with average harem size (r = .873), and their independent contributions plus the covariance between them accounted for over 75% of variance in mating success (table 21.6).

Determinants of Breeding Success in Hinds

Birth Weight and Date

Since the main cause of variation in breeding success among hinds was winter calf survival, our subsequent analysis of the causes of breeding success concentrated on this component. The overwinter survival of

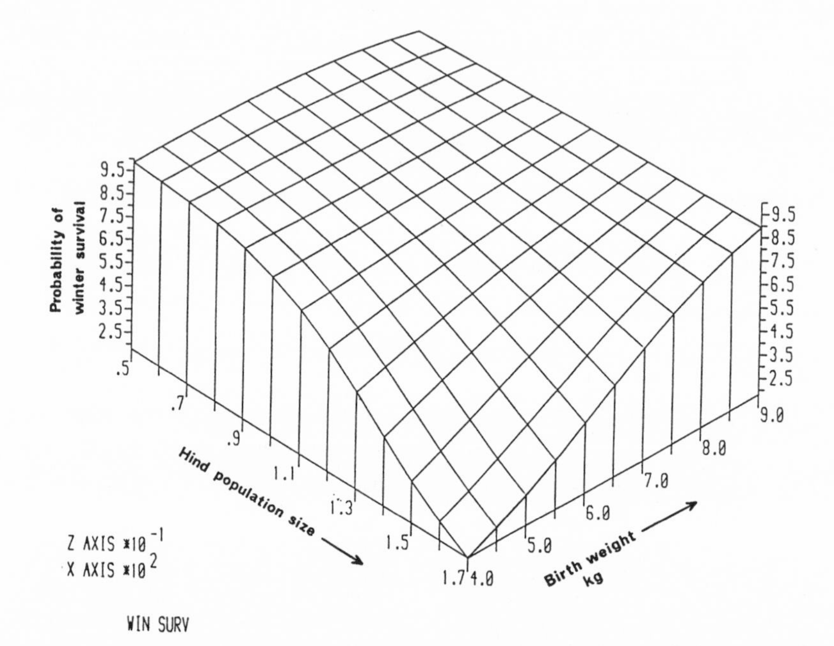

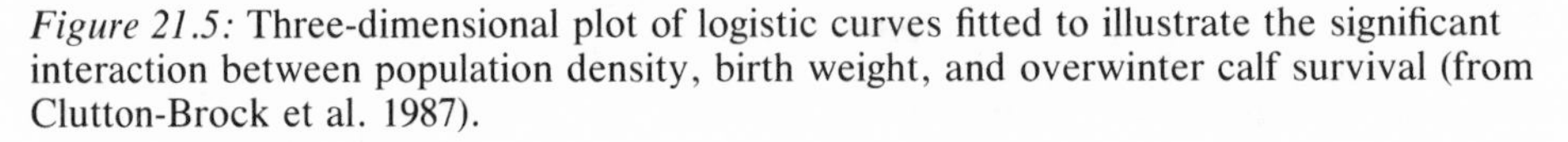

Figure 21.5: Three-dimensional plot of logistic curves fitted to illustrate the significant interaction between population density, birth weight, and overwinter calf survival (from Clutton-Brock et al. 1987).

calves was related positively to their birth weight and negatively to their birth date (Guinness, Clutton-Brock, and Albon 1978). These were presumably independent effects, since calf birth weight was positively correlated with birth date, while birth date was negatively correlated with calf survival.

During the period of our study, hind numbers in our study area increaed by nearly 200% (Clutton-Brock et al. 1987). As numbers rose, selection against light-born calves intensified (see fig. 21.5), though birth weight showed no consistent change. In contrast, birth date became gradually later, but selection against late-born calves remained approximately constant (Clutton-Brock 1987).

Average values of calf birth date and birth weight differed consistently between mothers. Hinds varied in their average deviation from the median date of parturition for the year from eight days before to eighteen days after. These differences were significantly correlated with the proportion of their calves that survived their first winter (fig. 21.6*a*). and with LRS ($r = -.356$, $t_{31} = 2.121$, $p < .05$; $r = -.362$, $t_{31} = 2.162$, $p < .05$).

The mean birth weight of calves varied between hinds from 4.1 kg to 8.1 kg, and this too was correlated with differences in calf survival (fig. 21.6*b*; $r = .466$, $t_{31} = 2.932$, $p < .01$). Multiple regression confirmed that birth date and birth weight had separate effects and that jointly these two variables accounted for 33.5% of the variance in overwinter calf survival and 31.0% of the variance in LRS (see table 21.7). The mean birth weight of a female's calves was correlated with her adult body weight ($r = .662$, $t_{16} = 3.18$, $p < .001$, $Y = 0.177X - 8.804$), which was, in turn, related to her own birth weight ($r = .622$, $t_{10} = 2.514$, $p < .05$).

Since mean calf birth weight varied between the years of the study from 5.95 kg to 7.30 kg in relation to temperature and food availability in spring (Albon, Guinness, and Clutton-Brock 1983), consistent differences in birth weight and calf survival were to be expected between cohorts of hinds. They proved to be substantial: mean offspring birth weight varied from 7.25 kg for the 1966 cohort to 5.60 kg for the 1970 cohort, while offspring survival calculated over the whole life span for the same two cohorts of females was 72% and 15% respectively (Albon, Clutton-Brock, and Guinness 1987). Analysis of these relationships in a larger sample of data confirmed that light-born cohorts produced light calves (fig. 21.7) and that the mean offspring survival of members of a cohort was related to the mean birth weight of calves that they produced (fig. 21.8). These results indicate that a substantial proportion of variation in the breeding success of hinds was of environmental origin.

Genetic Factors

The heritability of components of breeding success in hinds could not be estimated directly, since mothers and daughters usually shared a common home range whose quality affected their reproductive performance (see below). However, analysis of several polymorphisms indicates

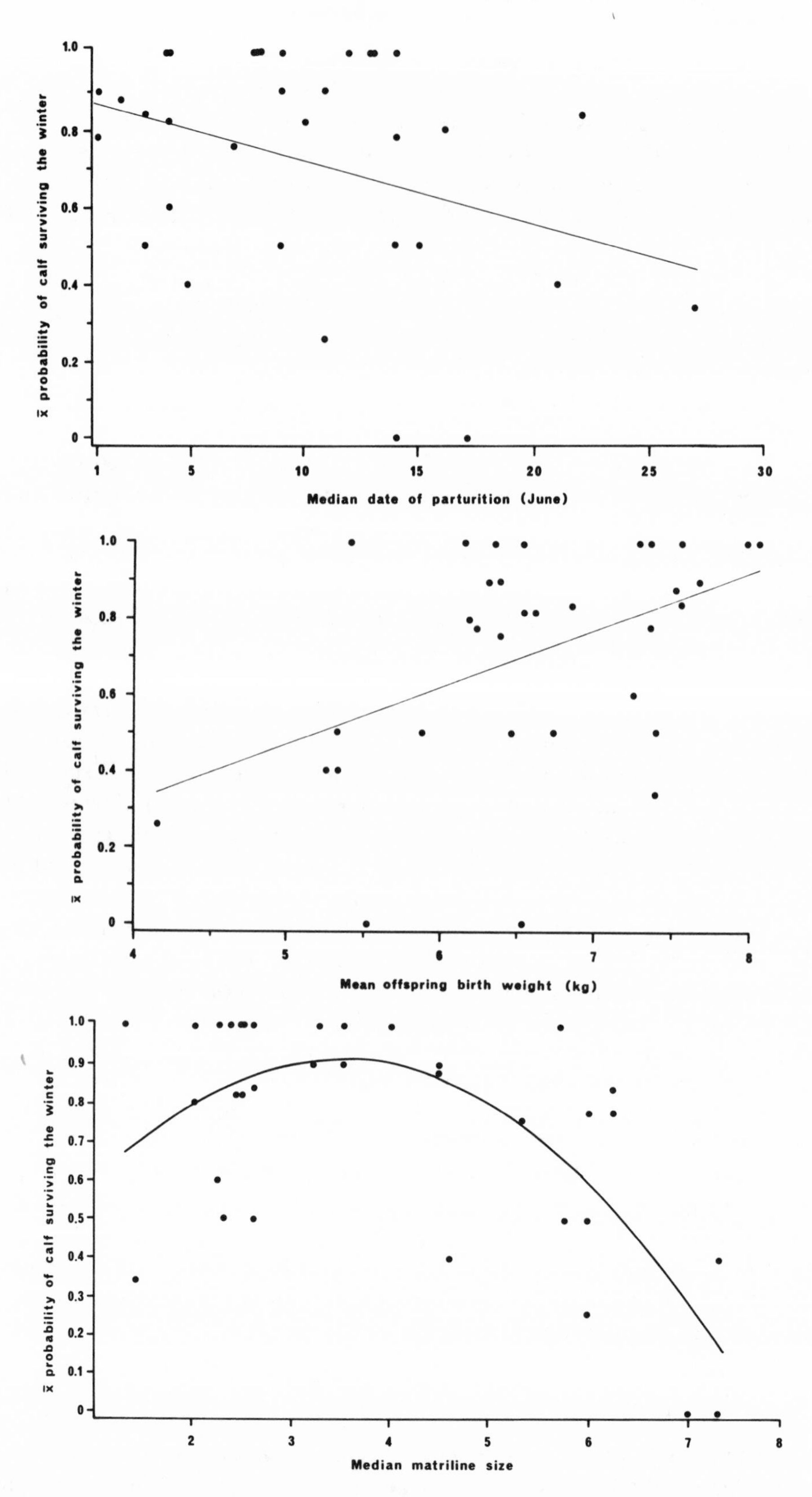

Figure 21.6: Average winter calf survival for thirty-three different hinds plotted against (*a*) the hind's median date of parturition, (*b*) the mean birth weight of her calves, and (*c*), the median size of her matriline.

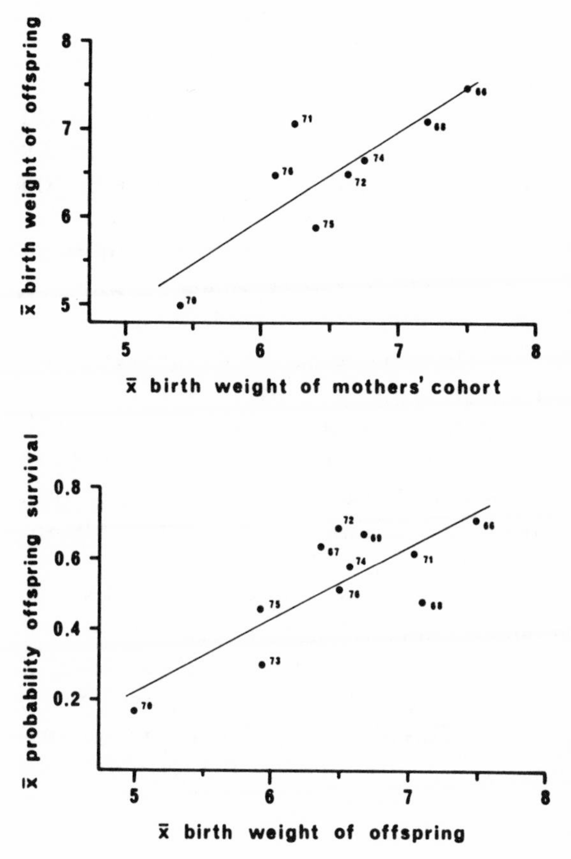

Figure 21.7: Mean offspring birth weight calculated across hinds belonging to different cohorts (1966–76) and plotted against the mean birth weight of the cohort to which they belonged. See Albon, Clutton-Brock, and Guinness 1987.

Figure 21.8: Mean offspring survival calculated across hinds belonging to different cohorts (1966–76) and plotted against mean offspring birth weight for the cohort to which they belonged ($r = .800$, $t_9 = 4.00$, $p < .01$; $y = .201x - .770$). See Albon, Clutton-Brock and Guinness 1987.

that part of the variation was of genetic origin. One allele at the MPI (mannose phosphate isomerase) locus was consistently associated with reduced survival of calves and appeared to be declining in the population (Pemberton et al., n.d.). The allele was unrelated to birth weight or date and did not appear to affect breeding success in calves that survived to adulthood.

Social Factors

Most environmental factors appear to affect calf survival through their influence on birth weight and date. One exception was the size of the hind's matrilineal group. As matriline size increased, competition for food intensified and breeding success fell (Clutton-Brock, Albon, and Guinness 1982). Large matriline size was associated with later birth dates and lower birth weights (ibid.) but also had independent effects on overwinter survival of calves (table 21.7). In addition, male yearlings born into large matrilines were less likely to survive than those born into small ones (Clutton-Brock, Albon, and Guinness 1982). The relationship between matriline size and LRS was nonlinear, and when matriline size was fitted as a two-degree polynomial, the independent effects of parturition date disappeared and a significantly greater proportion of the variance (47.1%) was accounted for (table 21.7, panel C).

The effects of matriline size operated partly or principally through their impact on the availability of food on the most heavily selected swards, the *Agrostis/Festuca* greens. When matriline size was replaced by the annual frequency with which matrilineal relatives were seen feeding on *Agrostis/Festuca* greens within a hind's range relative to the area of this community in their range, a stronger negative correlation with winter calf survival was found ($r = -.540$, $t_{26} = 3.271$, $p < .01$) compared with $r = -.486$, $t_{26} = 2.836$, $p < .01$).

A hind's breeding success was also related to her dominance rank. Dominance rank was positively correlated with live body weight in summer ($r = .495$, $t_{36} = 3.413$, $p < .01$) and with the animal's own birth weight ($r = .343$, $t_{63} = 2.895$, $p < .01$). Dominant hinds suffered less from feeding interference than subordinates (Thouless 1986), conceived earlier, and produced calves that were heavier than those of subordinates (Clutton-Brock, Albon, and Guinness 1984). Their sons (but not their daughters) were less likely to die as calves or as yearlings, and their LRS was nearly twice as high as that of subordinates (Clutton-Brock, Albon, and Guinness 1984). However, unlike the effects of matriline size, those of dominance were not independent of the effects of calf birth weight and parturition date, and including maternal dominance in the multiple regression shown in table 21.7 did not significantly increase the proportion of variance explained.

Table 21.7 Multiple Regression of Individual Differences in Winter Calf Survival and Lifetime Reproductive Success on Mean Calf Birth Weight, Mean Calf Birth Date, and Matriline Size

	Simple Correlation Coefficients				Cumulative Percentage of Variance Explained	
	(r)	T-values	p	Cumulative F	*Significance*	*Percentage of Variance*
		A. Winter Calf Survival				
Mean offspring birth weight	.466	2.224	.034	8.314	**	21.7
Mean date of parturation	−.353	−2.669	.013	7.319	**	33.5
Matriline size	−.511	−2.895	.007	8.915	***	48.9
		B. Lifetime Reproductive Success				
Mean offspring birth weight	.417	2.639	.013	6.310	*	17.4
Mean date of parturition	−.380	−2.396	.023	6.525	**	31.0
		C. Lifetime Reproductive Success				
Matriline size		3.348	.002	3.864	+	11.4
Matriline size^2		−3.658	.001	8.625	**	37.3
Mean offspring birth weight		2.283	.030	8.323	***	47.1

$^+p < .10$; $^*p < .05$; $^{**}p < .01$; $^{***}p < .001$.

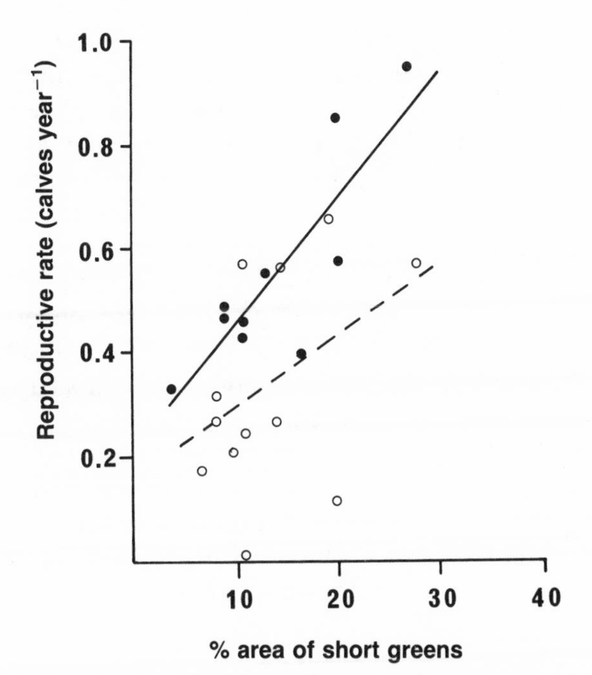

Figure 21.9: Reproductive rate (number of calves surviving to one year old produced per year of breeding life span) for hinds whose home ranges abutted the coast plotted on the percentage of observations of the mother on *Agrostis/Festuca* communities, 1973–81). *Solid circles*: hinds using gull-fertilized areas predominantly ($y = 0.202 + .0133x$; $F_{1,8} = 18.42$, $p < .01$). *Open circles*: Hinds not using gull-fertilized areas predominantly ($y = 0.16 + .0133\,x$, $F_{1,10} = 1.755$, n.s.). Comparison of slopes: $F_{1,18} = 0.929$, n.s.; comparison of elevations: $F_{1,19} = 7.12$, $p < .05$. See Iason, Duck, and Clutton-Brock 1986.

Home Range Quality

LRS was positively correlated with the size of the hind's range ($r = .409$, $t_{26} = 2.285$, $p < .05$) as well as with measures of the intensity of competition for preferred swards (see above). The presence of herring gull colonies on the north coast of Rhum provided a natural experiment on the effects of food quality on reproductive success (Iason, Duck, and Clutton-Brock 1986). *Agrostis/Festuca* swards around gull colonies received an estimated 15–20 kg of pure nitrogen and 2–3 kg of phosphorus per hectare per year, and the nitrogen content of vegetation on these patches was nearly twice as high as that in unfertilized areas. Among hinds whose ranges abutted the north coast, LRS was positively related both to the area of gull-fertilized greens in the range and to the extent to which they were used (fig. 21.9).

Determinants of Breeding Success in Stags

Fighting Ability

A stag's ability to acquire and defend a harem was closely related to his ability to win fights (Clutton-Brock et al. 1979). Fighting ability and breeding success in stags were both closely related to age, peaking between the ages of seven and ten in most individuals (fig. 21.2*a*). However, differences in fighting ability at a given age were also important, and there was a close correlation between mating success and our age-independent measure of fighting success ($r = .648$, $t_{31} = 4.137$, $p < .001$).

One way a stag's fighting ability affected the size of his harem was through its influence on his traditional rutting location. Stags could hold large harems only on large patches of greens, for on other vegetation communities the hinds strayed in search of food and the harem quickly fragmented. Competition for the areas of greens most heavily used by hinds was intense, and stags frequently began their rutting careers in peripheral sites, gradually moving to the center of one of the large patches as they matured, then moved back again to the periphery of the rutting area as they passed their prime (Clutton-Brock, Guinness, and Albon 1982). As a result of differences in hind density, the mating success of stags varied with the location of their rutting grounds.

Body Size and Breeding Success

A stag's body size had an important influence on his fighting ability and breeding success. Among twenty-two mature stags that were immobilized and measured, mating success was correlated with their average ranks calculated across three linear measures of body size (foreleg length, hind foot length, and back length, withers to base of tail ($r_s = .583$, $t_{20} = 3.206$, $p < .01$). Similarly, among stags measured after death, jaw length was correlated with fighting ability ($r = .472$, $t_{15} = 2.074$, $p < .10$) and with mating success ($r = .564$, $t_{17} = 2.814$, $p < .02$). The weight of cast antlers, another index of body size, was also correlated with mating success ($r = .709$, $t_{11} = 3.337$, $p < .01$). We found no evidence for selection against size at any stage of the life history, though it is possible that young males that adopt a rapid growth trajectory are more likely to die from starvation in poor years.

Early Development

As in hinds, the breeding success of stags was related to their early development. Among individuals born into the 1972 cohort whose birth date and birth weight were known, LRS was negatively correlated with birth date ($r = -.877$, $t_5 = 4.084$, $p < .01$), while the residual variation about this relationship was correlated with birth weight ($r = .779$, $t_5 = 2.774$, $p < .05$). Together these two variables accounted for more than 90% of the variance in LRS in stags belonging to this cohort.

The breeding success of stags was related to their mothers' rank and reproductive status, and the sons of dominant hinds showed consistently higher LRS than those of subordinates (Clutton-Brock, Albon, and Guinness 1986). When the effects of maternal rank had been taken into account, individuals weaned late because their mothers failed to conceive in the October following their birth showed higher LRS than those weaned at the beginning of their first winter.

Food Access

There was little evidence that consistent individual differences in food access had an important effect on the breeding success of stags. Stags lived for most of the year in large, unstable bachelor groups. Group members shared a common home range and varied little in the extent to which they used different plant communities. Though within some cohorts dominance rank was correlated with fighting success in the rut and with mating success, there was no consistent relationship between their rank in the bachelor group during their lifetime and LRS overall (r = .125, n = 28, n.s.).

21.3 Discussion

Researchers studying polygynous mammals have often been impressed by the obvious variance in male breeding success and have sometimes not stopped to consider that variance in female success may build up to similar levels through consistent differences in breeding success over a longer life span (Hrdy and Williams 1983). Although our results confirm that variance in lifetime breeding success is greater among males than among females, they show that the difference is not as large as is often supposed and is exaggerated by calculations within breeding seasons (Clutton-Brock 1983). In fact, differences in variance between the sexes would be further reduced if we were able to calculate in terms of grandoffspring, since maternal characteristics of environmental origin (such as body size) exercise an important effect on the breeding success of daughters (see above).

Though comparison of the causes of variation in reproductive success between the two sexes supports Trivers's (1972) contention that males are adapted to competing for females while females are adapted to competing for food, there are similarities as well as contrasts in the factors affecting breeding success in the two sexes. Competitive ability, adult body size, and early development are important to females as well as males. However, the effects of early development are stronger in stags than in hinds, partly because of the greater variance in male breeding success, but also partly because food shortage affects the growth and survival of males more than that of females. By comparison, food access in adulthood appears to be relatively more important in hinds and is related to the quality of their home range, the size of their social group, and their social rank. These differences in selection pressures are presumably responsible for the wide variety of differences in morphology, physiology, and behavior that exist between the sexes.

The relationship between early growth and adult breeding success contrasts with experiments on hill sheep that show that increasing the plane of nutrition and the rate of growth during the first winter of life has little or no consistent effect on an animal's subsequent fecundity or survival or on the survival of her offspring (Bradford, Weir, and Torrell 1961; Purser and Roberts 1964; Brierson, Ekern, and Homb 1960; Gunn 1968). However, these involved whole groups of animals that were treated in a standard fashion, and an early advantage in growth would not necessarily be associated with a persistent increment in resource access owing to improved competitive ability.

The precise causes of the relationship between early development and adult breeding success were unclear. While birth weight and birth date may have exerted a direct effect on the individual's growth and eventual size and breeding success, they may alternatively have set in motion a chain of separate causal relationships between growth at one stage and competitive ability and growth at the next. Yet again, it is possible that differences in offspring birth weight and birth date between hinds may be correlated with their milk yield, which could be the principal factor affecting the growth and survival of their calves.

Whatever their basis, the relationships between early development and subsequent survival and breeding success have three important implications. First, they show that estimates of female fitness based on the number of progeny born or weaned may exclude the single most important component of variation in reproductive success. Second, they suggest that a substantial proportion of variation in female breeding success may be of environmental origin. And third, they emphasize that strong selection pressures are likely to favor strategies of parental investment that minimize juvenile mortality and maximize offspring growth.

21.4 Summary

The extent and causes of variation in reproductive success were investigated in male and female red deer. Standardized variance in lifetime reproductive success was greater in stags than in hinds, though differences between the sexes were reduced when breeding success was calculated in terms of surviving offspring.

Differences in breeding success between hinds were caused principally by variation in offspring survival. These were closely related to differences in the average birth weight of their calves, which were related to the mother's body size as well as to her own birth weight. Annual variation in average birth weight produced marked differences in breeding success between successive cohorts of females. Other factors affecting the reproductive success of hinds included the size of their matrilineal group, their dominance rank within it, and the quality of their home range.

The mating success of stags depended on their fighting ability, which was related to their body size. As in hinds, adult breeding success was related to the individual's birth weight and birth date, but in contrast to hinds, differences in home range quality were not consistently related to breeding success.

21.5 Acknowledgments

We are grateful to the director of the Nature Conservancy (Scotland) for permission to work on Rhum and to Glenn Iason, Callan Duck, Martin Major, and Jerry Kinsley for assistance. The research is supported by the Science and Engineering Research Council, Natural Environment Research Council, and the Royal Society.

22 Lifetime Reproductive Success in Northern Elephant Seals

Burney J. Le Boeuf and Joanne Reiter

NORTHERN ELEPHANT SEALS, *Mirounga angustirostris* (fig. 22.1), offer numerous advantages for studying reproductive success in nature in a polygynous, sexually dimorphic, long-lived mammal. Polygyny in this species is extreme; as few as 5 out of 180 males may be responsible for 48% to 92% of the mating observed with up to 470 females during a breeding season. Adult males are three to seven and a half times heavier than adult females. Maximum age attained is

Figure 22.1: An adult male northern elephant seal mounts an adult female.

fourteen years for males and seventeen years for females. The animals mate annually on predictable, accessible islands, and individuals can be marked and their behavior on land recorded over their lifetimes (Le Boeuf 1974).

We present data on lifetime reproductive success (LRS) for each sex, the effects of age on breeding, variation in breeding success within each sex and its causes, and the principal components of breeding success. We compare the variance in reproductive success between the sexes, compare our results with those obtained on other animals, and discuss the implications of our data.

22.1 Study Site and Methods

Data are presented on marked, known-age animals born on Año Nuevo Island, California, USA, whose survival and reproductive performance were recorded on this island or on nearby Southeast Farallon Island (Le Boeuf, Whiting, and Gantt 1972; Le Boeuf and Kaza 1981). The study spanned the period 1964 to 1985. During this time the colony was increasing. Annual pup production on the Año Nuevo rookery increased from 52 to 1,685, a mean annual rate of 22.4%. This increase was due to the immigration of animals from southern California rookeries (Le Boeuf, Ainley, and Lewis 1974; Bonnell et al. 1978). Before 1974, breeding occurred at two major sites on the island. After 1975 the animals began breeding on the adjacent mainland as well (Le Boeuf and Panken 1977). By 1984 the animals bred in seven different locations on the mainland and two on the island.

Males always precede females on rookeries during the breeding season, and about 20% remain until all females have copulated and departed. Consequently, each breeding season the sex ratio on the entire rookery and around each group of females decreased rapidly from about 20–100:1 at the beginning of the breeding season, when the first females arrived, to 1:3 when female number peaked. After peak season, the sex ratio increased steadily to about 20–100:1 at season's end as females weaned their pups and departed from the rookery (Le Boeuf, Whiting, and Gantt 1972).

Subjects

The male sample consisted initially of 138 weaned pups out of 151 born in the years 1964 (n = 26), 1965 (n = 33), 1966 (n = 41), and 1967 (n = 38). At weaning (twenty-eight days of age), a single noncorrosive Monel Metal cattle ear tag (Kentucky Band Company) was attached to the interdigital webbing of a hind flipper (Poulter and Jennings 1966). All males were observed until they died; none lived to 1979.

The female sample consisted initially of 279 pups weaned out of 344 pups born in 1973 (n = 102) and 1974 (n = 177). At weaning, single plastic cattle ear tags (Dalton Jumbo Roto tags) were attached to the interdigital webbing of all subjects. During the 1984 breeding season, two females from the 1973 cohort (age eleven) and six females from the 1974 cohort (age ten) were still alive and breeding. We extrapolated the probable reproductive success of these females to age fourteen (see section on definitions and assumptions).

The final sample size for both sexes was reduced from initial values to compensate for loss of identity (see section below on tag loss).

Observations

At the beginning of each breeding season, we read the tags of known-age individuals and marked the animals with paint, dye, or a bleaching solution (Le Boeuf and Peterson 1969). Reproductive performance of individuals was recorded virtually every day throughout each breeding season from 1968 to 1985 at all sites on the Año Nuevo rookery (island and mainland) where females congregated. Observers monitored these sites from dawn to dusk on the island and from about 0900 h to 1500 h on the mainland. A similar procedure was followed by Point Reyes Bird Observatory researchers on Southeast Farallon Island, where 25% of the female subjects bred.

Each breeding season we recorded: (1) the daily location of marked males (e.g., near a group of females competing with other males or resting in nonreproductive areas; (2) sufficient daily aggressive interactions to allow calculation of a social hierarchy for the top ten to fifteen males at each site every ten days; (3) all copulations by marked males and females and unmarked males (Le Boeuf and Peterson 1969; Le Boeuf 1972, 1974); (4) the time, place of arrival, and date of parturition of each marked female; and (5) by marking pups at birth, offspring survival to weaning age for each female subject (Reiter, Panken, and Le Boeuf 1981).

Sources of Error

Two potential sources of error in estimating LRS are loss of identity and emigration. In the first case, we fail to account for individuals that are living because we cannot distinguish them from the dead; in the second case, we fail to account for individuals because they are not present where we are observing. We attempted to reduce these errors by employing the procedures and assumptions summarized below.

Tag Loss

We estimate that a maximum of 40% of the males lost their tags between weaning and appearance on the rookery at five years of age. We assume no tag loss after age five because additional plastic tags were attached to a hind flipper of each male when he was first observed on the rookery as an adult. We assume that no male lost both tags over the course of one year. The extra plastic tag provided insurance against tag loss and made the subjects easier to identify from a distance because they are larger than the Monel Metal tags and easier to read.

We assume an 11% tag loss rate between weaning and age two in females and 6% annually thereafter. We gave 33% of the identifiable survivors in the 1973 cohort and 23% of those in the 1974 cohort additional tags when they were first sighted on the rookery as adults. These estimates of tag loss are based on independent samples of males and females that were double-tagged at weaning and subsequently observed with only one tag as an adult (Le Boeuf, Reiter, and Sylvan, unpublished data). Estimates of survivorship, based on these tag-loss estimates, are lower than those obtained on southern elephant seals, *M. leonina*, using permanent branding marks (Carrick and Ingham 1962; see Results below).

We corrected for tag loss by reducing the original sample sizes by our estimates of tag loss. For males the original sample of 151 pups born is reduced by 40%, yielding $n = 91$. The adjusted sample size for females is $n = 204$. We assume that animals that lost their identity show the same distributions as the marked survivors. Our data consist of estimates of reproductive success (RS) for the identifiable survivors of both sexes each year that they were living and, by summation, throughout their lifetimes.

Emigration

All male subjects that survived to age five, except one, were observed on Año Nuevo and at no other rookery during the breeding season. Of the female subjects that survived to age three, 75% were observed breeding on Año Nuevo, and except for one female, the remainder were observed breeding on Southeast Farallon Island. We assume that marked males and females did not breed at any other rookeries during the study period because we saw none of them in searches of all other rookeries then in existence in southern California and in Mexico during virtually every breeding season from 1968 to 1978 (Le Boeuf 1974; unpublished observations) and because other investigators visiting island rookeries or working on them regularly (e.g., B. Stewart on San Miguel and San Nicolas Islands during the period 1979 to 1985) reported seeing none of our experimental subjects. Moreover, we know from tag/recapture studies (Bonnell et al. 1978) that only a few individuals from Año Nuevo (none in the present sample) migrated south during the study period and that males six years of age or older that appeared on a rookery during the breeding season did not skip years. Therefore we assume that marked adult males that were not observed on Año Nuevo or Southeast Farallon Island, after having first appeared there during a breeding season, were dead. Absentee females were treated differently (see section below).

Definitions and Assumptions

Our estimate of male RS is based on the percentage of copulations by a male with females in a certain harem multiplied by the number of females that gave birth and copulated in that harem. This ratio is an estimate of the number of females inseminated, I. A harem is defined as a group of females varying in number from two to about a thousand that give birth, nurse their pups, and copulate in a certain location; the highest-ranking male or males in the vicinity mate with these females and keep other males away from them.

We assume that all females were inseminated during copulations in the harem (where 95% of the copulations occur) or as they left the harem to return to sea (Cox and Le Boeuf 1977) and that at least 95% of the females that copulated were impregnated. We assume that the same males that dominated breeding during daylight hours did so at night (Le Boeuf 1972). We assume that copulatory success is correlated with fitness. However, we are aware that the ratio, I, may not be a perfect indicator of RS because most females copulate more than once over a period of two to three days and, in a large harem, a female may copulate with more than one male.

Since 70% of the females at Año Nuevo give birth where they copulated the previous year (Reiter, Panken, and Le Boeuf 1981), we devalued I for each male by the pup mortality rate in the area the following year to obtain a general estimate of the number of offspring each male sired that survived to weaning age, P. We assume that the pup-survival rate was equal across all males. Thus, I corresponds to success in obtaining mates and P approximates mating success by taking into account female fertility, or in this case pup survival.

We emphasize that estimates of male LRS were calculated from performance with all females present, most of whom were unmarked immigrants into the study area. In addition, the male subjects were born six to eleven years earlier than the female subjects and hence were observed at an earlier time in colony development.

Our estimate of female RS is B, the number of live births produced and W, the number of pups weaned in a healthy condition. When we did not know whether a female present on the rookery gave birth (after having done so in a previous year), we assumed we had overlooked her. We gave her a probability of .97 of giving birth based on the fecundity rate of marked females that we observe annually on the Año Nuevo rookery. This was done 15% of the times a female in the sample was present. When it was not clear whether a female successfully weaned her pup (41 out of 205 cases), we gave her a probability based on the performance of females of the same age in the same place and, when possible, the same time (see Reiter, Panken, and Le Boeuf 1981).

We assumed that females not observed on a rookery for two years in a row either were dead, were overlooked, or had lost their tags. For these two breeding seasons, we adjusted the life span, fecundity (B) and estimate of the pups weaned (W) for each female based on the probability of tag loss and the probability that a female might have been breeding unobserved. We assumed a 6% per annum tag-loss rate and a 33% per annum rate for failure to identify tagged females present. These probabil-

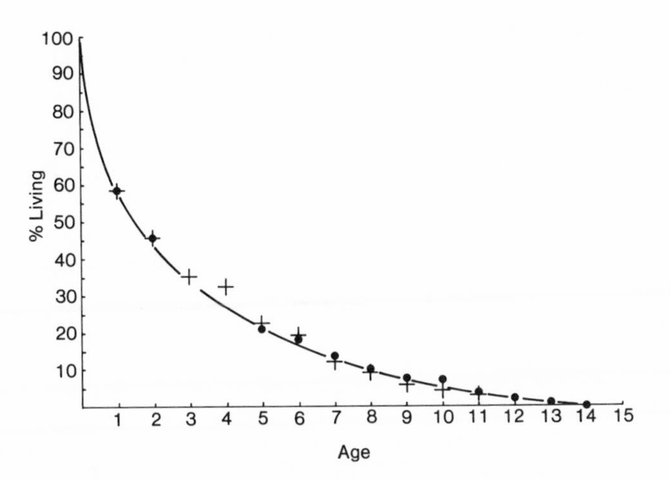

Figure 22.2: Estimated age-specific survival of males (*solid circles*) and females (*plus signs*) in this study. Estimates are based on a tag-loss rate for males of 40% from weaning to age five and zero tag loss thereafter; we assume a tag-loss rate for females of 11% from weaning to age three and a 6% annual rate thereafter. Data points for ages one and two are from Reiter, Stinson, and Le Boeuf (1978) and Reiter (1984).

ities were modified further to compute fecundity (B) by assuming a 97% probability of giving birth. The probability of a female's weaning a pup, W, was adjusted using the weaning rate for her age and for her harem in that particular year. The probability that we overlooked a female for a third straight year was so low as to be trivial. The eight females that were still alive in 1984 were given probabilities of giving birth and weaning pups in future years as described above, and life span was calculated from life history expectancies.

(At the close of the 1987 breeding season, four females were still alive and breeding, one from the 1973 cohort [fourteen years old] and three from the 1974 cohort [thirteen years old]. Each female gave birth and weaned her pup successfully in 1985, 1986, and 1987. Two additional females survived the 1984 breeding season. One gave birth in 1985, but we were not able to determine if she weaned her pup. The other was not observed in 1985 and 1986 but was seen during the summer of 1986. Further details on the methods used can be found in Reiter 1984.)

22.2 Results

Survival

Survival in both sexes is estimated to be about 58% to age one and 45% to age two (fig. 22.2). The survival rate of females is higher than males at age four, but thereafter the survival curves of the two sexes are similar. These estimates are similar to those obtained on southern elephant seals from the stable Macquarie Island population during the first year of life. However, the southern species has a 5% to 10% higher survival rate to age eight in males and females, respectively, and female survival is higher than male survival from age four to eight (Carrick and Ingham 1962).

Table 22.1 Estimates of Lifetime Reproductive Success in 19 Northern Elephant Seal Males Known to Have Survived to at Least Age Five out of 138 Tagged as Weaned Pups

Cohort and Male Number	Age in Years											I	P
	4	*5*	*6*	*7*	*8*	*9*	*10*	*11*	*12*	*13*	*14*		
1964													
53	0	n.s.	0	3	2	+						5	
				(3)	(1)								4
59	n.s.	n.s.	0	3	+							3	
				(3)									3
74	n.s.	n.s.	0	0	+							0	
													0
92	n.s.	n.s.	0	+								0	
													0
1965													
126	n.s.	n.s.	1	1	6	8	30	61	14	+		121	
			(1)	(1)	(5)	(7)	(25)	(47)	(7)				93
133	n.s.	n.s.	0	1	3	6	10	+				20	
				(1)	(2)	(5)	(8)						16
156	n.s.	0	0	0	1	10	11	+				22	
					(1)	(8)	(9)						18
162	n.s.	n.s.	0	0	1	30	66	+				97	
					(1)	(25)	(56)						82
177	n.s.	0	0	0	0	0	0	0	0	0	+	0	
													0
182	n.s.	0	+									0	
													0
1966													
229	n.s.	n.s.	n.s.	n.s.	n.s.*							0	
													0
236	n.s.	n.s.	1	1	0	3	2	10	+			17	
			(1)	(1)		(2)	(2)	(5)					11
247	n.s.	0	+									0	
													0
264	n.s.	n.s.	0	0	+							0	
													0
283	0	0	0	0	+							0	
													0
290	n.s.	0	0	+								0	
													0
1967													
319	n.s.	0	0	+								0	
													0
329	n.s.	0	0	0	0	9	8	20	26	+		63	
						(7)	(4)	(13)	(17)				41
424	n.s.	0	0	+								0	
I	0	0	2	9	13	66	127	91	40	0		348	0
P	0	0	2	9	10	54	104	65	24	0			268
Percentage I by age			0.6	2.6	3.7	19.0	36.5	26.2	11.5				
Percentage P by age			0.8	3.4	3.7	20.2	38.8	24.2	9.0				

Note: The numbers in the cells are estimates of the number of females inseminated (I) and, in parentheses, the number of offspring sired that survived to weaning age (P); n.s. denotes not seen and did not mate; + denotes death.

*Sighted in summer.

Most of the males in the sample of pups born (76.8%) did not survive to reproductive age (five years). The decline in survival rate is constant, going from 21% at age five to zero at about age fourteen.

Of the females, 65% did not survive to reproductive age (three years). The difference in the survival rates of females and males was greatest at age four, the time when most females were giving birth for the first time and the age of maximum pup productivity. Interestingly, the disparity in survival rates between the sexes in southern elephant seals at Macquarie Island is also greatest at the age of maximum breeding of females, age six (Carrick and Ingham 1962). The decline in the survival rate of northern elephant seal females between ages four and five appears to be linked to mortality factors associated with the onset of breeding (Reiter 1984).

Breeding Success: Males

Age at First Breeding

Nineteen males survived to age five (table 22.1). All but one of these males was observed on the rookery for the first time during the breeding season as a four-, five-, or six-year-old. Age at first mating was variable: two males copulated for the first time at age six, three at age seven, two at age eight, and one at age nine.

Effects of Age on Breeding

Beginning at age six, estimates of male RS increased with age up to eleven years and then declined to zero at age thirteen (fig. 22.3). Prime mating age was nine to twelve years; males in this age group are estimated to have inseminated 93% of the females present. Mean RS was highest in eleven-year-old males (I = 22.75 females) (fig. 22.3). A decline in RS after age eleven is indicated by the drop in performance of male 126 (table 22.1). Similarly, male 177 began showing a lack of reproductive effort at age ten by retiring to areas devoid of females. We estimate that no male inseminated more than one female before age seven or more than five females before age eight. Living to prime breeding age was not sufficient for achieving high RS; the male that lived the longest never mated. All males were dead by age fourteen.

Variation in Breeding Success

Only 8.8% of the sample mated during their lifetime; fewer than half of the survivors mated (table 22.1). We estimate that 8 males inseminated 348 females. A frequency distribution of LRS shows four groups of individuals among animals reaching breeding age: 3 males were extremely successful, inseminating 121, 97, and 63 females; 3 males inseminated 22, 20, and 17 females; 2 males inseminated fewer than 5 females apiece; and 11 males failed to breed. A distribution of LRS of the entire sample is plotted in figure 22.4.

Table 22.2 shows the means and measures of variance in male LRS for each cohort and for the entire sample. Overall, the mean number of females inseminated by males in the sample was 3.82 ± 17.61 and the mean number of pups produced to weaning age was 2.95 ± 13.74.

Both the means and the variances of male LRS varied greatly from year to year (table 22.2). Males in the 1965 cohort were far more successful than males in the other cohorts. We estimate that four males in this cohort (4.4% of the males) inseminated 75% of the females mated by all males in the entire sample.

Components of Breeding Success

The relative contributions of three multiplicative components to variance in male LRS were calculated using a FORTRAN program written

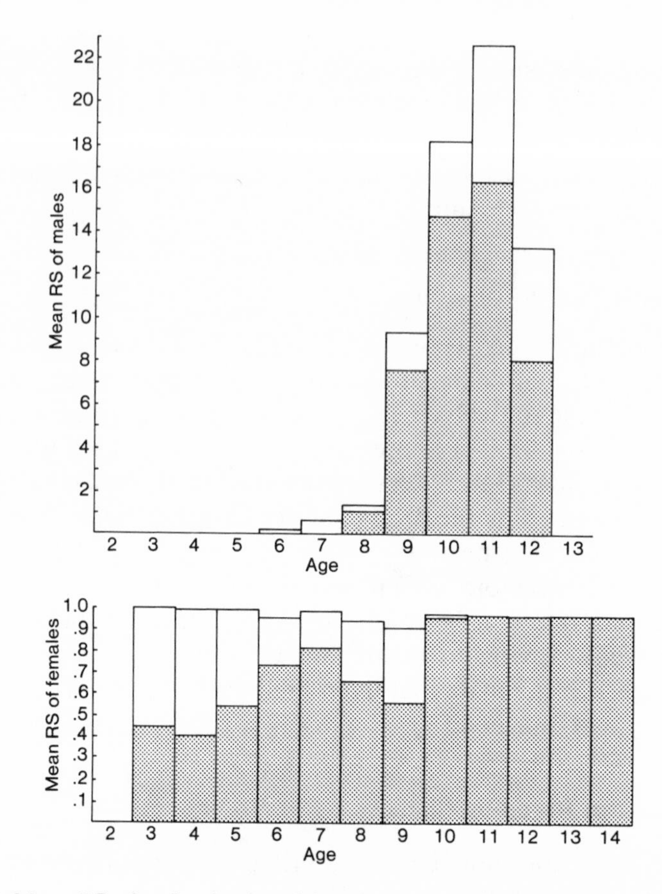

Figure 22.3: Mean RS of males (*top*) and females (*bottom*) as a function of age. Open bars refer to estimates of females inseminated (I) for males and pups born (B) for females. Shaded sections refer to estimates of pups sired and weaned for males (P) and pups weaned for females (W). A single female that gave birth and weaned her pup at age two is not shown. The values for eleven-year-old females are extrapolated for the 1974 cohort; the values for twelve-to fourteen-year-old females are extrapolated for the 1973 and 1974 cohorts.

Table 22.2 Lifetime Reproductive Success in Four Cohorts of Northern Elephant Seals: Means and Measures of Variation

Cohort	N	$\bar{x}$	σ^2	$\sigma^2/\bar{x}^2$
1964	17	1.35	14.24	7.78
		(1.18)	(10.65)	(7.70)
1965	21	12.38	1,085.75	7.08
		(9.95)	(693.65)	(7.00)
1966	26	.65	11.12	26.00
		(.42)	(4.65)	(26.00)
1967	27	2.33	147.00	27.00
		(1.52)	(62.26)	(27.00)
Total	91	3.82	309.95	21.19
		(2.95)	(188.79)	(21.77)

Note: The top figure is an estimate of the number of females inseminated (I); the figure in parentheses is an estimate of the number of offspring sired that survived to weaning age (P).

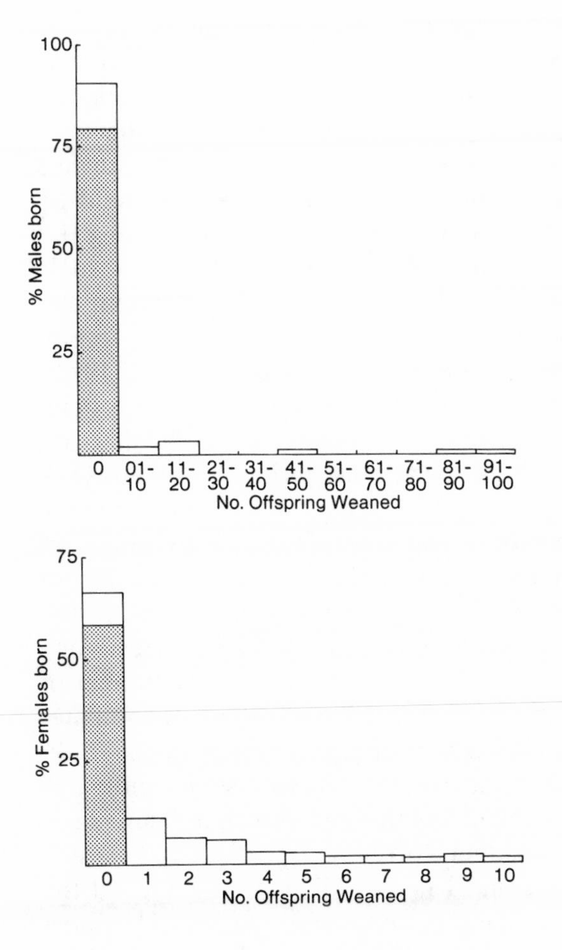

Figure 22.4: Estimated distributions of LRS in northern elephant seal males (*top*) and females (*bottom*) in this study. For individuals that weaned no offspring, the shaded section denotes the percentage of individuals of each sex that died before reaching breeding age; the remaining clear section shows the percentage of animals that reached breeding age but failed to produce a weaned pup. These distributions assume a 40% tag-loss rate for males from weaning to age five and a tag-loss rate for females of 11% from weaning to age three and 6% annually thereafter.

by David Brown. The variables for males were: L (life span), the number of years each male was capable of breeding (breeding seasons present on the rookery from age five to death); F (fecundity), the number of offspring sired (I) divided by L; and S (survival), the proportion of offspring that survived to weaning age (P).

The proportion of males that bred in the sample was .088. The overall variance was 188.79, 48.2% due to nonbreeders and 51.8% due to variation among the breeders. Variance in male LRS due to breeders can be broken down into 6% attributable to variation in L, 68% attributable to variation in F, and 2% due to variation in S. Less than 1% can be attributed to covariation in F and S and to covariation in L and S; about 27% is due to covariation in L and F. The contribution of the threefold covariation is .72.

Causes of Variation in Breeding Success

Among males that survived to breeding age, most of the variance in LRS is explained by differences in social rank achieved, the time in the breeding season when this rank was held, and harem size. Social rank was the most important cause of copulatory success or the lack of it, because rank determined access and proximity to females. The higher a male's rank, the more freedom he had to copulate without being disturbed by other males and the more frequently he prevented others from mating (Le Boeuf and Peterson 1969). Social rank refers to the position of a male relative to others competing in or near a harem. In aggressive encounters between two males, the one with the inferior rank retreats when threatened or chased by the other (Le Boeuf and Peterson 1969; Le Boeuf 1974).

Social rank achieved must be considered in relation to when it is held, how long it is held, and the size of the group of females for which males are competing. Males that achieved a high social rank during the seven-week period when the majority of females were in estrus (mid-January through the first week in March) dominated mating. As a rule, the five highest-ranking males held their positions throughout this period (see Le Boeuf and Peterson 1969); important shifts in rank usually occurred before the females came into estrus. In harems containing fifty females or less, one male, alpha, usually did all the mating (Le Boeuf 1974). As the number of females in a harem increased, more males were able to enter and copulate, but males continued to dominate mating in proportion to their social rank. Estimates of females inseminated by males with varying social ranks, as harem size increased to over a thousand females, are shown in figure 22.5.

In this study, the Pearson product correlation coefficient between highest rank achieved in a lifetime and I is .73. Males that achieved the highest social ranks in the largest harems dominated mating. The most successful male in the sample, male 126 (table 22.1), had social ranks of sixteen, seven, one, and five on the beach supporting the largest number of females when he was nine, ten, eleven, and twelve years old, respectively. In the year that he was the alpha male (1976), the harem in which he dominated other males contained 641 females at peak season, and he

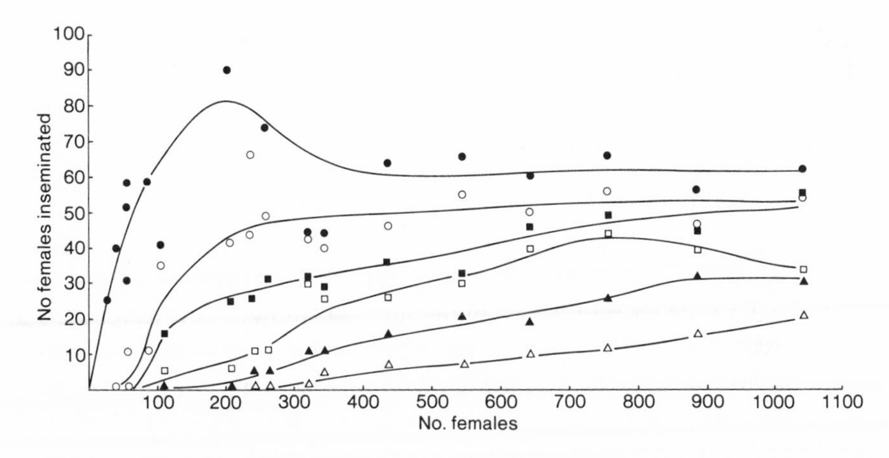

Figure 22.5: Estimated number of females inseminated (I) by males of varying social ranks (1, 2, 3, 5, 10, and 20) as the number of females in a harem increases. The data come from observations of two harems on Año Nuevo Island from 1968 to 1979.

did 10% of the mating observed, the highest proportion of any male present. Male 162 ranked third at age nine and second at age ten in the same large harem. As the beta male in 1975 when this harem contained 554 females, he did 12% of the mating observed, the highest proportion recorded that year. In contrast, male 177 lived longer than any other male in the sample but stopped competing when he reached prime age. Although he appeared on the rookery each breeding season when he was eleven, twelve, and thirteen and was in residence for one to two months each year, he was never seen on a breeding beach and never reached a sufficiently high social rank to allow him access to females. He never mated and was blind when last seen on the rookery.

Social rank is determined by bluffing and overt fighting between a pair of males; rank is maintained by threats, chases and biting (Le Boeuf and Peterson 1969; Le Boeuf 1974). Several variables are associated with winning fights and attaining high rank. These include time of arrival on the rookery, size, temporal patterning of challenges, conservation of energy, stamina, and fight strategy. We have observed over the years that males attaining high rank arrive early in the breeding season (during the interval 1–25 December), but not so early that their energy reserves are depleted by the time females begin coming into estrus in mid-January (males fast throughout the breeding season), and not so late that they have to fight many high-ranking males to achieve a high rank themselves (Le Boeuf 1974, 1981).

The highest-ranking males are among the largest males present. In 1984, the mean standard length (nose to tail in a straight line) of seven alpha males was 392 ± 12.2 cm, 15 cm longer than the mean of fifty-six adult males measured. In this sample, the only males longer than alpha males were older males past their prime and no longer competing actively for females. We have no measurements on lengths or weights of the males in the present sample. However, it was noted by all observers that the successful male 126 was unusually small for an alpha male. At weaning age, he was smaller than all the other males that survived in his cohort (table 22.1), measuring 114.3 cm in contrast to males 162 and 177, who both measured 162.6 cm. Clearly this male was an exception to the rule, an indication that size is but one of several variables important in determining social rank and RS.

Winning fights and achieving a high social rank is not easily predicted by focusing on a few variables that are easy to measure. We have seen males depose an alpha bull by challenging him immediately after he had fought and was exhausted. Some adult males sleep, avoid fights, and allow subadult males to approach nonestrous females early in the breeding season, apparently conserving their energy for the time when females come into estrus. The winner of a fight is not necessarily the male that does the most damage; we have seen males win fights without delivering a single butt or bite.

The main cause of failure to mate in nonbreeders was death before reaching social maturity. Two causes of death have been identified. White sharks, *Carcharodon carcharias*, kill and eat elephant seals of both sexes ranging in age from weaned pups to adults (Ainley et al. 1981; Le Boeuf, Riedman, and Keyes 1982). Every year on the Año Nuevo rookery, adult and subadult males are responsible for the deaths of a few weaned pups, yearlings, adult females, and adult and subadult males. Fatal injuries are inflicted during attempted mating or during fights. For example, in the 1984 breeding season, two alpha males and a challenger died from wounds inflicted by opponents. In addition, we have seen males retire from competition because of broken flippers, lost canine teeth, a broken os penis, or blindness. Finally, there are males that survive, compete, and fail, and most puzzling of all, males that survive but simply do not compete. We have seen some large, apparently healthy adult males spend the entire breeding season on a rookery but never compete with other males near harems.

Breeding Success: Females

Age at First Breeding

Most females gave birth for the first time at age four, the range being two to five (table 22.3). Most females, 64.7%, gave birth for the first time on Año Nuevo Island, where they were born, 11.8% gave birth on the Año Nuevo mainland, and 20.5% gave birth on Southeast Farallon Island (H. Huber, pers. comm.). Some females (17.6%) changed the location where they gave birth at least once in their lifetimes. For all females but one, the change was from the island to the mainland (less than $\frac{1}{2}$ km) or from Año Nuevo to Southeast Farallon or vice versa (a distance of 89 km). The exceptional female gave birth from age four to age eight on San Nicolas Island, 528 km south of Año Nuevo (B. Stewart, pers. comm.), after giving birth at age two and age three on Año Nuevo Island.

Effects of Age on Breeding

Table 22.4 shows that measures of reproductive success in females varied with their age. Young females four to five years old were most prolific at giving birth (42% of pups born). Females four to seven years of age weaned the most pups (62%). After a female gave birth for the first time, fecundity varied little with age; 97.6% of the females that appeared on the rookery gave birth. However, weaning pups in a healthy condition varied greatly with the age of the mother. Young females three to five years old weaned only 46% of the pups they produced. In comparison, females six to eight years old weaned 77% of their pups (table 22.4, fig. 22.3). The effect of age on breeding in females is treated in detail, using a larger sample, in Reiter, Panken, and Le Boeuf (1981) and Reiter (1984).

Variation in Breeding Success

Sixty-seven females gave birth to 243.2 pups, of which 152.4 were weaned in a healthy condition (table 22.4). All those that survived to reproductive age gave birth at least once in their lifetimes; however, 8.3% of the females that reached breeding age failed to wean a pup. Estimated

Table 22.3 Age at First Reproduction in Female Northern Elephant Seals

	Age				
Cohort	*2*	*3*	*4*	*5*	*6*
1973		6	11	1	
1974	1	18	25	3	2
Both cohorts	1	24	36	4	2
Percentage	1.5	35.8	53.7	6.0	3.0

Table 22.4 Reproductive Success as a Function of Age in Female Northern Elephant Seals from Two Cohorts, 1973 (n = 75) and 1974 (n = 129)

	Age (years)													
	2	*3*	*4*	*5*	*6*	*7*	*8*	*9*	*10*	*11*	*12*	*13*	*14*	Totals
Females present (n)	1	25	57.8	45.6	35.1	25.9	19.3	13.2	9.6	6.5	4.6	3.3	2.4	249.3
Pups born (B)	1	25	57.7	45.2	33.4	25.8	18.0	12.0	9.5	6.4	4.4	3.2	2.3	243.2
Percentage of females giving birth (B/N)	100	100	99.8	99.1	95.3	99.6	93.6	90.8	98.8	98.0	97.2	97.3	97.5	97.6
Pups weaned (W)	1	11.0	23.0	24.8	25.7	21.0	12.8	7.3	9.4	6.4	4.4	3.2	2.3	152.4
Percentage of females weaning pups (W/n)	100	44.0	39.8	54.4	73.3	81.1	66.4	55.6	98.5	98.0	97.2	97.3	97.5	61.1
W/B	100	44.0	40.0	54.9	77.0	81.4	70.9	61.3	99.7	100	100	100	100	62.7

Note: Statistics are extrapolated for ages eleven (1974 cohort only), twelve, thirteen, and fourteen.

Table 22.5 Means and Measures of Variation in Female Lifetime Reproductive Success in Two Cohorts

Cohort	n	$\bar{x}$	σ^2	$\sigma^2/\bar{x}^2$
		.92	4.31	5.12
1973	75	(.59)	(2.45)	(7.01)
		1.37	5.73	3.07
1974	129	(.85)	(3.28)	(4.58)
		1.20	5.23	3.62
Total	204	(.75)	(2.95)	(5.23)

Note: The top figure is pups born (B); the bottom figure is pups weaned (W).

LRS of survivors, measured as pups weaned (fig. 22.4), varied from zero to ten, with most survivors weaning one pup.

The means and measures of variance in female LRS for each cohort and for the combined sample are shown in table 22.5. The entire sample of females produced a mean of 1.20 ± 2.29 pups and weaned a mean of .75 ± 1.72 pups. The variance due to pup survival is 3.62/5.23 = 69.2%.

There was great year-to-year variability in female RS, an observation made in other studies (Reiter, Panken, and Le Boeuf 1981; Reiter 1984). Females in the 1974 cohort produced a higher proportion of pups than females in the 1973 cohort (1.17 vs. 0.92 per female), and they weaned a higher proportion of pups (0.85 vs. 0.59 per female).

Components of Breeding Success

The same FORTRAN program used to determine the relative contribution of multiplicative components to variance in male LRS was used for females. The variables were: L (life span), the number of years each female bred, F (fecundity), number of offspring/L, and S (survival), number of pups weaned/number of pups born. The proportion of breeders in the sample was 67/204 = .328. The overall variance was 2.98, 39% being due to the inclusion of nonbreeders and 61% to the variation among the breeders. The overall variance in female LRS due to breeders can be broken down into 36% attributable to variation in L, 0.1% attributable to variation in F, and 23% attributable to variation in S. None of the overall variance in breeders is due to covariation in F and S or covariation in L and F; 44% is due to covariation in L and S. The contribution of the threefold covariation is −1.1%.

Causes of Variation in Breeding Success

The principal causes of variation in breeding success in surviving females are factors important in weaning a pup successfully, a subject treated in detail for a larger sample of females, including the present sample, in Reiter, Panken, and Le Boeuf (1981) and Reiter (1984). These factors are summarized briefly. Breeding success, as measured by weaning a pup successfully, is higher in prime-age females (six years or older) than in young females (two to five years old) because the former are larger, are dominant in aggressive encounters, and have more maternal experience—

factors important in protecting and nurturing pups. Older females, being larger, wean larger pups. The pups of older females are bitten less frequently by neighboring females, and a higher percentage of them survive (Chistensen and Le Boeuf 1978). Older females breed earliest in the breeding season in the best location for rearing a pup, the center of the harem. Pups born early in the season have the option of stealing milk from other females once they are weaned (an important way for males to obtain a size advantage over competitors), they get more experience learning to swim and dive before departing from the rookery, and in the case of females, pups born early in the breeding season breed for the first time earlier in life than those born in mid- to late season. Females with maternal experience are more successful in weaning pups than similar-aged females without maternal experience. Since a female gives birth annually throughout her life, experience increases with age. The discrepancy in breeding success as a function of age is most extreme under conditions of high female and pup density.

Despite lower probability of a young female's weaning her pup successfully relative to a prime-age female, females in our sample with the highest estimates of LRS began breeding early in life. For example, females that weaned four or more pups in a lifetime began breeding at a mean age of 3.27 ± .65 years. Viewed from the other direction, females that bred for the first time at age two or three weaned a mean of 2.70 ± 2.56 pups; females primiparous at age four weaned 2.18 ± 2.35 pups; females primiparous at age five or six weaned 1.24 ± 1.21 pups. These mean differences in pups weaned as a function of the mother's age group are not statistically different ($F = 1.01$, d.f. = 2, 64, $p > .05$).

As is the case with males, the majority of females left no offspring because they did not survive to breeding age (fig. 22.2). Females are subjected to the same agents of mortality at sea as males. On the rookery, a few breeding females are inadvertently killed annually by males trying to mate with them as they leave the rookery at the end of lactation.

Variance in LRS of Males and Females Compared

As figure 22.1 shows, male breeding success is much more variable than female breeding success. We can calculate $I_m I_f$, the ratio of reproductive intensities in the two sexes, a measure of the relative intensity of selection on the two sexes (Wade and Arnold 1980). I_m from table 22.2 is 21.77; I_f from table 22.5 is 5.23, yielding an $I_m I_f$ value of 4.16 for the entire sample. $I_m I_f$ ratios, calculated from single cohort years for each sex, range from 1.10 to 5.90.

22.3 Discussion

The disparity in mean LRS between the sexes warrants comment. Why is mean LRS of males substantially greater than mean LRS of females? Why are the two values not equal? Why is mean LRS of females so low? This could be due to a variety of reasons that are not necessarily mutually exclusive:

1. We expect the mean estimates of RS to be equal if we had determined the RS of all males and females in the same population. This was not the case in our study. Male RS was calculated with all females present on the Año Nuevo rookery from 1971 to 1979; only a fraction of these were in the female sample. This was a period when many individual males and females immigrated to the Año Nuevo rookery from southern California rookeries; the operational sex ratio (the number of males present and in competition relative to the number of females that were present and in estrus) rose from 1:1.25 in 1971 to 1:4.15 in 1979.

2. The sample of males were breeding earlier in the population expansion than the sample of females, so density-dependent effects may have exaggerated the sex differences. The beginning of breeding by females in the sample coincided with the onset of crowded conditions on Año Nuevo Island, where most females in the sample gave birth. These adverse conditions for breeding are reflected by more than a twofold increase in the pup mortality rate after 1978 (Reiter, Panken, and Le Boeuf 1981). The pup mortality rate on Año Nuevo Island from 1978 to 1984 has ranged from 31% to 70%.

3. Estimates of male LRS may be biased toward successful breeders, or our estimates of female LRS may be biased toward unsuccessful breeders. This could arise if our assumptions of tag loss are far removed from reality.

4. The low mean RS of females, 1.20 pups born and 0.75 pups weaned, indicates a declining population or, possibly, a very atypical pair of cohorts. In fact, pup production of the Año Nuevo colony has been increasing annually since breeding was initiated in 1962 (Radford, Orr, and Hubbs 1965; Le Boeuf and Bonnell 1980). The explanation for this paradox is that the growth of the Año Nuevo population is attributable to a large annual influx of migrants from southern California rookeries, especially San Miguel Island, which produces ten times more pups annually than Año Nuevo (Bonnell et al. 1978; Reiter 1984). A high proportion of the breeding females at Año Nuevo consists of these southern immigrants. For example, among the 1976 cohort of females, only 23% of the three-year-olds and 42% of the four-year-olds were born on Año Nuevo; the remainder were from San Miguel and San Nicolas islands in southern California. Were it not for these immigrants, Año Nuevo–born females would not be replacing themselves, and as our data indicate, the colony would be declining. Reproductive data collected on additional female cohorts in the 1970s indicate that the reproductive data from females in this sample are not atypical (Reiter 1984).

Because some of the problems treated above are likely to influence estimates of the mean and variance in RS of both sexes, some of our results are necessarily speculative and must be interpreted with caution. We have observed individual seals through a window in time and space. A few cohorts of each sex have been examined that were born in one colony and lived during a time when colony numbers were increasing. Had the study been conducted on a large rookery in southern California, we would expect different estimates of LRS, especially for females. The latter from Año Nuevo are too low to be representative of the entire population. The growth of the population has increased exponentially from 1890 to 1980 (Le Boeuf 1981), and recently the growth has been estimated to be 14.5% per year (Cooper and Stewart 1983). The mean LRS of females at large colonies like San Miguel Island must be greater than two to be consistent with the well-documented population growth.

Although absolute estimates, like mean LRS, can be expected to vary from one colony to the next, the relative differences in our estimates of LRS between the sexes are large enough to give us confidence that they are real. Variance in LRS is much greater among males than among females. A few males enjoy spectacular mating success, and we assume this translates into reproductve success and fitness. We estimate that one male in the present sample inseminated over 100 females in his lifetime; a male in an earlier study was apparently more than twice this successful (Le Boeuf 1974). In contrast, the reproductive potential of females is low; few females bear the maximum of ten to fourteen pups in a lifetime. This result is consonant with sexual selection theory, which predicts that the sex with more variance in RS will evolve features important in reproductive competition. Northern elephant seal males differ from females in size, aggressivity, and appearance of secondary sex characteristics like the nose and neck, traits that play a role in male/male competition, the outcome of which determines access to mates. This difference in reproductive potential helps explain the greater risk taking of males early in life (Reiter, Stinson, and Le Boeuf 1978). The sex difference in variance in LRS among elephant seals is greater than that observed in some polygynous species for which data are available, for example, bullfrogs (Howard 1983) and lek-breeding birds (Payne 1984). However, comparisons across species must be interpreted with caution because of differences in definition, criteria, and methods.

Northern elephant seals share many similarities in sex differences in reproduction with polygynous, island-breeding pinnipeds such as southern elephant seals (Laws 1956; Carrick, Csordas, and Ingham 1962), gray seals, *Halichoerus grypus* (Anderson, Burton, and Summers 1975; Boness and James 1979), eared seals or sea lions, *Eumetopias jubatus*, *Zalophus californianus*, *Otaria byronia*, *Neophoca cinerea* and *Phocarctos hookeri* (Gentry 1970; Peterson and Bartholomew 1967; Vaz-Ferreira 1975; Marlow 1975), and fur seals, *Callorhinus ursinus* (Bartholomew and Hoel 1953), and with species in the genus *Arctocephalus* (e.g., Bonner 1968; Pierson 1978; McCann 1980). Most obvious among these is the great disparity in size between the sexes, the lack of parental investment in offspring by males, and the observation that a few males successful in male/male competition are involved in a disproportionate number of matings. We suspect further that these species are like northern elephant seals in that the variance in LRS of males is greater than that of females.

22.4 Summary

Lifetime reproductive success (LRS) was estimated in samples of 138 male and 204 female northern elephant seals, *Mirounga angustirostris*, marked at birth on Año Nuevo Island, California. The estimate for males was based on number of copulations with different females; the estimate for females was number of pups weaned in a healthy condition. Males began competing to mate at age five and achieved peak success at age nine to twelve. Females gave birth for the first time at age two to six, four being the mean age. Fecundity was high (97%) in females of all ages, but females six years of age and older were more successful in weaning their pups than females five years old or less. Males showed great variability in LRS. Only 8.8% of the sample mated, but some males were very successful, inseminating as many as 121 females; 33% of the female sample mated. Most breeding females weaned one to three pups in their lifetime; a few females weaned as many as ten. The most important component of LRS in males is the number of offspring sired. Access to mates is determined by living to maturity and achieving high social rank among males competing in a large harem. Large size is one of several factors important in fighting and in achieving high social rank. Important components of LRS in females are life span and rearing pups successfully. In surviving females, variables associated with advancing maturity—size, dominance, and maternal experience—help in nourishing, protecting, and weaning pups. Breeding success in male elephant seals is much more variable than in females, a sex difference found in other polygynous mammals like red deer and predicted by sexual selection theory.

22.5 Acknowledgments

We thank our colleagues, graduate and undergraduate students, research assistants, teaching assistants, and friends who helped to collect the data, and Clairol, Inc., for animal-marking solutions. This study was supported by grants GB-16321, DEB 77-17063, and BNS 74-01363 A02 from the National Science Foundation.

23 Reproductive Success of Lions

Craig Packer, Lawrence Herbst, Anne E. Pusey, J. David Bygott, Jeannette P. Hanby, Sara J. Cairns, and Monique Borgerhoff Mulder

African lions face fundamentally different ecological problems than most other mammals do, and they also depend more on cooperative behavior than virtually any other vertebrate. Lions are territorial but prey primarily upon large herbivores that are often migratory (Schaller 1972). Thus in some habitats the lions' food supply is highly erratic and ephemeral, whereas in others it is more stable. Lions resemble other social carnivores and many higher primates in being long-lived and highly social. However, they differ in that they do not typically form dominance hierarchies (Schaller 1972; Bertram 1978; Bygott, Bertram, and Hanby 1979; Packer and Pusey 1982, 1985). The lack of social dominance is perhaps an essential feature of lion sociality, because each individual may therefore be equally affected by any factor that raises or lowers the reproductive success of the entire group (ibid.; also see Caraco and Wolf 1975; Vehrencamp 1983). In this chapter we examine the major ecological and social factors influencing reproductive success in lions, with special reference to the effects of group size. We contrast populations living in three widely differing habitats and confirm that both sexes derive inherent advantages from living in groups.

23.1 Methods

Long-term records are maintained on all the lions in Ngorongoro Crater and in a 2,000 km^2 area of the Serengeti National Park, Tanzania (fig. 23.1). The Serengeti study area includes two distinct habitats: the eastern plains and the woodlands around Seronera. All the lions in these areas have been studied continuously since 1974 or 1975 (Hanby and Bygott 1979; Packer and Pusey 1982), and two of the woodlands prides have been studied since 1966 (Schaller 1972; Bertram 1975). These areas typically contain a total of about three hundred individuals residing in fifteen to twenty social groups ("prides").

A *pride* is a fission/fusion social unit of one to eighteen adult females and their dependent offspring. Membership in the pride is stable, but pride females are usually in smaller subgroups scattered throughout the pride range (Schaller 1972; Packer 1986). All females of the same pride are genetic relatives in our study areas: females either join their mothers' prides or form new ones with members of their natal cohort (Pusey and Packer 1987). All females in the same pride breed at a similar rate (Schaller 1972; Bertram 1975; Packer and Pusey 1983b). A *coalition* of one to seven adult males maintains residence in the pride for about two years before being replaced by another coalition (Bygott, Bertram, and Hanby 1979; Packer and Pusey 1982). Both females and males are considered adult at four years. Four is the median age when females have their first surviving litter (see below) and the median age when males first become resident in a pride (Pusey and Packer 1987).

All lions resident in the study areas are individually recognizable from

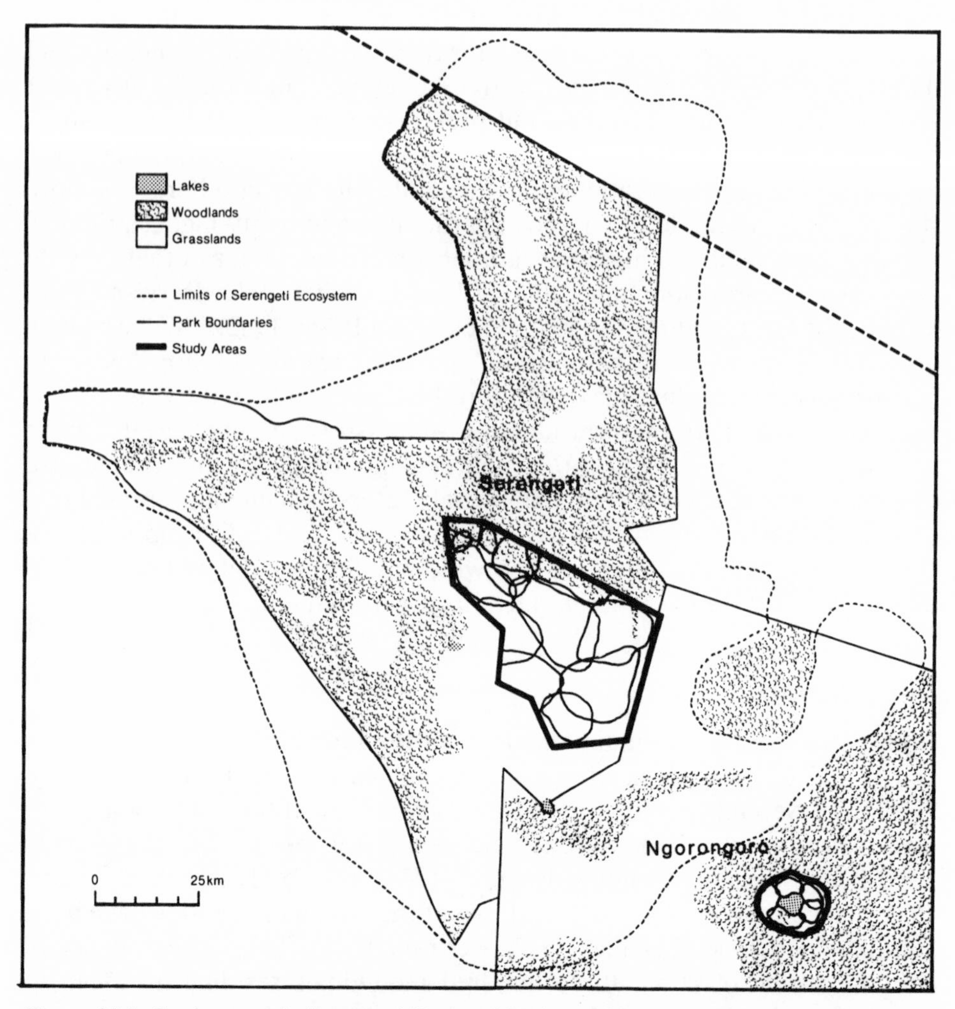

Figure 23.1: Study areas in Serengeti National Park and Ngorongoro Conservation Area, Tanzania. Approximate pride ranges are drawn in the study areas. The woodlands prides are those whose ranges are primarily in the woodlands habitat at the northern edge of the study area.

natural markings (see Pennycuick and Rudnai 1970), and we also identify all nomadic lions that enter the areas. We attempt to locate every resident at least once every two months and to record their reproductive condition (see Packer and Pusey 1983b). Birth dates of cubs can be estimated, but initial litter size can never be known with certainty, since females hide their cubs until they are four to six weeks old. Estimated litter size is one to six, and 98% of litters are one to four (Packer and Pusey 1987). Litters that are lost before the cubs are seen are assumed to have consisted of only one cub, because females often abandon single cubs (Rudnai 1973; Packer and Pusey 1984). Cubs that disappear before the age of eighteen months are assumed to have died, unless their mothers disappeared at the same time and were believed to have emigrated. Older individuals are assumed to have died if they disappear singly; if two or more individuals disappear simultaneously, it usually indicates emigration (see Pusey and Packer 1987).

Demographic data in this chapter come from the authors' observations and from data given in Bertram (1975). In spite of the continuity of these studies, there have been numerous gaps of over six months in observations of each pride, and thus certain data have not been collected with precision. Females may have given birth unnoticed if the entire litter died at an early age, and males may have briefly become resident unnoticed. We have not interpolated birthrates of females during these gaps, but no gap was long enough for us to miss cubs that reached one year of age (our primary measure of reproductive success).

If females of the same pride give birth synchronously, they typically pool their cubs and rear them communally. If "communal litters" are formed during gaps in observations, then we cannot assess maternity. In these cases we award each mother an equal proportion of the communal litter, since females often do have equal numbers of cubs in such litters and certainly do have more cubs than females that did not contribute to it. About 70% of the 1,237 cubs born during the study could be attributed to specific mothers, 15% to a pair of females, and the remainder to three to eight females. Thus our estimates of variance in reproductive success between individual females may be slightly low, but only a small proportion of each female's total cubs were in such litters, and thus the bulk of her reproductive success was accurately recorded.

Our data on the lifetime reproductive success of females are based on the performance of fifty-nine females that died after reaching four years of age. This sample was biased toward younger females, and we therefore calculated the expected age distribution at death of a cohort of fifty-nine from the survival data given in figure 23.4. We thus discarded at random fourteen of the females that had died between four and twelve years of age and replaced them with fourteen living females of thirteen to seventeen years. We extrapolated the reproductive success of six of these living females for one to two years and added cubs both according to the age-specific data given in figure 23.3 and according to each individual's deviation from the mean reproductive success for her age.

Data on mating activity by individual males are far too sparse to use as an index of reproductive success within breeding coalitions, and our estimates of male reproductive success make two assumptions. First, we assume that all cubs conceived during a coalition's tenure were fathered by members of that coalition. This cannot be verified until the paternity exclusion analysis currently in progress is completed. However, we have no behavioral data to suggest that nonresident males ever father cubs, except in the rare cases where fathers are still resident when their daughters reach maturity. Females typically avoid mating with their fathers and will temporarily leave their prides to mate with unfamiliar males (see Pusey and Packer 1987). The current analyses remain essentially unchanged if we assume that either all or none of the cubs were conceived by these females' fathers. Second, we assume that coalition partners have equal reproductive success during their lifetimes. This assumption will also be tested by biochemical analysis. However, observations of mating activity and the fact that females often show synchrony in estrus suggest that coalition partners often have comparable reproductive success (Bertram 1975, 1976; Bygott, Bertram, and Hanby 1979; Packer and Pusey 1982, 1983b). A coalition may remain in residence after the death of one or more of its members, and individual reproductive success is therefore calculated by awarding each male an equal proportion of the cubs conceived in his lifetime.

Results are presented separately for each habitat whenever there is no significant concordance between habitats. If trends for each habitat differ in elevation but not in slope for a particular analysis, then pooled data are graphed and the habitat differences noted. In some analyses, statistics have had to be based on the number of cubs that were born or died in particular circumstances. Cubs of the same litter cannot be treated as statistically independent because they are born and often die at the same time. We have therefore divided the number of cubs by two (the approximate average litter size) and calculated chi-square values from the number of "litters." Similarly, except for the analysis of individual reproductive success in table 23.1 and figure 23.2, statistical tests on males are based on average values for each coalition, since coalitions act as a unit.

Annual rates are calculated according to weather year (the beginning of one rainy season until the end of the following dry season—for example, November 1976–October 1977). Because of the effect of rainfall on the movements of the migratory herds in the Serengeti (see below), we have relied on rainfall as an indicator of prey availability. Although nearly thirty rain gauges are ready every month in the Serengeti study area, no good rainfall data are available for the crater floor. However, because the rainfall on the eastern Serengeti plains generally comes from the east, we have used those data to estimate rain in the crater.

23.2 Study Sites

The Ngorongoro Crater is a large volcanic caldera that contains one of the highest lion population densities in Africa (van Orsdol 1981). The crater floor is mostly open grassland but includes a number of swamps and marshes. These wetlands flood during the rainy season (November–May), and thus green grass is continuously available around their edges as the water recedes during the dry season. Although most of the larger ungulates show daily and seasonal movements around the crater floor, only a small proportion of wildebeest and zebra leave the crater in the wet season, and most of the buffalo and eland leave in the dry season (Kruuk 1972; Estes and Estes 1979). Thus there is only minor seasonal variation in the overall herbivore biomass in the crater (van Orsdol 1981), and the resident biomass of the lions' preferred prey species is the highest in Africa.

The crater is surrounded by highland forest around two-thirds of its circumference and by arid scrub/grasslands occupied by Masai tribesmen around the remaining third. Resident lion prides occasionally move up the crater walls, and there is evidence that a considerable proportion of subadults permanently emigrate from the crater (Pusey and Packer 1987). However, the surrounding area appears to support a far lower lion density than the crater floor, and no immigration into the crater has been recorded in the past ten years. Thus the crater is essentially a self-contained ecological unit, and the lions are a distinct and isolated population. The crater lion population crashed in 1962 (Fosbrooke 1963), reached its current level by 1975, and has since been stable (Pusey and Packer 1987).

In contrast, the Serengeti ecosystem is defined by the vast migration of its dominant herbivores (figure 23.1). The wildebeest, zebra, and gazelle all prefer to graze on the eastern plains, where the soil is mineral-rich volcanic ash from the crater highlands to the east (Kreulen 1975). Rainfall is sufficient to stimulate grass growth on the plains only during a few months in each rainy season (November–May; McNaughton 1979), and rainfall levels are lower on the plains than elsewhere in the Serengeti ecosystem (Sinclair 1979). The lion population in the plains thus has access to plentiful prey only during these rainy months. During the dry season, only a few warthog, topi, oryx, and Grant's gazelle remain on the plains (Hanby and Bygott 1979). However, if there is sufficient rainfall during the normally dry months of June–October, wildebeest and zebra will return briefly to the northern and western edges of the plains, and the plains lions temporarily move to the edge of the woodlands when prey is scarce on the plains. Several prides on the Serengeti plains are known to have originated from adjacent areas of woodlands (Hanby and Bygott 1979). The size and number of prides on the plains increased between 1966 and 1976 (Hanby and Bygott 1979), and the number of adult females living on the plains has increased continuously from 1973 to 1984 ($r_s = .846$, $n = 12$ years, $p < .01$).

The woodlands area around Seronera is both a dry-season refuge for many herbivores and also a "holding area" for the migratory species as they go to and from the plains. The area is on the migratory route during the changes between seasons and is visited by some of the migratory herds during dry spells in the rainy season and wet spells in the dry season (see Maddock 1979). When the wildebeest and zebra are absent in the dry season, these lions prey primary upon buffalo and gazelle (Packer 1986). Rainfall is heavier here than on the plains at all times of year, but in the wettest part of the rainy season prey may become scarce, and these lions will make temporary forays more than 25 km onto the plains in extremely wet years (Schaller 1972; pers. obs.). The lion population in this area also increased between 1966 and 1974 (Hanby and Bygott 1979), but it has since stabilized.

23.3 Results and Discussion

In the following sections we first compare the lifetime reproductive success (LRS) of females and males and specify the causes of variance in the reproductive success of each sex. We then describe in detail the causes of mortality of cubs and of adults, since these are important determinants of reproductive success. Finally, we show how the reproductive success of both sexes depends on group size.

Lifetime Reproductive Success

"Reproductive success" is defined as the number of offspring reaching twelve months of age. A large proportion of subadults disperse from the study areas between the ages of one and one-half and four years (Pusey and Packer 1987), and thus data on the number of offspring reaching reproductive age are incomplete.

Females

Figure 23.2 shows the frequency distribution of lifetime reproductive success of breeding females. Age-specific measures of female reproduction are given in fig. 23.3. Female lions typically have their first litter at three to four years of age, and their reproductive performance starts to decline at eleven years (fig. 23.3). Data on birthrates and litter size are only approximate (see Methods, above) but data on the number of cubs surviving to twelve months are accurate. These show that female reproductive success also declines at eleven years and virtually stops at fifteen years. Survival of cubs does not change significantly with maternal age. The average interbirth interval for females whose previous cubs survived is twenty-four months (Pusey and Packer 1987), and the peaks in reproductive success at three, five, seven, and ten years presumably reflect this interval.

Tables 23.1–23.3 show the relative importance of life span, fecundity, and cub survival on the lifetime reproductive success of the breeding females that reached eight years of age. We restrict this analysis to females that reached eight years because of the statistical artifact that arises from excluding nonbreeders from the sample. Variance in fecundity is highest shortly after maturity, since late first breeders have zero fecundity at their death, whereas early first breeders have a high apparent fecundity at death because of the brief duration of their reproductive life span. Consequently there was a significantly negative correlation between life span and fecundity among all *breeders* that reached four years of age ($r_s = -.3346$, $n = 54$, $p < .05$), but not over all females that reached four years ($r_s = -.0538$, $n = 59$, n.s.).

Tables 23.1–23.3 suggest that although all three variables make a substantial contribution to the lifetime reproductive success of breeding

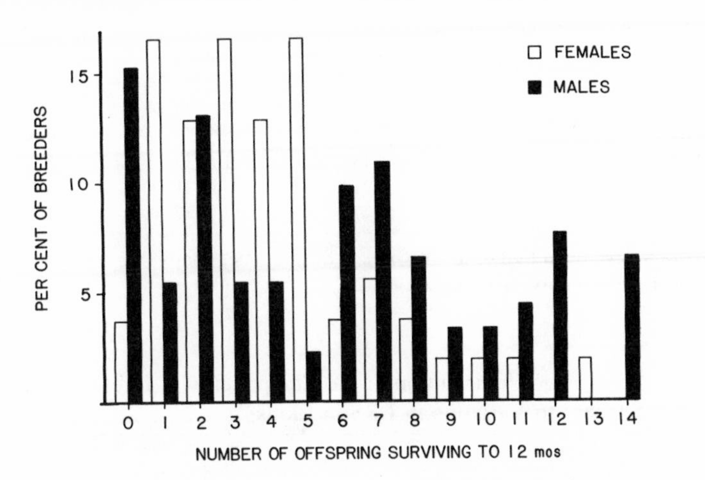

Figure 23.2: Lifetime reproductive success of males and females. Frequencies are given in terms of the percentage of breeders of each sex (males: $n = 91$; females: $n = 54$; see text). Any possible effects from differences across habitats have been ignored.

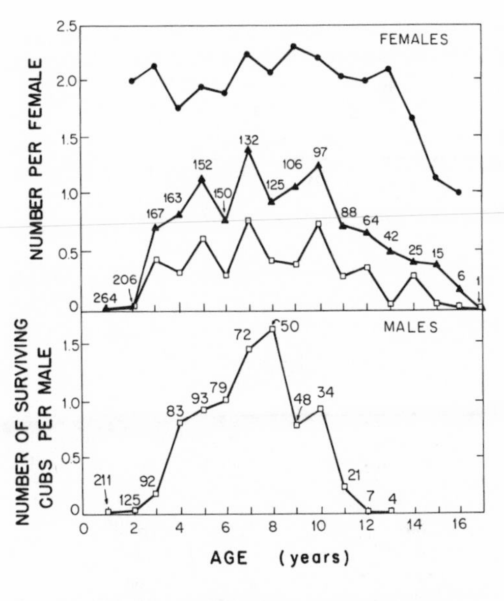

Figure 23.3: (*a*) Age-specific reproduction of females. Numbers above birthrates give sample size for each age. There were no habitat differences in any of these measures. (*b*) Age-specific reproductive success of males. Data include nonbreeding males as well as those that were members of resident coalitions.

Table 23.1 Mean and Variance of the Components of Lifetime Reproductive Success in Lions

Component	Original *Mean*	Original *Variance*	Standardized Variance
Females (*n* = 54)			
L	7.7374	10.3595	0.1730
F	1.2968	0.5299	0.3151
S	0.4374	0.0621	0.3246
LF	9.0035	17.8392	0.1772
FS	0.5474	0.2419	0.7520
LS	3.4400	5.4318	0.4742
LFS	3.8100	7.8029	0.4051
Males (*n* = 31 coalitions)			
L	4.4731	3.8675	0.1933
F	2.2757	2.7177	0.5247
S	0.3843	0.0725	0.4908
LF	9.3420	48.3257	0.4665
FS	1.1337	1.4939	1.9531
LS	1.7137	2.0331	0.6880
LFS	4.4796	17.1047	1.1175

Source: Program developed by David Brown (see this volume, chap. 27).
L = life span; F = fecundity; S = cub survival.

Table 23.2 Percentage Contribution of the Components of Lifetime Reproductive Success to Variation in LRS in Lions

Component	L	F	S
Females[a]			
L	21.71		
F	−25.17	39.45	
S	−15.05	42.60	84.57
LFS	−48.10		
Males[b]			
L	17.30		
F	−22.51	46.95	
S	0.33	83.88	43.93
LFS	−69.89		

Source: Program developed by David Brown (this volume, chap. 27).
L = life span; F = fecundity; S = cub survival.
[a]Forty-seven breeding females that survived to eight years.
[b]Breeding male coalitions.

Table 23.3 Partition of Lifetime Reproductive Success in Lions: Inclusion of Nonbreeders

	Females (number of breeders = 54)	Individual Males (number of breeders = 91)
Proportion of breeders	0.3293	0.2218
Overall variance (OV)	5.8607	8.7288
Percentage of OV due to nonbreeders	55.74	52.32
Percentage of OV due to breeders	44.26	47.68

Source: Program developed by David Brown (this volume, chap. 27).

females, cub survival is the most important component. However, life span is probably more important than the tables indicate. We have had to include only females over eight years of age, whereas a number of adult females died before that age. When all females that reached four years of age are included, life span accounts for 43% of the variance. We will discuss causes of mortality in both cubs and adults in subsequent sections of this chapter.

The importance of fecundity is difficult to assess. Fecundity should be *inversely* related to cub survival in lions, because females that lose their cubs resume breeding much more quickly than those whose cubs survive (see Packer and Pusey 1984). Thus a female that lost every litter shortly after birth might have a litter once every four to six months (and have a high fecundity), whereas a female that successfully reared each litter would have a litter only once every eighteen to twenty-four months (and have a relatively low fecundity). The positive correlation between survival and fecundity in tables 23.1–23.3 probably results from our inability to detect many litters that die at an early age (see Methods), and thus our data greatly underestimate the fecundity of females with low cub survival. Therefore the apparent relationship between fecundity and reproductive success may not be valid.

Males

Figure 23.2 shows the estimated lifetime reproductive success of ninety-one males that became resident in the study prides. No attempt has been made to correct the age structure of this sample as was done for females (see Methods), because the average reproductive life span of males is only thirty-three months (range 5–130 months, n = 31 coalitions). However, these data include a number of males that are still breeding and an unknown number of males that may have been resident outside the study areas before or after they resided in a study pride (see note to table 23.5). Furthermore, we have assumed that there is no variance in reproductive success within coalitions except that due to male mortality (see Methods). Thus these estimates are too low and should be considered rough approximations. We include them in order to compare variance in male reproductive success with that for females and because more complete data would probably not substantially alter the shape of the overall distribution.

As would be expected in a polygynous species, variance is higher in male reproductive success than in that of females (tables 23.1–23.3), and the true magnitude of this difference is probably greater than these data indicate.

Reproductive success is confined to a narrower range of ages in males than in females (fig. 23.3*b*) Male coalitions must compete successfully against other coalitions in order to gain and retain residence in prides (Schaller 1972; Bertram 1975; Bygott, Bertram, and Hanby 1979). Larger coalitions oust smaller ones from prides and chase nomadic coalitions

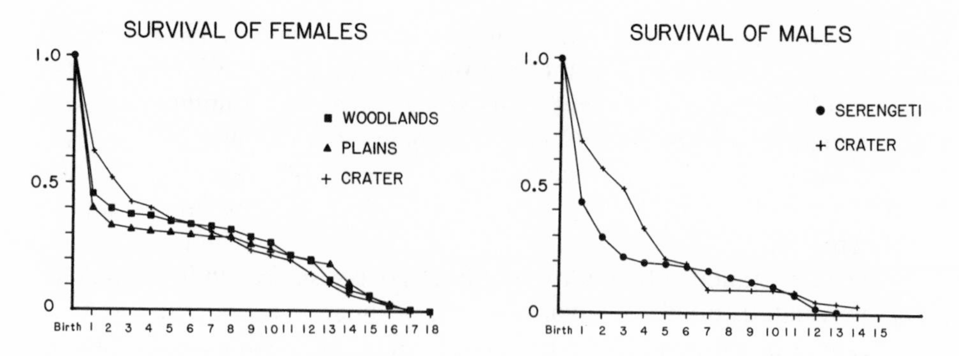

Figure 23.4: Survival curves for each sex in each habitat. Mortality rates from birth to one year of age were significantly higher in the Serengeti plains (1974–84) than in the woodlands (1966–84) ($\chi^2 = 6.887$, $n = 426$ "litters," $p < .01$), and mortality in the woodlands was higher than in Ngorongoro Crater (1975–84) ($\chi^2 = 9.320$, $n = 412$, $p < .01$). Data include all individuals alive when records were begun in each area and all males that immigrated into the study prides from elsewhere. Initial age estimates of these individuals are made by comparing photographs with those of known-age individuals. Individuals that "disappeared" rather than died (see Methods) were included as survivors for each age before their disappearance. Because males are more likely to disperse from the study areas than females (Pusey and Packer 1987), and because substantial rates of disappearance result in underestimates of mortality, we have corrected for this sex bias by estimating true mortality of males from the sex ratio at each age.

from their prides' ranges. Although such encounters typically involve only chasing, gang attacks also occur, and severe injuries may be inflicted. Thus males must be old enough and healthy enough to compete successfully for residence in a pride.

Successful males become resident in their first pride when they are about four years old (Pusey and Packer 1987), and their first cubs are born about six months later (Packer and Pusey 1983a). Male coalitions remain in individual prides for an average of only twenty-six months, but some large coalitions gain residence in a succession of prides (Pusey and Packer 1987). However, even males in these coalitions look conspicuously old by ten years of age, being covered with scars and wounds, and very few survive to twelve years (see fig. 23.4). After an aging coalition loses residence, the males' small cubs are usually killed by the replacement coalition (see below).

Tables 23.1–23.3 show the relative contributions of life span, fecundity, and cub survival to male lifetime reproductive success. Overall, these components have effects similar to their effects in females, except that cub survival appears somewhat less important to male reproductive success. Fecundity and cub survival are highly correlated, probably because the cubs of males with high "fecundity" were born synchronously to many females shortly after those males took over their pride. Such synchronous litters are subsequently reared communally and have higher survival than cubs born asynchronously (Bertram 1975). It is doubtful that coalitions actually do vary in fecundity; such variation probably results from conditions affecting the females in the prides where they gain residence.

Lifetime reproductive success of males is also significantly correlated with coalition size (r_s = .4110, n = 31 coalitions, $p < .05$), time spent in residence in prides over their lifetime (overall tenure length) (r_s = .7963, $p < .001$), and the total number of females in the prides held during their lifetime (r_s = .5819, $p < .001$). Of these, only coalition size relates directly to differences in competitive ability; the other two are consequences of competitive ability. (Note that these analyses include only coalitions that maintained residence long enough to father cubs; many coalitions never do so; see table 23.5.)

Tenure length is one of the most important components of lifetime reproductive success of breeding males, but tenure is highly correlated with coalition size (r_s = .7372, $p < .001$). Males that maintain residence long enough are able to protect their young from infanticide by subsequent coalitions (tenure vs. cub survival: r_s = .5797, $p < .001$). Males that remain in the same pride long enough to sire successive litters also gain from the fact that females conceive three months sooner when mating with males that have sired their previous cubs than when mating with males that have just entered their pride (Packer and Pusey 1983a). Tenure and fecundity are significantly correlated (r_s = .5505, $p < .01$).

The number of females gained by each coalition is also correlated with coalition size (r_s = .5745, $p < .001$). This is because larger coalitions gain more prides during their careers (coalition size vs. number of prides held: r_s = .7782, $p < .001$). They do not preferentially become resident in large prides (coalition size vs. average size of prides held: r_s = .0076, n.s.).

Coalition size in itself is not as highly correlated with reproductive success as are many other variables, because the advantage of belonging to a coalition of a particular size depends on the average coalition size in the population. This varies from year to year, and when no large coalitions are present, smaller coalitions can maintain residence for long periods. Coalitions of four or more males are always composed of members of the same natal cohort (see below). Natal cohort size depends on cub survival within a pride, and cub survival is synchronized over large areas of the park by weather patterns (see next section). Thus the median coalition size in the Serengeti increased from two in the late 1960s to three in the mid-1970s, following a general increase in the lion population, and has since declined again to two (also see Packer and Pusey 1983c).

Cub Mortality

Mortality is highest during the first year of life (Schaller 1972; Bertram 1975), and this early mortality varies across habitats: nearly two-thirds of cubs born on the plains die before one year of age, whereas only one-third of crater cubs die by that age (fig. 23.4). Since cubs' deaths are rarely witnessed, it is usually difficult to determine the exact cause. Nevertheless, there are two contexts in which the cause of death can be reliably inferred: male takeovers and season of low prey availability.

Incoming males have been observed to kill unweaned cubs seven times, and cub mortality rates are much higher in the first months following the arrival of a new set of males than at other times (Bertram 1975; Packer and Pusey 1983a, 1984). Cubs almost never survive more than two months into the tenure of a new male coalition in their mothers' pride. By killing the cubs, males speed up the females' return to sexual receptivity by an average of eight months (Packer and Pusey 1984). All cub mortality within two months of a male takeover is therefore considered here to be due to infanticide. The proportion of deaths attributed to infanticide is similar across all three habitats, and 27% of all mortality before twelve months of age occurs in this context. Infanticide is mostly restricted to cubs of six months or less, though they continue to be at risk until about twenty months (fig. 23.5; also see Pusey and Packer 1987). Cubs older than eighteen months are generally evicted from their natal pride by the new males (Hanby and Bygott 1987; Pusey and Packer 1987).

Schaller (1972) suggested that 50% of cub mortality in the Serengeti was the result of starvation, but though many malnourished cubs have been observed, the death of a starving cub has been seen only once (Packer and Pusey 1984). However, the prevalence of starvation can be inferred from a seasonal pattern of cub mortality that is correlated with seasonality of prey availability. In the Serengeti plains, cub mortality other than that due to male infanticide is highly seasonal: 68% of the remaining cub deaths occur during the dry season, which constitutes only 43% of the year ($\chi^2 = 18.70$, 1 d.f., $n = 64.5$ "litter" deaths [see Methods], $p < .001$). During the dry season, the estimated prey availability on the plains is about one-twenty-fifth that of the rainy season (van Orsdol 1981), and cubs in plains prides are obviously malnourished during dry-season months. In

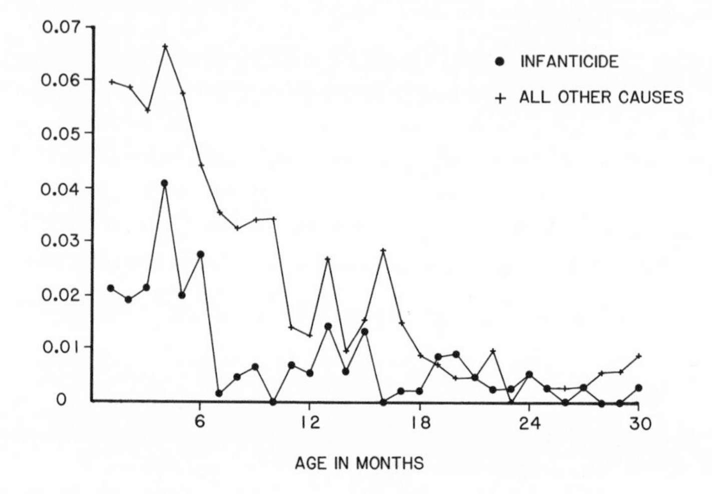

Figure 23.5: Age-specific mortality owing to infanticide. The overall proportion of deaths that were due to infanticide was similar across all three habitats (Ngorongoro Crater: 22.8%; Serengeti plains: 33.5%; woodlands: 24.9%; $\chi^2 = 2.75$, 2 d.f., n.s.), and there was significant concordance in the age at which cubs were most at risk from infanticide (Friedman test: $\chi^2 = 45.26$, 30 d.f., $p < .05$), as well as from other causes ($\chi^2 = 62.91$, $p < .001$). Sexes are lumped because male and female cubs showed a similar age-related pattern of mortality from infanticide ($r_s = .392$, $n = 30$, $p < .05$).

Table 23.4 Correlations between Environmental Conditions in Different Years and Annual Lion Cub Recruitment Rates within Each Habitat

Correlation of Cub Recruitment with	Serengeti Woodlands	Serengeti Plains	Ngorongoro Crater	χ^2	$\hat{\rho}$	z
Percentage of females exposed to male takeovers	−.4386	−.7341**	−.6273*	0.98	−.6135	3.300***
Dry-season rainfall	.2363	.3614	.3333	0.09	.3105	1.670#
Wet-season rainfall	−.7636**	−.6932*	−.0061	4.59*		
Number of females in habitat	−.3182	−.3386	−.4061	0.05	−.3529	1.898#
Dry-season rain in woodlands	—	.8735***	—			

Note: Data from the Serengeti are for cubs born in 1973–83 (cubs born in 1973 reached one year of age in 1974, the first year of data for most of the study prides). Data from the crater are from 1974–83. When there is no significant heterogeneity in correlations across habitats (given by the χ^2 with 2 d.f.), an overall correlation coefficient ($\hat{\rho}$) and its associated z value have been calculated (Kraemer 1975).

# $p < .05$, one-tailed test (see text); * $p < .05$, two-tailed test; ** $p < .02$; *** $p < .001$.

contrast, cub mortality is only slightly higher during the dry season in the Serengeti woodlands (49%, $\chi^2 = 2.49$, $n = 110$, n.s.), and there is no indication of seasonality of mortality in Ngorongoro Crater (39%, $\chi^2 = 0.07$, $n = 35.5$, n.s.). In the Serengeti woodlands there are obvious cases of cub starvation in some dry seasons, but this is probably balanced by starvation during the wet season in exceptionally wet years (see below). As outlined above, seasonality of prey availability in this area varies strikingly from year to year. In the crater, we have never seen a conspicuously malnourished cub, and the crater shows virtually no seasonality in prey availability (van Orsdol 1981). Mortality of individuals older than one year does not show significant variation across seasons in any of the three habitats.

Van Orsdol (1981) found that cub survival to twelve months was highly correlated with poor-season biomass across six habitats that included our three sites and that by themselves are in accordance with this trend. This suggests that food availability is a major determinant of cub survival in the most seasonal areas but that other factors must be acting to limit populations in nonseasonal areas.

Schaller (1972) saw cubs being killed by other species of carnivores, and there is also evidence that cubs die owing to accidental maternal neglect and intentional abandonment (Packer and Pusey 1984). However, we have no way of assessing the relative importance of these causes of mortality, since we cannot define a context when such mortality is likely to occur. More extensive observations of death are required.

We have examined the major variables that are likely to affect the annual cub recruitment rates in each of our study areas (table 23.4). These variables are directly related to the two major causes of cub mortality outlined above: infanticide and starvation during the season of lowest prey availability. Cub recruitment rate is the number of cubs reaching one year of age per adult female, and high recruitment rates indicate low cub mortality. Across all three habitats, there is a strong negative relationship between cub recruitment and male takeover rate (table 23.4), as expected because of infanticide by the incoming males. Cub recruitment is predicted to be positively correlated with dry-season rainfall within each of the three habitats, because such rainfall attracts more herbivores to the lions' ranges (see Study Sites, above). However, the overall trend is only weakly significant (table 23.4).

The relation of cub recruitment on the Serengeti plains to dry-season rainfall in the woodlands is far higher than its relation to rainfall on the plains (table 23.4). The plains prides often move to the edge of the woodlands during the dry season, and thus this area appears to be their critical refuge during harsh conditions. Hanby and Bygott (1979) had attributed the population increase on the plains in the early 1970s to higher dry-season rainfall there. However, dry-season rainfall in the two habitats is positively correlated (r_s = .5941, n = 16 years (1969–84), $p < .05$), and it is the rainfall in the adjacent woodlands that appears to be critical. The soil on the eastern plains is so friable and the dry-season rainfall there is so slight that the resultant growth of green grass is usually insufficient to attract the migratory herbivores for any appreciable period outside the true rainy season.

Cub recruitment is inversely related to wet-season rainfall in both Serengeti habitats (table 23.4), but probably for different reasons in each. The migratory herds return to the woodlands less frequently in the rainy season during extremely wet years, and even resident prey move out to the plains. We have seen emaciated cubs in the woodlands during periods of heavy rain, and thus the woodlands cubs may be more likely to starve in these years. On the plains, greater numbers of nomadic males are attracted to the area during extremely wet years because the migratory herds are present more persistently, and thus takeover rates in the plains prides are highest in wet years (r_s = .800, n = 11, $p < .01$). Such nomads pass through the woodlands study area on their way to and from the plains, and thus takeover rates there also tend to be higher in wet years (r_s = .452, n = 11, n.s.). Note that wet-season rainfall is not correlated with dry-season rainfall in either area (woodlands: r_s = −.053; plains: r_s = .021, n = 16 years).

It is also possible that higher rainfall may increase mortality through flooding or disease; but if this were the case, recruitment rates in the crater should also be lowest in the wettest years. Following extremely heavy rainfall in the wet season of 1962, there was a "plague" of bloodsucking *Stomoxys* flies in Ngorongoro Crater that reduced the lion population from seventy-five to six to fifteen (Fosbrooke 1963). However, the population had returned to its former level by 1975 (Pusey and Packer 1987), and subsequently there has been no detectable effect of variation in wet-season rainfall on cub recruitment (table 23.4).

Finally, Bertram (1973) suggested that pride sizes might be density dependent. There is some indication that cub recruitment rates are highest in years when the number of females in a habitat is lowest, although this effect is small (table 23.4). This correlation is not due to increased frequencies of infanticide: the takeover rate in each habitat does not increase significantly with increasing density of females in that habitat. There may be some effect of higher density on cub starvation, but our data are not adequate to detect it.

Adult Mortality

Females

Although cub mortality is higher in the Serengeti than in Ngorongoro Crater, mortality of females is lower at each age from three to ten years (fig. 23.4). In spotted hyenas, a similar contrast between the two areas was deduced from the differing age structures of the two populations (Kruuk 1972). Kruuk suggested that the Serengeti hyenas were limited by prey availability, which resulted in higher cub mortality. In contrast, the crater hyenas were subject to greater territorial aggression and hence to greater adult mortality because of higher population density. Lion density in the crater is four times the density in the Serengeti plains and twice that in the woodlands. Although our findings in figure 23.4 are in broad agreement with Kruuk's, we have no direct evidence that crater females are more frequently killed by intraspecific aggression. One female in the Serengeti was known to be fatally wounded by other females, whereas the only known cause of mortality in adult females in the crater is injury inflicted by Cape buffalo ($n = 4$). However, a much larger proportion of subadult females disperse from their natal pride in the crater than in the Serengeti (Hanby and Bygott 1987, Pusey and Packer 1987), and we therefore suspect that females are subject to higher levels of aggression in the crater.

Males

After one to three years of age, mortality of males is higher than that of females, and after five years it appears more similar across habitats (fig. 23.4). Males from the two Serengeti habitats have been combined, since males often move back and forth between the two areas. Although a higher proportion of males survive to one year of age in the crater, mortality from ages two to five is far higher, apparently owing to attacks by adult males. Young males often become nomadic after their second year, but whereas young males in the Serengeti can often move to areas unoccupied by residents and can follow the superabundant migratory prey species, young crater males are in more constant contact with resident prides (see Hanby and Bygott 1987) and are often seen with severe wounds. The similarity of male life expectancy in the two areas after five years may be because levels of adult male/male competition are similar regardless of female density.

Group-Size-Specific Reproductive Success

Lions are the only social cat, and it is therefore of considerable importance to determine whether individuals derive a reproductive advantage from living in groups. We analyze separately the effects of grouping on the reproductive success of each sex.

Males

As described above, male lions form stable coalitions that compete intensely against other coalitions for access to female prides. Bygott, Bertram, and Hanby (1979) first showed that per capita male reproductive success increases with coalition size, and their analysis indicated that larger coalitions do better because they are more likely to gain access to a pride, remain in residence longer, and gain access to more females. These findings are all confirmed by our analyses.

Figure 23.6*a* shows the precise payoff to individual males from belonging to coalitions of different sizes. This provides more complete information on the lifetime reproductive success of the coalitions included

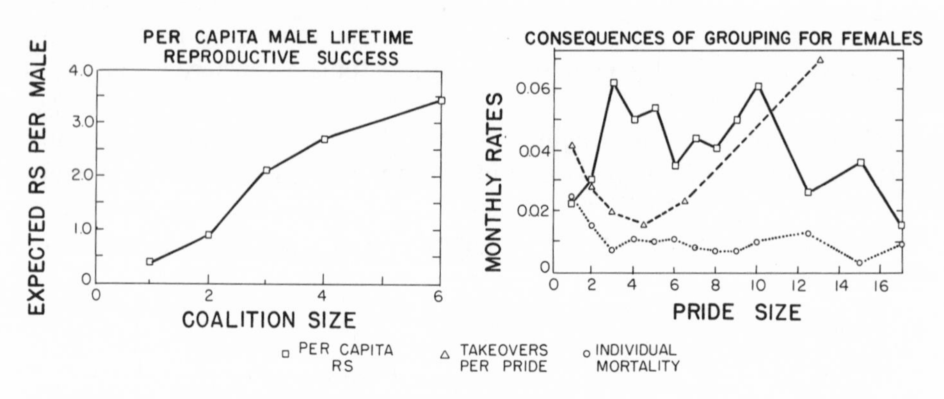

Figure 23.6: (*a*) Per capita reproductive success versus male coalition size. Calculations are based on data given in table 23.5. There were insufficient data to test for concordance between the Serengeti and the crater, but they showed a similar increase with increasing coalition size. (*b*) Pride-size-specific reproductive rates, mortality, and frequency of male takeovers. Data from adjacent cells have been combined to provide sufficient sample size in each cell to allow use of chi-square tests. Monthly reproductive rates per female showed significant overall heterogeneity ($\chi^2 = 21.039$, $n = 272.5$ surviving "litters," 12 d.f., $p < .01$). Prides of one or two females had significantly lower reproductive rates than did prides of three to ten ($\chi^2 = 6.585$, $n = 259.5$, 1 d.f., $p < .01$); prides of eleven to eighteen also did worse than those of three to ten ($\chi^2 = 5.250$, $n = 250.5$, $p < .02$). There was no significant heterogeneity among pride sizes of three to ten ($\chi^2 = 8.166$, $n = 237.5$, 7 d.f., n.s.). Mortality of females in prides of one or two females was significantly higher than in larger prides ($\chi^2 = 10.885$, $n = 123$ deaths, 1 d.f., $p < .01$). Heterogeneity across pride sizes in frequency of male takeovers approaches significance ($\chi^2 = 10.48$, $n = 77$ takeovers, 5 d.f., $p < .10$, two-tailed test). Solitary females and those in large prides suffered more frequent takeovers than those in moderate sized prides (solitaries vs. moderate: $\chi^2 = 5.20$, 1 d.f., $p < .05$; large vs. moderate: $\chi^2 = 18.00$, $p < .001$). Although there was a significant difference in per capita reproduction rate between the crater and the plains ($p < .05$), and though the crater females also suffered higher mortality than their Serengeti counterparts (fig. 23.4), there was reasonably high concordance between habitats on the effects of pride size for all three variables.

Table 23.5 Reproductive Success of Male Coalitions in Lions

Coalition Size	Proportion That Stay Together	Proportion That Gain Residence	Average Observed RS per Resident Male (*n*)	Expected RS per Male
1	—	1/16	6.500 (2)	0.406
2	all	6/14	2.136 (11)	0.915
3	all	6/12	4.249 (12)	2.125
4	all	3/4	3.625 (2)	2.719
5–7	3/8	3/3	9.206 (4)	3.452

Note: Data on the proportion of coalitions remaining together and on those that became resident in prides are restricted to males originating within the study area. Coalitions that fragmented before gaining residence are included in subsequent columns according to their sizes after the split. Males that acquired unrelated companions during thier nomadic phase are included in estimates of the proportion that gain residence according to the size of their coalition both before and after the acquisition. Pairs are significantly more likely to gain residence than are singletons ($p < .05$, Fisher test). Data on per capita reproductive success of resident males include all males regardless of origins but include only the coalitions for which there are data on at least half of their reproductive output. Several other coalitions dispersed into or out of the study areas or only recently gained residence.

in Bygott, Bertram, and Hanby's paper (1979) and includes additional coalitions that became resident after 1977. Per capita reproductive success is estimated by taking the proportion of coalitions of each size that gain exclusive access to a group of females, multiplied by the number of cubs surviving to one year of age that were conceived during the coalition's tenure, divided by the initial number of males in the coalition. Figure 23.6*a* thus gives an estimate of the *expected* reproductive success of each male when the coalition is first formed; some coalition members may die before the remainder have lost tenure (see above). This estimate assumes that coalition partners have equal reproductive success during their lifetimes, and although this may be true in most coalitions (see Methods), males apparently do have disparate success in some (see Packer and Pusey 1982). We discuss the implications of unequal reproductive success below.

Values for each of these group-size-specific factors are listed in table 23.5. It can be seen that the probability of gaining residence increases continuously with increasing coalition size, but the difference is most striking between solitaries and pairs. Per capita reproductive success of resident coalitions also increases with coalition size, but this shows considerable variance (see preceding sections).

Note that we have calculated the proportion of coalitions gaining access to prides in a slightly different way than Bygott, Bertram, and Hanby (1979), who simply divided the number of resident coalitions of each size by the total number of coalitions of that size in the population. We now know that very large coalitions often break up shortly after leaving their natal pride (Pusey and Packer 1987) and thus would not be seen as a large nonresident coalition.

The tendency for solitary males to form coalitions of two to three is expected from the large mutual advantage of cooperation (Packer and Pusey 1982), and coalitions of unrelated males have reproductive success similar to that of comparably sized coalitions composed entirely of relatives. However, it is puzzling that unrelated males join only up to a maximum of three: nine of fourteen pairs and trios of known origins include nonrelatives, but all six coalitions of four or more males are composed entirely of relatives ($p < .05$, Fisher test). Although males might continue to gain by acquiring further unrelated partners, the gain beyond three may be too small to compensate for the time lost in seeking further companions. Males acquire unrelated companions only while they are not yet resident, and Pusey and Packer (1987) found that males with unrelated partners gained initial residence at a later age than did cohorts of relatives.

Another possible explanation is that although larger coalitions enjoy higher average reproductive success, the temporal patterning of the availability of estrous females may be such that the males in a coalition of four or more are never all able to mate simultaneously, whereas the males in a pair or trio can often do so. This would lead to inherently higher levels of variance in mating success in the large coalitions. Thus males should join them only either when they could predict that they would have a higher reproductive success from joining or when their companions were their close kin. In the latter case a subordinate might still increase his inclusive fitness by enhancing the reproductive success of his dominant brothers even if he had a lower individual mating success than he would in a smaller coalition.

Females

It is not possible to calculate group-size-specific reproductive success for females in the same way as for males. Whereas a male coalition starts at its largest size and then declines through mortality, female pride size may fluctuate widely over time. For example, since 1966 the Masai pride has split into four separate groups, and the number of adult females in the main pride gradually increased from six to eighteen, then slowly declined to seven. Therefore, rather than estimating the per capita lifetime reproductive success for each pride size (as was done in fig. 23.6*a*), we found the number of surviving cubs that were born in each pride size each month. The number of cubs born each month in each pride size divided by the number of adult females in that pride gives the monthly per capita reproductive success for each size.

Figure 23.6*b* shows that there is considerable variation in per capita reproductive success of females in different-sized prides. Females in prides of three to ten had significantly more surviving cubs per month than those in either smaller or larger prides.

What is the cause of this relationship? Although most authors assume that the basic advantage of sociality to female lions is cooperative hunting, reanalysis of all the available data do not support this hypothesis (Packer 1986). Caraco and Wolf (1975) developed a model for estimating daily food intake by different-sized hunting groups from Schaller's (1972) data, and their analysis suggested that groups of two or three gain the highest rates of food intake per day. However, their model erroneously assumed that all groups of two or more would have the same probability of capturing multiple prey in a single hunt and that hunting rate is independent of food intake. Schaller (1972) had found that groups of four or more made the highest proportion of multiple kills, and subsequently both Elliott, Cowan, and Holling (1977) and van Orsdol (1981) found that lions resume hunting more quickly after a small meal and thus that individuals with a lower success rate or smaller average meal size might compensate by hunting more often. When these assumptions were corrected, the model suggested that *solitary* females gain the highest food intake *per hunt* (Packer 1986). The model cannot be used to estimate *daily* food intake without data on group-size-specific hunting rates. An empirical assessment of daily food intake showed that females constrained to remain in groups of three or four adult females were thinner than those in groups of one or two, but that there were no differences across all pride sizes (Packer 1986). Females in large prides spend considerable time either alone or in small subgroups and are thus apparently able to forage as efficiently as members of smaller prides.

Further studies of hunting behavior are currently being conducted to test alternative hypotheses about the advantages of cooperative hunting. However, any advantages that may exist will have to be far greater than has so far been suggested to account for the higher per capita reproductive success of females in prides of three to ten. Both of the analyses above predict that rates of food intake will decline rapidly as hunting group size increases above one to three.

The results in figure 23.5 and table 23.4 show that infanticide is a major cause of cub mortality, and data on the frequency of male takeovers across all pride sizes suggest an important advantage to females from living in moderate-sized groups. Figure 23.6*b* includes the monthly probability of a male takeover in each pride size, and the relationship mirrors that of group-size-specific reproductive success. Not only do male takeovers result in the loss of a female's unweaned cubs, but females also take twice as long to conceive when mating with a new set of males (Packer and Pusey 1983a,b).

The higher takeover rate in smaller prides probably results from two factors: first, smaller groups of females are less able to defend their cubs against alien males (Packer and Pusey 1983a). Second, males generally annex small prides of females while they are still resident in a larger adjacent pride and continue to spend more time with the larger pride. The smaller prides are therefore more vulnerable to subsequent invasion. The takeover rate in larger prides may be higher because very large numbers of females attract many more males to their range. Larger groups of gelada baboons are similarly subject to higher takeover rates, and Dunbar (1984) showed that males preferentially select larger groups for their takeover attempts.

Female mortality is also higher for solitaries and pairs (fig. 23.6*b*), and this may be due in part to their higher rate of male takeovers. Several females have died in defense of their cubs (Packer and Pusey 1983a), and females in small prides may be especially vulnerable during male attacks. Solitaries and pairs may also be subject to harassment from females in larger prides, but we lack direct evidence of this. Note that because of the high survival rates of females in each pride size, the relationship between female fitness and pride size is virtually identical to that between reproductive rate and pride size.

Although females apparently gain the same advantage from living in groups as males do, there is a striking contrast in their tendency to join unrelated companions. About half of all male coalitions include nonrelatives, but there is no case of unrelated females forming prides in these populations (Packer 1986). There appear to be greater constraints on females than on males in finding suitable partners. More than 33% of females leave their natal pride, and in comparison with females that remained there, dispersing females suffer lower reproduction rates in the Serengeti and higher mortality in Ngorongoro Crater (Pusey and Packer 1987). It is not known whether these costs of dispersal are directly due to the disadvantages of leaving a familiar area, but it is striking that dispersing females almost always settle in a portion of their natal range or an adjacent area (Pusey and Packer 1987). Familiarity with an area is likely to be especially important for successful reproduction in female lions through knowledge of suitable denning sites and good hunting areas; and such familiarity is most readily achieved by natal philopatry (Waser and Jones 1983). Thus solitary females may be unable to profit from a move to a new area to join another solitary or a pair. It is noteworthy that the only cases ever reported of unrelated females forming new prides followed a severe drought in the Kalahari Desert in which the natal areas of several females had become uninhabitable and they joined forces in a new area over 40 miles from their natal ranges (Owens and Owens 1984). Thus these females accepted unrelated companions when they were forced to range well away from their natal areas.

Why are lions the only social felid? The only advantage of group living that we have so far been able to detect is that moderate-size prides suffer lower rates of infanticide. Infanticide is an important source of cub mortality in lions, but it also appears to be common in tigers and mountain lions, and both of these species are solitary (Packer and Pusey 1984). Although female lions may not gain a strong positive advantage from cooperative hunting, the available empirical data suggest that the consequences of group hunting are at worst only mildly negative and are probably not as disadvantageous in lions as they would be in other species (Packer 1986). Thus in no other species might the advantages of group defense of cubs against infanticidal males be greater than the disadvantages of group foraging.

23.4 Summary

Long-term demographic data are presented on lions living in the Ngorongoro Crater and in Serengeti National Park, Tanzania. Variance in lifetime reproductive success is higher in males than in females, and successful reproduction is confined to a narrower age range in males. Cub survival appears to be the most important variable affecting lifetime reproductive success of females. In males, cub survival and "fecundity" contribute equally to lifetime reproductive success. However, successful intercoalition competition is a necessary prerequisite for male reproduction. Considerable cub mortality is due to infanticide by incoming males and to starvation. Within each habitat, more cubs are reared in years with lower levels of male replacements, and hence less infanticide. Cub recruitment rates are higher where prey availability is greater. Cub recruitment is also higher when there are fewer adult females in each habitat. Adult females and subadult males in the area of highest population density, the Ngorongoro Crater, suffer higher mortality than their counterparts in either Serengeti habitat, but mortality of adult males is similar in all areas.

Males in large coalitions gain higher individual reproductive success. Solitary males readily join to form coalitions of two or three, but coalitions of four or more are always composed of relatives. Prides of three to ten adult females have higher per capita reproduction rates than smaller or larger prides, and these moderate-sized prides suffer fewer male takeovers. Females in prides of one or two females also suffer higher mortality. Unlike males, solitary females never form larger prides with unrelated companions. The evolution of female kin groups in lions is discussed.

23.5 Acknowledgments

We are very grateful for the continued encouragement of the Tanzanian government, and in particular the Serengeti Wildlife Research Institute, for our research. Over the past eleven years the Serengeti lion study has been generously supported by grants from the H. F. Guggenheim Foundation (to Packer and Pusey), National Geographic Society (Packer and Pusey), Royal Society of Great Britain (Pusey), Science Research Council of Great Britain (Bygott and Hanby), New York Zoological Society (Bygott and Hanby), Eppley Foundation (Pusey), American Philosophical Society (Packer and Pusey), Chapman Fund (Cairns), Sigma Xi (Packer), Hewlett-Packard (Packer), NIMH grant no. MH15181 (Packer), Graduate School of the University of Minnesota (Packer, Pusey, and Herbst), and NSF grant no. BSR-8406935 (Packer and Pusey). The African Wildlife Leadership Foundation funded collection of the rainfall data by the Serengeti Wildlife Research Institute. We thank Tim Clutton-Brock for many useful comments on earlier drafts of this chapter.

24 Reproductive Success in Vervet Monkeys

Dorothy L. Cheney, Robert M. Seyfarth, Sandy J. Andelman, and Phyllis C. Lee

ALTHOUGH THE RELATIVE IMPORTANCE of food supply and predation on population growth has been a topic of much theoretical debate, the debate has suffered from the inability to compare these factors empirically. Food supply is known to have an important effect on reproduction and survival in nonhuman primates. To date, however, most long-term demographic studies have been conducted on provisioned populations or in areas where predators are rare, with the result that the possible effects of predation have of necessity been ignored. This is an important omission, because predation and food supply are likely to act differently on different reproductive parameters and individuals. Fecundity, for example, is clearly affected more by food supply than

Figure 24.1: Vervet monkeys in Amboseli National Park.

by predation, and while mortality due to food competition is often concentrated among low-ranking animals, there is no a priori reason to expect predation to act more strongly on individuals of some dominance ranks than others.

In this chapter we summarize the results of seven years' research on the factors affecting reproductive success in one population of East African vervet monkeys, *Cercopithecus aethiops* (fig. 24.1). Unlike many terrestrial primates that range over large areas (e.g., baboons: Altmann, Hausfater, and Altmann, this volume, chap. 25), vervet groups inhabit small territories that differ markedly in the availability of food and water over even short distances. The effects of habitat differences on reproductive behavior can therefore be measured directly. Moreover, perhaps because of their relatively small size, vervets suffer high rates of predation. It is therefore possible to compare the relative importance of predation and food competition on reproductive success. Our study cannot pretend to offer a definitive analysis of the factors determining reproductive success, which in any primate species would take years of observation. Instead, we attempt to provide some comparative data on the ways food supply and predation affect individuals of different age, sex, and dominance rank.

24.1 Habitat, Study Groups, and Methods

Habitat

Research is conducted on three social groups with adjacent territories in Amboseli National Park in southern Kenya. The park consists of arid savanna and savanna-woodland (Struhsaker 1967; Altmann and Altmann 1970; Western and van Praet 1973; Lee 1981; Western 1983). The dominant tree species are *Acacia xanthophloea*, which grows primarily near water holes, and *Acacia tortilis*, which is found away from water. Rainfall is light, averaging 300 mm per year (Dunne and Dietrich 1980; Lee 1981). Water holes are the only major source of water during the dry seasons of June–November and January–April.

Vervets in Amboseli depend heavily on *A. xanthophloea* for both refuge and food; they eat the tree's flowers, fruits, leaves, thorns, and gum. Unlike the gum of *A. tortilis*, xanthophloea gum is rich in carbohydrates and low in tannins (Hausfater and Bearce 1976) and is an important food source during the dry seasons, when other foods are scarce (Lee 1981). Over the past three decades, however, a gradual rise in the saline water table in the Amboseli basin has killed large numbers of xanthophloea trees, as well as numerous shrubs vervets feed on (Western and van Praet 1973; Western 1983). Tree mortality has been exacerbated by elephant damage, which has curtailed regeneration (Western 1983). The decline in xanthophloea trees has been correlated with a dramatic decrease in the number of vervet monkeys (table 24.1; see also Struhsaker 1973, 1976).

Although vervets feed less on *A. tortilis* than on *A. xanthophloea*, they do eat its flowers, leaves, and fruit. Annual variation in the produc-

Table 24.1 Changes in Group Size and Habitat Quality in Vervet Monkeys over a Five-Year Period

	March 1978	March 1983	Percentage Change
Group A			
Group size	30	14	−53.3
Range (km^2)	0.27	0.21	−29.6
Density (per km^2)	103.45	28.57	−72.4
Number of *A. xanthophloea*[a]	14	6	−57.1
Group B			
Group size	17	19	+11.7
Range	0.12	0.19	+58.3
Density	77.0	45.0	−41.6
Number of *A. xanthophloea*	22	27	+22.7
Group C			
Group size	27	13	−51.2
Range	0.24	0.16	−33.3
Density	79.0	56.0	−29.1
Number of *A. xanthophloea*	27	5	−81.5

[a]Number of *A. xanthophloea* refers only to mature trees.

tion of tortilis pods appears to have a profound effect on survival in groups with no permanent surface water, which rely heavily on tortilis pods during the dry season (Lee 1981). The failure of tortilis pods in 1978, for example, was correlated with the deaths of 41% of the twenty-nine members of one study group (group A), which has no access to permanent water holes (see below).

Each study group occupies a territory that averages 0.23 km^2 in size and is aggressively defended against other groups (Cheney 1981). Despite their close proximity, there are distinct differences in territory quality among the three study groups. Although group A's territory is adjacent to that of group B and is separated by less than half a kilometer from two permanent water holes, it has no permanent surface water. For ten months of the year group A obtains water only from tortilis tree crevices, where moisture gathers and is retained in small amounts (an average of 5 ml; Wrangham 1981). Access to such trees is the source of much intragroup competition (see below). The majority of trees in group A's range are *A. tortilis*; the range includes only a small number of mature xanthophloea trees, all of which are dying (table 24.1). In contrast to that of group A, group B's range includes two permanent water holes, over twenty tortilis and mature xanthophloea trees, and numerous immature xanthophloea trees. Group C also has access to two water holes and a number of immature xanthophloea trees. However, by 1983 its range included only a few tortilis and mature xanthophloea trees (table 24.1). Group C spends a great deal of time feeding in a large papyrus swamp, particularly during the dry season.

Comparison of habitat quality in the three territories has shown that group B's has the most abundant food supply and group A's the least, with group C's territory being intermediate (Lee 1983a, 1984).

Since 1981, as the number of xanthophloea trees has continued to decline, group A has expanded its range to include more tortilis trees (table 24.1), while group B has expanded its range at the expense of groups A and C. Although group B has gained access to more mature xanthophloea trees as a result, all of these are dying. The decline in group C's range has been correlated with encroachment by other groups and has occurred primarily since 1981, during a period of high adult female mortality (from eight adult females in January 1982 to four in October 1983).

Study Subjects

We have observed all three study groups without interruption since March 1977. Group A was also studied by D. Klein (1978) from 1974 to 1976. Study animals are well habituated to humans and can be observed at close range by observers on foot. The data presented here are complete as of October 1983.

Since 1977 the three study groups have ranged in size from ten to thirty-one individuals, with one to eight adult males and two to eight adult females per group. Males emigrate to adjacent groups at about sexual maturity, often in the company of brothers or other age-mates (Cheney and Seyfarth 1983). With rare exceptions, females remain in their natal group throughout their lives. Infants are defined as animals under one year, while males aged one to five years and females aged one to four years are considered juveniles. Males begin to copulate with ejaculation at about five years of age, and though most females first give birth at age five, three four-year-old females have given birth to live offspring.

Like baboons and macaques, adult female vervets can be ranked in a linear dominance hierarchy that is generally stable over time and accurately predicts access to resources such as food, water, and social partners (Struhsaker 1967; Seyfarth 1980; Cheney, Lee, and Seyfarth 1981; Wrangham 1981; Whitten 1982). Offspring acquire ranks immediately below those of their mothers (Cheney 1983; Lee 1983b). Although adult males can also be ranked in linear dominance hierarchies, these seem to depend less on maternal rank than on such factors as age, size, and fighting ability. Males' dominance ranks are also less stable than those of females. Over a three-year period, males changed ranks at a rate of 0.75 ranks per male per year, while the comparable figure for females was 0.11 (Cheney 1983). In dyadic interactions, adult males are usually dominant over adult females. However, perhaps because sexual dimorphism in size is not great (females' body weight is approximately 80% that of males; Whitten 1982), females regularly form successful aggressive coalitions against males. The threat of such coalitions appears to allow females to dominate males in some dyadic interactions.

Most of the results described here are restricted to births occurring since 1977. Thus female reproductive success is calculated only in terms of infant survival between 1977 and 1983, even in those cases when females are known to have produced surviving offspring between 1974 and 1976.

This is done to control for the probability that some females may have had nonsurviving offspring during that period. Including only the surviving members of previous birth cohorts would therefore have biased our analysis.

Of the thirty-three adult females included in our sample, fourteen were multiparous as of 1976. In estimating the ages of thirteen of these older females, we were conservative and assumed that any known offspring of mature females were those females' first offspring, born when the female was five (see below). Thus, for example, a female with a year-old offspring in 1977 was assumed to be six years old. (One additional female was estimated to be over ten years of age in 1977 judging by physical appearance and tooth wear.) While this procedure no doubt underestimated some females' ages, it ensured that any difference in reproductive success due to life span were minimum estimates. Thirteen females are known to have reached at least ten years of age, and two others at least thirteen.

During the breeding season, female vervets show no apparent external morphological changes correlated with the occurrence of ovulation, nor do males and females form sexual consortships like those found among baboons. As a result, the reproductive success of male vervets can only be estimated, by recording the frequency and timing of copulations by each individual. Such data provide, at best, a very rough estimate of male reproductive success, since captive studies of macaques using known genetic markers have shown that there is not always a strong positive correlation between paternity and estimates of male reproductive activity (e.g., Stern and Smith 1984).

Data on copulations were collected opportunistically during the 1977–80 and 1983 breeding seasons. In 1981 and 1982 mating activity was studied intensively by Andelman (1984), who also collected urine from some females in an attempt to relate sexual activity more precisely to the timing of ovulation (Andelman et al. 1984). In estimating male reproductive success, we used the proportion of all copulations obtained rather than only those that occurred during periods of probable conception, for two reasons. First, the two measures were significantly positively correlated, and the selection of a particular measure had little effect on results. Second, in many cases only a small number of copulations were observed during fertile periods, because sexual activity in general occurred at low rates. Restricting analysis to copulations that occurred during fertile periods would thus have severely reduced our sample size.

Causes of Mortality

Mortality was ascribed to a number of different categories. *Confirmed predation* occurred whenever an observer either actually witnessed a kill or, in the case of three leopard attacks, saw a predator carry off a monkey who had been observed to be healthy within the previous twenty-four hours. Confirmed predators in Amboseli are leopards (n = 3 predations in this study), crowned hawk eagles, martial eagles (n = 1), pythons (n = 3), and baboons (n = 4) (Struhsaker 1967; Cheney and Seyfarth 1981). Suspected predators include small carnivores such as servals and caracals (n = 1), marsh mongoose, and Verreaux's eagle owls. *Disappear healthy* (n = 37) occurred when an animal disappeared within twenty-four hours of having been seen in good condition. In twenty-six cases animals disappeared overnight, and observers typically arrived in the morning to find the monkeys either giving alarm calls or remaining unusually long in their sleeping trees. They often appeared nervous throughout the day, rarely descending to the ground and running into trees at the slightest disturbance. The remaining eleven cases involved disappearances by animals in group C in the large papyrus swamp that borders the group's range and seemed to be due to predation. In most cases swamp disappearances were accompanied by alarm calls, and two unsuccessful attacks—one by a leopard on an adult male and another by either a python or a marsh mongoose on a yearling female—have been observed in the swamp.

Illness (n = 14) occurred whenever an animal disappeared within twenty-four hours of having been observed to be weak, listless, or suffering from a particular disease. Two diseases have been observed on more than one occasion, both most common in group A during the dry season (see below). One is a staphylococcic infection of the eye (J. Else, pers. comm.), causing swelling and reddening of the eyelid, often such that one or both eyes are swollen shut. A second, unknown disease has been observed only in males; symptoms are blackened, shriveled testicles. Especially in the case of infants, the coat color of ill animals also often changes from its usual olive gray to pale whitish, a change that may be related to protein deficiency (S. Altmann, pers. comm.). Predation and illness were mutually exclusive categories; animals who disappear following signs of illness are not considered to have died of predation.

If the mother of an unweaned infant died (n = 5), the subsequent infant death was classed as *die following mother's death*. Finally, twenty-eight animals disappeared while the observer was absent for more than twenty-four hours, and the cause of the disappearance could not be determined. All subsequent analyses that consider causes of mortality exclude animals that disappeared in this last context.

24.2 Results

Fecundity

As in other populations of East African vervets (Whitten 1982), births in Amboseli have been highly seasonal, with 87% (n = 75) occurring during the months of October, November, and December (see also Struhsaker 1967). Although there was no difference in the timing of births across groups, there were intergroup differences in the age at which females first gave birth. Group A females, who had no access to permanent water holes, first gave birth at a mean age of 5.69 (s.d. = ±1.13) years, whereas females in groups B and C gave birth earlier, at mean ages of 4.40 (±0.58)

Table 24.2 Summary of Reproductive Differences among Three Study Groups of Vervet Monkeys

Measure	Group A	Group B	Group C
Infant female mortality	0.63	0.47	0.63
Infant male mortality	0.52	0.60	0.67
Juvenile female mortality	0	0.06	0.08
Juvenile male mortality	0	0.11	0.28
Adult female mortality	0.17	0.10	0.22
Female age at first birth	5.7 ± 1.1	4.4 ± 0.6	5.1 ± 0.6
Interbirth interval	21.3 ± 5.0*	16.3 ± 5.6	13.8 ± 3.5

Note: All figures given are mean annual rates.

*Significant difference between group A and groups B and C ($F_{1,43} = 11.40$, $p < .01$). No other between-group comparisons yielded significantly different results.

and 5.08 (±0.64), respectively (table 24.2). Although not significant, these differences suggest that habitat quality, and in particular the availability of water, may influence age at first reproduction. Of the fifteen females known to have reached at least ten years of age, none showed any reduction in the rate of live births with age, suggesting that any decrease in fecundity due to age becomes evident only after at least thirteen years. In this respect our data differ from those of Strum and Western (1982), who report that fecundity in an expanding population of baboons was highest among individuals of midreproductive age (see also Dunbar 1980). This difference may occur because few Amboseli vervets survive to old age.

Fecundity differed significantly between groups with and without permanent water holes. The mean interbirth interval in group A was almost two years, whereas in groups B and C it was significantly shorter (table 24.2). Within each group, interbirth intervals were also affected by the survival of immediately preceding offspring. Females whose offspring did not survive their first twelve months were significantly less likely to skip a birth season than those whose offspring survived ($\chi^2 = 8.01$, d.f. = 1, $p < .05$).

Within groups, there was no correlation between female dominance rank and either age at first reproduction or interbirth interval. Similarly, there was no consistent tendency for high-ranking females to give birth earlier in the birth season than low-ranking females, as has been reported for one group of vervets in northern Kenya (Whitten 1983).

Over six breeding seasons, 69% ($n = 16$) of live births in group A and 67% ($n = 30$) of births in group C were of male infants. In contrast, only 37% of births in group B ($n = 27$) were males (group B compared with groups A and C: $\chi^2 = 4.41$, d.f. = 1; $p < .05$). Intergroup sex-ratio differences were not obviously related to habitat differences, since groups B and C shared a more similar habitat with each other than with group A. These intergroup differences may simply have reflected small-sample biases, and more data are needed to examine the possible long-term effects of such trends.

Within each group, a female's rank had no apparent effect on her offspring's sex (table 24.3). This study therefore differed from some studies of macaques and baboons that have reported rank-related sex-ratio biases (Altman 1980; Silk et al. 1981; Simpson and Simpson 1982; Meikle, Tilford, and Vessey 1984). Similarly, in contrast to some captive rhesus monkeys and bonnet macaques (Simpson et al. 1981; Silk et al. 1981), the sex of her offspring did not affect the length of a female's subsequent interbirth interval.

Mortality

As in other populations of both capitve and free-ranging Old World monkeys, most vervet mortality occurred in the first twelve months, with an average of 57% of all infants dying in their first year (fig. 24.2; see also Drickamer 1974; Sade et al. 1976; Altmann et al. 1977 and this volume, chap. 25; Mori 1979; Silk et al. 1981; Sugiyama and Ohsawa 1982; Fairbanks and McGuire 1984). In contrast to captive vervets, however, where most deaths are concentrated in the first six weeks of life (Fairbanks and McGuire 1984), deaths in Amboseli occurred throughout the first year, showing peaks at four months (19% of all infant deaths) and at eight to nine months (30%). The cause of infant mortality was often difficult to determine precisely, since most disappearances occurred overnight. However,

Table 24.3 Summary of Reproductive Differences among Female Vervet Monkeys of Different Dominance Ranks

Measure	Dominance Rank (quartile)			
	First	*Second*	*Third*	*Fourth*
Percentage of infants surviving				
Group A	20 (n = 5)	60 (n = 5)	50 (n = 4)	0 (n = 3)
Group B	43 (n = 7)	50 (n = 6)	60 (n = 10)	50 (n = 10)
Group C	38 (n = 8)	17 (n = 6)	33 (n = 9)	43 (n = 7)
Overall	35 (n = 20)	41 (n = 17)	44 (n = 18)	40 (n = 20)
Percentage of males born				
Group A	60 (n = 5)	60 (n = 5)	100 (n = 4)	50 (n = 2)
Group B	43 (n = 7)	40 (n = 5)	20 (n = 5)	40 (n = 10)
Group C	75 (n = 8)	50 (n = 6)	67 (n = 9)	71 (n = 7)
Overall	60 (n = 20)	50 (n = 16)	61 (n = 18)	53 (n = 19)
Percentage of males surviving				
Group A	0 (n = 3)	67 (n = 3)	50 (n = 4)	0 (n = 1)
Group B	33 (n = 3)	100 (n = 2)	0 (n = 1)	50 (n = 4)
Group C	17 (n = 6)	33 (n = 3)	50 (n = 6)	20 (n = 5)
Overall	17 (n = 12)	63 (n = 8)	45 (n = 11)	30 (n = 10)
Percentage of females surviving				
Group A	50 (n = 2)	50 (n = 2)	—	0 (n = 1)
Group B	50 (n = 4)	33 (n = 3)	75 (n = 4)	50 (n = 6)
Group C	100 (n = 2)	0 (n = 3)	0 (n = 3)	100 (n = 2)
Overall	63 (n = 8)	25 (n = 8)	43 (n = 7)	56 (n = 9)

Note: All survival data are to one year of age. Overall, no between-quartile comparisons yielded significantly different results. Similarly, no differences between females above and below median rank were significant. Two infants whose sexes were not known were omitted from some analyses.

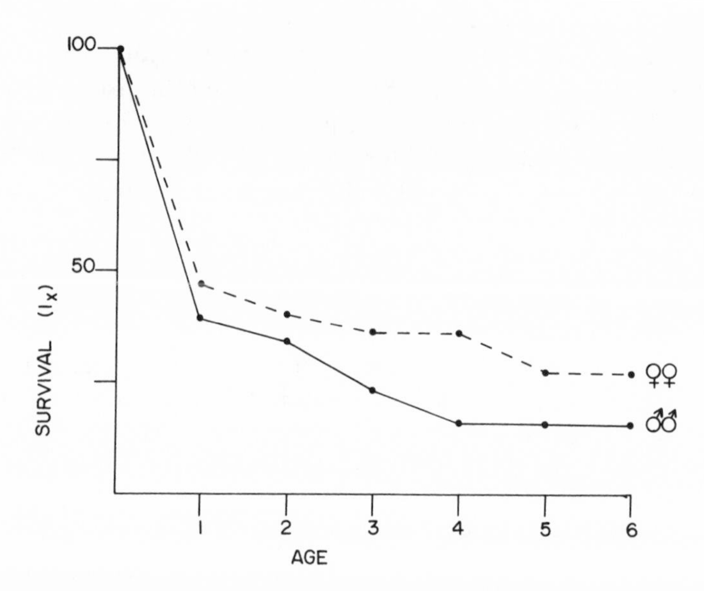

Figure 24.2: Age-specific survival rates for male and female vervet monkeys during their first six years of life. Data from the three study groups have been pooled. At no age was there a significant difference between males and females in rates of survival.

the first peak in infant mortality coincides with a reduction in carrying by the mother, when infants may become more vulnerable to predation, and the second with weaning during the dry season, when infants may be more susceptible to illness (see also Whitten 1982; Lee 1984). There was no relation between infant survival to twelve months and group size or between infant survival and the number of infants present in a group. Similarly, the presence of older siblings had no apparent effect on infant survival. Finally, unlike captive bonnet macaques and vervets (Silk et al. 1981; Fairbanks and McGuire 1984), primiparous females (n = 17) did not have fewer surviving infants than did multiparous females ($n = 58$; $\chi^2 = 0.24$, d.f. = 1, $p > .50$).

While mortality among males and females occurred at approximately the same rates through two years of age, mortality was greater among males between the ages of two and four, although not significantly so (fig. 24.2). This mortality occurred before sexual maturity and was not obviously associated with migration or increased intrasexual competition (see, e.g., Koford 1965; Dittus 1975; Otis, Froehlich, and Thorington 1981). All juvenile male mortality occurred in groups B and C and appeared to be due to predation (see below). It is possible that some behavioral factor increased the vulnerability of these animals to predation, though the cause of such susceptibility is not known.

In none of the three groups was there a significant correlation between dominance rank and infant survival. Rank-specific comparisons of infant survival revealed no differences among females in the four rank quartiles (table 24.3). Similarly, an individual-specific analysis in which each female was assigned the average rank she occupied during her breeding years revealed no correlation between rank and infant survival (group A: r_s = .137, n = 10; group B: r_s = .180, n = 8; group C: r_s = .279, n = 14; two females that never bred were eliminated from analysis). There was also no indication that female life span was influenced by dominance rank. Finally, high- and low-ranking females showed no consistent differences in the proportion of sons or daughters that survived (table 24.3). For example, although the sons of females in the top-ranking quartile survived at a lower rate than did their daughters, the reverse was the case in the second-ranking quartile. The lack of any consistent relation between dominance rank and survival was at least partly due to the high rate of predation, which affected all rank quartiles.

Confirmed or suspected predation is the major cause of mortality in Amboseli vervets and accounts for at least 69% of all deaths. On average, approximately 15% of the population is estimated to die of predation each year. Between 1977 and 1983, predation and illness affected animals at different times of the year (fig. 24.3). Whereas illness is most common during the dry-season months of July–October, when food is most scarce, predation occurs most often during the rainy seasons and the short dry season between the rains (January–June). The increased vulnerability of vervets during these periods may occur because many ungulate species disperse from the park during wet periods, possibly causing predators to concentrate more exclusively on vervets. That predation is not concentrated on animals who are already ill may exacerbate the effects of predation on the population.

Illness was a more important cause of mortality than predation in group A (illness = 11; confirmed and suspected predation = 8), but predation was far more important than illness as a cause of mortality in groups B and C (group B: illness = 1, predation = 17; group C: illness = 2, predation = 24). The high rate of predation in these groups appeared to be due to their proximity to water holes and the large swamp, where both prey and predator species were more abundant. Predation was particularly high in group C, which regularly fed in the swamp.

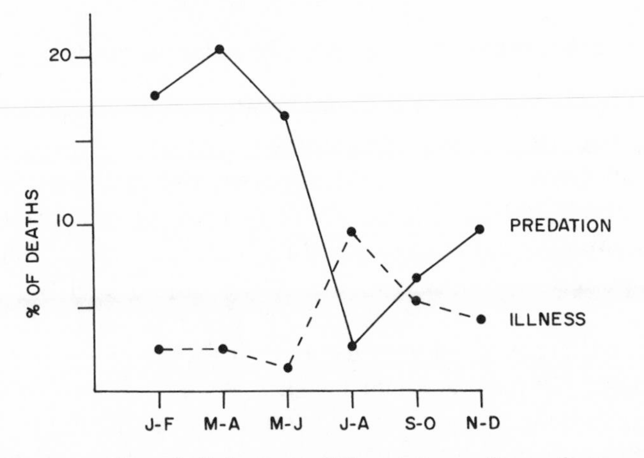

Figure 24.3: The proportion of all deaths at different times of year that were due to illness and predation (confirmed plus suspected; see text).

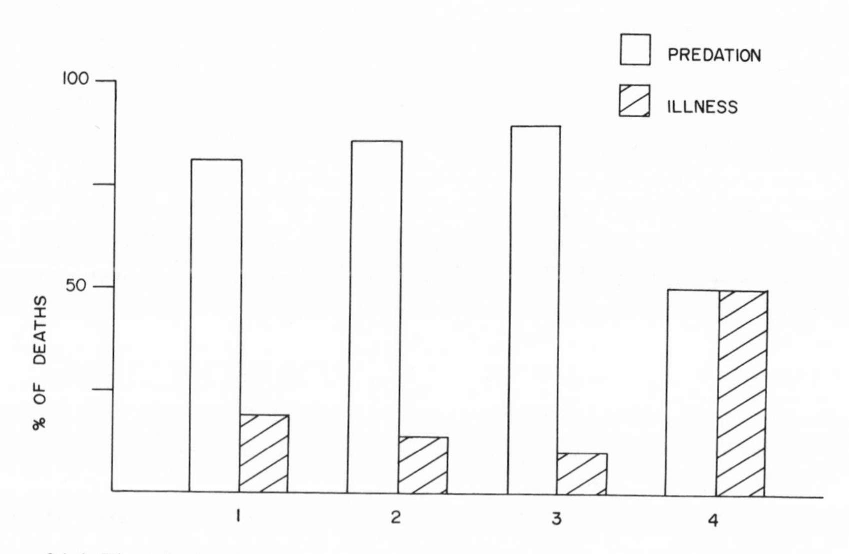

Figure 24.4: The relative frequency with which adult females and juveniles in the top, second, third, and fourth dominance rank quartiles died of predation as opposed to illness. See text for sample sizes.

Overall, confirmed and suspected predation caused significantly more adult female, juvenile, and infant deaths than did illness (two-tailed binomial test, $n = 50$, $x = 11$, $p < .001$). Although predation and illness were equally likely to strike individuals in all rank quartiles, a female's rank did seem to influence the probability that she or her offspring would die of predation rather than illness. Figure 24.4 divides adult females and their immature offspring into quartiles based on their dominance rank and shows the relative frequencies of deaths due to illness and predation. In the first, second, and third rank quartiles, females and their offspring were significantly more likely to die of predation than of illness (binomial test, first quartile: $n = 16$, $x = 3$; second: $n = 14$, $x = 2$; third: $n = 10$, $x = 1$; all $p < .03$). In the fourth quartile, animals were as likely to die of illness as of predation ($n = 10$, $x = 5$, $p > .50$). Thus predation exerted a significantly stronger effect than illness in all but the lowest-ranking quartile. That predation was the major cause of mortality, and that it acted strongly on animals of all ranks, explains at least in part the lack of correlation between female dominance rank and offspring survival.

Deaths from illness were particularly common among low-ranking animals in group A, which has the poorest habitat. Most such deaths coincided with the failure of tortilis pods in 1978 (see above). Of the eight females and infants who died during this period, six were below median rank. Competition over access to the limited supplies of water in tortilis tree crevices may have contributed to the mortality of low-ranking animals, since there were numerous signs of food and water stress during this period (Lee 1981; Wrangham 1981).

Results were similar for a small sample of adult male deaths. Nine of eleven (82%) confirmed and suspected predations affected males above median rank, whereas two of three (67%) illnesses affected males below median rank. As was the case for females, deaths due to illness among group A males occurred during the dry season of 1978. During that period, four of seven males either disappeared or died of illness. Of the four, three were below median rank.

Factors Affecting the Reproductive Success of Females

Tables 24.4 and 24.5 summarize the relative contributions of breeding life span, fecundity, and infant survival on female reproductive success (see Brown, this volume, chap. 27). Ideally, the analysis should consider only females whose entire breeding life is known. Owing to small sample sizes and the relatively short duration of the study, however, our analysis includes both older females of estimated age and younger females who were still breeding. As a result, that portion of a female's breeding years

Table 24.4 Mean and Variance of the Components of Lifetime Reproductive Success in Female Vervet Monkeys

Component	Original		Standardized Variance
	Mean	*Variance*	
L	4.258	4.465	0.246
F	0.795	0.049	0.078
S	0.474	0.101	0.448
LF	3.260	3.564	0.311
FS	0.377	0.074	0.522
LS	2.243	3.123	0.765
LFS	1.746	1.970	0.764

Note: Analysis includes thirty-one breeding females. The proportion of breeders was estimated from mortality rates. See also text.
L = breeding life span; F = Fecundity; S = proportion of offspring surviving to one year.

Table 24.5 Percentage Contribution of the Components of Lifetime Reproductive Success to Overall Variance in LRS among Female Vervet Monkeys

Component	L	F	S
All females			
L	32.24		
F	−1.65	10.16	
S	9.23	−0.52	58.67
LF	−8.12		
Breeders only			
L	14.84		
F	−0.76	4.68	
S	4.25	−0.24	27.01
LF	−3.74		
Breeders and nonbreeders:			
Proportion of breeders, .267			
Overall variance (OV), 1.116			
Percentage of OV due to nonbreeders, 53.96			
Percentage of OV due to breeders, 46.04			

Note: Analysis includes thirty-one breeding females. The proportion of breeders was estimated from mortality rates. See also text.
L = breeding life span; F = fecundity; S = proportion of offspring surviving to one year.

that occurred between 1977 and 1983 is treated as representative of her entire breeding life. Given the variation in both fecundity and infant survival across different years and groups, this assumption may not be justified. The short duration of our study has also forced us to define offspring survival in terms of survival to one year. Although most mortality does occur during infancy, it is not restricted to this period, and the procedure therefore misrepresents true rates of offspring survival. Despite these liabilities, however, the analysis does permit some consideration of the relative contributions of life span, fecundity, and infant survival to lifetime reproductive success.

Only 27% of female vervets in Amboseli are estimated to survive to breeding age, with the result that the proportion of variance in reproductive success due to nonbreeders is higher than that due to breeders (table 24.5). Among females who survive to breed, infant survival accounts for 59% of the variance in lifetime reproductive success, while life span accounts for 32% and fecundity for 10%. There is greater individual variation in infant survival than in either life span or fecundity, apparently because infant mortality in Amboseli is so high that even females who survive to breed for a number of years cannot be assured of leaving any surviving offspring. Thus females whose breeding life is long do not necessarily have greater lifetime reproductive success than those whose breeding life is short. These results can be contrasted with those that might be expected in provisioned groups, where infant mortality is generally low and where almost all females that reach breeding age produce at least some surviving offspring. In such groups, variation in reproductive success would probably result more from differences in fecundity or breeding life span than from infant survival.

Factors Affecting the Reproductive Success of Males

Males vary in the age at which they disperse from their natal group (Cheney and Seyfarth 1983), and as a result the sexually mature males within a given group may be either natal or nonnatal. During seven breeding seasons, natal males over five years of age (n = 12) occupied significantly lower dominance ranks and achieved a significantly lower proportion of copulations than did nonnatal males (n = 65). On average, natal males were dominant to 17.4% of all males, whereas nonnatal males were dominant to 54.1% ($F_{1,75} = 10.27$, $p < .01$). Natal males achieved an average of 9.3% of all copulations, while the average for nonnatal males was 30.2% ($F = 6.68$, $p < .02$). As among baboons (Packer 1979a), the low copulatory success of natal males appeared to derive in large part from their lack of sexual attraction to females. With the exception of one individual who became the dominant male in his natal group and bred successfully with several females, natal males seldom competed for access to sexually receptive females. Moreover, most emigration by natal males occurred during the breeding season, and the proximate cause of such migration appeared to be sexual attraction to females in neighboring

Table 24.6 Mean Percentage of Copulations Accounted for by Male Vervet Monkeys of Different Dominance Ranks and Lengths of Tenure in the Group

Rank and Tenure	Group A	Group B	Group C
Above median rank	27.4	40.2	43.4
(*n*)	(17)	(9)	(11)
Below median rank	18.7	39.2	29.4
(*n*)	(11)	(6)	(7)
U	53.5	26.5	26.0
Tenure less than twelve months	18.8	32.7	62.0
(*n*)	(5)	(3)	(3)
Tenure more than twelve months	26.0	39.9	29.6
(*n*)	(19)	(10)	(12)
U	29.0	12.5	13.0

Note: Data from seven breeding seasons are combined. All statistical comparisons were insignificant. *U* = Mann-Whitney *U* test.

groups (Cheney 1983; Cheney and Seyfarth 1983). Because most natal males were low ranking and seldom attempted to copulate, including natal males in an analysis of male copulatory behavior might artificially bias results against low-ranking individuals. The following analysis therefore concerns only nonnatal males.

Although many studies of baboons and macaques have documented a positive correlation between male rank and mating success, others have shown no such relation (see reviews by Berenstain and Wade 1983; Silk 1985). Thus, although dominant males usually have greater access to resources than subordinate males, they do not always achieve greater copulatory success. In this study dominant males also did not consistently achieve more copulations than low-ranking males. In some years there was a significant positive correlation between rank and copulation frequency; in other years the correlation was negative. Overall, in seven breeding seasons, the correlation in the three study groups between male rank and copulatory success averaged .006, and it varied from +1.00 to −1.00. On average, males above median rank accounted for a larger proportion of copulations than males below median rank, but this difference was not significant (table 24.6). Moreover, except in those years when there was only one adult male in a group (twice in twenty-one group breeding seasons), the correlation between rank and copulation frequency did not appear to be influenced by the number of males in the group. For example, in one year when there were seven adult males in group A, high-ranking males copulated significantly more than low-ranking males. During the same year, when there were only two adult males in group B, the lowest-ranking male achieved the most copulations.

In contrast to Amboseli baboons (Altmann, Hausfater, and Altmann, this volume, chap. 25), the mating success of male vervets was also unaffected by length of tenure in the group. The copulation frequency of males who had been resident in the group for less than one year did not differ from that of longer-term residents (table 24.6). Similarly, the limited data available on males of known age revealed no relation between age and copulation frequency. Insufficient long-term data precluded a more refined analysis of the effects of tenure and age on mating success.

Because infant mortality in Amboseli is high and dominance rank does not always ensure mating success, it is possible that a male who mates at even low frequencies over a number of years may leave more surviving offspring than a male who achieves high mating success during only one or two years. As a result, male reproductive success may be influenced more by whatever factors contribute to longevity than by traits commonly associated with intrasexual competition, although such traits are obviously not mutually exclusive. Female choice probably also strongly affects male mating success. Andelman (1984) presents data showing that female vervets reject as many at 50% of all male attempts at copulation and that females exert strong mate choice. Some preferences may remain stable from year to year while others change. Similarly, some male vervets who are of long tenure in their group can maintain copulatory success despite falling in dominance rank, whereas for others copulatory success declines with declining dominance rank. Clearly, the reproductive success of males is more difficult to measure, and its causes are more difficult to explain, than that of females. At this stage in our analysis, no single factor satisfactorily explains variation in male mating behavior.

24.3 Discussion

Factors Affecting Individual Variation in Reproductive Success

Several studies of free-ranging primates have documented correlations among preferential access to food, female dominance rank, and reproductive success. Dittus (1979) showed that a decrease in food supply increased both intragroup competition and mortality among low-ranking toque macaques. Similarly, in one unprovisioned population of Japanese macaques, the offspring of high-ranking females were heavier and survived at higher rates than did the offspring of low-ranking females (Mori 1979; but see Hazama 1964; Sugiyama and Ohsawa 1982). Whitten (1983) also found that high-ranking female vervets in northern Kenya were heavier than low-ranking females, presumably owing to preferential access to high-quality food. High-ranking females also gave birth earlier in the birth season and had shorter interbirth intervals. These effects were limited to one of two study groups, however, and in neither group was there a correlation between rank and infant survival.

Most studies showing a correlation between female rank and fecundity or offspring survival have been conducted on provisioned or captive groups (Fairbanks and McGuire 1984; Drickamer 1974; Wilson, Gordon, and Bernstein 1978; Wilson, Walker, and Gordon 1983; Silk et al. 1981; but see Gouzoules, Gouzoules, and Fedigan 1982; Wolf 1984). Indeed, in one population of Japanese macaques the relation between female rank and reproductive success disappeared when food provisioning ceased (Sugiyama and Ohsawa 1982). One possible explanation for this finding is that high-ranking females are better able to take advantage of increases in food supply than are low-ranking females.

The decline in xanthophloea trees in Amboseli might be expected to have increased intragroup competition for food and thus to have exaggerated rank differences in reproductive success among vervet monkeys. Such competition might also be predicted to have lowered fecundity among low-ranking animals, as has been postulated to occur in some primate populations (Dunbar and Dunbar 1977; Silk et al. 1981). In Amboseli, high-ranking females do spend more time than low-ranking females feeding on preferred food items (Wrangham and Waterman 1981; see also Whitten 1983) and are also able to exclude low-ranking females from preferred tree water holes (Seyfarth 1980; Wrangham 1981). Perhaps as a result, illness during the dry season is concentrated among low-ranking animals. There is no evidence, however, that such competition lowers the fecundity of low-ranking animals. Moreover, despite high mortality among low-ranking animals during periods of food scarcity, we have been unable to document a correlation between dominance rank and reproductive success. This is because predation occurs at a higher rate than illness and is high among animals of all dominance ranks. Predation therefore overrides the competitive advantage gained by high-ranking females and obscures the effect of illness on animals of low rank.

The precise reason for the variation in susceptibility to predation is not known. Although high-ranking animals alarm call more than low-ranking animals, there is no evidence that alarm calls increase the caller's vulnerability to predation. Similarly, there is no consistent tendency for some animals to lead group progressions or to precede others into new and potentially dangerous areas (Cheney and Seyfarth 1981, 1985). Whatever its cause, predation does appear to account for the low correlation between rank and reproductive success in this study, and it may also contribute to the lack of such a relationship in other free-ranging populations (e.g., Amboseli baboons: Altmann, Hausfater, and Altmann, this volume, chap. 25). To date, of all studies conducted in areas of high predation, only one on baboons has reported a correlation between female rank and infant survival (Busse 1982). The precise cause of this rank-related difference in infant survival was not known.

Factors Affecting Sex-Ratio Biases

Trivers and Willard (1973) have hypothesized that because maternal rank or condition may usually have a greater effect on the reproductive success of sons than of daughters, the sex ratio of offspring among high-ranking females should be biased toward sons, while that among low-ranking females should be biased toward daughters. In primates this hypothesis has received support from one study of semicaptive rhesus macaques (Meikle, Tilford, and Vessey 1984; see also Clutton-Brock, Guinness, and Albon 1982). In contrast, several studies of captive macaques (Silk et al. 1981; Simpson and Simpson 1982; Silk 1983) and free-ranging baboons (Altmann 1980; Altmann, Hausfater, and Altmann, this volume, chap. 25) have found offspring sex ratios among high-ranking females to be female biased rather than male biased and have attempted to explain this result on the grounds that when females remain in their natal groups and assume their mothers' ranks, high-ranking females will be able to confer a competitive advantage on their daughters and will therefore invest more in daughters than will low-ranking females.

High- and low-ranking female vervets in Amboseli do not differ in the sex ratio of their offspring, and there are several possible (admittedly post hoc) explanations for the lack of such biases. First, in contrast to Meikle, Tilford, and Vessey (1984), we find no evidence that maternal rank influences the rank or reproductive success of adult sons. Sons of low-ranking females are known to have assumed high ranks following transfer from the natal group, and sons of high-ranking females are known to have assumed low ranks. Second, it is possible that owing to the lack of correlation between mating success and rank, male reproductive success in Amboseli is not more variable than female reproductive success.

There are also several possible explanations for the lack of a female sex bias in the offspring of high-ranking females. First, the local resource competition model demands that there be some relation between female rank and reproductive success. In Amboseli, however, high rates of predation appear to swamp the competitive advantage gained by high-ranking animals and may reduce the benefits of investment in daughters. Second, the reproductive suppression of low-ranking females may depend in part on the ways inter- and intragroup competition affect reproductive success. Facultative sex-ratio adjustment may be advantageous only under conditions when a group has grown large enough to compete successfully with its neighbors. Among Amboseli vervets, severe reduction in group size (for example, in group C) appears to diminish the group's ability to defend its territory against other groups and adversely affects females of all ranks. Thus, when mortality among all females is high, it may not be in the interests of high-ranking females to suppress the reproductive output of low-ranking females or to limit their recruitment of daughters.

Factors Affecting Population Growth Rates

The reduced fecundity of group A and the high predation rates in group C emphasize that the decline in the Amboseli vervet population is occurring for different causes and at different rates in each group. Group C, which has suffered both the highest predation and the greatest mortality of mature xanthophloea trees, has declined at a faster rate than group B, which has expanded its range and increased in size as its mature xanthophloea trees have died. Group B's female-biased sex ratio has also contributed to its relatively slow rate of decline. The decline of group A, which has a longer interbirth interval but suffers lower rates of predation than the other two groups, has been intermediate between those of groups B and C.

Since mortality in groups B and C is due primarily to predation, we can hypothesize that predation is the primary factor regulating growth rates in these groups. This may not be true, however, for group A, where food supply seems to influence female reproduction more strongly. In extreme years the failure of one plant species can result in high mortality. In other years group A females experience slower maturation rates and longer interbirth intervals than females in groups B and C.

Competition for food also mediates reproductive success, both within and between groups. During periods of food and water shortage, high-ranking animals survive at a higher rate than low-ranking animals, apparently because they are able to gain preferential access to such resources. Moreover, intergroup aggression may prevent groups without surface water from gaining access to water holes (Cheney 1981, 1983; Wrangham 1981; see also Dittus 1975, 1979). Food shortages may also contribute indirectly to the high predation rates experienced by Amboseli vervets, since the decline in mature xanthophloea trees appears to have forced animals to forage in areas where they are most vulnerable to predation. Thus, for example, with the death of most of its xanthophloea trees, group C has increased the frequency with which it feeds in the swamp, even though a high proportion of animals have died there.

In short, food supply and predation interact to affect reproductive success in Amboseli vervet monkeys. These factors may vary in relative importance not only over brief periods, but also over very small geographical distances. Unitary explanations that suggest an exclusive role for one of these factors may therefore greatly oversimplify this complex interaction.

24.4 Summary

Groups of vervet monkeys in Amboseli National Park, Kenya, inhabit territories that differ markedly in quality over small geographical distances. Food supply, in particular the availability of *Acacia xanthophloea* trees and permanent water holes, affects both fecundity and mortality. Groups without surface water show longer interbirth intervals and suffer higher mortality from illness during the dry season than do groups whose territories include water holes. Predation also strongly influences the vervet population: over a seven-year period, predation accounted for at least 69% of vervet mortality. Predation accounted for the majority of deaths in the three highest-ranking quartiles, while animals in the lowest-ranking quartile were equally likely to die of predation and illness. Infant survival and life span were the two most important factors influencing female reproductive success, but neither of these factors was correlated with dominance rank. Similarly, there were no rank-related differences in the sex ratio of infants. Among males, neither dominance rank nor tenure was correlated with copulatory success.

Although food supply apears to be the most important factor affecting survival in groups without surface water, predation appears to be a more important factor in groups whose ranges include water holes. Food supply, however, may also indirectly affect predation rates, since the recent decline in the number of feeding trees appears to have caused vervets to forage in areas where they are more vulnerable to predation. The relationship between food supply and predation is therefore complex, and both interact to affect vervet numbers.

24.5 Acknowledgments

We are grateful to the Office of the President and the Ministry of Tourism and Wildlife of the Republic of Kenya for permission to conduct long-term research in Amboseli National Park. We also thank J. Kioko and B. Oguya, the wardens of Amboseli during this research, for their cooperation and assistance. This work would not have been possible without the help of J. Else and M. Buteyo of the Institute of Primate Research in Nairobi. We are also grateful to D. Klein, for his background information on group A, and to S. Milgroom and B. Musyoka Nzuma, who helped collect data. S. Altmann, R. Dunbar, M. Hauser, B. Smuts, T. Struhsaker, and R. Wrangham made helpful suggestions on an earlier draft of this manuscript. We also thank T. Clutton-Brock and two anonymous reviewers for their suggestions. This chapter was prepared while Cheney and Seyfarth held fellowships at the Center for Advanced Study in the Behavioral Sciences, Stanford, California, and were supported by grants from the National Science Foundation (BNS 76-22943), the Alfred P. Sloan Foundation (82-2-10), and the Exxon Education Foundation. Finally, we wish to express particular thanks to the foundations that have supported our research: the Harry Frank Guggenheim Foundation, the Wenner-Gren Foundation, the L. S. B. Leakey Foundation, the National Institutes of Health (NS 19826), and the National Science Foundation (BNS 80-08946 and 82-15039).

25 Determinants of Reproductive Success in Savannah Baboons, *Papio cynocephalus*

Jeanne Altmann, Glenn Hausfater, and Stuart A. Altmann

BABOONS (FIG. 25.1) ARE AMONG the largest, most sexually dimorphic, and most terrestrial of monkeys. The presence in baboons of externally visible correlates of reproductive physiology (Gillman 1935; Gillman and Gilbert 1946; Kriewaldt and Hendrickx 1968; S. Altmann 1970) aids studies of their reproductive success, and their ground-dwelling habit in the African grasslands facilitates observation. Conversely, other characteristics of baboons—their slow maturation, litter size of one, long period of parental care, behavioral flexibility, and long adulthood that spans changing ecological and social conditions—all complicate attempts to measure and account for reproduc-

Figure 25.1: Savannah baboons, *Papio cynocephalus*, in Amboseli National Park,

tive success. However, based on over thirteen years of data from known individual savannah baboons, *Papio cynocephalus*, in Amboseli National Park, Kenya, we can now provide fairly good estimates for some reproductive and survival parameters and tentative estimates for others.

Despite the emphasis on male behavior and mating success in the first studies of this and other primate species, knowledge of males' life histories remains much less complete than knowledge of females'. Male baboons take several years longer to reach full maturity. They are the sex that disperses: they leave their natal group near maturity and sometimes change groups once or more during adulthood. One often cannot determine what has happened to a male that disappears from a group or determine his reproductive success in a new group even if one knows that he successfully transferred. For the same reason, one does not usually have much prior information on males that immigrate into study groups. In addition, information about paternity, particularly in noninterventive studies but also in studies that have included attempts at genetic analysis, remains much poorer than for maternity. Consequently, our analyses of female reproductive success are more complete than those of male success.

25.1 Background

The data on which this chapter is based come from almost daily records on all individuals of "Alto's Group," a group that varied in size from about thirty-five to sixty-five during the study period, averaging approximately forty-five animals including an average of fourteen adult females and nine adult males. Records on each day of observation include presence or absence of each individual, checks for wounds and other pathologies, the condition of each female's sexual and paracallosal skin and the presence or absence of menstruation, partners involved in the exclusive sexual consortships that are characteristic of this species (Rasmussen 1985), copulations, and ad libitum records of agonistic interactions to determine dominance relationships. (Details of the methodology can be found in Hausfater 1975; J. Altmann 1980; Altmann, Altmann, and Hausfater 1981; Hausfater, Altmann, and Altmann 1982.)

In baboons, because the time of ovulation can be well determined from changes in sexual skin swellings and because menstruation is externally visible, delayed menstruation, indicating a possible pregnancy, can readily be detected. In addition, pregnancy can be confirmed within a few weeks of a missed menstruation through reddening of the paracallosal skin (S. Altmann 1970); moreover, the skin loses its pinkness if the fetus is later miscarried. Consequently, we can identify fetal wastage as well as later offspring loss. Because pregnancy can be detected shortly after implantation and because few miscarriages have occurred thereafter, we can determine infant sex ratios not only at birth but near the time of conception.

Once an infant is born, its fate can be followed through age six, because group transfer is extremely rare before that age. Because dead infants are carried for several days by their mothers, the cause of death can sometimes be estimated and its occurrence confirmed for infants, though not often for juveniles.

Approximately one hundred conceptions occurred in Alto's Group during 1971–83 and have been used here for analysis. We do not have information on all variables of interest for all conceptions, so sample sizes vary from analysis to analysis. For example, age of mother is known or well estimated only for those mothers that reached maturity since 1971, whereas dominance rank of mother is known for all conceptions. Season of conception and birth is known for all, rainfall at that time for most. For conceptions starting with those in 1979, we have determined probable paternity based on mating and consortship records during the days of most likely conception; in addition to mating records, dominance rank and length of residency in the group was known for each male during this five-year period.

Reproductive success involves more than conception, of course. We know offspring survival through infancy for all conceptions, through age six (age of first reproduction for females and age of attaining the subadult stage for males) for a fraction of these.

The present report includes data on twenty-eight adult females and thirty-eight adult males. However, for most of the subjects the study period overlaps only a portion of their life span or even of their reproductive span. Similarly, only a fraction of the infants they produced during this period have had time to reach maturity. Thus our reconstruction of reproductive success necessarily involves both cross-sectional as well as longitudinal analyses. Wherever possible, we have done both types of analysis to verify that the results are consistent. We have also been able to replicate some of the findings during a shorter, more recent study of a second group (Hook's Group) in a slightly different habitat within the same park.

Unfortunately, because many of the life spans are incomplete, because the factors determining reproductive success in our animals are not independent (see below), and because we cannot yet evaluate most factors that determine life span, the results of our studies cannot yet be translated directly into quantitative estimates of contribution to variance in lifetime reproductive success. Moreover, because primate populations are age structured, with overlapping generations and overlap of reproduction, we have not been able to take advantage of analytical techniques developed by Arnold and Wade (1984a,b) for evaluating selection.

We begin with an investigation of factors affecting conception success first in females, then in males. Next we examine sources of variability in offspring survival. Finally, we consider determinants of the length of an individual's reproductive span.

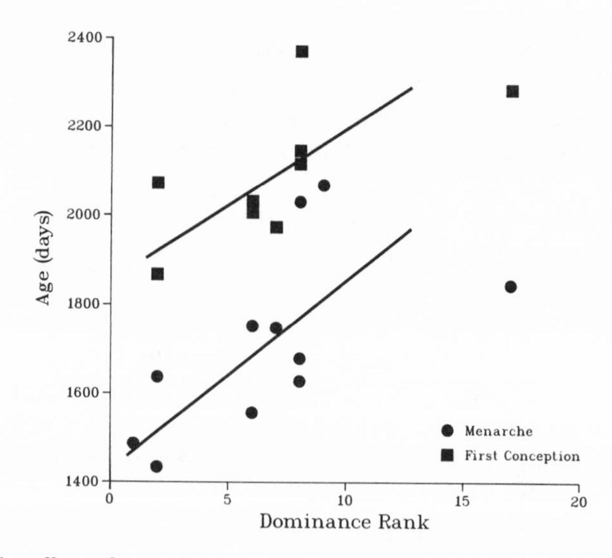

Figure 25.2: The effect of a mother's dominance rank when her daughter was conceived on the age at which the daughter reaches menarche ($y = 23.5x + 1525$; $r^2 = .37$, $p < .05$). The highest-ranking female is, by convention, indicated as having the lowest rank number, one.

25.2 Results

Conception Success of Females

Attainment of full reproductive capacity is a gradual process in female baboons, and its first externally visible stage is menarche. In captivity, a female reaches menarche at three to three and a half years of age and conceives her first infant within the next year (Snow 1967; Gilbert and Gillman 1960). In Amboseli, however, menarche occurs when females are approximately four and a half to five years old, and conception first occurs at age six (Altmann et al. 1977; Altmann, Altmann, and Hausfater 1981; also Nicolson 1982 for olive baboons). Daughters of high-ranking females attain reproductive maturity younger than do daughters of low-ranking females: they reach menarche about three hundred days earlier, and they conceive their first infant two hundred days earlier on the average (fig. 25.2). This provides a high-ranking female with an additional half-infant advantage—roughly a 7% adult life span advantage—relative to an average of seven offspring conceived (two surviving to adulthood) per female.

Once a female reaches age six, her fertility remains relatively constant over at least the next ten or twelve years. If a female lives to an old age, nonreproductive pathologies, such as extreme tooth wear and stomach tumors, as well as reproductive pathologies, particularly tumors, will increasingly interfere with reproduction (Lapin et al. 1979, based on captive animals). In the absence of pathologies, most nonhuman primate females, including baboons, continue to produce offspring until death (see summary in Graham, Kling, and Steiner 1979 for data on the existence of menopause in a few individual captive macaques and its absence in other nonhuman primates, and see Strum and Western 1982 for reproductive decline from unknown causes in several free-ranging olive baboons estimated to be very old). In Amboseli, the variance but not the mean in reproductive parameters, including interbirth interval, seems to increase after age fifteen, but the sample of known-age elderly females is limited, and no data are available on factors that might be affecting age changes in reproduction.

The probability of conception is strongly affected by the survival of the immediately preceding offspring. If an infant dies, its mother's postpartum amenorrhea terminates within a few weeks, and she conceives in only one or two cycles, whereas if the infant survives, postpartum amenorrhea lasts approximately twelve months, and three or four menstrual cycles usually occur before conception (Altmann, Altmann, and Hausfater 1978). Thus the absence of strong birth seasonality means that baboons have a greater ability than seasonally breeding primates (e.g., vervets, Cheney et al., this volume, chap. 24) to recoup losses from infant mortality. In addition, fewer females will simultaneously be in estrus or giving birth. As a result, female/female reproductive competition may be less than in primates that are highly seasonal and thus more synchronous in their reproduction.

Finally, ecological factors affect conception rates (J. Altmann 1980; Strum and Western 1982). In our population this occurs through effects on the probability of cycling. In Alto's Group, the proportion of females that are cycling in the group is higher during wet seasons than during dry ones, because then it is easier both for females to attain menarche and for postpartum females to resume menstrual cycling (unpublished data). Once females are cycling, however, we can detect no effect of season on the probability of conception.

Although overall conception rates are not related to dominance rank, infant sex ratios shortly after conception are correlated with maternal rank. The highest-ranking third of females produce three to four females to each male offspring, and the lowest-ranking third produce two males to each female (fig. 25.3). We suggest that differences in offspring sex ratios are an immediate effect of social stress on behavior or physiology, as through timing of conception with respect to ovulation or changes in vaginal pH, and that this stress is a function of the number of females that dominate the conceiving female. Offspring sex ratio is not a function of maternal age or parity. As we describe below, the initial offspring sex-ratio bias is further exaggerated by an interaction between maternal rank and sex-specific offspring survivorship.

Conception Success of Males

When females first conceive at age six, the males of their cohort are just reaching subadulthood. Although physiologicaly capable of inseminating a female, six-year-old males are about half the weight of fully adult males and rank below them in the dominance hierarchy. It takes several

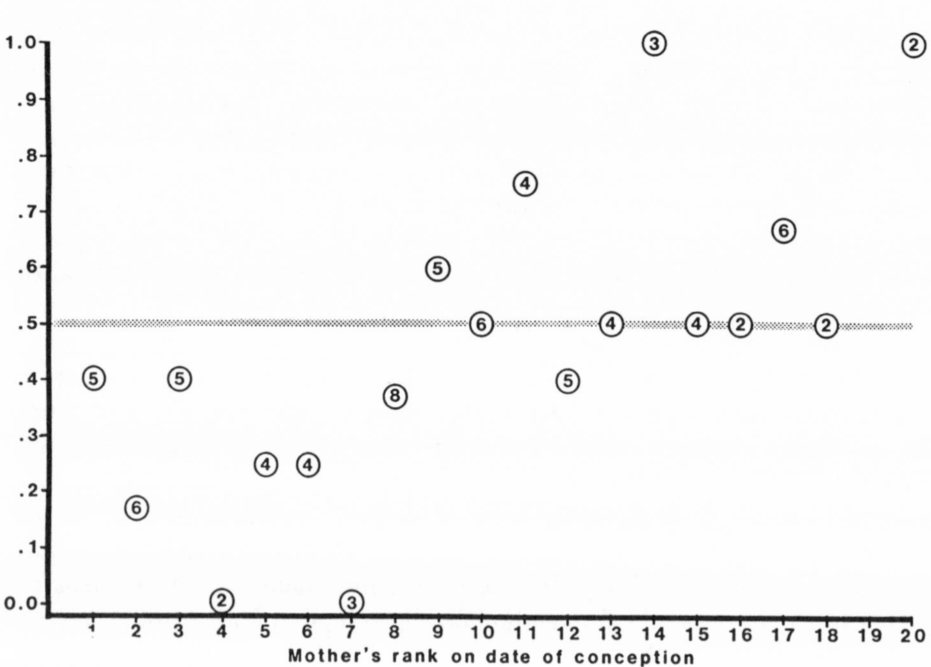

Figure 25.3: The effect of a female's dominance rank at the time she conceives on the probability that the offspring will be a male (July 1971–June 1981). The sample size of conceptions at each dominance rank is indicated within the circle marking the sex ratio for that rank. We determined a female's absolute dominance rank at the time of each conception; deaths, maturations, and rarely, rank changes among adults sometimes result in a female's having different rank numbers at the conception of successive offspring.

more years before these young males attain the size and social relationships that facilitate their fathering infants in the group.

Only rarely do matings during the days of likely conception occur outside the exclusive consortships maintained by adult females and males. Sometimes a single male is a female's only consort throughout these days, at other times consortships are maintained by two or more males successively within and between days (Hausfater 1975; Rasmussen 1985).

Juvenile males occasionally appear in the mating (but not consortship) records for females during the days of likely conception, but subadult males are virtually absent from both records: in this situation, juveniles but not subadults are tolerated by adult males (Hausfater 1975). Males reach subadulthood at approximately six years (Altmann, Altmann, and Hausfater 1981) and by seven and a half years of age they are large and strong enough to rise in the adult male dominance hierarchy, often precipitously, commonly becoming second or third ranking. During this age, usually after the rise in rank, the males often leave their group for a few days at a time, but of nineteen males who reached age six in Alto's Group, only one is known to have transferred groups successfully before age eight. One other subadult male died, two almost surely died, two others either emigrated or died, and the remaining thirteen stayed in the group through their eighth birthday. After rising in rank, these males sometimes mate with fertile females in their natal group. However, these matings are rarer than for fully mature males of the same rank. In sum, most males (68%) survive subadulthood in their natal group but are unlikely to produce offspring there during this period.

The situation changes dramatically after age eight. Males of this age are usually high in rank and well within the size range of fully adult males even though their growth is not yet complete. Of the thirteen males who reached both subadulthood and adulthood in Alto's Group, eight are known or thought to have emigrated from the group between ages eight and nine. Two others died or emigrated in their tenth year. At that same age, two males that emigrated a year earlier returned to their natal group. Finally, one male in his tenth year is still in Alto's Group.

At least four males who reached subadulthood in Alto's Group—known or presumed to be their natal group—have stayed or returned there during adulthood. If males do stay or return as adults, they become fully participating mating adults within the group. The small sample available thus far does not suggest any pattern of maternal rank, male rank, or presence of adult female relatives that distinguish the "leavers" from the "stayers" and "returners." Approximately 20% of males reaching subadulthood or almost 30% of those reaching adulthood in Alto's Group may have spent their entire reproductive career there, but this is apparently a less common pattern in other baboon populations (Packer 1979a).

The remainder of this section focuses on the reproductive success of adult males who leave their natal group and reproduce elsewhere and is based on an analysis of patterns of residency, rank occupancy, and consortship for males that immigrated into Alto's Group or were there at the start of the study.

We first consider the data on residency duration, using actuarial estimates for incomplete durations (Cutler and Ederer 1958). Of residency durations for males that did not mature in Alto's Group, 32% were ten months or less, and 3% lasted for eleven to twenty months, 8% each for twenty-one to thirty months and thirty-one to forty months, none for forty-one to fifty months, 14% for fifty-one to sixty months, and the remaining 35% for more than sixty months. Males who were recorded as having more than one period of residency in the group had two periods, usually consisting of a short period (less than ten months) and then a second residency period of several years. Overall, the median time from when a male was first associated with the group to the last date of association (ignoring gaps due to absences) was thirty-four months for males who immigrated into the group during the study, and males who were with the group at the start of the study stayed for a median of fifty-one months subsequently.

Within thirty days of entering the group, most (12/18) immigrant males had obtained the highest position in the dominance order of adult males that they would ever achieve, and virtually all males (17/18) had done so by the middle of their first year of residency in the group. Furthermore, the initial rank position of immigrant males proved to be a

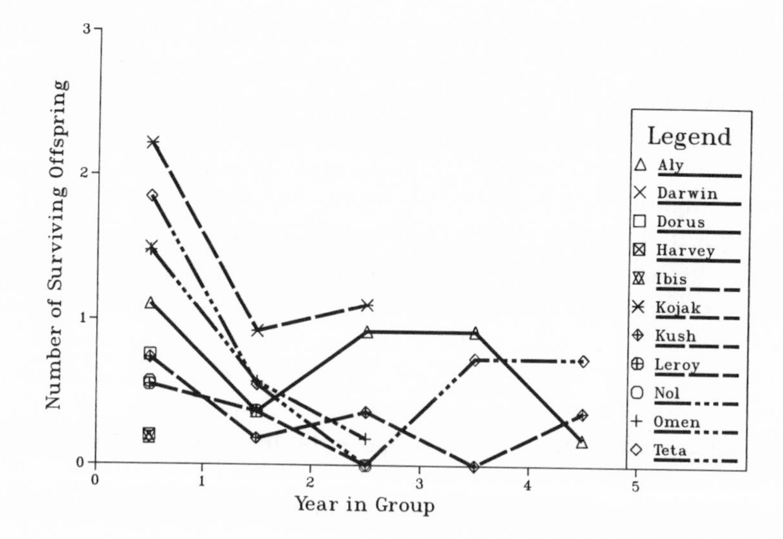

Figure 25.4: The yearly reproductive success of males as a function of the length of time they have been resident in a group. Each symbol represents a different adult male. For methods used to calculate number of surviving offspring for males see text.

good predictor of their subsequent duration of residency in the group: rank occupancy patterns and residency duration were not independent attributes of an adult male's life history. Specifically, immigrant males who obtained only an initial rank position in the lower half of the hierarchy among adult males remained with the group for a median of eighteen months (range 2–90+). In contrast, those males that initially obtained a rank position in the upper half of the adult male dominance order subsequently remained with the group for a median of sixty-seven months (range 44–89).

For each infant conceived during the five-year period 1979–83, we examined our daily monitoring records for its mother's matings and consortships on the days of likely conception. A male who appeared in those records for several matings or was recorded as a consort was considered a "likely father" of that infant. Almost all infants had one to three likely fathers, and overall the average was two likely fathers. Consequently, for the purpose of the present chapter, the number of infants a male was considered to have fathered was taken as half of those for which he was one of the likely fathers. Beyond that, the number of surviving offspring for males was taken as the number of infants fathered multiplied by the average rate of survival to age six.

A linear regression of cumulative reproductive success on duration of residency accounted for 56% of the variance in male reproductive success (fig. 25.4). Note that in this analysis the duration of residency for some males represents their complete stay in the group, while for others it is how long they had been in the group as of the end of 1983, the cutoff time for these data. Thus, at any duration shown in figure 25.4 the values include the cumulative reproductive success both of males who departed after staying that long and of those who had been there that long as of the end of 1983 but were still in the group. In particular, the first-year results include both low-success males who left after a brief residency and high-success males who entered in 1983 and stayed for several years thereafter. Long-association males were the likely fathers of more infants in their *first* year of residency in the group than was the case for short-association males.

Recently completed analyses of lifetime rank occupancy sequences for the adult males of Alto's Group have demonstrated that the average rank occupied by males declined steadily with each additional year they associated with the group. This slow decline in male rank over time may be the main factor responsible for the finding that the per annum reproductive success of males was substantially reduced in their second and later years with the group compared with their first year with the group (fig. 25.5) Nevertheless, males that stayed with the group even for very long periods were likely fathers of infants in almost all years of their residency. The total number of offspring attributable to each male varied considerably both across year of residency and when summed over all years with the group (figs. 25.4, 25.5). Year-to-year fluctuations in the number of conceptions in the group, duration of occupancy in the top few dominance ranks during the first years of residency, and development of mating-partner preferences by the females all probably contribute to this variance.

Females exercise mate choice by the extent to which they contribute to or thwart consort formation or maintenance. At one extreme, they frequently approach (follow) the male, present sexually, and stand for male mountings. At the other extreme, they move away at most approaches by the male, present rarely and only at male instigation, and walk out from under the male as he starts to mount. Consequently, the time and

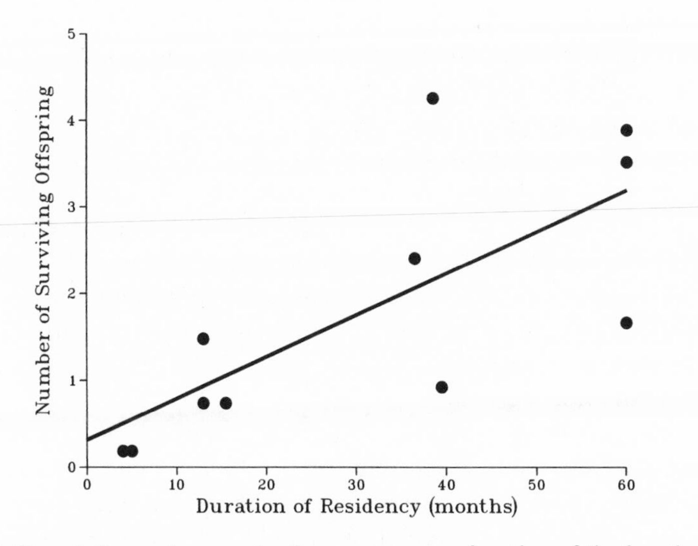

Figure 25.5: Cumulative male reproductive success as a function of the length of time the male has been resident in the group. See text for method used to calculate number of surviving offspring for males ($y = 0.05x + 0.26$; $r^2 = .56$, $p < .01$).

energy a male must expend per copulation and ultimately the number of copulations he achieves and his likelihood of fathering an infant will vary considerably as result of female behavior. Although quantitative data are not yet available, it appears that a low-ranking male will fight higher-ranking males for access to consort females primarily if these are females with whom the low-ranking male has mutually cooperative relationships. The nondominance basis for female choice remains to be determined (see also Rassmussen 1980; Smuts 1985; Bercovitch 1985).

Survival of a Female's Offspring

To attain high reproductive success, not only must individuals conceive offspring, but those conceptions must result in offspring that survive and reproduce. In most primates, few offspring are produced in a lifetime, usually one at a time. Consequently the fate of each greatly affects a parent's ultimate reproductive success.

In Alto's Group, 10% (n = 95) of pregnancies result in miscarriage or stillbirth. Mortality is approximately 25% in each of the first two years of life and declines in the next two years to virtually zero for the later juvenile period (J. Altmann 1980 and below). In this section and the subsequent one, we consider, first for female and then for male parents, what factors are related to variability in offspring mortality. We examine survivorship through the first two years, because infancy lasts between one and two years, and through age six, because that is the age of first conceptions in females and attainment of subadulthood in males.

The probability that an infant will survive is affected by maternal parity, but only in separating first pregnancies from later ones. The rate of miscarriage and stillbirth is the same for both parity groups, but live-born infants of primiparous females are half as likely (.29, n = 17) as later ones (.63, n = 70) to survive to two years of age. Survival success for second conceptions is indistinguishable from that for higher parities (.62, n = 16 vs. .63, n = 54). Variability in survival of first offspring is not accounted for by variability in age of the mothers at first conception.

Social dominance rank of mothers bears no simple relationship to offspring survival in Amboseli. Survivorship is .43 through two years of age and .35 through age six for the conceptions of the top-ranking females, .45 and .31 for those of the low-ranking ones. Because further analysis requires partitioning of our limited data sets, we will need more data and data on other groups to pursue complex interactions with confidence. However, several suggestive patterns, discussed below, emerge in the data now available.

Offspring survival is an interactive function of infant sex and maternal dominance rank: for both high- and low-ranking females, the offspring of the sex toward which birth ratios are biased have higher probabilities of survival. Considering the upper third (ranks 1–7) and lower third (ranks 14–20) of the dominance hierarchy, those in which sex ratio is biased, daughters of high-ranking females have a better chance of surviving to age six (.47, n = 15) than do sons of high-ranking females (.00, n = 5) or daughters of low-ranking females (.25, n = 4). Likewise, sons of low-ranking females are more likely to survive (.33, n = 9) than are daughters of low-ranking females (.25, n = 4) or sons of high-ranking females (.00, n = 5). The ranks that have the most divergent sex ratios, the highest third of the ranks, have the greatest survival advantage conferred on the preferred sex.

Sex bias in offspring mortality affects the offspring of low-ranking females at a different age than it does offspring of high-ranking females. The daughters of low-ranking females are not at greater mortality risk during infancy, perhaps as a result of their mothers' greater protectiveness (J. Altmann 1980), but they are more vulnerable as juveniles, as also suggested for macaques (Dittus 1979 for wild toque macaques; Silk et al. 1981 for captive bonnet macaques). In contrast, so far only one son of a high-ranking female has survived the first two years (n = 7).

Because the sample of "wrong sex" offspring is small, we have not yet been able to identify likely sources of the mortality differences, particularly in the case of sons of high-ranking females. However, available data reveal that "right-sex" offspring have gestations that are a few days longer, their mothers experience a postpartum amenorrhea that is one month longer, and the mothers' cycling time to next pregnancy is also one month longer, suggesting prolonged maternal care. Gestation length is positively correlated with probability of survival in Amboseli baboons (unpublished data) as it is in humans (Van Valen and Mellin 1967). The differences in postpartum amenorrhea and in the subsequent time spent cycling are of the order of 10% to 25%. These several differences do not yet reach statistical significance. However, each is consistent with the initial sex-ratio biases and with the survival differences. The complete set of results is consistent with the hypothesis that the observed offspring sex biases are adaptive.

Survival of a Male's Offspring

A male's reproductive success, like that of a female, will depend not only on achieving conceptions but also on any factors that lead to the survival of those offspring he does father. Males provide both direct and indirect care for infants (DeVore 1963; Ransom and Rowell 1972) and for young juveniles, primarily for those that are likely to be their own (see, e.g., J. Altmann 1980; Nicolson 1982; Stein 1984). Such care is sometimes dramatic in life-threatening situations, but it also occurs more subtly on a daily basis. Data are not yet available on the effectiveness of this care.

Males differ in the extent of care they provide. These differences have not yet been quantified, however, nor is it clear to what extent they represent life-stage differences or whether they result in lifetime differences in reproductive success. It appears that differences are partially related to life stage and, though not independently, to dominance rank: a male who is top ranking does not seem to provide as much care as those who are middle and low ranking.

The survival of a male's offspring will also depend on those maternal factors that affect successful gestations and infant and juvenile survival, so that mate selection by males may be a key component of their reproductive success. Several studies document selectivity among baboons (e.g., Hausfater 1975; Rasmussen 1985; Smuts 1985), but it is still unclear to what extent the selectivity results from female choice versus male choice. Moreover, much of the basis of choice remains to be elucidated, as noted above. These studies failed to find any indication that males are selective among females by dominance rank. Two studies report that adult males select multiparous females as consorts in preference to nulliparous ones (Rasmussen 1985; Takacs 1982). Unfortunately, these studies pooled data from females at all stages of adolescence; our (unpublished) data and those of Scott (1984) suggest that there is a major shift in adult male interest late in adolescence, when the females are likely to conceive. Judging by the poor survival chances for a female's first offspring, males should mate with nulliparous females only if it does not reduce their chances of fathering an infant by a female who has already had at least one infant or if doing so appreciably increases the likelihood that the young female will choose him as a mate in subsequent years.

Adult Reproductive Span and Survival

Age emerges as a major factor affecting cumulative female reproductive success, accounting for 79% of the variance in the linear regression (fig. 25.6). However, in a number of cases the data points are based on adult spans that are not yet complete. Consequently we cannot yet estimate the variance in life span or specify the proportion of lifetime reproductive success accounted for by life span. Because the expected survival of a newly adult female is approximately ten to twelve years, and because no known factors seem to have a very strong influence on adult span, we cannot yet predict individual differences in female reproductive success based on predictors of longevity.

Our data do indicate, however, that mortality rates are higher for females who are caring for young infants. Females spend approximately half their adult years in postpartum amenorrhea, caring for an infant (Altmann et al. 1977), and we would therefore expect that half the females who die would be in this reproductive state if probability of death is independent of reproductive condition. Three females who died were suspected of having pathologies that interfered with reproduction and were excluded from the analyses, although their inclusion would not have changed the results: one cycled continuously for the eight years before her death, the other two had extended periods of postpartum amenorrhea until death. Of the remaining thirteen females that died, eleven were in postpartum amenorrhea (chance $p = .01$) at the time of their deaths. The remaining two were pregnant. We conclude that there is a mortality cost of

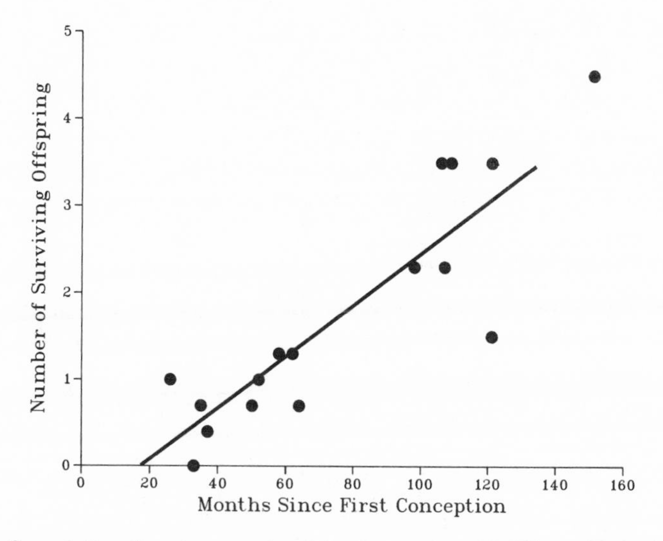

Figure 25.6: Cumulative female reproductive success as a function of months since first conception. The actual survival of each offspring was used for this graph. For any offspring that was alive and less than six years of age at the time the calculations were done, we used survival probabilities from that age through age six based on the longitudinal data for Alto's Group ($y = 0.03x - 0.59$; $r^2 = .79$, $p < .005$).

reproduction (J. Altmann 1980, 1983). Females that repeatedly lose young infants and therefore produce more infants in compensation for this loss may have a higher mortality rate, which could emerge as a source of individual differences in life span and consequently in reproductive success.

For males, we rarely know if we have observed their whole reproductive span, even if data are complete for their time in a single group. Some reproduction may occur in the natal group before a male leaves, and a male may live and reproduce in more than one nonnatal group before he dies (unpublished Amboseli data; also Packer 1979a; Strum 1982). We do not know whether males that are unsuccessful in one nonnatal group and stay only briefly subsequently make a successful transfer into another group. Moreover, there is considerable variability in total reproductive success, at least in the short term, among groups within the same population (Altmann, Hausfater, and Altmann 1985), so that attaining high relative reproductive success in different groups may have different absolute values. Thus, though the present data suggest that age, or at least length of residency in a group, is a major determinant of reproductive success for males as well as females, factors that we cannot yet evaluate may confound the relationship.

25.3 Summary and Discussion

Agonistic dominance among males has long been recognized as a feature of baboon society and as a factor affecting differential access to fertile females (see Packer 1979b and references therein). A quantitative study by Hausfater (1975) demonstrated that dominance accounted for approximately half of the short-term variance among males in the frequency of matings with females around the time of ovulation. It took longer to appreciate the importance of female dominance relationships, even though these are much more stable than those of males (Hausfater 1975), not only within generations but between them (Hausfater, Altmann, and Altmann 1982). However, a relationship between female dominance status and reproductive success remained uninvestigated (but see, e.g., Dunbar and Dunbar 1977 and Dunbar 1984 for geladas; and Drickamer 1974 and Gouzoules, Gouzoules, and Fedigan 1982 for macaques), partially as a result of the expectation from evolutionary theory that females in sexually dimorphic mammals with a somewhat polygynous mating system would have low reproductive variance and might choose among mates but not compete reproductively.

We have presented longitudinal data that demonstrate the effect of dominance on the reproductive success of female baboons, and also the limitations of that effect for both males and females. With respect to female reproductive success, daughters of high-ranking females attained reproductive maturity earlier than did daughters of low-ranking females. However, no dominance effect on conception rates could be demonstrated beyond first conception, nor could we discern any simple relationship between dominance and adult survival or offspring survival, the other main factors affecting lifetime reproductive success.

Dominance did affect sex-specific conception rates and survival of offspring. High-ranking females produced more daughters than sons. These daughters appeared to garner somewhat greater maternal investment, and they were more likely to survive infancy than were sons of these high-ranking mothers. In contrast, low-ranking females produced offspring sex ratios that were biased toward males. These males appeared to garner more maternal investment, and although the males did not have higher chances of surviving infancy, they were more likely than daughters of low-ranking mothers to survive the juvenile years and reach sexual maturity.

Questions regarding parental biasing of offspring sex ratios and subsequent investment have been of considerable interest in the evolutionary literature. Conditions that might favor the production of one sex over the other include environmental/nutritional stress, social stress, and the ability of the parent of each sex to affect its own reproductive success and that of each sex of offspring. Authors rarely treat these factors separately or consider that for any given species or population these factors may work in opposing directions rather than the same direction. Adequate information on these effects would result, for different species, in predictions of sex-ratio biases that differ in direction—or no bias—when the several factors are taken together. The absence of the necessary empirical evidence often results in a posteriori explanations of observed sex-ratio bias or of their absence.

Several recent studies of cercopithecine primates, including the two in this volume, differ in the relationships found between maternal dominance rank and offspring sex ratio (see J. Altmann 1980; Silk et al. 1981; Simpson and Simpson 1982; Paul and Thommen 1984; van Schaik and van Noordwijk 1983; Meilke, Tilford, and Vessey 1984; for other cercopithecine data). Whether these differences represent methodological differences, small-sample differences, or real differences remains to be seen. Likewise, in the absence of relevant measurements of those variables that are predicted to favor selection for parental influence on sex ratio, we cannot evaluate whether any or all of these sex ratios are those that would be predicted on the basis of adaptive arguments even if the results are "real." Finally, despite suggestive results on the relationship between timing of conception and offspring sex ratios in several mammalian species, neither the relevant normative physiological nor behavioral evidence is available for nonhuman primates.

Adult male immigrants who stayed at least a year in their new group tended to be high ranking and to produce the most infants during that first year. However, high rank is at least partially a temporary life-history stage for males rather than a stable trait, thus diluting over a lifetime the short-term reproductive advantages (Hausfater 1975; Saunders and Hausfater 1978). Elapsed residency in a group, inversely correlated with dominance rank, affected the estimated per annum number of offspring for males. Nonetheless, males continued to mate with fertile females, subsequent to rank loss, for the remaining years they were in the group, resulting in a strong correlation between length of residency and estimated reproductive success.

Instability of dominance for males, biased offspring sex ratios at conception and in offspring survival for females, yet-to-be-determined roles of mate choice in males and females, and undetermined contributions of fathers to offspring survival left adult survival as the apparent predominant factor affecting within-sex variance in reproductive success in our investigations thus far. It remains for future studies to determine the magnitude of these unmeasured effects as well as the extent of variance in adult survival and the identification of those factors that affect adult survival.

Studies of artificially fed macaques (e.g., Mori 1979; Sugiyama and Ohsawa 1982) demonstrate greater differential effects of female dominance on reproductive success under conditions of artificial feeding than otherwise. Because the artificial feeding resulted in a more concentrated food resource as well as a higher average nutritional plane, it is hard to disentangle these two factors (see J. Altmann 1986). However, it is clear that the proportion of variance in reproductive success attributable to various factors may vary with ecological and demographic conditions. This is further illustrated by the fact that infant survival is the major factor affecting conception rates in Amboseli (Altmann, Altmann, and Hausfater 1978 and unpublished data), where infant mortality is appreciable, whereas food availability has been identified as the major factor in olive baboons at

Gilgil, Kenya, where infant mortality is negligible (Strum and Western 1982). Any rank-related variability in infant survival or in food availability might have different effects on conception at these sites.

Male and female baboons appear to have alternative strategies for increasing their rates of conception and the survival of their offspring, thereby reducing the variance in reproductive success. Biasing sex ratios, adapting different styles of maternal care, choosing mates, and taking advantage of potential social resources all seem to be utilized by baboon females. Males differ in choice of group in which to mate, decision to emigrate, timing of one or more migrations, choice of and extent of competition for mates, and care of potential offspring. We have much yet to learn about the causes and consequences of these various behavioral and life-history patterns.

25.4 Acknowledgments

Many individuals and organizations have contributed to the success of the longitudinal project. We appreciate the hospitality and cooperation provided by the Institute of Primate Research, the Office of the President, the National Council of Science and Technology, the Zoology Department of the University of Nairobi, and the National Parks, all in Kenya. To M. Buteyo, J. Else, J. Kioko, B. Oguya, and D. Sindiyo we are particularly grateful. Financial support has been provided by the H. F. Guggenheim Foundation, the National Science Foundation (BNS 78-09178), and the United States Public Health Service (MH 19617, HD15007). For contributions to the monitoring data set we thank J. Friedman, R. Jansma, R. Mututua, R. Noë, B. Noë, M. Pereira, F. Saigilo, A. Samuels, C. Saunders, J. Scott, J. Shopland, J. Silk, S. Sloane, P. Stacey, D. Stein, J. Stelzner, D. Takacs, and J. Walters. Assistance with data analysis, graphic plots, or manuscript preparation was provided by S. Alberts, M. Bell, S. Galhotra, C. Johnson, and D. Sherman.

26 Reproductive Success in Three Kipsigis Cohorts

Monique Borgerhoff Mulder

THE STUDY OF HUMAN demography has shown that populations vary greatly in their rates of fertility and mortality (Wrigley 1966), but there have been few systematic attempts to describe variation in the components of reproductive success between individuals within populations (Driver 1981). This chapter analyzes the contributions of life span, fertility, and offspring survival to individual variation in lifetime reproductive success within males and females for a subpopulation of Kipsigis agropastoralists and examines factors determining reproductive variability and its major components.

26.1 Ethnographic Background to the Study

The Kipsigis are a Nilohamitic people of Kericho District, southwestern Kenya (fig. 26.1). Originally migrants from northeastern Africa, they are thought to have moved to this area in the late eighteenth century, ousting Masai, Sirikwa, and Kisii tribes (Orchardson 1961). Until the 1930 they

Figure 26.1: Six Kipsigis brothers.

were largely pastoralists, but after Europeans settled in the area the Kipsigis adopted maize as a subsistence crop, a development that led to the emergence of individual landholdings (Manners 1967). Like cattle, land is passed from father to son. Surplus maize is sold for cash by richer individuals, who buy livestock (and occasionally land), meet emergency medical expenses, invest in the education of their children, and start small businesses (for example, cattle trading or carpentry). Few individuals have been employed outside the area.

All Kipsigis men belong to an *ipinda* (age-set), delineated by the year of their circumcision (Prins 1953). This chapter focuses on the reproductive careers of three Kipsigis cohorts: Nyongi, men who were circumcised before 1921, Maina men (1922–30), and Chuma men (1931–38). Women do not have age-sets but are assigned to parallel cohorts (see Methods).

26.2 Reproductive Life Histories

Marriage is arranged between the parents of the prospective spouses and requires the payment of bridewealth (approximately six cows and four goats in the period under analysis; see Borgerhoff Mulder 1988a). Bridewealth constitutes a major expenditure (approximately one-third of an average man's herd), and a man's first marriage is financed by his father. Some men reaching adulthood fail to marry, usually because of poverty.

Through marriage a man acquires exclusive rights to the sexual and domestic services of his wife for the rest of his life. Adultery is extremely rare and recognized divorce almost unknown (Peristiany 1939), although maltreated wives may desert. In most other cases husbands are, by all accounts, rather certain of the paternity of their wives' children (see Borgerhoff Mulder 1987c).

Men in this sample marry between eighteen and forty-six years of age (mean 26.4). After approximately ten years of marriage a Kipsigis man commonly takes a second wife, occasionally at the instigation of his first wife (Borgerhoff Mulder 1985). Of the Nyongi, Maina, and Chuma cohorts, 31% have one wife, 34% have two wives, and 35% have three wives; one man has twelve wives. Some men continue to marry into their seventies and reportedly father offspring, although a very old polygynyist will allow sons by previous wives rights to his younger childbearing wives.

Accurate measures of mortality and life span of the full cohort from which the sampled individuals are drawn cannot be determined. The high mean life span of 68.4 years for all recorded deceased and living married members of the oldest cohort of men probably results from undersampling males who died in their early married years rather than from age exaggeration (Kenya Fertility Survey 1980), because most men know their age at circumcision and their circumcision ceremonies could be accurately dated to the year (see below).

All girls undergo clitoridectomy at the onset of menarche, after which they spend one to three years in seclusion huts under instruction (Peristiany 1939) and, by custom, being fattened. On leaving these huts, practically all women are married within a few months. The average age at circumcision is 15.4 years, ranging from 12 to 19 years, and the average age of marriage is 16.4, ranging from 12 to 20 years in this sample. Illegitimate motherhood is rare, and before missionary intervention, infants born to uncircumcised girls were throttled at birth. After marriage a wife moves to her husband's family's land and becomes entirely dependent on his herd; she has no property rights of her own and shares her husband's resources with her co-wives on a strictly equal basis (Peristiany 1939; Borgerhoff Mulder 1987e).

Women continue to perform agricultural, domestic, and husbandry work throughout pregnancy and give birth in their own huts. For the cohorts under consideration infants were breastfed for two to three years, with millet porridge introduced at seven months; no commercial milk supplements were used. By eighteen months the infant is cared for almost exclusively by sibling caretakers (Borgerhoff Mulder and Milton 1985). No individual reported using traditional or modern methods of birth control, and there is a low incidence of venereal disease, as reported to local clinics.

Among all reported live births, 24% fail to reach adulthood. Mortality between birth and eighteen months (infant mortality) is 12% and is generally due to congenital defects and perinatal factors in the first few weeks and to hazards associated with weaning (Mundot Bernard 1975; Conde and Boute 1971). Mortality between nineteen and sixty months (childhood) is 7%, probably due to poor nutrition and disease (Grounds 1964; Cantrelle 1971). Mortality between five years and marriage (immature) is 5%, caused mainly by disease (notably malaria) and infections resulting from circumcision. These figures are similar to those for parallel cohorts reported in the Kenya Fertility Survey (1980).

26.3 Methods and Samples

The Study Site

The study site of 35 square km incorporates a former ''Reserve'' area (Kipkelat Location) and adjacent land in Narok District where Kipsigis live as squatters on Masai land. The area has been little influenced by the modernization experienced in many parts of rural Kenya. Two missionary hospitals were established 30 km away in the 1950s but are not heavily used, and more recently primary health care has become available locally.

Data Collection

All households in the study area were censused between June 1982 and December 1983. Questioning each individual in Kipsigis, I determined, wherever possible, his (her) dates of birth, circumcision, and marriage, numbers of live births, and surviving offspring. Discrepancies between reports of husbands and wives were subsequently checked with both parties, and life-history details on decreased or absent spouses were recorded. Reproductive histories of polygynously married wives not resident in the study area were taken from their husbands and verified with co-wives or relatives.

Eighteen months of residence in the area, use of the Kipsigis language, thorough checking of interview responses, and the readiness of Kipsigis to talk about their own family life and that of neighbors make it highly unlikely that deceit could have been consistently maintained in reporting life-history events. Furthermore, results have subsequently been replicated (Borgerhoff Mulder 1987c), restricting analyses to an intensively studied sample of one hundred families (see Borgerhoff Mulder and Caro 1985). Finally, past life-history events could be accurately dated to the year by cross-referencing with male circumcision ceremonies, severe droughts, and other events of known date.

Age, Life Span, and Married Life Span

Most men know the year of their circumcision. If they also know their age at circumcision, life span (L) could be calculated directly; if not, it was estimated by assuming a man was circumcised at the median age for that cohort. Proportion of life span spent married (M) was calculated as the years elapsed since a man's first marriage divided by life span.

It was not possible to determine absolute life span for most women, and estimates of reproductive life span were used. A woman's reproductive lifespan (L_r) begins at marriage and is assumed to have ended after thirty years have elapsed, at the approximate age of forty six, which falls centrally in the range of menopausal age recorded in developing countries (Bongaarts 1980). Subsequent use of a finer measure of reproductive life span among menopausal women, age at last live birth minus age at marriage (see Borgerhoff Mulder 1987b), does not affect the results based on *estimated* reproductive life spans reported here.

A woman was classed as monogamous or polygynous according to the number of wives ever married by her husband rather than the number of currently surviving co-wives. To measure the degree of co-wife competition (Borgerhoff Mulder and Caro 1983), I calculated the mean number of co-wives a wife experienced during her reproductive years.

Fertility

A man's fertility depends on the number of his wives and the number of children they produce per year. For each man the number of potentially reproductive wife-years was calculated by multiplying the number of wives by their reproductive life span. The mean number of reproductive wives per married year (W) was calculated by dividing total reproductive wife-years by the husband's married life span, thus providing a measure of polygyny. Fertility (F) was calculated as the number of live births per reproductive wife-year. In some cases the only data available were the number of surviving offspring per reproductive wife-year (O), calculated as the number of offspring surviving to twenty-one years (or, if born after 1963, to the present) divided by the total number of the father's reproductive wife-years.

For women fertility (F) was measured as the number of live births divided by reproductive life span (L_r). Where information on offspring mortality was not available, surviving offspring per year (O) was measured as the number of surviving offspring divided by L_r.

Offspring Survival

Offspring survival (S) was calculated as the proportion of live births surviving to twenty-one years, or to the present if born after 1963. Where the age of the deceased offspring was known, the probabilities of surviving through infancy (S_1), childhood (S_2) and immaturity (S_3) were calculated.

Lifetime Reproductive Success

For both sexes lifetime reproductive success (LRS) was measured as the number of surviving offspring produced over the life span. For males, this is a product of life span (L), proportion of the life span spent married (M), mean number of reproductive wives per married year (W), and number of surviving offspring per wife-year (O). For females, it is the product of reproductive life span (L_r), live births per reproductive year (F), and proportion of livebirths surviving (S).

Sample Selection

Data for three age-sets (*ipinda*) were analyzed separately. The complete male sample contained thirty-four Nyongi, fifty-two Maina, and forty-four Chuma men, and from this sample comparative indexes of selection (Wade and Arnold 1980) were calculated. For all other analyses, the sample was restricted to twenty-nine Nyongi, forty Maina, and thirty-eight Chuma men for whom there were sufficient details for analyzing the components of LRS. Of the Nyongi cohort twenty-four out of twenty-nine individuals were deceased by 1983, and all were reproductively dead; deceased (group 1) and living (group 2) members of the Maina cohort were analyzed separately; and analyses for the Chuma men were restricted to living individuals. Among Maina and Chuma men approximately 30% still had currently childbearing wives. Truncating Maina and Chuma samples only slightly underestimates terminal LRS, because reproductive rates drop off in both samples after forty-five years of marriage, and relatively few men take subsequent wives after this period (Borgerhoff Mulder 1987d).

Women were allocated to the male cohort whose median date of first marriage fell closest to their own date of marriage, resulting in three female "cohorts": those married between 1918 and 1929 are termed "Nyongi," 1930–39 "Maina," and 1940–53 "Chuma" (n = 61, 73, and 218, respectively). The last cohort included some wives of the subsequent male cohort. Offspring survival was not available for all individuals in these cohorts.

Mortality before marriage was the major cause of breeding failure. Because only very rough estimates (30%–40%; Peristiany 1939) were available for such mortality during the early colonial period, analysis was restricted to breeding individuals.

Socioeconomic Measures

The number of cows and number of acres held by a man in 1982–83 were used to measure his wealth. These measures provided a coarse but robust index of relative wealth differences between men throughout their reproductive years (see Borgerhoff Mulder 1987c). The oldest (Nyongi) men had already given out their wealth to their sons, and their wealth status could not be determined. Women do not own wealth, but an estimate of resources for wives (wealth per wife) was attained by dividing the husband's assets (cows and acres) by the number of women he married.

26.4 Results

Breeding Success and Reproductive Life Span

Age-related changes in breeding success differed between the sexes; the numbers of currently surviving offspring born per year to men and women in the Nyongi cohort by five-year intervals are shown in figure

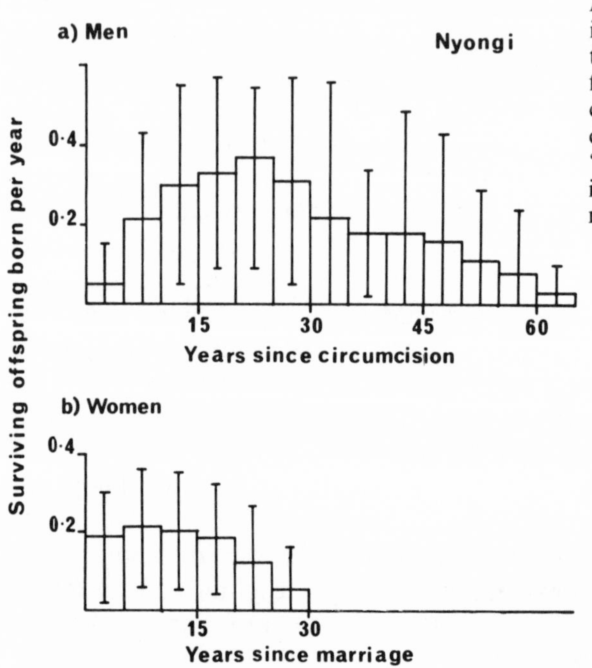

Figure 26.2: Numbers of surviving offspring per year for (*a*) twenty-nine Nyongi men by five-year intervals since circumcision (median circumcision date 1914), and (*b*) sixty-one "Nyongi" women by five-year intervals since marriage (married before 1930).

26.2. In men, breeding success peaked between twenty and thirty years after circumcision, and in women within the first ten years of marriage. Whereas no woman gave birth to a surviving offspring after thirty years of marriage four Nyongi men fathered surviving offspring up to sixty years after their circumcision, and two individuals did so up to sixty-five years after circumcision. Differences between the sexes in the length of reproductive life span reflect the polygynous marriage system (Borgerhoff Mulder 1987d).

Variation in Lifetime Reproductive Success and Its Components

Men

Mean lifetime reproductive success varied between 11.1 (Nyongi and Chuma cohorts) and 14.62 (live and dead Maina samples combined), with a few very successful individuals having over twenty-five (and in one case eighty) surviving children (tables 26.1 and 26.2). Standardized variance ($\sigma^2/\bar{x}^2$) in LRS for the complete sample (not shown in table 26.1) ranged from .32 (Chuma) to .79 (Maina live and dead combined). Dividing these values by the standardized variances for the full sample of females gave indexes of the comparative intensity of selection (I_m/I_f) of 1.61 (Nyongi), 2.63 (Maina), and 1.88 (Chuma) (Wade and Arnold 1980).

Distribution of LRS and its main components, number of reproductive wives per married year (W), number of surviving offspring per wife-year (O), and proportion of live births surviving (S), are shown for the full Maina cohort in figure 26.3. Values for Nyongi and Chuma were very similar (see table 26.1). Within cohorts the mean number of reproductive wives per married year varied from less than one to more than six. The proportion of life span spent married (M) and life span (L) varied little

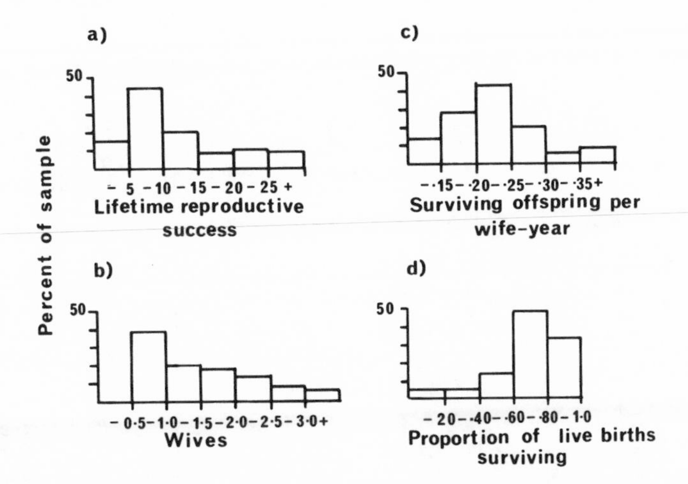

Figure 26.3: Distribution of the variation in (*a*) lifetime reproductive success (LRS) among forty Maina men and its main components: (*b*) mean number of reproductive wives per married year (W); (*c*) number of surviving offspring per reproductive wife-year (O); and (*d*) the proportion of live births surviving (*S*) (n = 18). Results were similar for all cohorts.

Table 26.1 Mean and Variance of the Components of Lifetime Reproductive Success in Three Cohorts of Kipsigis Men

	Nyongi (Deceased and Living)			Maina						Chuma (Living)		
				Deceased (1)			Living (2)					
	Original		Standardized	Original		Standardized	Original		Standardized	Original		Standardized
Component	*Mean*	*Variance*	*Variance*	*Mean*	*Variance*	*Variance*	*Mean*	*Variance*	*Variance*	*Mean*	*Variance*	*Variance*
L	73.97	226.68	0.04	58.33	112.95	0.03	74.24	14.36	0.00	64.37	11.00	0.00
M	0.59	0.02	0.05	0.50	0.01	0.05	0.62	0.01	0.02	0.60	0.01	0.03
W	1.42	0.47	0.23	1.66	0.48	0.17	1.59	1.80	0.71	1.14	0.25	0.20
O	0.17	0.01	0.18	0.21	0.01	0.13	0.23	0.00	0.03	0.26	0.01	0.10
LRS	11.07	54.03	0.51	9.73	24.63	0.23	17.56	286.85	0.97	11.13	35.84	0.28

Note: Calculated using the method developed by David Brown (this volume, chap. 27) for twenty-nine Nyongi men, fifteen deceased (1) and twenty-five living (2) Maina men, and thirty-eight living Chuma men.

L = life span; M = proportion of life span married; W = reproductive wives per married year; O = surviving offspring per wife-year.

Table 26.2 Percentage Contribution of the Components of Lifetime Reproductive Success to Variation in LRS in Three Cohorts of Kipsigis Men

	Nyongi (Deceased and Living)				Maina								Chuma (Living)			
					Deceased (1)				Living (2)							
Component	*L*	*M*	*W*	*O*	*L*	*M*	*W*	*O*	*L*	*M*	*W*	*O*	*L*	*M*	*W*	*O*
L	8.15				14.31				0.27				0.96			
M	4.28	9.33			21.93	20.60			−0.01	2.48			0.23	9.49		
W	−7.35	−4.21	46.07		−13.17	−7.82	75.09		5.06	27.30	72.86		2.91	10.31	70.98	
O	−1.41	−3.54	−5.03	35.72	−26.12	−31.79	−50.46	56.90	−0.44	−4.32	−10.37	7.45	−2.96	−1.30	−25.90	34.84
Three- and four-way interactions	17.50				49.53*				−0.29				0.44			

Note: Calculated using the method developed by David Brown (this volume, chap. 27) for twenty-nine Nyongi men, fifteen deceased (1) and twenty-five living (2) Maina men, and thirty-eight living Chuma men.

L = life span; M = proportion of life span married; W = reproductive wives per married year; O = surviving offspring per wife-year.

*High negative two-way and positive three- and four-way interactions may result from the small sample size.

between individuals: most individuals were married for between 50% and 70% of their life span, and life span varied largely as a function of sample partition and sample truncation. The components of reproductive success in men were not highly intercorrelated (shown for the entirely reproductively dead Nyongi cohort, table 26.3), with the exception of life span and the proportion of life span spent married, a statistical effect that merely reflected the fact that men who die early spend a lower proportion of their lives married.

The mean number of reproductive wives married to a man during his married life contributed more to variance in LRS than any other component, although the number of surviving offspring born to a man's wives was also important (see table 26.2). Life span, the proportion of the life span spent married, and the proportion of offspring surviving to adulthood contributed little to reproductive variance among men in any sample.

Women

Mean LRS for complete "Nyongi," "Maina," and "Chuma" cohorts is 4.6, 4.8, and 6.8, with standardized variances ($\sigma^2/\bar{x}^2$) of .28, .30, and .17, respectively (see table 26.4 for a subset of the data). Distribution of the

Table 26.3 Correlation between the Components of Lifetime Reproductive Success for Twenty-nine Nyongi Men

Component	L	M	W
M	.95***		
W	−.08	−.01	
O	.22	.34	.03

Note: Pearson correlation coefficients. Results were similar for all cohorts.
L = life span; M = proportion of life span married; W = reproductive wives per married year; O = surviving offspring per wife-year.
***$p < .001$.

Table 26.4 Mean and Variance of the Components of Lifetime Reproductive Success in Three Cohorts of Kipsigis Women

	"Nyongi"			"Maina"			"Chuma"		
	Original		Standardized	Original		Standardized	Original		Standardized
Component	*Mean*	*Variance*	Variance	*Mean*	*Variance*	Variance	*Mean*	*Variance*	Variance
L_r	29.52	6.26	0.01	25.02	15.69	0.02	28.90	21.53	0.03
F	0.26	0.01	0.10	0.28	0.01	0.10	0.31	0.01	0.12
S	0.69	0.04	0.08	0.67	0.04	0.10	0.79	0.03	0.05
LRS[a]	5.67	3.62	0.11	5.51	5.38	0.18	6.96	6.76	0.13

Note: Calculated using the method developed by David Brown (this volume, chap. 27) for 27 "Nyongi" women, 49 "Maina" women, and 184 "Chuma" women.
L_r = reproductive life span; F = fertility per reproductive year; S = offspring survival.
[a]Because information on offspring survival is more usually available for women alive in 1983, and because only individuals with one or more surviving offspring were used, these samples are biased toward successful women; compare with mean LRS for the complete cohorts cited in text.

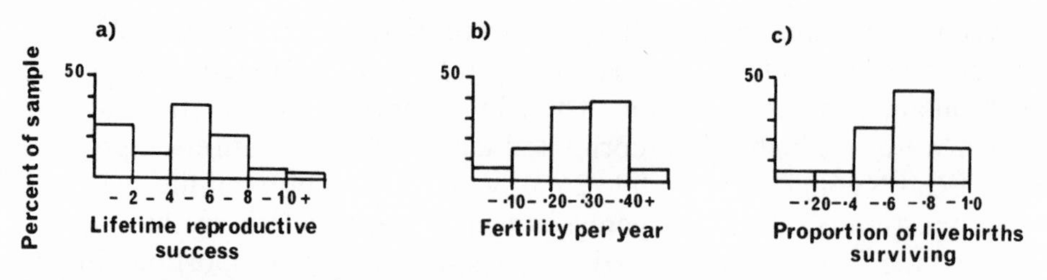

Figure 26.4: Distribution of the variation in (*a*) lifetime reproductive success (LRS) among seventy-three "Maina" women and its main components: (*b*) fertility per reproductive year (F) (n = 49) and (*c*) proportion of live births surviving (S) (n = 49). Results were similar for all cohorts.

variance in LRS, live births per reproductive year (F), and proportion of live births surviving (S) are shown in figure 26.4 (as with males only for the "Maina" cohort). Intercorrelations between all the components of LRS are shown only for "Chuma" women in table 26.5 and were similar in all three cohorts: live births per reproductive year were negatively correlated with infant survival (S_1) in all cohorts, reflecting the shorter interbirth intervals following infant deaths. In addition, a consistent negative relation was found between reproductive life span and surviving offspring per year, probably mainly due to the decrease in breeding success with age. Despite this negative relation between L_r and O, there was no evidence that women with higher breeding success were more likely to die during their reproductive years: age-specific breeding success did not differ between women who died within thirty years of marriage and those who survived. Reliable information on maternal age at infant mortality was not available for these cohorts, so the reasons for the positive correlation between reproductive life span and infant survival could not be determined from this sample.

Partitioning variance in LRS into its components showed that both fertility per reproductive year and the proportion of live births surviving made substantially higher contributions to variance in female LRS in all cohorts than did reproductive life span (table 26.6). Further analyses on a smaller sample for which age-specific offspring survival was available showed that differential offspring survival during the first eighteen months (S_1) accounted for more than twice as much variance as did survival through childhood (S_2) and almost three times as much as survival through the immature period (S_3). The percentage of variance explained by differential survival during infancy and childhood decreased in later cohorts, but the contribution of immature survival remained relatively constant over time.

Determinants of Reproductive Success and Its Components in Men

In all cohorts LRS was significantly correlated with the size of a man's plot and the size of his herd (table 26.7). This result was independent of male education and employment (Borgerhoff Mulder 1987c). The positive effects of wealth on lifetime reproductive success in men were due to the strong association between plot size and number of wives in all three samples (Maina (1): $r = .70$, $n = 11$, $p = .016$, Maina (2): $r = .91$, $n = 25$, $p < .001$, Chuma: $r = 49$, $n = 37$, $p = .004$), plotted for the Maina cohort

Table 26.5 Correlation between the Components of Lifetime Reproductive Success for 160 "Chuma" Women

Component	L_r	F	S_1	S_2
F	−.49***			
S_1	.43***	−.20**		
S_2	−.09	.01	−.07	
S_3	−.05	.19*	−.19*	−.11

Note: Pearson correlation coefficients. The sample is a subset of the sample used in tables 26.4 and 26.6 for which age-specific offspring survival was available. Results were similar for all cohorts.

L_r = reproductive life span; F = live births per reproductive year; S_1 = proportion of live births surviving infancy; S_2 = proportion of live births surviving childhood; S_3 = proportion of live births surviving immature period.

*$p < .05$; **$p < .01$; ***$p < .001$.

Table 26.6 Percentage Contribution of the Components of Lifetime Reproductive Success to Variation in LRS in Three Cohorts of Kipsigis Women

	"Nyongi"			"Maina"			"Chuma"		
Component	L_r	*F*	*S*	L_r	*F*	*S*	L_r	*F*	*S*
L_r[a]	6.76			10.44			19.19		
F	−45.39	97.36		−1.20	56.97		−34.02	86.39	
S	5.97	−79.78	78.02	−1.26	−17.29	54.37	1.22	−2.12	39.86
LRS		37.06			−2.03			−10.52	

Note: Calculated using the method developed by David Brown (this volume, chap. 27) for 27 "Nyongi" women, 49 "Maina" women, and 184 "Chuma" women.

L_r = reproductive life span; F = fertility per reproductive year; S = offspring survival.

[a]Variance due to L_r is underestimated in these samples because information on offspring survival is more usually available for women alive in 1983. Percentages of variance due to L_r in the complete samples are 16.28, 22.15, and 16.75, respectively.

Table 26.7 Correlation between Wealth and the Major Components of Reproductive Success in Kipsigis Men

	Maina (1)	Maina (2)		Chuma	
Component	*Acres*	*Acres*	*Cows*	*Acres*	*Cows*
LRS	.72***	.92***	.86***	.42**	.50**
W	.70*	.91***	.84***	.49**	.62***
O	−.04	.05	.02	−.14	−.17
S		.36	.37	.05	.08
n	11	25	25	37	34

Note: Pearson correlation coefficients.

LRS = lifetime reproductive success; W = reproductive wives per married year; O = surviving offspring per wife-year; S = proportion of live births surviving.

*$p < .05$; **$p < .01$; ***$p < .001$.

(fig. 26.5). Although exceptionally wealthy men were generally highly polygynous, the strength of the correlation coefficients of these results was not affected when the six wealthiest men with more than seventy cows or one-hundred acres were excluded. In addition, wealthier men reported a lower age at marriage in the Chuma cohort ($r = -.49$, $n = 17$, $p = .043$). There was no effect of wealth on the number of live births per reproductive wife-year, offspring survival, life span, or proportion of life span spent married.

Wealth may also affect reproductive chances of offspring: the number of acres owned by a man was positively correlated with the proportion of his sons married polygynously ($r_s = .45$, $n = 25$, $p = .026$) in the Maina (2) cohort.

Determinants of Reproductive Success and Its Components in Women

Wife's Wealth

Lifetime reproductive success in the "Chuma" cohort was positively correlated with the number of acres a husband owned divided by the number of his wives (acres per wife) ($r = .22$, $n = 182$, $p = .003$), although the percentage of variance explained was small. This relationship probably results from a positive correlation between acres per wife and offspring survival ($r = .19$, $n = 175$, $p = .014$). In the "Maina" cohort acres per wife had positive effects only on infant survival ($r = .43$, $n = 24$, $p = .037$). In

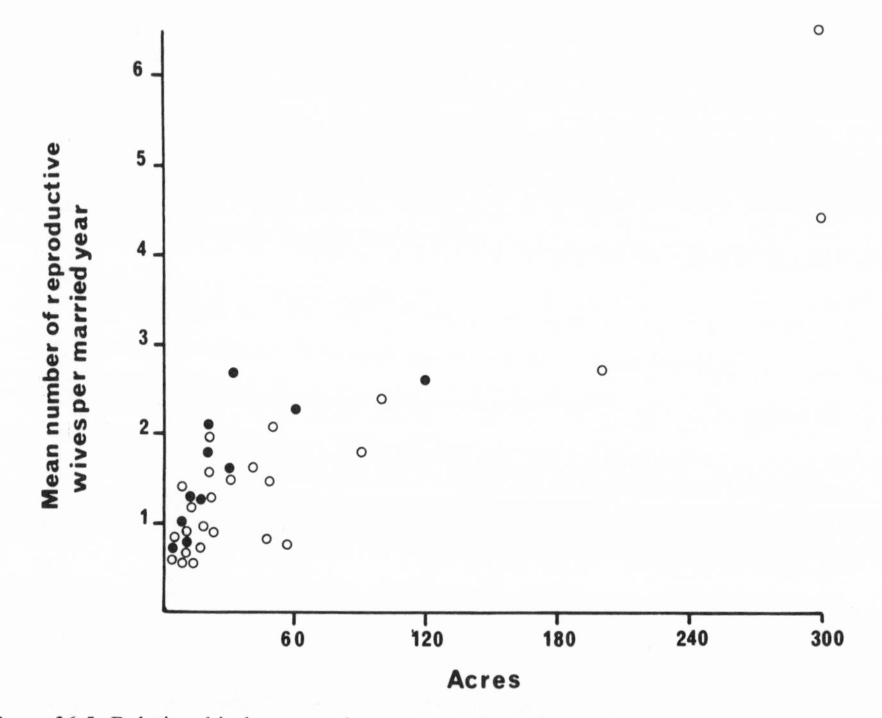

Figure 26.5: Relationship between the mean number of reproductive wives per married year (W) and plot size for Maina men (solid circles are deceased individuals).

no cohort was wealth per wife significantly related to fertility (but see Burgerhoff Mulder 1987e for recent changes). Husband's employment and education had no marked effects on any component of female reproductive success.

Marital Status

Polygynously married women were no poorer in terms of cows or acres per wife than were monogamously married women, yet the survival of offspring of polygynously married women was .16 lower in the "Maina" cohort ($t_{39,9} = -3.00$, $p = .007$) and .07 lower in the "Chuma" cohort ($t_{136,43} = -2.54$, $p = .012$). In the "Maina" cohort the negative effects of polygyny on offspring survivorship were attributable to the older age of polygynous husbands (see below), but this could not account for the differences observed in the "Chuma" cohort. The mean number of co-wives a woman experienced during her reproductive years had no effects on reproductive success or offspring survival.

Husband's Age Relative to His Wife

Women married to men of an older cohort than themselves had markedly lower LRS (1.5–2 offspring) than did women married to men of the same cohort in the "Nyongi" and "Maina" cohorts ($t_{11,48} = -2.39$, $p = .020$ and $t_{15,54} = -2.23$, $p = .029$); the difference was not significant in the "Chuma" cohort. Husband's relative age was negatively related to offspring survivorship in all three cohorts: "Nyongi" women married to Kablelach men (the preceding age set) experienced .61 lower offspring survivorship than those married to Nyongi men, but the sample was small ($t_{2,25} = -3.68$, $p < .001$); "Maina" women married to Nyongi men experienced .32 lower offspring survival than those married to Maina husbands ($t_{12,36} = -2.96$, $p = .010$), and "Chuma" women married to Maina men experienced .09 lower offspring survival than those married to Chuma men ($t_{48,131} = 2.34$, $p = .021$). In each cohort these effects were independent of marital status.

Bride's Age

Age of bride at marriage was available only for a subset of the "Chuma" cohort and was negatively related to her LRS ($r = -.36$, $n = 39$, $p = .02$). A marked decrease in LRS occurred among women married between eighteen and twenty years old; these older brides had approximately 2.7 fewer surviving offspring than those married between thirteen and seventeen ($t_{10,29} = 3.57$, $p = .001$), because they had fewer live births per reproductive year (.08) ($t_{10,29} = 2.77$, $p = .009$) and lower offspring survival (.10) ($t_{10,29} = 2.48$, $p = .018$). These results were retained when controlled for measures of wife's wealth and for education and have subsequently been shown to be associated with age at menarche (Borgerhoff Mulder 1987b).

Cohort Membership

The full sample (of which table 26.5 is a subset) showed that "Chuma" women had approximately two more surviving children (6.78) than did women in the "Maina" (4.84) and "Nyongi" (4.59) cohorts ($F_{2,338}$ = 22.06, $p < .001$). This reflected an overall increase over time in offspring survival, not fertility: offspring born to "Chuma" mothers had a 10% higher chance of surviving than those born to "Maina" or "Nyongi" mothers ($F_{2,252}$ = 9.210, $p < .001$) owing to significant reductions in mortality during infancy ($F_{1,193}$ = 4.91, $p = .028$, using "Chuma" and "Maina" samples where data on age-specific mortality were available). The effects of cohort were independent of wealth, as measured in this study, and of employment, education, marital status, and husband's age.

26.5 Discussion

In contrast to males of most polygynous mammals (e.g., Clutton-Brock, Guinness, and Albon 1982), men are reproductive for much of their life span. This difference may result from the pattern of resource acquisition and ownership characterizing many human societies (Borgerhoff Mulder 1987d): resources are accumulated over a man's lifetime through inheritance and individual endeavor, and a man's rights to these resources (and the wives he can buy) are largely protected by tribal legal sanction (e.g., Peristiany 1939); thus most men can acquire and defend a resource base without resorting to force or risking physical injury. Moreover, the resources are in themselves buffers against starvation and hardship, which may help reduce the rate of physical deterioration.

Age is strongly associated with a reduction in breeding success among women in both developing and developed countries (Bongaarts and Potter 1983) and probably results from the higher incidence of anovulatory cycles, longer postpartum amenorrhea, higher abortion rates, and lower frequency of intercourse among older women (Gray 1979).

Variance in male success is largely caused by variation in the number of mates, as in other polygynous breeding systems (Bateman 1948) and in some human populations (Wood, Johnson, and Campbell 1985; Chagnon 1979). More specifically in this context, the high variance in numbers of wives among Maina men is most probably attributable to the sudden injection of British economic interests in rural areas: after 1930 the Maina cohort, young men at the time, were able to convert earned wealth from newly salaried chieftainships and estate management into cattle, land, and wives.

Variance in female reproductive success is due principally to differences in fertility and infant survival, although the importance of the latter has been decreasing over time, most probably because of improved medical facilities and transport (see Borgerhoff Mulder 1987e for analyses of recent cohorts). The most likely explanation for the strong negative relation between fertility and infant survival is that anovulatory cycles are curtailed after the loss of an infant (Short 1976; Konner and Worthman 1980) and that mothers seek to replace lost infants (Preston 1975), but deleterious effects of short birth intervals on infant survival cannot be excluded (Boerma and Vianen 1984; Scrimshaw 1978). The length of the reproductive life span explains little reproductive variance among women, suggesting unusually low levels of death in childbirth, and may in part result from a sampling bias against reporting wives who died during their reproductive years.

The comparative intensity of selection (I_m/I_f) as measured in this study is probably an underestimate: the higher average breeding success of men than of women suggests that poor men, who leave the community to become plantation laborers and marry late in life, if at all, achieve low LRS. Were it possible to include these emigrants in the analysis, the comparative intensity of selection would be higher.

As regards the determinants of reproductive success, the close association between wealth and number of wives is not surprising, given that almost all men want several wives (Peristiany 1939). Polygyny is best predicted by number of cows in the younger cohort and acres in the older cohort, reflecting increased tendencies for men throughout their life span both to spend cattle as bridewealth (for themselves and their sons) and to invest in land. This is a new development; before the 1930s wealth could be held only as cattle.

The problem of causality between wealth and reproductive success cannot be precisely determined without longitudinal data on wealth changes over a man's reproductive and marital lifetime (but see Borgerhoff Mulder 1987c); however, wealthier men marry earlier, and their sons are more likely to be polygynous, suggesting that problems in amassing resources curtail and delay marriage opportunities among poor men (see also Wrigley 1966).

The positive relation between wealth and male reproductive success has been shown in other developing societies (e.g., Hawthorn 1970; Shaikh and Becker 1985; Irons 1979; Wood, Johnson, and Campbell 1985; Turke and Betzig 1985), but it could simply be an epiphenomenon of modernization (Borgerhoff Mulder 1987a), with richer men having greater access to Western commodities such as medicine, education, and commercial infant supplements, all of which tend to increase fertility and offspring survival (Nag 1980). In the Kipsigis case, the cause of the relationship is the number of wives married to a man, irrespective of their fertility or the survival chances of their offspring, thereby excluding the possibility that moderization accounts for the relationship, and this is corroborated by the consistency of the relation between wealth and male reproductive success across all cohorts circumcised between 1922 and 1964 (Borgerhoff Mulder 1987c).

The positive association between wealth per wife and female reproductive success, particularly offspring survival, occurs because in richer households there is sufficient food throughout the year (Borgerhoff Mulder 1987f), favoring successful pregnancy outcomes (Mosley 1977), helping offset the costs of lactation (Harrison, Boyce, and Platt 1977; Jelliffe and Jelliffe 1978), and increasing the survival chances of offspring (Grounds 1964; Mundot Bernard 1975). Moreover, both rich women and their children are less likely to be sick than poor women and their children (Borgerhoff Mulder 1987e). However, the high infant mortality of polygynously married women in the "Chuma" cohort cannot be attributed to wealth shortages and may reflect other aspects of co-wife competition and rivalry (Le Vine 1962; Brabin 1984). Two overt cases of co-wife aggression (poisoning and child battering) were observed.

In general, however, polygynously married women do not differ from monogamously married women either in overall reproductive success or in wealth per wife, contrary to some previous ethnographic findings (Brabin 1984), suggesting that mate choice among Kipsigis may follow an ideal-free distribution with regard to husbands' wealth (Fretwell 1972), as the polygyny threshold model predicts (Verner 1964; Orians 1969b).

The lower reproductive success of women married to relatively old husbands has previously been attributed to lower fertility of old men (Benedict 1972), but in the Kipsigis case, differences in offspring survival account for the effect. This may be attributable to the low contributions of relatively old husbands to household, agricultural, and husbandry work (pers. obs.). Alternatively, lower offspring survival of older husbands may occur because older men are less willing to use medical facilities.

Reproductive advantage of young brides and young primiparous mothers has been noted in previous studies (Aghajanian 1981; Bumpass, Rundfuss, and Janosik 1978; Gubhaju 1983; Kenya Fertility Survey 1980). Among the Kipsigis it is a consequence of both higher fertility and higher offspring survival of mothers marrying young. Early marriage per se is unlikely to account for this effect, since adolescent breeding is often characterized by subfecundity (Jain 1969), high fetal loss (Leridon 1977), and infant death (Miller and Stokes 1985; Srivastava and Saksena 1981). It may be that girls who were well nourished in early development (in the case of the Kipsigis, the daughters of rich men) reach menarche early (Borgerhoff Mulder 1988b; see also Frisch and McArthur 1974; Burrell, Healey, and Tanner 1961), marry young and are also more successful at childbearing and child rearing. This is supported by the negative relationship between fertility and age of menarche found both in the Kipsigis (Borgerhoff Mulder 1987b) and in Romanian women (Critescu 1975) and more peripherally by the associations between fat indexes and fertility (Frisch 1978) and completed family size (Mueller 1979).

Cultures are highly variable, and no claim should be made for the generality of the Kipsigis case. Nevertheless, similar examination of the major components of reproductive success in other societies will allow us to investigate whether variation in the major determinants of breeding success is associated with differences in behavior directed at enhancing reproductive success. The attempt to find links between the sociocultural environment and the principal determinants of reproductive variance avoids the danger of trying to distinguish between cultural and biological determinants of behavioral diversity (Borgerhoff Mulder 1983) and opens the possibility of a behavioral ecology of cultural diversity.

26.6 Acknowledgments

I thank all the people of Abosi, Tabarit, and Kamerumeru for their warm welcome and cooperation and Philip arap Bii and his family for their special hospitality. Angelina Chemngetich Mosongik, Lena Cherono Talam, and Sostin arap Mibei were very helpful in introducing me to the families. Thanks also to Tim Clutton-Brock and members of the Large Animal Research Group, Department of Zoology, Cambridge, for providing such a stimulating environment in which to work, to Steve Albon for enormous help with statistics, and to Tim Caro for valuable advice throughout the study. They, David Coleman, Mark Elgar, Geoffrey Hawthorn, and Roger Schofield all commented on the manuscript. Research and anlysis were supported by the National Geographic Society and the Graduate School of Northwestern University.

5 General Issues

In the first chapter of this section, Brown describes the basis and rationale of his method for partitioning variance in lifetime reproductive success into its component parts and discusses alternative methods (chap. 27). As he stresses, all methods of partitioning lifetime reproductive success have their disadvantages. Multiple regression of raw data is unsuitable, since it assumes that relationships between components are additive when they are multiplicative and, even if the data are logged, involves estimating relationships between lifetime success and its components that are known exactly. The approach suggested by Arnold and Wade (1984a,b) has the disadvantage that the reference population changes between stages of the life history and that it does not incorporate the contribution of covariation between stages separated in time. Finally, Brown's own method calculates average values for components of fitness over the life span, disregarding age changes. For example, where fecundity declines with age, short-lived individuals will appear to have high average fecundity while long-lived ones will be scored as having low fecundity. As a result, the method can produce misleading estimates of covariation.

In the second chapter, Grafen reviews the uses and limitations of data on lifetime reproductive success. Grafen's review emphasizes the important distinction between studies of selection in progress and adaptation, concentrating primarily on the latter. In studies of adaptation, the central aim is to discover why a particular trait is advantageous, and measures of the opportunity for natural or sexual selection are of limited relevance. Evidence that behavioral or morphological traits are associated with components of fitness will not always help to reveal why these relationships occur, in some cases because correlations between particular traits are caused by associations with other variables. Few field biologists would disagree with Grafen's emphasis on the need for experiments, but in most circumstances a combination of observational and experimental studies provides more reliable answers than either type of evidence on its own.

In the last chapter, I review the generalizations that have emerged from the twenty-five empirical studies, covering six principal questions: How widely does breeding success vary among females? How widely does it vary among males? Are fitness components positively or negatively related to each other? Does variation in breeding success occur by chance? What phenotypic variables affect reproductive success in each sex? And to what extent is variation in fitness heritable?

27 Components of Lifetime Reproductive Success

David Brown

In all animal populations studied so far, individuals differ widely in the number of offspring they produce during their life span. Understanding the extent and distribution of these differences is relevant both to studies of population dynamics (Lomnicki 1978) and to studies of selection (Wade and Arnold 1980). Different stages of the life span commonly contribute disproportionately to variation in lifetime reproductive success (LRS), and in both contexts it is often important to identify the contributions made by different stages.

In theory, the number of stages into which an animal's life span can be divided is almost infinite. In a long-lived, iteroparous species, it would be possible to examine the contributions to variance in LRS of survival and fecundity during each year of the life span, and in several parous species the contributions of individual days could be investigated. However, it is generally more practical to restrict analysis to a smaller number of stages. In particular, several workers have found it convenient to partition variance in LRS into four components: survival to breeding age, reproductive life span (L), average fecundity per year of the reproductive life span (F), and offspring survival between birth or hatching and reproductive age (S). One reason for partitioning variation in breeding success in this way is that there are a priori grounds for believing that different ecological and behavioral factors are likely to affect these four components (Clutton-Brock, this volume, chap. 1).

What statistical techniques are appropriate to examine the contribution of different components to variance in lifetime reproductive success? Multiple regression, on the raw or log-transformed data, has been used by some workers (e.g., Howard, this volume, chap. 7), but it involves estimating a relationship between LRS and its components that is known exactly (e.g. LRS = L F S in the wolf example, sec. 27.2) and is therefore unsuitable. A partition based on the relationship log(LRS) = log(L) + log(F) + log(S) is possible, but the log-transformation is not necessary and therefore requires some justification other than mathematical convenience. Ostensibly Arnold and Wade's (1984a,b) method for partitioning the opportunity for selection into components is suitable. They achieve a partition into single-component terms and terms due to covariances between components. There are some statistical arguments against its use in this situation, which are considered in section 27.7. McCauley (1983) previously gave an approximate method for partitioning the opportunity for selection, but his method made no provision for covariance between components.

This chapter outlines a method for partitioning variance in LRS, utilizing an exact relationship between V(LRS) and the variances and covariances of its components, which a number of chapter authors have found useful. The method is presented for a simple case of two components, and more generally for *n*, in section 27.1, and applied to a simple numerical example in section 27.2. How terms in two components or more include effects of covariation and independent variation is explained in section 27.3, and methods for determining the effects of sampling error are discussed in section 27.6. An allowance for nonbreeders is made in section 27.5, and different ways of taking into account covariation are covered in section 27.4. The discussion (sec. 27.7) covers further comparison with other methods and also aspects of interpretation of the results. Nonmathematical readers are advised to omit the second part of section 27.1—discussing the general case of *n* components—and all of section 27.4 from a first reading.

27.1 The Statistical Method

The main part of the analysis involves only those animals that succeeded in producing at least one offspring, which might or might not have survived to maturity. For these parent animals, data is available on all the stages that contribute to LRS, and this is necessary for the main part of the analysis. Those animals that did not produce any offspring—the nonbreeders—are dealt with in a separate adjustment in section 27.5.

Before giving a general definition, let us consider the case of just two variables, L and F, reproductive lifetime and fecundity. Let m_L and m_F be the means, and V(L) and V(F), the variances of L and F respectively. Goodman (1960) gives the variance of the product LF as

$$V(LF) = m_L{}^2V(F) + m_F{}^2V(L) + Q(L,F).$$

The three terms in the expansion are: two terms that are the products of the square of the mean of one component and the variance of the other; and Q is a complex term involving simultaneous variation in L and F, and the covariance between L and F, but also covariances of squares of the deviations of L and F from their means. For the moment Q can be simply regarded as a joint variation term. In any particular case, whether we are referring to theoretical population values, or sample estimates, Q can be calculated as the difference between V(LF) and $m_L{}^2V(F) + m_F{}^2V(L)$. So V(LF) can be partitioned into a contribution due to variation in L, one due to variation in F, and one due to joint variation in L and F. The formula can be simplified by division of both sides by $m_L^2 \, m_F^2$, obtaining

$$V(LF)/(m_L m_F)^2 = V(L)/m_L^2 + V(F)/m_F^2 + Q(L,F)/(m_L m_F)^2,$$

which can be written

$$V(LF/m_L m_F) = V(L/m_L) + V(F/m_F) + Q(L/m_L, F/m_F).$$

Thus the simplest expression of the relationship is obtained in terms of the variables divided by their means. This relationship can be used to determine what percentage of the variation in LF is due to variation in L, to variation in F, and to joint variation in both variables.

Now we consider the general case of *n* component variables. Then the lifetime reproductive success can be expressed as a product of component variables $X_1X_2 \ldots X_n$. First the individual X-variables are scaled so that all of them have a mean of 1. These scaled variables are denoted by the subscript s:

$$X_{si} = X_i/E(X_i) \text{ or } X_i/\text{mean}(X_i).$$

The method is based on the result given by Goodman (1962) that the variance of the product $X_{s1}X_{s2} \ldots X_{sn}$ is given by

$$\sum H_i + \sum_{i<j}\sum H_{ij} + \sum_{i<j<k}\sum\sum H_{ijk} + \ldots + H_{1\ldots n}$$

where the H's are defined as follows.

$$H_i = V(X_{si})$$
$$H_{ij} = V(X_{si}X_{sj}) - H_i - H_j$$
$$H_{ijk} = V(X_{si}X_{sj}X_{sk}) - H_i - H_j - H_k - H_{ij} - H_{ik} - H_{jk}.$$

This enables the variance of a product of any number of variables to be split up into additive components involving single variables (H_i or $H(X_i)$), pairs of variables (H_{ij} or $H(X_i, X_j)$), triples of variables (H_{ijk} or $H(X_i,X_j,X_k)$), and so on. It is convenient now to introduce some further notation:

$$G(X_i) \text{ or } G_i = V(X_{si})$$
$$G(X_iX_j) \text{ or } G_{ij} = V(X_{si}X_{sj}).$$

More generally, $G(X_i \ldots X_m)$ or $G_{i\ldots m} = V(X_{si} \ldots X_{sm})$.

This notation applies only when the variables $X_i, X_j \ldots$ are the basic component variables of the product. G values are dimensionless. To simplify the interpretation, the G and H values can be expressed as a percentage of the variance of the overall product of the scaled variables. These percentages are denoted by G′, H′, and so forth, a prime being added to denote division by $V(X_{si} \ldots X_{sn})/100$. So $G'_{ij} = 100\ V(X_{si}X_{sj})/V(X_{s1} \ldots X_{sn})$; the G and H terminologies overlap for single components so that $H_i = G_i$ and $H'_i = G'_i$.

27.2 A Numerical Example

The analysis is illustrated using data from an artificial population of wolves. The variables measured were: L(breeding life span), F(fecundity, or number of offspring born per year of the breeding life span), and S(survival to sexual maturity of the offspring), which when multiplied together give the lifetime reproductive success. A summary of the data for the sample of thirty breeding animals is given in table 27.1. The average breeding life span of this sample of wolves was about eight years, during which on average they produce three offspring each per year, and about half of these survived to maturity. The average LRS was about 11. The variances of the individual variables and the two- and threefold products are given in the next column. The column headed G gives the variances of the individual variables and the products after the individual variables have been divided by their means; so for the individual variables, the values of G are the squares of the coefficients of variation of the variables, whereas for the products this is not so (this occurs because the mean of a product of two or more variables is usually different from the product of the individual means). The last column gives the G values expressed as a percentage of the G value for LRS: this G′ column gives the percentage of V(LRS) accounted for by the individual variable, or the product of variables indexing the row. Thus only 6.9% of V(LRS) is accounted for by variation in S (and S only); 87.3% of V(LRS) is accounted for by variation in the product FS.

Then tables 27.2 and 27.3 give the partition of V(LRS) in two scales: table 27.2 gives a breakdown in terms of the variances of the original components after scaling by their means (G and H); table 27.3 gives the same breakdown but expressed as a percentage of V(LRS) (G′ or H′). A generalization of the two component formula in section 27.1 to three components is easily made:

$$H'(L) + H'(F) + H'(S) + H'(L,F) + H'(L,S) + H'(F,S) + H'(L,F,S).$$

The partition is composed of terms of the following types: H′(F)—the contribution of the variation in F and in no other components; H′(L,F)—the contribution of variation in L and F and in no other components, which is in excess of the sum of the single component contributions, H′(L) + H′(F); H′(L,F,S)*M* the contribution of variation in all three components, which is in excess of the sum of the contributions due to variation in the single components, H′(L) + H′(F) + H′(S), and in all the constituent pairs of components, H′(L,F) + H′(F,S) + H′(L,S).

The partition into single component and two component contributions can conveniently be drawn up in a lower triangular matrix, with the diagonals containing the single component contributions, H'_1 or H′(L), H'_2 or H′(F), H'_3 or H′(S), and the values below the diagonals containing the two component contributions H'_{12} or H′(L,F), H'_{23} or H′(F,S), and H'_{13} or H′(L,S). The rows and columns are labeled by the appropriate component-variable labels. This is supplemented in this case by a single three component contribution, H'_{123}. From table 27.3, it can be seen that about 42% of V(LFS) can be accounted for by variation in L, about 72% by variation in F, and only about 7% by variation in S. Thus we might say that the most important single component is fecundity, in the sense that if

Table 27.1 Summary of an Artificially Generated Sample of Wolf "Data": Means and Variances of the Individual Components and the Products

Component	Mean	Variance	G	G′
L	7.80	8.23	0.135	42.4
F	3.07	2.17	0.229	71.8
S	0.47	0.0048	0.022	6.9
LF	24.33	260.80	0.454	142.2
LS	3.45	0.65	0.050	15.5
FS	1.44	0.57	0.279	87.3
LFS	10.8	39.73	0.319	100.0

Note: Sample size = 30. G = variances of components and products after individual components have been divided by their means; G′ = G expressed as a percentage of V(LRS).
L = breeding life span; F = fecundity; S = survival of offspring.

Table 27.2 Wolf "Data": Upward Partition

Component	L	F	S
L	0.13535		
F	0.08920	0.22934	
S	−0.10784	0.02740	0.02211
LFS[a]	−0.07625		

Note: See sections 27.2 and 27.4. Contributions to $V(LFS)/(\bar{L}\bar{F}\bar{S})^2$: The diagonals give the single-component contributions (e.g., H(F)), and the off-diagonal elements give the joint-variation terms corresponding to row and column labels.
L = breeding life span; F = fecundity; S = survival of offspring.
[a]Joint-variation term for variation between the three components.

Table 27.3 Percentage Contribution of the Components of Lifetime Reproductive Success to Variation in LRS: Wolf "Data"

Component	L	F	S
L	42.39		
F	27.94	71.82	
S	−33.77	8.58	6.93
LFS	−23.88		

Note: Results from table 27.2 expressed as a percentage of V(LFS).

the other two components L and S were set equal to their means (i.e., the only variation was in F), the product LFS would have a variance that is 72% of the full variance when variation in all three components is allowed. So if all the wolves had a reproductive life span of exactly 7.80 years and an average offspring survival of exactly 0.47, their lifetime reproductive success would be 72% as variable as it is when these two variables are allowed to take their real values. In the same sense, L is moderately important accounting for 42% of V(LRS), and S is relatively unimportant, accounting for only 7% of V(LRS). The two component contributions account for a moderate amount of the overall variation, although some of the contributions are negative: (L,F) 28%, (L,S) −34%, (F,S) 9%. What interpretation can we put on these? The extra variance induced in the product by joint variation in L and F, over and above the sum of their single-component contributions, 42% + 72% or 114%, is 28%, giving a grand total for these two component, L and F, of 142%. You will notice that this is also in accord with the value of G′(LF) given in table 27.1. This means that if the average offspring survival of each wolf were set to exactly 0.47, the LRS would have a variance of 142% of what it would be if all three components took their real values; and this 142% is greater than the sum of the individual component contributions by 28%. We will see in the next section that this 28% can be further broken down into terms that are more meaningful, related to the degree of correlation between L and F. Similar interpretations can be put on the three-component contribution, which is about −24%, but this is not pursued here.

An important feature of the partitions above is that the contribution of any subset of variables is the same irrespective of the other variables that are included in the product. This enables simultaneous examination of the breakdown of the variance of any subproduct along with the complete product. For example, in the artificial example above, we could examine the bottom right-hand corner involving only F and S completely separately from the rest. The total percentage in that subtable is 87.33 and this is the percentage of V(LFS) accounted for by variation in the subproduct FS. This can be further broken down into 71.82% of V(LFS) (or 71.82 × 100/87.33 = 82.24% of V(FS)) due to variation in F alone; 6.93% of V(LFS) (or 7.94% of V(FS)) due to variation in S alone; and the remainder, 8.58% of V(LFS) (or 9.82% of V(FS)) due to simultaneous variation in F and S.

Two further qualifications ought to be made of these results. Some will be subject to large sampling errors; for example, an approximate 95% confidence interval on the percentage due to variation in L is 22,90, whereas the point estimate is 42%. For further details see section 27.6. Also, there are other ways of carrying out the partition, which are outlined in section 27.4 and discussed further in section 27.7.

27.3 Independent Variation and Covariation

This section deals with the important question whether two components vary together in such a way as to reduce or, alternatively, reinforce their individual variations. If all the component variables in the product vary independently of each other, the two and more variable terms in the partition take a particularly simple form:

$$H_{ij} = G_iG_j \qquad H_{ijk} = G_iG_jG_k \qquad H_{ijkl} = G_iG_jG_kG_l$$

These elements are said here to be due to *simultaneous independent variation* in the particular component variables. What this means is that, even if a pair of variables contributing to LRS vary independently of each other, there is still a corresponding joint variation term in the partition. For example, consider a simple case of two variables: the total number of offspring, surviving or not, born to one parent, is a product of L and F only. If L and F are completely independent in the population (i.e., there is no connection, either causal or otherwise, between breeding life span and fecundity), the variance of the number of offspring, LF, could be partitioned into a term due to variation in L only, one due to variation in F only, and a further term due to joint variation in L and F. To see how these terms arise, let us write $L = m_L + \delta_L$ and $F = m_F + \delta_F$ where m_L and m_F are the means, and δ_L and δ_F are the variations about the means of L and F respectively, then

$$LF - m_Lm_F = m_L\delta_F + m_F\delta_L + \delta_L\delta_F.$$

When we consider the variation in LF about its mean, which is approximately given by m_Lm_F (exactly so when L and F are independent), we see that there are in fact three components: that due to $m_L\delta_F$, which is the mean of L times the chance component of F; that due to $m_F\delta_L$, which is the mean of F times the chance component of L; and that due to the two chance components together, δ_L times δ_F. The first is the component due to variation in F only, the second due to variation in L only, and the third is the component due to joint variation in L and F. Even when L and F vary independently, we still get this third component. This also illustrates a further feature of the partition. When we say that a component is due to variation in one variable alone, we mean that it is a product of the variation in that variable and the means of the other variables: thus the importance of variation in one variable is enhanced when the mean of any other variable is increased and everything else remains unchanged.

Let us now consider the situation when two variables are not independent. It can be seen intuitively that if two variables vary together, that is, are positively correlated, the variance of the product will be increased over what it would be if they were independent. Conversely, if they are negatively correlated it is possible for one variable, so to speak, to vary in such a way as to cancel out the variation in the other variable, and so the variance of the product will be much decreased. This paragraph deals with the quantification of this feature. If the component variables in the product are not independent, then each two- or more variable term can be split into a part, I, due to simultaneous independent variation in the

Table 27.4 Partitioning of Variance in Lifetime Reproductive Success: Off-Diagonal Entries for Wolf "Data"

Component	L	F	S
L	42.39	18.21	−34.71
F	9.72	71.82	6.99
S	0.94	1.59	6.93
LFS	independent, 0.21; covariation, −24.09		

Note: The off-diagonal terms in table 27.3 are partitioned into simultaneous independent terms (below the diagonal of this table) and covariation terms (above the diagonal). The three-variable term is partitioned at the bottom of the table.

component variables, and a part due to the departure of the variables in question from independence, J, termed the contribution due to *covariation* among the complete set of variables in the particular term. So we get

$$H_{ij} = I_{ij} + J_{ij} \text{ where } I_{ij} = G_iG_j \text{ and } J_{ij} = H_{ij} - I_{ij}.$$

$H_{ijk} = I_{ijk} + J_{ijk}$ where $I_{ijk} = G_iG_jG_k$ and J_{ijk} is determined by difference as above. Similarly for terms involving four or more components. It should be mentioned here that J_{ijk} and higher-order J's are rather complex quantities. They include some contributions due to the amplification of pairwise covariation by independent variation in other variables, as well as contributions due to covariation involving all the variables.

If we apply this to the wolf sample, we get the results in table 27.4. In this the simultaneous independent terms are placed below the diagonal and the terms due to covariation above the diagonal. In this case the simultaneous independent contributions are quite small: (L,F) 10%, (L,S) 1%, and (F,S) 1.6%. The contributions due to pairwise covariation between the variables are a little more substantial: (L,F) 18% (F,S) 7%, and (L,S) being −35%. Thus breeding life span and offspring survival of wolves are so negatively correlated that the variance of their product LS is less, by 35% of V(LFS), than it would be if they were independent. Some of the advantage of breeding for longer is lost as those animals that breed longer have offspring that do not survive as well. By contrast, the variations in breeding life span and fecundity tend to slightly reinforce each other in this sample.

The three-variable component can be similarly split: virtually zero due to simultaneous independent variation in the three variables, and about 24% due to variance reducing covariation between the three variables.

Caution should be exercised in interpreting the correlations that might be revealed by the analysis above, apart from the need to take into account sampling variation. For example, in some cases high negative correlations were obtained between L and F. This could arise as follows. Some animals breed poorly or not at all when they are rearing the previous year's offspring, and this induces a cycle of alternating good and poor breeding. This feature would induce a higher fecundity for those animals that breed only once, and if there were a moderate number of such animals, the result could be a negative correlation between L and F.

27.4 Upward and Downward Partitions

The previous sections have dealt with the change in variance of a product during the process of construction from single variables, pairs of variables, triples of variables, and so forth. This results in what is called here an *upward* partition of V(LRS). It reflects the change in variance as LRS is built up from any combination of constituents. It also has the advantage that any subgroup of two or more variables can easily be examined separately from the overall analysis.

The elements of the upward partition corresponding to variation in single variables give what the variance of LRS would be if *all the other variables* were set to constants equal to their means; and two- and three-variable components can be interpreted similarly. An alternative approach is to consider how V(LRS) changes as the variable or pair or triple of *variables under consideration* are set to constant values equal to their means but other variables retain their values; in other words, to examine the change in the variance of the product as variation in the variables in question is deleted from the overall product. This *downward* partition of V(LRS) involves calculating, in the expression for V(LRS), the sum of all the terms involving the variance of the variable or variables in question and also any terms involving covariation of those variables and other variables. This gives a direct measure of the contribution of the set of variables in question to the variance of the overall product.

Considering our sample of wolves again, the downward partition is given in table 27.5. The variable that changes V(LRS) the most by being set to a constant value equal to its mean is F, reducing V(LRS) by about 84%; L on being thus removed from the product causes a reduction of only about 13%; removal of S, on the other hand, causes an increase in V(LRS) of 42%. What this last figure means is that S is so negatively covarying with the other two variables that its effect on the product is to reduce its variance by a substantial 42% rather than increasing it. The two-variable terms in this downward partition give the effect of deleting two variables beyond the effect of deleting them separately. For example, the effect of deleting L and F together is only slightly different from the sum of the effects of deleting them separately, the difference being only −4%. On the

Table 27.5 Wolf "Data": Downward Partition

Component	L	F	S
L	12.68		
F	−4.07	84.46	
S	57.64	15.29	−42.14
LFS	−23.88		

Note: See section 27.4. Expressed as a percentage of V(LFS).
L = breeding life span; F = fecundity; S = survival of offspring.

contrary, deleting L and S together has quite a different effect compared with the sum of the effects of deleting them separately.

27.5 Contribution of Nonbreeders

If the last two variables are fecundity and offspring survival, then the individuals which cannot be included in the analysis above because of incomplete data are those that did not successfully breed. Suppose that as well as the n_b animals included in the above analysis, there are n_f that fail to breed. V(LRS) and $\overline{LRS}$ are the sample variance and mean respectively of LRS, calculated of course only for the breeders. Then the proportion of breeders is given by $p = n_b/(n_b + n_f)$. The overall variance in LRS among the breeders and nonbreeders is given by

$$pV(LRS) + p(1 - p)\ \overline{LRS}^2.$$

The first term in the expression above is the proportion due to variation among animals that successfully breed, which can be further partitioned using the proportions already derived. The second term is the proportion of the overall variation due to failure to breed. This expression is due to Alan Grafen.

Suppose besides the sample of thirty wolves that bred successfully, there were ten that did not; then inserting numerical values in the expression above we get $(0.75)\ (39.37) + (0.75)\ (0.25)\ (10.8^2) = 29.53 + 21.87$. Thus 57% of the variance in overall breeding success can be attributed to breeders, and further partitioned down as above, and 43% due to nonbreeding.

27.6 Assessing Sampling Error

The theory above applies to a statistical population—the totality of animals we wish to make inferences about. The statistical population might be somewhat notional: a construct necessary to use the tools of statistical inference. In some studies included in this volume, all the animals within a defined area over a particular period of time have been studied and data from them included in the analysis—a whole cohort of animals. In other studies, only a cross-sectional sample of the whole biological population has been studied. In the former case, we are certainly interested in how the particular biological population variance of LRS can be partitioned into its components; but we are also interested in the extent to which we might have been deceived had the population studied been slightly shifted in time or space—that is, if the "chance" variables affecting the outcome had taken slightly different values within the observed range of variation. This interpretation defines the statistical population of interest, even though only vaguely so. In the second case we actually have some explicit sampling variability to contend with, but it makes no difference to the mechanics of the computation and inference. In a sense, therefore, we need to apply the results above to "sample data." For simplicity, we use exactly the same functions of the observations in the sample as we would have used had we measured the whole population. In this way the theory can be applied to samples. This means that we use slightly biased estimates of some quantities (for example, in sample variance calculations we divide by n, the number of observations, rather than $n - 1$, but except for very small samples this is unlikely to produce substantial biases in the results).

Some method is therefore required for assessing whether the properties observed in our "sample" reflect real properties of the statistical population. Because most of the methods make no assumptions about distributions and the component variables in the product will not all follow the same class of distributional form, we need some robust methods for assessing the magnitude of the likely sampling variation and for testing hypotheses about particular components.

Efron and Gong (1983) give a review of methods available for the nonparametric estimation of statistical error (not depending on distributional assumptions). Of the methods described there we mainly use the *bootstrap* here. This is effectively using the sample, whether univariate or multivariate, as an estimate of the population and assessing sampling variability by randomly sampling this estimated population (i.e., a population containing only the sample values in the same proportions as in the sample) using a computer pseudorandom sampling procedure. To take a simple example, suppose our sample of three components of LRS consists of the following measures of L, F, and S for five animals only.

L: 10, 9, 7, 8, 11; F: 2, 3, 4, 2, 3; S: 0.5, 0.4, 0.6, 0.3, 0.4.

The percentage of V(LRS) accounted for by variation in reproductive lifetime, L, in this sample is 19.6%. The bootstrap distribution of this percentage is obtained by repeatedly taking random samples of five from a population consisting of a large number, say one thousand, of individuals with (L,F,S) = (10,2,0.5), an equal number, one thousand, with (L,F,S) = (9,3,0.4), and so on, and for each of the resulting samples of five calculating in the usual way the percentage of V(LRS) due to variation in L. From these a histogram can be drawn that approximates the sampling distribution of the particular statistic. The standard deviation gives an estimate of the standard error of the estimated percentage. If the histogram consists of estimated percentages from each of four hundred samples of five animals, then the 2.5 and 97.5 percentiles (i.e., the 10th and the 390th value) give a 95% confidence interval on the percentage in the population. This uses the "uncorrected percentile" method for confidence intervals. Figure 27.1 gives the bootstrap distributions for some of the component variables and some of the covariation terms in the partition for our sample of wolves, with 95% confidence intervals. The excessive spread of some of these sampling distributions and confidence intervals should be noted.

Another technique available for testing certain hypotheses—for ex-

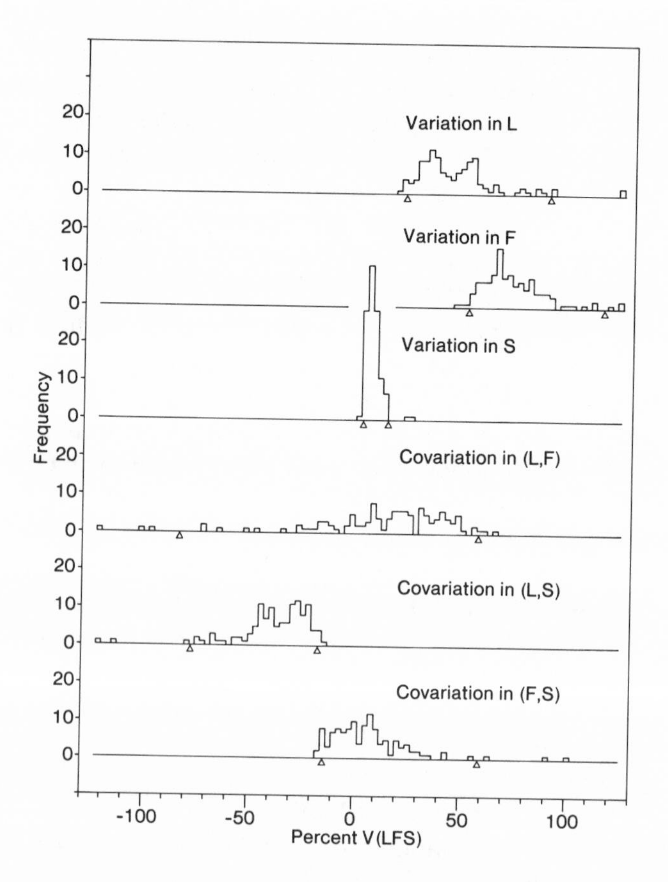

Figure 27.1: Estimated bootstrap sampling distributions of the percentage of V(LFS) accounted for by variation in L, variation in F, and variation in S, and covariation in the pairs (L,F), (L,S) and (F,S). Approximate 95% confidence intervals on the percentages are indicated by triangles beneath the appropriate axes.

ample, whether a covariation contribution is zero—is *randomization testing*. Instead of drawing randomly from a population consisting of the sample values, using a computer we calculate the distribution of the statistic under repeated randomization. For example, if we were interested in the covariation term J(L,F), we could repeatedly allocate at random the L values in our sample to the F values and obtain a histogram of the resulting values of J(L,F). If the observed value of J(L,F) for the original sample is in the tail of this randomization distribution, we say that the contribution of the covariation between L and F is significantly different from zero at the particular significance level obtained.

27.7 Discussion

Care should be taken interpreting the results of the partition of the product variance not to use experience derived from the analysis of multiple regression. To illustrate the differences, consider a case of two components whose product is w. Most investigators would take logarithms of these variables before doing regression, in order to make log(w) the sum of the logged components. One immediate difference is that the present method includes explicitly a contribution due to covariation between the two components, whereas in multiple regression this contribution usually inflates the contribution of whichever is the first component to be included in the multiple regression. Considering the wolf data, the correlation between log(L) and log(F) is only .07, but the joint variation contribution of the upward partition is about 20% of V(LF), and the single variation components are 30% for L and 50% for F. In contrast, log(L) when entered first into the regression accounts for 43% of V(log(LF)), and log(F) 64%.

If the two components are highly correlated, then the regression of log(w) on the log of either component could account for a large part of the variation whether the correlation is negative or positive, only the absolute magnitude of the correlation affecting the result. With the methods of the present chapter, if the correlation is positive, then the proportion of V(w) accounted for by either single component would be reduced, possibly to a very small value, and the dominant term in the partition might be that due to covariation. If the correlation is negative, the contribution of a single variable will be increased possibly to well over 100% of V(log(w)). As an example, let log(w) equal the sum of log(L) and log(S) in the wolf data, which are highly negatively correlated ($r = -.96$). The joint variation term of the upward partition is now −217% of V(LS), the single component terms being 273% for L and 45% for S. In comparison, log(L) when entered first into the regression accounts for 97% of V(log(LS)), and log(S) for 80%; and when these components are entered second in the regression, the terms account for only 20% and 3% of V(log(LS)) respectively.

Multiple regression results can be presented in more complex ways that explicitly allow for the correlation between the explanatory variables: there is a long history of such methodology ranging from Wright (1921, 1934), via Kempthorne (1957), to Newton and Spurrell (1967a,b), and recently Whittaker (1984). Newton and Spurrell's (1967b) presentation is similar to that in this chapter, but the underlying model is totally different, in their case being an unknown multiple regression equation in which coefficients are estimated, in ours a specific product function that is known. The objectives of the two analyses are different: in one case to produce an economical predictive model; in the other, to assess the influence of each variable in a known equation.

This last example illustrates another feature of the present method: the occurrence of contributions of components well in excess of the variance of the original product. This means, in the case of single-component contributions, that the component, relative to its average, is more variable than the overall LRS. This is not an artifact of the analysis but a real feature of many sets of data. A necessary accompaniment of such a feature is a substantial contribution due to negative covariation between some of the components; this negative covariation is, in a sense, compensating for the excessive variation in the single component.

It is not necessary that LRS or its components be normally distributed for the partition to give useful results; the partition holds whatever the distribution. However, the bootstrap sampling distributions given in figure 27.1 indicate the considerable potential errors in the estimates of the contribution of each component to V(LRS) for a sample size (thirty) that in other circumstances might be regarded as adequate, from reasonably near normal distributions. If the distribution of any component departed substantially from normality in the direction of excessive skewness or kurtosis (peakedness), much larger samples would be required to obtain results of acceptable accuracy. The same safeguards should be applied in this analysis as in any parametric analysis of variability or of correlation. For example, the data should be scanned for outliers (individuals with values of one or more components very far from the body of the data), and the analysis should be repeated excluding these values to check if it is disproportionately affected by them.

Arnold and Wade (1984a,b) interpret variance in LRS as the opportunity for selection and provide a partition of V(LRS) within that context. There are two aspects of their mathematical formulation that make it inappropriate in this situation.

1. Their emphasis throughout is on the opportunity for selection in the population that actually exists at any stage. Considering their analysis of Howard's (1983) data, given in Arnold and Wade (1984b), the overall variance of LRS is partitioned into components due to variation in number of mates (M), number of eggs per mating (E), and hatching success of those eggs (H). But the variance in number of eggs per mating is estimated over all matings—it is no longer considered a feature of the list of males that were started with; similarly the variance of H is estimated over the population of eggs. Thus the reference population changes from stage to stage. In the methods of this chapter all the variances are unweighted functions of the initial population or sample of parents.

2. Their method does not allow the contribution of covariation between a single early stage and a late stage to be evaluated. In the example above, the contribution of covariation between M and H is not calculable directly, and feedback that might be reflected in a large contribution of this covariation could be an important feature of the population dynamics.

There are two disadvantages of the present method that cannot be solved by simple modifications. The first is that the method uses averages over the whole lifetime of an individual, when it is well known that many components change as an animal gets older. Preadult mortality can be allowed for by including survival to maturity as an explicit component. It would also be possible to allow for variation in performance between a small number of periods of an individual's life, but the analysis would quickly become very complex and is beyond the scope of the present chapter. The second is that the adjustment for nonbreeding, necessitated as a separate adjustment by the resulting incomplete data for the later components in the reproductive cycle, ignores the stage at which failure to breed occurred. The result would in general be an underestimation of the contribution due to variation of those components whose zeroes had been excluded. It would in theory be possible to make such an adjustment at each stage at which failure occurs, but this also would complicate the analysis and interpretation considerably.

The expression (1) above gives the partition of V(LRS) into constituents of the form H_i, H_{ij}, up to $H_{1 \ldots n}$ if there are *n* components of LRS. These H terms as they stand are the elements of the upward partition of V(LRS); sums and differences of exactly the same H terms form the elements of the downward partition of V(LRS). So the upward and downward partitions are different expressions of the same mathematical decomposition of V(LRS). Which should be used in practice? If all the components act independently, the two forms will be essentially the same except for sampling error, and then of course the choice is unimportant. An example of this is the Rhum red deer study (Clutton-Brock, Albon, and Guinness, this volume, chap. 21). When there are substantial covariances between components, they will give quite different results. In this case the answer lies in the way either partition is interpreted. When assessing the single-component terms, for example, in an upward partition, it is vital to consider also the joint-variation and covariation terms with other components as well as the single-component terms, since the importance of the single-component terms could be increased by positive covariation with other components or reduced by negative covariation. The downward partition, in a sense, does this automatically: for a single component, the contribution of covariation and simultaneous variation with other components is included in each single-component term. So provided the covariation terms are taken into account directly or indirectly, it does not much matter which form of the partition is used.

In summary, this chapter provides formulas for partitioning the variance of lifetime reproductive success into additive contributions due to variation in single components and to variation and covariation among groups of components. Two main ways of presenting and interpreting this partition are available, depending upon whether the contribution of a component, or group of components, is considered in isolation or in conjunction with variation in other components. A contribution due to those animals that fail to breed can also be estimated. These methods are useful for identifying major sources of variation in lifetime reproductive success, and hence for directing attention to the environmental and behavioral factors that most influence it. In this, however, it is important to take into account sampling error, which might be substantial and for which some nonparametric techniques are suggested.

27.8 Acknowledgments

I am grateful to Tim Clutton-Brock and Steve Albon for drawing my attention to this problem, and to them, Peter Rothery, and Martin Major for very helpful discussions; also to Robert Marrs, who efficiently programmed some of the calculations.

28 On the Uses of Data on Lifetime Reproductive Success

Alan Grafen

THERE ARE NOW MANY studies in which data on lifetime reproductive success (LRS) are available, and in this chapter I wish to consider the kind of problems that arise in drawing inferences from such data. By LRS I mean direct observation of the total number of offspring produced in a lifetime by each member of a set of known individuals. The most complete proposed set of analyses is that of the Chicago school, namely Wade and Arnold (1980), Lande and Arnold (1983), and Arnold and Wade (1984a,b). These analyses will be scrutinized with a view to the analysis of adaptation. Various topics will be relevant. I will emphasize two distinctions: the first is between the analysis of adaptations and the detection of selection in progress, and the second is between natural and artificial variation. I will discuss the likely causes of natural variation and stress the necessity of entertaining and relying on hypotheses beyond the limits of the data, considering particular examples of the truism that correlations do not establish cause. Two general conclusions will, I hope, remain with the reader: that LRS data are not a panacea for epistemological complaints, and that a theory-free analysis is likely to be an interest-free analysis.

Another distinction is important in understanding my purpose in this chapter. On the one hand is the general idea of studying animals intensively for a long period, so that whole lifetimes are observed, and attaching importance to the number of offspring individual animals produce. I believe that the application of this idea is a genuine advance in technique and that we understand animals better for it. On the other hand are particular proposals about the use of LRS data and about what inferences may be drawn from variances, correlations, and regressions involving LRS. Wade and Arnold (1980), Clutton-Brock (1983), Lande and Arnold (1983), Arnold (1983a,b), Arnold and Wade (1984a,b), and others contrast the advantages of the LRS approach with disadvantages of other methods. This advocacy is natural for a new and pathbreaking approach to field studies and new methods of analysis. However, as the publication of this book shows, the LRS method is becoming increasingly popular. It is therefore timely to look hard at its *disadvantages* and think about what *cannot* be inferred from it. My chapter takes the value of the general approach for granted and aims to provide a critical appraisal of some specific ways of using LRS data.

Subsequent sections are organized as follows. Section 28.1 explains the distinction between adaptation and selection in progress and illustrates that different kinds of variation are best used to study them. Section 28.2 explores the likely causes of natural variation and shows how LRS regressions can become difficult to interpret as the hypothesized cause of natural variation becomes more complex. In section 28.3 the value of LRS variance decompositions will be considered. Building on the discussions in the previous sections, section 28.4 evaluates the Chicago school's proposed techniques for analyzing adaptations. In section 28.5 measures of fitness are discussed, in particular the relationship between LRS and Darwinian fitness. Section 28.6 discusses the application of inclusive fitness to data. Brief conclusions are given in section 28.7.

28.1 Adaptation versus Selection in Progress

The distinction between adaptation and selection in progress is simple yet important. An adaptation in the sense of Williams (1966a) is a feature of an organism that can reasonably be said to serve a purpose and is the result of natural selection in the past. Selection in progress is gene frequencies changing now as a result of differences in design between genetically different individuals. An organism may have an adaptation even if selection is not operating on it now. I do not know whether genetic variation is currently affecting the eye in humans, and I do not need to know in order to recognize the eye as an adaptation, to study its function, and to analyze its adaptive value. On the other side, selection in progress may be modifying an existing adaptation, creating a new adaptation, or simply changing the value of a quantitative trait back and forth as generations proceed. A paleontological analogy can be made between adaptations and the bulk of existing fossils, on the one hand, and between selection in progress and present-day corpses, some of which are currently being turned into fossils, on the other.

I am sure this distinction has been widely appreciated before, but I do not think it has been made explicitly in print, presumably because there was no need. Darwin was interested in selection in progress—partly because it was evidence for the mutability of species—and in adaptation. His theories of natural (1859) and sexual (1871) selection are still our only explanations for the existence of organic complexity and adaptation. Wright's four-volume treatise on evolution (Wright 1967–78), is almost entirely concerned with selection in progress and is a fund of information about balanced polymorphisms, selection coefficients, effective population sizes, rates of gene substitution, linkage disequilibrium, dominance, and epistasis. Fisher's book (Fisher 1958), in contrast, is mainly concerned with understanding adaptation and treats selection in progress as an important but logically subsidiary topic. Its topics include mimicry, sex ratios, extravagant male characters, infanticide, and the heroic virtues. Kimura's neutral theory (Kimura 1983) is about what fraction of genetic variability can be attributed to selection in progress, as opposed to random drift and mutation pressure, and is all but irrelevant to the study of adaptation.

The study of adaptations begins with trying to answer the question why. Why do male red deer have antlers? Why are kingfishers brightly colored? Why are black grouse polygynous? These are the kinds of questions that have always been asked about animals, and the key to them was provided by Darwin (1859). The comparative approach, the theory of evolutionarily stable strategies, and functional morphology are methods of studying certain kinds of adaptations. I suspect that most authors and most readers of this book are interested in explaining adaptations.

The study of selection in progress is also fairly old. Animal breeders who keep track of their stocks are interested in selection in progress. The school of ecological genetics is devoted to the study of selection in progress in nature and has made many fascinating discoveries (Ford 1975). To distinguish between different hypothesized modes of evolution—for example, the Fisher-Haldane mode of a succession of more-or-less independent gene substitutions and the Wrightian shifting balance—it is important to study selection in progress. I believe that the authors and readers of this book are less interested in currently changing gene frequencies than in adaptations.

To illustrate the differences in the kind of study necessary to investigate these two distinct problems, I shall use as an example the spot number on the hind wing of *Maniola jurtina*, a character much studied by ecological geneticists whose work is reviewed by Ford (1975) and Brakefield (1984). The present book is about measuring the reproductive success of individuals, a technique the ecological geneticists did not use. The experiments I propose will therefore be hypothetical, and I do not wish to suggest that they are superior to those in fact used.

Suppose first that we wish to discover the adaptive significance of spot number. The obvious experiment is to paint spots on or off the hind wings and to compare (for example) predation, mating success, and thermoregulation in the groups with different numbers of spots. If we found that spottier butterflies were eaten less often but had the same mating success and temperature control, we could conclude that the function (or better, *a* function) of the spots was to avoid predation. LRS is not very useful here because it is too all-encompassing a measure. We wish to know *why* the butterfly has the spots, not *how much* more successful more spotted individuals are than less spotted. To understand the adaptive significance of spot number, we want to pin down more exactly the mechanism of advantage. We can make a start by finding if spot number correlates with components of LRS, hoping to find where it is useful to look more closely for the reasons behind the advantage of spots.

If, on the other hand, we are interested in selection in progress on spot number, then we are looking for evidence of gene-frequency changes at the loci that affect spot number. It would be pointless in this case to create variation by painting spots. Ford and colleagues measured the frequency of adult morphs in successive generations and sought to exclude the other possible causes of the observed changes. With data on LRS of individuals, we could find the covariance between LRS and natural spot number. According to the "secondary theorem of natural selection" of Robertson (1966, 1968), the selective change in a character is equal to its genetic covariance with LRS divided by mean LRS. The genetic covariance is equal to the phenotypic covariance (the one we observe) multiplied by the heritability of the character. So by showing that the heritability was not zero (Brakefield 1984), which is to say there is additive genetic variance for spot number, the covariance between spot number and LRS could be used to demonstrate that there was selection in progress at the loci affecting spot number. It would not demonstrate that spot number was part of the causal chain from genes to differential success of individuals, for it could be another, pleiotropic, effect of the genes that determine spot number. In fact, Ford reports that the selection they detected by measuring spot numbers was probably the result of differential parasitism of the butterfly larvae by a hymenopteran. The reason for the correlation between spot number as an adult and susceptibility to parasite attack as a larva is not known.

We have examined distinct ways of studying the two distinct problems. One particularly important difference between them is the kind of variation exploited. It was simplest, though not necessary, to use artificial variation in the study of adaptation, and it was necessary to use natural variation to study selection in progress.

The distinction between adaptation and selection in progress does not mean there are no connections between them. One obvious connection is that current adaptations are the result of selection that was in progress at some time in the past. Another connection arises in some modern theories of sexual selection and the maintenance of sexual reproduction. Hamilton and Zuk (1982) proposed that sexual selection is a defense against certain kinds of parasitism, in which females choose comparatively unparasitized males so that their offspring will in turn be comparatively resistant to parasites. This is a good case for illustrating the distinction between adaptation and selection in progress, because an adaptation in one character (female choice) is based on the continuing existence of selection in progress not in itself, but in another character (resistance to parasites).

28.2 Using Natural Variation to Study Adaptation

It is a truism that correlation does not prove causation, and this is the reason artificial variation was used in the proposed experiment to discover the adaptive significance of spot number. If we control the variation ourselves, we know its cause and can eliminate the possibility that any other variable correlated with spot number is the true cause of differences in LRS. Had we used natural variation in spot number, it would have been much more difficult to eliminate this possibility. The gain from using natural variation is that we need only measure (components of) LRS and ten characters in order to investigate the adaptive significance of ten characters: to conduct ten experiments to do the same thing would be much more work. It is prudent to be suspicious of bargain offers. Of course all methods of investigation have their drawbacks, and in analyzing LRS data I would certainly perform the LRS analyses first and unleash my suspicions on the results afterward.

That natural variation may mislead us does not mean that it will. In this section I consider some plausible causes of natural variation and ask how they would affect an attempt to discover the adaptive significance of spot number by means of correlations using the natural variation in spot number and LRS (or components of LRS).

Why might spot number vary in nature? The first candidate component of variation in spot number is mutational variance held in check by stabilizing selection. This is variation caused by the occurrence of slightly deleterious mutations that are eliminated so slowly that they remain and contribute to variance for some time. If this is the cause, then the natural variation is just as good as the same quantity of artificial variation. Apart from mutations, no other differences exist between many-spotted and few-spotted butterflies.

If the population is in equilibrium between mutational variance and stabilizing selection, LRS should be higher at the intermediate values of spot number and lower at each extreme. The slope of a linear regression of LRS against spot number would therefore be zero or very small. That LRS is lower at both ends would produce a negative quadratic term in a polynomial regression. Lande and Arnold (1983) discuss using regressions in this way. As before, however, this regression with LRS tells us how advantageous different spot numbers are but says nothing about why. To discover the adaptive significance, it would be of more interest to know the effect of spot number on particular causes such as predation or mating success.

Simple mutational variation checked by stabilizing selection is unlikely to be the sole source of variability in spot number. (Spot number has now become simply a less abstract way of saying "the character of interest." Ford [1975] and Brakefield [1984] should be consulted for facts about the real spot number.) Let us consider the effect of purely environmental variation. Suppose a warm pupation site is an unplannable piece of good luck and that it has an effect on spot number that no genetic combination could reliably produce. Suppose further that those individuals that pupate in a slightly warmer place have more spots while those that pupate in a cooler place have fewer, and that no other variables are affected in this way. Now the linear component of a regression of LRS on spot nubmer need no longer be zero, since it is possible even in equilibrium that the possession of more spots is consistently advantageous. The effect of spot number on predation could be investigated by regressing predation rate on spot number, and it could be established that the adaptive significance of spot number was as an antipredator device. With this cause of natural variation, correlation with components of LRS is a reliable guide to adaptive significance.

I was careful to say that simple mutational variance was as good as *the same quantity* of artificial variance. We cannot choose how much mutational or environmental variance there is, and the ability to detect effects will be roughly proportional to the amount of variance. Experiment then still has the advantage that the quantity of variance can be chosen. The experimenter of course has problems of his own, including the naturalness of his manipulations and the disturbance they cause.

With these two benign causes of natural variation, the existence of correlation is evidence for causation. However, we cannot be sure that these are the only two causes acting in a particular case. We consider next other possible causes of natural variation that lead to less welcome conclusions. We have discussed additive genetic variation and environmental variation that affect spot number only. Problems arise when the cause of natural variation means that other characters are correlated with spot number.

The first malignant cause is pleiotropic mutational variance. If mutations affect a number of characters, then those that are disadvantageous in all their effects will be more quickly eliminated than those that are advantageous in some effects and slightly disadvantageous on balance. This implies that the mutations that persist will tend to produce a negative correlation between those characters that are positively associated with fitness (Falconer 1981, 300). Consequently an individual who has a disadvantageous spot number will tend to have, say, an advantageous wingspan, and similarly in reverse. By using natural variation in spot number we will measure the combined effects of spot number and the correlated part of wingspan. The consequence of this is to diminish the strength of associations between spot number and LRS and, to a lesser extent, with its components. Wright (1968–78, 1:61) states that "the available evidence indicates that pleiotropy is virtually universal." We define characters according to our interests, and we can hardly expect the same divisions to be observed by the biochemistry of development.

The second factor is the "silver spoon" effect, a particularly likely example of a common environmental cause that influences many characters. I define the silver spoon effect as positive correlations between characters in the adult that are positively associated with fitness, brought about by the common underlying cause of favorable or unfavorable environmental events during development. It differs from the example of warm pupation sites only in that many characters are affected, and this is exactly what causes problems. Let us switch examples and think of juvenile red deer. There are many accidents of childhood, including the extent of parental care, the severity of winters, and the chance occurrence of epidemics. They are likely to produce some individuals that are vigorous and fit in both senses of the word and others that are weak and unfit. The syndrome of characters called "quality" may often be determined by the silver spoon effect. If this does occur, we will observe a positive correlation between LRS and a whole host of characters, most of which will be correlated with each other. The components of LRS would also tend to be intercorrelated. To discover the adaptive significance of antler size, we would have to partial out the effects of all the correlated characters. The effect of early food availability is documented for domesticated animals by Sadleir (1969). Lande (1982) discusses the likelihood of negative genetic correlations but positive phenotypic correlations between characters positively associated with fitness.

The third factor has been called "making the best of a bad job" (Dawkins 1980). This applies particularly to behavioral characters, and the idea is that the observed differences may be a response by individuals to their particular circumstances. These preexisting circumstances are then confounded in the analysis with the character. This effect is fairly subtle, and I can think of no plausible animal example. I believe this is because we would have to understand a species rather better than we generally do to recognize it. As a human analogy, consider a group of children who have to catch a bus to school in the morning. Some will walk and some will run. Will the walkers or the runners be more likely to catch the bus? I would bet heavily on the walkers. For any given child on a given morning, running would increase the chance of arriving on time. However, those who choose to run will be those who got up late and are at risk of missing the bus, and those who choose to walk are those who got up early and will catch the bus with no dfficulty. A correlation of speed from home to bus with catching the bus, naively interpreted, would therefore reveal that running made a child less likely to catch the bus. The difference in behavior is the result of a decision that produces a correlation between the behavior and an underlying variable. Particularly for reproductive strategies, we must suspect that differences in behavior are "adaptive reactions" of this sort of preexisting relevant physical and social circumstances.

These adaptive reactions are in one sense just correlations between characters, but I believe they have a special power to mislead, for three reasons. First, the individual animal is likely to have much more information about its state of health and nutrition than we have. Second, in deciding between fighting or reproductive strategies, a small difference in state can cause a large difference in the strategy chosen, particularly if the options available are discrete. Third, it is natural for us to ascribe differences in success to correlated differences in behavior, perhaps because we think of our own behavior as causing but uncaused. Important but poorly observed information about an animal's state may therefore be converted into easily observed behavior by the animal's adaptive reaction to that state. We would then assign to the behavior what are really consequences of the underlying state.

This completes my catalog of malignant causes of natural variation that would invalidate simple inferences about adaptive significance, though the catalog is by no means exhaustive. In no case was a correlation with LRS itself of much importance. If the purpose of the exercise is to understand the adaptive significance of a character, correlations with predation rate or harem size or fighting ability are more to the point. This is because an adaptive explanation is about why something is advantageous, not just how advantageous it is. The main conclusion of this section is that what a correlation with LRS or its components tells you depends on the causes of the natural variation in the character. It will not always be easy to discover those causes.

28.3 The Partitioning of Natural Variation in LRS

The stages between which variation may be divided can be illustrated using the example of *Maniola jurtina*. An egg has a certain chance of surviving to become a larva, which has a certain chance of surviving to become a pupa, which has a certain chance of surviving to become an adult. Once adult, its reproductive success is the product of longevity and fecundity per unit time. Individuals may differ to varying degrees in their success in these five stages.

There are a number of proposals for partitioning variance in LRS between different stages of the life cycle and reproductive cycle (Wade and Arnold 1980; Arnold and Wade 1984a,b; Brown, this volume, chap. 27), similar to the key factor analysis of Varley and Gradwell (Varley and Gradwell 1960; Southwood 1978) and an extension of work by Crow (1958). Crow suggested a partition of his "index of selection," and Jacquard (1974, 322–30) gives a number of applications of this partition to human populations. The idea is that by decomposing the variance between survival as egg, larva, and pupa, adult longevity, and fecundity, we can identify the stages at which individuals differ greatly. The total variance in LRS, its difference between the sexes, or the variance among males in mating success are sometimes thought to be useful in deriving a measure of sexual selection (Wade 1979; Payne 1979; Wade and Arnold 1980). My aim in this section is to discuss what significance we can attach to these variances under different possible sets of causes of natural variation, when our purpose is to explain adaptation. Since the various causes are explained in the previous section, I shall not explain them again.

The first candidate cause of variability is mutational variation held in check by stabilizing selection. Surprisingly, the amount of genetic variability in fitness in this case seems not to have been investigated directly. Turelli (1985), reconciling conflicting results of Latter (1960) and Bulmer (1972, 1980) on the one hand and Lande (1976) on the other, shows that the variance of an ordinary character will be greater with higher mutability of the loci involved and with weaker stabilizing selection. Fitness is no ordinary character, and it appears likely from Turelli's finding that genetic variance in fitness will be greater with higher mutability and strong stabilizing selection. Thus the genetic variance in fitness at a stage does increase with the strength of stabilizing selection, but comparisons between stages are vitiated unless we know the mutabilities of the sets of loci affecting fitness at the different stages.

Once environmental variation is allowed, however, the inference from variance at a stage about the strength of selection acting at that stage becomes even weaker. The larger part of variance in success in the various stages of life may well be environmental, as suggested by the effect of food availability during youth on subsequent success in domesticated animals (Sadleir 1969) and the low heritabilities generally reported for characters strongly related to fitness (Falconer 1981).

The claim made by Wade and Arnold (1980) is that the variance assigned to a stage sets an upper limit to the selection that can go on at that stage. (They use language differently: By "selection" they mean the phenotypic change within a generation, and by "response to selection" they mean the phenotypic change between generations.) This upper limit is to directional selection in progress, as distinct from stabilizing selection in progress and as distinct from adaptive value. Upper limits are likely to be good guides in two kinds of circumstances. One is when the upper limit is likely to be nearly attained, as would be the case if the internal volume of a glove were used as a guide to the volume of the hand that wears it. (It is a poor guide to the quantity of air trapped inside the glove.) The other case is where the upper limit contains two elements in roughly constant proportions, one of which is the value of interest. For example, the volume of a container filled with air is a good guide to the volume of free oxygen it contains because fluctuations in pressure and in relative concentrations of the different atmospheric gases are comparatively small. We have seen that the limit to selection in progress is probably not very nearly attained in general, and I know of no reason to believe that genetic and environmental variation should be of roughly constant proportions in different characters or life stages. But for the purposes of understanding adaptation, the most important point is that the variance at a stage, if it measures anything at all, measures something to do with selection in progress, not with adaptive value.

Let us now turn to the significance of variance in LRS within sexes, and in particular to whether it can be used to detect or measure sexual selection, as has been suggested by Payne (1979), Wade (1979), and Wade and Arnold (1980). In this section I consider questions about the adaptive significance of character, not about selection in progress. The kind of question we are trying to answer is if measuring variance in LRS can help us decide whether antlers in red deer are the product of sexual selection. I shall argue that the answer is no, that the natural variance in LRS is entirely irrelevant to this problem. The substance of the argument follows from the distinction between adaptive significance and selection in progress. Variance now in LRS may be relevant to sexual selection in progress, but it need have nothing whatever to do with the nature of selection in the past.

Suppose that male red deer live for ten years without mating, in their eleventh year develop large antlers and defend a harem, and then die. Suppose what is even more unlikely, that each male gains exactly the same number of matings in its life. We cannot conclude that antlers are not sexually selected, despite the zero variance in LRS between males. There will also be zero variance in components of LRS such as number of mates. It may be that all males have the same size antlers and that any variant would be punished by being unable to hold a harem (say because females would leave a small-antlered male, while a large-antlered male would be too encumbered to fight). In this case the adaptive significance of antlers is to attract females, and antlers are certainly the result of sexual selection. The point of this extreme example is to show that current variance in LRS in males and sexual selection are in principle quite separate.

It is interesting to compare these arguments with those of Arnold (1983a, 67), who discusses the analysis of sexual selection using variances in LRS and its components. He states that the aim of the method is "merely to characterise sexual selection by its statistical effects on phenotypic characters within a generation," that this "of course, tells us nothing about how selection actually worked in the past," and that "the goal is to understand the process of sexual selection by direct measurement of its contemporary impact." It seems fairly clear that the focus of interest here is measurement of selection in progress, not the analysis of adaptation.

I believe that most evolutionists and behaviorists would say they were primarily interested in adaptation, as opposed to selection in progress, once the distinction is brought to their attention. Their primary concern is why male red deer have such big antlers, not whether there are genes now changing in frequency that affect antler size. Of course the gene-frequency changes are interesing, just as the physiological mechanisms of antler creation and the morphological modifications for bearing the antlers are interesting. But they are not central.

In principle, therefore, variance in LRS and analysis of adaptations due to sexual selection are separate. It is likely, though, that variance among males in LRS will be higher in sexually selected species. This is because monogamous species tend to be less strongly sexually selected and will probably have a lower variance in LRS. A positive correlation between variance in LRS and degree of sexual selection would therefore not be surprising, and it would not refute the logical point that one is neither an explanation nor a measure of the other.

28.4 The Proposed Methods of the Chicago School

In this section I will consider the methods proposed by Wade and Arnold (1980), Lande and Arnold (1983), and Arnold and Wade (1984a,b). Three problems with their methods arise from earlier discussion, but it is not my intention to discourage the application of their methods. (Experiments of course have problems of their own.) With suitable data I would certainly use them, because selection in progress is interesting and because they provide hints about adaptation. My aim is only to give some cautions about the interpretation of their results.

In making claims for their methods, Arnold, Wade, and Lande do not always distinguish clearly between the analysis of adaptation and the detection of selection in progress. It is clear, however, that the design of their methods is to detect selection in progress. Witness the connection, of which they make much, between their methods and the dynamic equations of evolutionary change, and the idea that they are measuring the *potential for selection*, of which only a certain fraction becomes the *response to selection*. Their technique could be compared with other, more direct, techniques for detecting selection in progress, such as those described by Dobzhansky (1970) and Ford (1975). More relevant here is that some of their proposed methods may be useful for the study of adaptation while others may not, and those that are useful for this distinct purpose will require different caveats and face different problems of interpretation.

The variance partitions, magnitudes of variances in LRS, and regressions of LRS on characters are valuable additions to our knowledge of the natural history of a species. However, they tell us little about selection in progress unless we can estimate the environmental causes of variation at each stage, and they reveal little if anything about adaptation. Like the existence of sexual dimorphism, they may *suggest* that males in a species are sexually selected but cannot be used to *show* that they are. The technique that *can* tell us about adaptation is the regression of components of LRS on characters.

Let us consider the multivariate regression techniques proposed by Lande and Arnold (1983) and developed by Arnold and Wade (1984a,b). Using natural variation, they estimate a number of quantities for each character that are relevant to selection in progress. Three of them are the opportunity for selection (the variance in the character), the selection differential (the covariance of the character with an LRS component), and the selection gradient (the slope of the best-fitting straight line relating the LRS component to the character, holding other variables constant). The first two are not likely to be relevant to adaptation, but the third is. The meaning of a selection gradient may be seen as follows. Suppose antler length is measured in inches and LRS is measured in percentage of the mean LRS. Then a selection gradient for antler length of three means that if two stags are the same in every respect except that the first one's antlers are longer by one inch, then the first will have an LRS that is (on average) higher by three. "The same in every respect" means the same in every other character included in the analysis, not the same in every other character that affects fitness, which is of course the unattainable ideal. It will be seen that the technique avoids many of the problems of the correlation of characters.

How valuable is this selection gradient in the analysis of adaptation? The problems are two we examined in section 28.2, namely, correlation of characters and the amount of natural variation and the additional problem of *incompleteness*.

One strength of the Lande and Arnold (1983) analysis is that correlation between characters that are included in the multivariate regression will not confound the estimates of selection gradients, but as they point out, the estimates may be confounded by missing charactrs. How likely is it that all or most relevant characters will be included? In *Maniola jurtina*, the relevant correlate of adult spot number was thought to be parasitism as a larva. A character may be influenced by pleiotropic loci with (to a student of behavior) obscure physiological effects. The silver spoon effect described in section 28.2 is likely to be the result of unobserved environmental effects. To the extent that we can find surrogates for these effects, such as weight at one year old, their influence on the selection gradients of other characters can be removed; but to the extent that those surrogates are imperfect, that influence will remain. The "adaptive reaction" of section 28.2 is also a problem for the reasons explained there, so long as we remain ignorant of the behavioral rules the animals use. Clear and observable differences in behavior may be consequences of hard to observe underlying differences in health or vigor that are the true causes of variation in success.

The next problem in the interpretation of selection gradients is the quantity of natural variation. If this is small, then it will be difficult to detect an effect that genuinely exists. The variation that is relevant in the Lande and Arnold (1983) analysis is not simply the variance in a character, but only the part of it that cannot be predicted by variation in other characters. A selection gradient insignificantly different from zero may occur because the character is selectively irrelevant or because the character does not have enough independent variation. Andersson (1982) found that male widowbirds with larger tails were more successful than those with smaller tails, using an experiment. The experimentally manipulated tails were about five standard deviations from the mean. To detect the same size of difference using natural variation, even supposing it to be benign, would have required much larger sample sizes.

The third problem is that the data set may be incomplete. The most likely cause is that early mortality may prevent the measurement of important variables. In *Maniola jurtina*, the spot number of individuals parasitized as larvae by hymenopterăns is believed to be different from those that escape parasitism. By using natural variation, we cannot discover this, because the spot number can be measured only in adults. The penalty for a male red deer planning to have a large body size may be that by allocating resources to growth instead of to defense against disease, it is more likely to die as a juvenile. So the "largest" males die young and cannot have their adult size measured. There will often be an "invisible fraction" of individuals who do not appear in the regression but should. The close connection between the multivariate regression and the dynamic equations of evolutionary change depends on the completeness of the sample of individuals in the regression. Sex-limited characters may be genetically correlated with other characters in the sex in which they are not expressed, and if this is so, then a whole sex may belong to this "invisible fraction."

In conclusion, the Chicago school's methods are interesting analyses to perform on suitable data. They provide clues about the adaptive significance of characters and are interesting in their own right as a contribution to the natural history of the species. For the reasons given above, the regression results must be treated with caution when applied to the analysis of adaptation. The variance partition is not a measure of the extent to which the behavior or morphology of a species may be attributed to sexual selection.

28.5 Measurements of Fitness

In this section I discuss ways of measuring fitness. A fieldworker who has measured the LRS of a number of individuals may like to feel that the bulk of the work is over, that it is necessary only to find the correct statistical technique, and a convenient computer program to implement it, before the biological significance of the study will be revealed. The feeling is tempting because fitness is the central concept of evolutionary biology, and LRS seems to be fitness. I hope to convince the reader that LRS is in an important sense not Darwinian fitness. In the next section the important case of inclusive fitness is discussed.

Following Williams (1966a, 158), we may interpret Darwinian fitness as a property of a design, not of an individual, and particularly not as the number of offsrping an individual happens to produce. Rather it is the number of offspring that a given design of animal will on average produce. Suppose a new mutation occurs, and the reproductive success of its few bearers fluctuates greatly between generations because of random factors peculiar to individuals, such as being struck by lightning. If Darwinian fitness *were* LRS, we would be forced to say that the Darwinian fitness of the new design fluctuated greatly between generations. Williams would want to say this only if the environment changed so much that a bad design in one year was a good design in the next. Too much should not be made of this distinction, but it is important that we are trying to estimate something we cannot directly observe. With a large enough sample, we could get as close as we like.

LRS is clearly a very important measurement in estimating Darwinian fitness, but this does not mean they are the same thing. Darwinian fitness is the property of a design, while LRS is a property of an individual. LRS can be known exactly, while we estimate Darwinian fitness with error. The difficulties encountered in previous sections with ascribing differences in LRS to differences in a character are difficulties in estimating Darwinian fitness from LRS. Any difficulty in discovering LRS is purely observational.

The crucial extra stage in estimating Darwinian fitness from LRS comes in deciding what character is to be studied. Plotting mean LRS against size for range of size-class intervals will estimate the Darwinian fitness of different-sized animals. The Lande and Arnold (1983) regression is a more sophisticated version of this, and their selection gradient will estimate the marginal effect on Darwinian fitness of an increase in size. Special problems arise when we try to apply this method to social behavior, as a special case of the "invisible fraction" discussed in the previous section.

Suppose we wished to determine whether an altruistic act was favored by selection or not, by means of the LRS method. The obvious way to proceed is to find the average LRS of altruists and compare this with the average LRS of nonaltruists. The problem here is that altruism will often be conditionally expressed, and thus many of the apparent nonaltruists may be altruists for whom the appropriate circumstances did not arise. If the "genuine" nonaltruists and the "altruistic" nonaltruists benefited from the altruism of others in the same way, this would be unimportant. However, if altruism is directed at relatives, then "altruistic" nonaltruists will be the recipients of help more often than "genuine" nonaltruists. The result will be that the success of nonaltruists will be overestimated, and so the selective advantage of altruism will be underestimated. I have discussed this elsewhere (Grafen 1984, 1985). In the next section I discuss how the advantage of altruism could be estimated. The main point for now is that LRS fails to give the right answer when we cannot identify the sets of individuals whose average LRS would give the correct estimates of Darwinian fitness.

Next in this section on measurements of fitness, a few comments on the problem of overlapping generations. By fitness I have meant (expected) lifetime number of offspring. Fisher (1958) and, in more detail, Charlesworth (1980) have pointed out that offspring further in the future should be discounted by a factor that depends on the growth rate of the population. Selection favors early reproduction in a growing population and late reproduction in a shrinking population, if the total number of the individual's offpsring must remain constant. Equally, in a static population, selection is indifferent between early and late reproduction if the total number of offsping remains constant.

Populations that are neither extinct nor continually increasing must have a long-run average growth rate of nearly zero. If individuals can detect and respond in their behavior to the growth or shrinking of the population, and if the number of offspring they can produce is independent of when they are produced, then they may indeed be selected to value differently offspring produced at different times. If, on the other hand, they lack this adaptive flexibility, or the population never grows or shrinks very fast anyway, then they will tend to behave appropriately to the long-run average growth rate of nearly zero. This is the justification for the simplifying assumption that expected number of offspring is the proper measure of fitness.

Finally, at what stage should fitness be measured? I have used expected number of *offspring*. Why not grandoffspring, or descendants at some hypothetical time in the infinite future? Suppose our aim is to understand the evolution of a character, which for simplicity we shall assume to take one of two possible forms. Then our choice of fitness measure must satisfy two properties: that we can work out the fitness measure from our knowledge of or assumptions about the character, and that if one form has a higher fitness than other, it will increase in frequency in the population. Subject to these conditions, the simpler the fitness measure the better. The use of number of offspring is suitable if the character *in the adult* does not affect its offspring's survival or fitness after the age of counting.

A very convenient assumption in measures of fitness is independence of control. When we count offspring as a measure of the parent's fitness, we implicitly assume that up to the stage of counting the survival of an offspring does not depend on its own phenotype. This independence of control is unlikely to be fulfilled by a simple measure of fitness that extends beyond offspring to future generations. This assumption is not as restrictive as it seems, for the independence of phenotypic effects applies only to the character being studied, not to the phenotype as a whole. Cheverud (1984) presents a model with cross-generational effects.

The best measure of fitness is then determined by the character we have in mind rather than by any deep principles. This is another illustration of the fact that Darwinian fitness and LRS are different, because Darwinian fitness should be computed differently for different characters according to the stage of reproduction at which they take effect. For example, a character that determines survival as a juvenile will be best studied using survival to breed as a measure of fitness. Using number of offspring surviving to breed would be assessing the phenotype of the offspring rather than the phenotype of the adult. Contrariwise, we might study a character that affected the survival to first breeding of the individual's offspring. Then number of surviving offspring is the appropriate measure of fitness because it assesses the individual's phenotype and because the offspring's own value of the character does not confound the analysis. Using number of offspring born as a measure of fitness does justice to neither of the examples, although it could be used. One individual's fitness would then depend on another individual's phenotype, which is the circumstance for which Hamilton (1963, 1964, 1970) developed his theory of inclusive fitness. Simplicity suggests using an individual measure where possible.

One much-discussed character for which a fitness measure would have to go beyond offspring is sex ratio. In simple models, grandoffspring are best counted. The effect of an individual's phenotype on *number* of descendants is not apparent until the grandoffspring generation, but independence of control is maintained because the offspring's phenotypes do not take effect until the generation after that. The point, then, is that the best place to measure fitness depends on the character being studied. If no place can be found that satisfies independence of control, then more complicated modeling is necessary.

I conclude this section by repeating its important points. There is an important sense in which LRS is not Darwinian fitness, but rather Darwinian fitness is an abstraction that may be estimated, not always straightforwardly, from LRS. LRS data are not, as it may seem at first thought, an infallible oracle for answering important evolutionary questions about a species. Indeed, we have yet to come across a question for which they are decisive, and in every case it would be valuable to have other kinds of data as well.

28.6 Inclusive Fitness

In this section I discuss the application of inclusive fitness (Hamilton 1964, 1970) to data. Elsewhere I have discussed errors in alleged definitions of inclusive fitness and also worked out an example from data as an illustration of how inclusive fitness should be applied (Grafen 1982, 1984). Here I shall repeat the most salient points and explain their relation to the dichotomy of natural versus artificial variation. First, it is better to apply Hamilton's rule (shortly to be defined) than inclusive fitness to data. The two are formally the same, but as historical fact Hamilton's rule has been correctly applied, whereas inclusive fitness has been incorrectly applied to data (Grafen 1984). The reason is that applying Hamilton's rule suggests to the user's mind the correct logic of differences.

To apply Hamilton's rule, we need to estimate the three terms that appear in it. The rule is that "a social action is favored by natural selection if $rb - c > 0$": r is the relatedness of the donor to the recipient, b is the benefit to the recipient of the action and c is the cost of the action to the donor. I have discussed relatedness and derived and defended Hamilton's rule as an evolutionary principle at some length in a recent paper (Grafen 1985). In field studies of the sort in this book, relatedness will usually be known through observed common ancestry, or sometimes it may be estimated from electrophoretic data (Pamilo and Crozier 1982). Here I will concentrate instead on the estimation of b and c.

The benefit and cost are differences in number of offspring that are caused by the performance of the social action. The meaning of b therefore involves something we do not observe; b is the number added to the recipient's number of offspring by the social action. We need to know how many the recipient would have had with the social action, and how many without. Obviously we cannot measure both in the same individual, but this does not mean we are helpless. We must use our understanding of the animal and its actions to estimate what that difference is. If, for example, the social action is help in surviving the winter, then we may know the reproductive value of an animal that survives the winter of a given age and the chance of survival with and without help by observing the survival of individuals with different fat reserves.

The complications involved, and the assumptions that need to be made, may at first seem a disadvantage in applying Hamilton's rule. This should really be turned on its head. If your aim is to understand the adaptive significance of a social action, then these "complications" are exactly the constituents of that understanding. If you do not know the effect on survival of help, then you cannot possibly understand the adaptive significance of that help. The value of Hamilton's rule is that it draws attention to three essential quantities, and says: If you can estimate r, b, and c, then you understand the adaptive significance of the social action. If you cannot estimate them, it is unlikely that you understand the adaptive significance of the social action.

The conclusion of more general interest is that Hamilton's rule involves working out what would have happened in circumstances we do not observe. It is therefore an example of an artificial variation technique, though not necessarily an experimental one. These workings out are exactly what is necessary to discover the adaptive significance of the social action studied.

28.7 Conclusions

The primary purpose of the contributors to this book seems to me to be the analysis of adaptation rather than the detection or measurement of the changes in gene frequency that constitute selection in progress. The logically most straightforward way to understand an adaptation is to perform an experiment, as Andersson (1982) did with his widowbirds. An alternative is to use LRS data. Problems arise in this LRS method because it relies on natural variation in characters, when the cause of that natural variation is usually unknown. Correlations with relevant characters not included in the analysis, the quantity of variation, and the omission of dead individuals from the sample of those measured are three such problems.

The intensive kind of study that produces LRS data is likely to uncover at the same time many interesting facts about the species. The LRS data themselves add to our knowledge of the species' natural history in an important way. They are not, however, the answers to a theoretician's prayer. The methods of analysis of LRS data proposed by Wade and Arnold (1980), Lande and Arnold (1983), and Arnold and Wade (1984a,b) seem primarily designed to study selection in progress—that is to say, gene frequencies changing now rather than adaptation. Their multivariate regression method is a very interesting one to perform even for the study of adaptation but is often difficult to interpret because of its use of natural variation.

It is necessary to understand the causes of natural variation in order to draw conclusions from natural LRS data about adaptations and about their attribution to natural or sexual selection.

28.8 Acknowledgments

I am grateful to T. H. Clutton-Brock, M. Ridley, M. S. Dawkins, J. A. Endler, S. J. Arnold, M. A. A. Otronen, W. B. Sutherland, N. E. Pierce, J. R. Krebs, and S. A. Frank for helpful comments and discussions. The work was supported by a Julian Huxley Junior Research Fellowship at Balliol College, Oxford, and by a Royal Society 1983 University Research Fellowship.

29 Reproductive Success

T. H. Clutton-Brock

THE TWENTY-FIVE EMPIRICAL CHAPTERS in this book provide an opportunity to compare the results of longitudinal studies of reproductive success in different species. In this final chapter, I return to the questions raised in the Introduction, briefly reviewing some of the contrasts and generalizations that have emerged and assessing the implications of these studies for future work.

29.1 How Widely Does Breeding Success Vary among Females?

The small range of clutch or litter size that is found within seasons in most birds and mammals has led to suggestions that variation in female breeding success is necessarily low and that competition between females is consequently unlikely to be intense (see Hrdy and Williams 1983). For example, "Most adult females in most animal populations are likely to be breeding at or close to the theoretical limit of their capacity to produce and rear young. Among males, by contrast, there is always a possibility of doing better" (Daly and Wilson 1978, 59). Or, "About 20% of the [rhesus macaque] males (all high in the status hierarchy) performed about 80% of the copulations, whereas all estrous females tended to be impregnated. These data make it clear that only males are directly involved in differential selection among rhesus and most probably all the terrestrial and semiterrestrial primates" (Freedman 1979, 33).

The available information on female breeding success provides no justification for arguments of this kind, and all four parts clearly show that variation in female breeding success is substantial, both in species that show extended care (and consequently have small clutch sizes) and in those that do not (see fig. 29.1). Moreover, there appears to be a consistent tendency for estimates of standardized variance in female breeding success based on the number of zygotes produced to underestimate variation in the number of offspring recruited into the breeding population, in some cases by as much as an order of magnitude (see fig. 29.1). No field study has yet been able to incorporate variation in offspring breeding success as

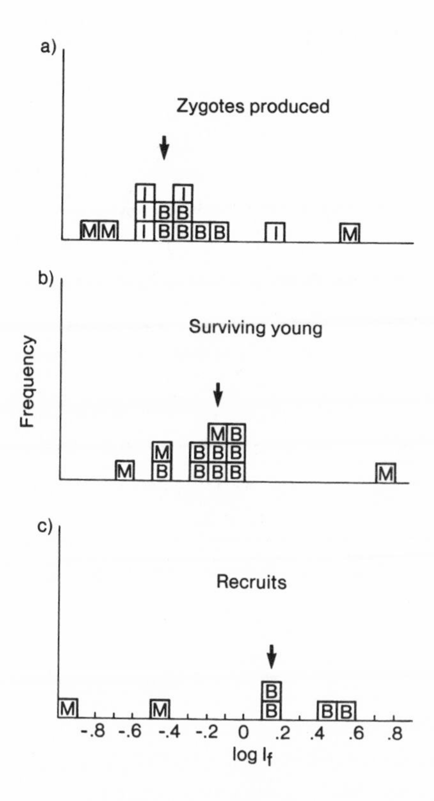

Figure 29.1: The distribution of (log) estimates of the total opportunity for selection among breeding females (I_f): (*a*) Standardized variance in the number of zygotes produced over the life span (data from chaps. 1, 3, 4, 5, 8, 10, 11, 13, 17, 20, 21, 22, 24, and 26). (*b*) Standardized variance in the number of young reared to hatching or fledging (data from chaps. 8, 10, 11, 13, 14, 16, 17, 20, 21, 22, 23, and 24). (*c*) Standardized variance in the number of young contributed to the breeding population (data from chaps. 8, 9, 10, 20, 21, and 26). I, insects; B, birds; M, mammals; arrow indicates median.

well, but the existence of consistent relationships between early growth and breeding success in adulthood (see chap. 21) indicates that their inclusion would raise estimates of variation in female breeding success further still.

Comparisons of variation in female success between studies are hampered by differences in sampling techniques and methods of analysis. None of the eighteen studies that examined variation in female success over the breeding life span were able to follow entire samples of individuals throughout their breeding lives, and all suffer to a greater or lesser degree from inaccuracies in the measurement of fecundity or juvenile mortality, problems of immigration and emigration, disappearance of individuals for unknown reasons, biases toward individuals dying young, and the inclusion of truncated life spans. Moreover, comparisons between studies were complicated by variation in the definition of breeding adults, the stage at which reproductive success is calculated, and the duration of the study relative to the species' life span as well as by demographic factors affecting mean reproductive success.

29.2 How Large Is the Contribution of Different Components of Fitness in Females?

One of the most obvious generalizations arising from the analysis of fitness components in females is that individual differences in offspring survival are one of the most important components of variation in lifetime reproductive success among breeding females in many birds and mammals (see parts 3 and 4). In particular, differences in offspring survival *after* fledging or weaning were the principal source of variation in reproductive success among breeding adults in several species (see chaps. 8, 20, and 21). Studies of invertebrates have not yet been able to measure individual differences in offspring survival in natural populations (but see Prout 1965), though evidence that egg size differs between females and is related to the growth and survival of offspring suggests that differences may occur (van den Berghe and Gross, n.d.; Wall and Begon, n.d.).

The magnitude of individual differences in offspring survival raises the question of how fitness should be measured (see Endler 1986). The omission of offspring survival may produce misleading estimates of the strength of selection. Conversely, its inclusion can cause the effects of heritable factors that influence survival to be included twice (see Prout 1965; Endler 1986). As Grafen points out (chap. 28), there is no single answer to this problem, and fitness may need to be defined in different ways in different circumstances. One important criterion in deciding what stages of the life history to include is the extent to which the survival or breeding success of progeny depends on their own phenotype versus that of their parent(s) (see chap. 28). In the first case, it may usually be more convenient to limit the definition of fitness to a single generation, but in the second, the inclusion of offspring survival or breeding success may be necessary.

The magnitude of variation in offspring survival between fledging or weaning and recruitment suggests that we might expect parents to extend care beyond fledging or weaning where they can increase the survival of their progeny by doing so, even where the costs of extended care to their subsequent reproductive potential are high. In several mammals, mothers appear to help their daughters acquire territories (Montgomery and Sunquist 1978) or high dominance rank (Silk 1983), and they may even help suppress animals born into the group that are likely to compete with their own offspring in the future (Dittus 1979, 1980; Silk 1983). Indeed, it may be that the evolution of female-bonded groups in many birds and mammals represents an adaptation for maximizing the survival of offspring after fledging or weaning (see Goss-Custard, Dunbar, and Aldrich-Blake 1972).

29.3 How Widely Does Breeding Success Vary among Males?

Though the eighteen studies that measured variation in breeding or mating success among adult males suffered from the same problems as estimates of variation in female success (see above), their results indicate that the lifetime breeding success of males is most variable in species where direct conflict over mating access is common and least variable in species where males compete indirectly and conflict is unusual (see chaps. 2, 3, 7, 18, and 22). High variance is often associated with prolonged defense of territories (as in *Megaloprepus*) or harems (as in elephant seals), but it can also occur where males compete for single females (as in *Jalmenus*). Whether

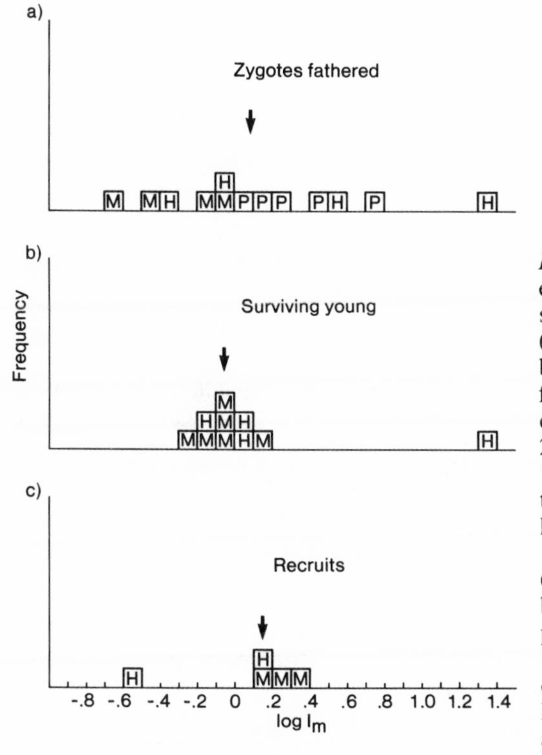

Figure 29.2: The distribution of (log) estimates of the total opportunity for selection among breeding males (I_m): (*a*) Standardized variance in the number of zygotes matings or of zygotes fathered over the life span (data from chaps. 3, 4, 5, 10, 11, 17, 18, 20, 21, 22, 23, and Banks and Thompson 1985). (*b*) Standardized variance in the number of offspring reared to hatching or fledging (data from chaps. 10, 11, 14, 16, 17, 20, 21, 22, 23, 26). (*c*) Standardized variance in the number of offspring recruited to the adult population (data from chaps. 9, 10, 17, 21, and 26). M, monogamous (including scrub jays); P, promiscuous; H, harem-holding (including lions); arrow indicates median.

variance in male breeding success is consistently higher in species where breeding access depends on the individual's dominance rank than in species where males defend territories (see Owen-Smith 1977) or in lek-breeding animals compared with other polygynous species is not yet clear (see Payne 1984; Trail 1985). The compression of male reproductive life spans in highly polygynous species should make us wary of any arguments based on estimates of variation in male success within particular days or breeding seasons.

Differences between monogamous and polygynous species are smaller than might have been expected. Although variance in male success in monogamous species is never as great as in the most strongly polygynous ones (see parts 3 and 4), there is considerable overlap between the two groups (see fig. 29.2), and when offspring survival is included in measures of male success, there is no consistent difference between monogamous and polygynous species (see figs. 29.2*b*,*c*; see also Clutton-Brock 1983).

In both groups, standardized variance in male success commonly exceeds variance in female success, though differences between the sexes in polygynous species are smaller than estimates based on short-term data would suggest (see Payne 1979). Ratios of the total opportunity for selection in the two sexes (I_m/I_f: see Wade and Arnold 1980) are an unsatisfactory basis for comparison, since they vary as a result of species differences among *females* and are strongly affected by local differences in the sex ratio as well as in mean fitness (see chaps. 3, 4, and 7). However,

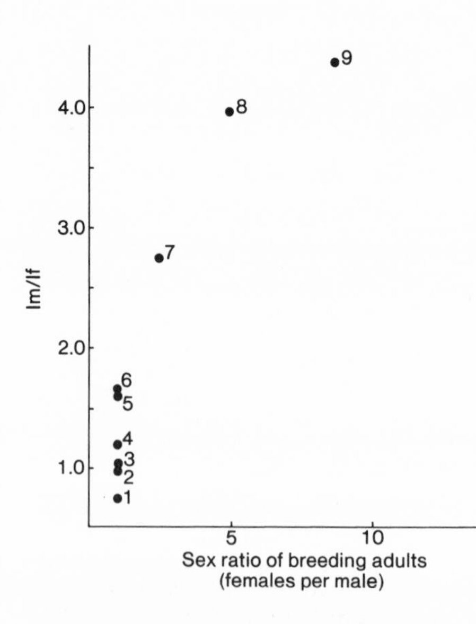

Figure 29.3: The comparative opportunity for selection in males and females (I_m/I_f; see Wade and Arnold 1980), plotted against the degree of polygyny (sex ratio of breeding adults in birds and nonhuman mammals: (1) great tit; (2) Bewick's swan; (3) scrub jays; (4) kittiwake; (5) song sparrow; (6) house martin; (7) lion; (8) red deer; (9) elephant seal.

the available data do suggest that, across species, the relative opportunity for selection in males increases with the number of females that males can monopolize (see fig. 29.3). The higher variance in male mating rate in the most polygynous species may be partly caused by the increased opportunity for chance to influence mating rate in species where the maximum mating rate is high (see Sutherland 1986b). However, the evidence that mating success depends on phenotypic characteristics in these species indicates that this is unlikely to be the only reason for the association.

As Grafen emphasizes in the preceding chapter, it is important to be clear about the implications of these results. Neither high variance in male breeding success, nor high variance in mating success, nor any ratio of the two measures (such as the ratio of I_s/I_m recommended by Wade and Arnold 1980) indicates that direct competition between males is necessarily intense or that selection pressures will necessarily favor the evolution of secondary sexual characters. Not only may much of the observed variance in mating success be unrelated to phenotype (see Clutton-Brock 1983; Sutherland 1985a,b), but selection may favor traits that improve the male's longevity or his success in scramble competition for mates. As Howard argues (see chap. 7), sex differences in morphology or behavior should instead be associated with differences in the extent to which particular behavioral or morphological traits affect fitness in the two sexes. The common correlation between polygyny and sexual dimorphism (see Alexander et al. 1979; Payne 1984) does not invalidate this argument: it presumably occurs because selection pressures in males and females are most divergent in polygynous species.

29.4 How Large Is the Contribution of Different Components of Fitness in Males?

The studies of insects (part 1) and monogamous birds (part 3) show that differences in life span are a major cause of variation in the fitness of males that reach breeding age, as they are in females. In polygynous mammals (part 4), variation in mating rate is the single largest component of fitness variation among breeding males in most species studied. Since a substantial proportion of the variation in mating rate appears to be caused by individual differences in phenotype (see chap. 21, for example), selection pressures favoring traits that improve mating success are likely to be strong. As Darwin originally noted (1871), secondary sexual characters are especially highly developed in these species.

Comparisons of the contributions of different components of reproductive success to fitness in males are commonly used to measure the intensity of sexual selection (see chap. 1). At least five indexes of the intensity of sexual selection are in use (see Clutton-Brock 1983; Sutherland 1985b):

1. The extent of variation in male breeding success (see Payne 1979, 1984).
2. The ratio of variation in breeding success in the two sexes (Ralls 1977).
3. Variation in the total number of matings in a season or over the life span.
4. Variation in mating rate (McCauley 1983; Koenig and Albano 1986).
5. The ratio of standardized variance in mate number to standardized variance in total breeding success (I_s/I_m) (see Wade and Arnold 1980; Arnold and Wade 1984a,b; chap. 2).

In addition, three "environmental" measures that are likely to affect the opportunity for sexual selection are also sometimes used to estimate its strength. These are the operational sex ratio (Emlen 1976), relative parental investment by males and females (Trivers 1972), and the proportion of time males spend searching for mates (Sutherland 1985b).

It is important to appreciate that all five indexes as well as the three environmental parameters attempt to measure the *opportunity* for sexual selection (see chap. 1), assuming that this reflects the intensity of sexual selection on traits affecting mating success. This will not necessarily be the case (see chap. 28), for it is not likely that the proportion of variance in mating success caused by phenotypic differences is the same in all species. Moreover, the three environmental measures will not always be closely related to the opportunity for sexual selection.

A second objection to all five indexes is that none of them precisely reflects Darwin's original concept of sexual selection. The first three indexes include variation in breeding or mating success due to differences in longevity (which is usually regarded as a component of natural selection), while the fourth and fifth exclude the effects of consistent differences in mate quality, which may be an important source of selection pressures favoring secondary sexual characters in monogamous species (see also Trail 1985).

A final objection is the difficulty of comparing any of these eight measures between species. The problems of obtaining similar samples of reproductive data and of deciding precisely which components of fitness should be allocated to sexual selection and which to natural selection are well illustrated by any attempt to compare Fincke's and McVey's studies of odonates (see chaps. 3 and 4). The apparent simplicity of the three environmental variables disappears as soon as any attempt is made to define them in a way that could apply to a wide range of animals.

The difficulties of measuring the opportunity for sexual selection should lead us to reconsider why we need to do so. All the indexes listed above are generally used for one of three purposes: to show that sexual selection is operating (see Sutherland 1985a,b; Koenig and Albano 1986); to compare the opportunity for selection at different stages of the reproductive process (Wade and Arnold 1980; see chaps. 3 and 4); or to explain the distribution of sexually dimorphic traits (Clutton-Brock, Harvey, and Rudder 1977).

None of these aims requires a precise distinction between "natural" and "sexual" components of selection. The most direct test of *whether* sexual selection is operating is to investigate whether differences in phenotype affect mating rate or mate quality. If they do, sexual selection is occurring, however small the observed variance in mating success (see below). If not, it will seldom be possible to be sure that sexual selection is occurring, however large the variance in mating success (see Section 29.7). Where the aim is to compare the opportunity for selection at different stages of the reproductive process, an analysis of fitness components will generally be more useful in indicating where and when selection may be acting (see chap. 1) than comparisons between two gross categories. Finally, where the aim is to explain the distribution of sex differences, we need to understand the selection pressures that produced and maintain them in different species (Clutton-Brock 1983).

29.5 How Are Fitness Components Related to Each Other?

The studies in this book confirm that positive correlations between fecundity and longevity are common in both sexes (see chaps. 2, 4, 5, 9, 16, 21, and 23). The most likely explanation of these associations is that they occur because variation in phenotypic quality obscures the negative consequences of breeding success (see Clutton-Brock and Harvey 1979; Reznick 1985). Since variation in phenotypic quality is often large, these results need not suggest that the reproductive costs of breeding are low.

Negative correlations between components of breeding success were found in a minority of studies (see chaps. 11, 17, and 20), and even these may have been artifacts of reduced breeding success in animals with long life spans or of artificially high estimates of fecundity in individuals that die young. The most obvious conclusion to be drawn is that comparisons of the subsequent fate of successful and unsuccessful breeders or correlations between components of reproductive success tell us little about the costs of breeding (Partridge and Harvey 1985). Comparisons that control for individual variation in reproductive potential (Clutton-Brock, Albon, and Guinness 1984), experimental manipulation of reproductive costs (Nur 1984a; Boutin 1984), and selection experiments (Reznick 1985) are needed.

29.6 Does Variation in Breeding Success Occur by Chance?

One aim of this book was to determine whether random variation in access to mates commonly accounts for the observed variation in mating success (Sutherland 1985b). One way of approaching this question has been to compare observed distributions of mating success with those generated by random models (e.g., Banks and Thompson 1985; Hubbell and Johnson 1987). However, though this may tell us whether random variation *could* account for the observed distribution of breeding success, it cannot tell us whether it *does* do so. As Sutherland (1985a) points out, failure to exceed a random model does not indicate that differences in mating rate have arisen by chance, while a significant deviation from the model may merely indicate that one or more of its parameters is unrealistic. Instead, the question must be answered by investigations of whether differences in breeding success are related to variation in phenotype.

The studies suggest that chance seldom accounts for the observed variation in breeding success in either sex. All but one of those that investigated the causes of variation in female breeding success found relationships with phenotypic variables, though these were not always consistent (see also Price and Grant 1984; Price et al. 1984). The exception was Smith's study of song sparrows, where intercohort differences in breeding success were large and could have masked the effects of phenotype. Similarly, all but one of the studies that investigated the causes of variation in male mating rate in promiscuous or polygynous species found that differences were related to phenotypic factors, even though the observed distribution of mating success did not always exceed the binomial (see chap. 3 and Banks and Thompson 1985). The one exception was McVey's study of dragonflies (chap. 4), where no phenotypic correlates of mating rate were identified. However, her elegant demonstration that individual differences in mating rate were consistent showed that some phenotypic variable must have been involved. These results obviously do not indicate that chance was unimportant. Instead, they emphasize that the question we should be asking is *what proportion* of observed variation in mating success is attributable to chance. So far, no studies have been able to answer this question satisfactorily (but see Houck, Arnold, and Thisted 1985).

In addition to the effects of pure chance, short-term environmental changes unrelated to phenotype can have a major effect on all components of breeding success, especially in short-lived or semelparous animals. Environmental factors that affect breeding success directly include climatic variation within, or between, breeding seasons (see especially chaps. 8 and 12), local differences in habitat quality (see chap. 6) or in the sex ratio (see chaps. 4 and 7), and changes in population density (see chap. 21). In addition, in many social species the size of the group an individual belongs to can affect its reproductive success (see chaps. 19 and 21), while changes in group membership can also have important effects on survival or fecundity (see chap. 23).

29.7 What Factors Affect Reproductive Success in Males and Females?

Five phenotypic characteristics were commonly related to reproductive success in the studies represented here:

Age

An initial improvement in breeding success with age was found by most studies, and in some cases the improvement continued through several breeding attempts. In fulmar petrels, for example, breeding success continues to improve until the bird's tenth year (chap. 17). The causes of improvements in breeding success with age evidently vary widely and can include physical maturation (chap. 2), experience of reproduction (chap. 12), the age or experience of the individual's breeding partner (chaps. 9 and 26), and changes in dominance rank (chap. 20), breeding site, or breeding time (chap. 16; see also Coulson and White 1959). In addition, young breeders may show reproductive restraint (Curio 1983), though there is little firm evidence of this.

A decline in breeding performance and survival with increasing age may be a feature of all animals (see chap. 2), but the extent to which it is expressed in natural populations appears to increase with the species' average longevity. The decline in survival and breeding success with increasing age is particularly pronounced in males of polygynous vertebrates, where direct competition for mates is intense (see chaps. 18, 21, and 22) and is commonly associated with an obvious decline in physical condition.

The prevalence of age effects has important implications for studies of selection and demography. If age variation is ignored, both the opportunity for selection and the intensity of selection on age-related traits may be seriously overestimated, especially in males (see chap. 1; see also Clutton-Brock 1983). Omitting relationships between maternal age and offspring survival in the calculation of life tables may also lead to inaccurate estimates of the demographic consequences of changes in age structure (see part 4). Finally, the existence of age effects raises important questions about their physiological and evolutionary causes. Do the cumulative effects of reproduction advance senescence, or are the costs of breeding limited to the period of breeding activity, as Partridge's work on *Drosophila* suggests (chap. 2)? What are the causes of the rapid deterioration in the condition of males with increasing age, and why has selection not retarded them? And is it true, as some recent studies suggest (see Clutton-Brock 1984), that although reproductive performance commonly declines with age, reproductive effort may increase?

Body Size

Among vertebrates as well as invertebrates, female body size is commonly correlated with many components of breeding success, including fecundity (see chaps. 2, 3, and 11), weight of individual eggs or offspring (chaps. 3, 13, 21; see also Jones 1973; Ojanen, Orell, and Väisänen 1979), milk yield (Hanwell and Peaker 1977), and offspring survival (chap. 21). Among males, larger individuals commonly live longer (chap. 2) and obtain priority of access to resources and mating partners (Ferguson and Fox 1984; Borgia 1982; Ridley and Thompson 1979; Price 1984; Payne 1984; see chaps. 3, 5, 20, and 21), though this is not always the case (Payne 1984), especially in species where males form competitive coalitions or where male success depends on female choice (see chap. 18).

These relationships should prompt us to ask what limits adult size (Clutton-Brock and Harvey 1983). In some cases exceptional environmental conditions may select against larger adults (for example, see Bumpus 1899; Price and Grant 1984). However, large adult size is usually associated with faster growth rates or longer growth periods in juveniles, and it may be that large juveniles are consequently more susceptible to starvation (see Case 1978; Price and Grant 1984).

Unfortunately, only Howard's study of bullfrogs and woodfrogs was able to compare the effects of body size on males and females (see chap. 7). In bullfrogs, where the two sexes are similar in size, variation in body size has similar effects on reproductive success in males and females, while in woodfrogs, where females are substantially larger than males, size has a greater effect among females. To test current explanations of the adaptive significance of sexual dimorphism (see Clutton-Brock, Harvey, and Rudder 1977; Alexander et al. 1979), direct comparisons of the effects of size on reproductive success are needed in polygynous species where males are the larger sex or where the sexes are similar in size. In addition, we need to understand substantially more about the *reasons* size affects reproductive success in different species.

Dominance Rank

In social species, dominance rank is a common correlate of breeding success in both sexes (see chaps. 6, 14, 19, 21, and 25), though studies of primates show that these relationships can be inconsistent (Robinson 1982; Fedigan 1983). Relationships between dominance and breeding success are frequently complex (Chapais 1983; Chapais and Schulman 1983). For example, in vervet monkeys the offspring of high-ranking females are apparently less likely to die from illness than those of subordinates but more likely to die from predation (chap. 24), while in Altmann's baboons the sons of high-ranking mothers and the daughters of low-ranking ones are less likely to survive than offspring of the opposite sex (chap. 25).

Correlations between breeding success and dominance rank, like those between breeding success and body size, can have several causes. For example, correlations between female rank and breeding success may occur because a female's rank affects her access to resources or to superior partners (see Whitten 1983; Wrangham 1981; chap. 14) or her ability to protect her offspring from dominant conspecifics (Dittus 1977, 1979; Silk 1983). Alternatively, these correlations may arise because phenotypically superior females show high reproductive success and are also likely to be high ranking. For example, in red deer, where a female's rank is related both to her body size and to her breeding success, it is not yet clear whether dominance adds to the proportion of variance accounted for by her body size (chap. 21). To isolate the effects of social rank itself, future studies will need to control for variation in age and size.

Mate Choice

Several of the studies of monogamous birds provide evidence that the characteristics of an individual's mate can exert an important effect on the breeding success of the pair. Experience of breeding together is evidently important: among established breeding pairs, remating is commonly associated with a reduction in breeding success (chaps. 9, 16, 20), and in Bewick's swans the breeding success of pairs also increases with the duration of the pair-bond (chap. 14).

In house martins, flycatchers, and Bewick's swans there is evidence that the mate's age and social rank can affect an individual's breeding success (see chaps. 11, 12, and 14). These effects can differ between the sexes as well as between different components of reproduction. In Bewick's swans, the body size of the male has a stronger effect on the dominance and breeding success of the pair than the size of the female (chap. 14), and in great tits the female's age predicts clutch size and laying date while the male's age predicts fledging success and recruitment (chap. 9). The breeding success of females can also be strongly influenced by the resources the male is able to control (Pleszczynska and Hansell 1980; Halliday 1983; Partridge and Halliday 1984; see also chap. 26) or by the protection he can provide for eggs or young (see chap. 24). Variation in the genetic qualities of their mates may also contribute to females' breeding success, but as yet there is little direct evidence that it does so in natural populations (see Partridge and Halliday 1984).

In many polygynous species, as well as in monogamous ones, males have the opportunity to select among females, and there is some evidence that they choose females whose reproductive potential is above average

(Johnson 1982; Thornhill and Alcock 1983). If so, variation in mate quality may represent an important component of breeding success to males too, and selection would be expected to favor choosiness in both sexes.

Early Development

The studies in this book provide extensive evidence that in species as diverse as *Drosophila* (chap. 2), great tits (chaps. 8 and 9), red deer (chap. 21), and man (chap. 26) an individual's survival and breeding success are closely related to its early development. Birth or hatching date is often correlated with subsequent growth and survival (see chaps. 8, 12, and 21). In addition, studies of insects (Begon, n.d), fish (van den Berghe and Gross, n.d), birds (see chap. 8), and mammals (see chap. 21) have shown that variation in egg weight or birth weight is related to the size and survival of juveniles. Differences in juvenile growth are commonly related to variation in adult size (Begon, n.d.; van den Berghe and Gross, n.d.; Clutton-Brock, Albon, and Guinness 1986), for although compensatory growth occurs in domestic and laboratory animals, in natural populations competition commonly prevents compensation, maintaining or exaggerating initial differences between individuals (see Rubenstein 1981). Differences in adult size not only affect fecundity or mating success (see above), but can also influence the reproductive performance of the next generation through their influence on egg size, birth weight, or early growth (see chap. 21).

Since climatic differences between years often affect birth weight and juvenile growth (see above), a substantial proportion of variation in breeding success may be of environmental origin. One line of evidence supporting this is that in long-lived animals average breeding success calculated across the life span often shows pronounced differences between successive cohorts (see chaps. 21 and 22).

The close associations between early growth and reproductive success in adulthood have several important implications. First, they suggest that changes in climate or population density may have protracted effects on reproduction and on population dynamics (see Albon and Mithan, n.d.). Second, they complicate attempts to investigate heritability in natural populations (see below). And third, they show that selection is likely to favor flexible strategies of development and reproduction that enable individuals to make the most of their opportunities (see Davis 1982).

29.8 Are Components of Breeding Success Heritable?

Only two of the twenty-four chapters examined the heritability of reproductive success or its components. Van Noordwijk (chap. 8) found heritabilities of about 40% for clutch size, and there was evidence that it was under stabilizing selection, while in song sparrows Smith (chap. 10) found no evidence of heritable variation in any of the components of breeding success that he examined. In addition, a recent study of collared flycatchers (*Ficedula albicollis*) has found heritabilities close to zero in components of lifetime reproductive success but higher heritabilities in traits less closely related to breeding success (Gustafsson 1986), confirming Fisher's (1930) prediction that there should be a negative association between the heritability of a character and the proportion of variance in fitness it explains. In contrast, research on red deer indicates that juvenile survival is related to genetic differences (Pemberton et al., n.d.).

Unfortunately, the relationship between a mother's phenotype and the growth of her offspring (see above) complicates most ways of estimating heritability in natural populations. Maternal effects will not necessarily be removed by cross-fostering eggs or neonates (e.g., Smith and Dhondt 1980), since the mother's phenotype can affect the size or content of the egg or the birth weight of the neonate, and this is likely to influence the juvenile's subsequent development and breeding success (Wall and Begon, n.d.; van den Berghe and Gross, n.d.). Where fathers are involved in the rearing process, father/offspring regression may overestimate heritability for similar reasons, and even in species where fathers do not assist in rearing, this technique may still overestimate genetic heritability if assortative mating occurs or males tend to breed with mates belonging to the same cohorts as themselves.

29.9 Prospects

The studies published here have implications for the measurement and interpretation of variation in reproductive success in natural populations and provide some indication of future research priorities. First, they demonstrate the potential pitfalls of relying on cross-sectional data alone. Estimates of variation in breeding success at particular points in time or over small parts of the life span may either under- or overestimate variation in lifetime reproductive success (see chaps. 2, 3, and 21). In particular, ignoring the effects of age on breeding success can lead to gross errors in estimates of the comparative opportunity for selection in the two sexes (see chaps. 18 and 22). Consistent relationships between phenotypic quality, reproductive performance, and longevity are common and can lead to changes in the distribution of phenotypes included in successive samples from the same population, which may either reduce or exaggerate estimates of the effects of age or previous reproductive history on breeding success (see chaps. 4 and 14). And short-term environmental changes commonly exert a strong influence on breeding success that can easily obscure the effects of variation in phenotype (see chaps. 5, 10, and 21).

Second, the studies reveal the complexity of measuring the costs or benefits of particular traits and emphasize the need to use different measures of reproductive success to answer different questions. Most differences in phenotype probably affect several components of fitness (see chaps. 9, 14, and 19), and some, such as effects on longevity or offspring survival, are harder to measure than others. This appears to argue for comparisons of lifetime reproductive success. However, it is frequently more difficult to identify the effects of phenotypic variation on lifetime success than its effects on particular components of fitness, partly because different phenotypic traits affect different components of fitness (see chaps. 9 and 14), and partly because of the large contribution of chance to some components of breeding success (e.g., life span). As a result, it is often advisable to investigate the effects of phenotypic variation on particular fitness components rather than on lifetime success.

Third, the studies emphasize the importance of statistical or experimental methods of controlling for the effects of confounding variables. Few of the studies that investigated the causes of variation in reproductive success attempted to control for other factors, and none of them used the techniques for calculating selection gradients successfully employed in a number of previous field studies (Price and Grant 1984; Price et al. 1984). Though these techniques have mostly been used for investigating relationships between simple anatomical variables and survival, there is no reason they should not be used to examine relationships with other components of fitness (Gibson 1987). Since there is ample evidence that all components of reproductive success are affected by a wide range of environmental and phenotypic factors, both these techniques and field experiments have an important part to play (see chap. 28).

The studies provide some indication of future research priorities. Their results should reassure us that variation in fitness is usually substantial in both sexes and should encourage future work to concentrate more on the causes of this variation in both sexes. They show that we need to know considerably more about the extent and causes of prereproductive mortality and the role of early development in determining breeding success in adulthood. Finally, they underline the importance of distinguishing clearly between studies of selection and adaptation.

Research on selection, adaptation, and demography will continue to rely on the measurement of reproductive success and survival. Longitudinal studies of individuals will never entirely replace short-term, cross-sectional data, nor is it desirable that they should, for some important questions can be answered only by measuring the immediate consequences of variation in phenotype or behavior. Nevertheless, it is clear that research on the life histories of individuals has an important role to play in the development of population biology that has only begun to be explored.

References

Aghajanian, A. 1981. Age at first birth and completed family size in western Malaysia. *J. Biosoc. Sci.* 13:197–201.

Ainley, D. G., and DeMaster, D. P. 1980. Survival and mortality in a population of Adélie penguins. *Ecology* 61:522–30.

Ainley, D. G.; Strong, C. S.; Huber, H. P.; Lewis, T. J., and Morrell, S. H. 1981. Predation by sharks on pinnipeds at the Farallon Islands. *Fish. Bull.* 78:941–45.

Alatalo, R. V.; Carlson, A.; Lundberg, A.; and Ulfstrand, S. 1981. The conflict between male polygamy and female monogamy: The case of the pied flycatcher *Ficedula hypoleuca*. *Am. Nat.* 117:738–53.

Alatalo, R. V.; Gustafsson, L.; and Lundberg, A. 1984. High frequency of cuckoldry in pied and collared flycatchers. *Oikos* 42:41–47.

Alatalo, R. V.; Lundberg, A.; and Stahlbrandt, K. 1982. Why do pied flycatcher females mate with already mated males? *Anim. Behav.* 30:585–93.

Albon, S. D.; Clutton-Brock, T. H.; and Guinness, F. E. 1987. Early development and population dynamics in red deer. 2. Density-independent effects and cohort variation. *J. Anim. Ecol.* 56:69–81.

Albon, S. D.; Guinness, F. E.; and Clutton-Brock, T. H. 1983. The influence of climatic variation on the birth weights of red deer (*Cervus elaphus*). *J. Zool.* 200:295–98.

Albon, S. D., and Mithan, S.n.d. Population fluctuation in red deer: The importance of cohort differences. In preparation.

Alcock, J. 1983. Territoriality by hilltopping males of the great purple hairstreak, *Atlides halesus* (Lepidoptera, Lycaenidae): Convergent evolution with a pompilid wasp. *Behav. Ecol. Sociobiol.* 13:57–62.

Alexander, R. D. 1974. The evolution of social behavior. *Ann. Rev. Ecol. System.* 5:325–83.

Alexander, R. D.; Hoogland, J. L.; Howard, R. D.; Noonan, K. M.; and Sherman, P. W. 1979. Sexual dimorphism and breeding systems in pinnipeds, ungulates, primates and humans. In *Evolutionary biology and human social behavior: An anthropological perspective*, ed. N. A. Chagnon and W. Irons, 402–35. North Scituate, Mass.: Duxbury.

Altmann, J. 1980. *Baboon mothers and infants*. Cambridge: Harvard University Press.

———. 1983. Costs of reproduction in baboons. In *Behavioral energetics: Vertebrate costs of survival*, ed. W. P. Asprey and S. I. Lustick, 67–88. Columbus: Ohio State University Press.

———. 1986. Adolescent pregnancies in non-human primates: An ecological and developmental perspective. In *School-age pregnancy and parenthood: Biosocial dimensions*, ed. J. Lancaster and B. Hamburg. Hawthorne, N.Y.: De Gruyter Aldine.

Altmann, J.; Altmann, S. A.; and Hausfater, G. 1978. Primate infant's effects on mother's future reproduction. *Science* 201: 1028–29.

———. 1981. Physical maturation and age estimates of yellow baboons, *Papio cynocephalus*. *Am. J. Primatol.* 1:389–99.

Altmann, J.; Altmann, S. A.; Hausfater, G.; and McCusky, S. 1977. Life history of yellow baboons: Physical development, reproductive parameters, and infant mortality. *Primates* 18:315–30.

Altmann, J.; Hausfater, G.; and Altmann, S. A. 1985. Demography of Amboseli baboons, 1963–1983. *Am. J. Primatol.* 8:113–25.

Altmann, S. A. 1962. A field study of the sociobiology of rhesus monkeys, *Macaca mulatta*. *Ann. N.Y. Acad. Sci.* 102:338–435.

———. 1970. The pregnancy sign in savannah baboons. *Lab. Anim. Digest* 6:7–10.

Altmann, S. A., and Altmann, J. 1970. *Baboon ecology*: *African field research*. Chicago: University of Chicago Press.

Andelman, S. J. 1984. Ecology and reproductive strategies of vervet monkeys. Ph.D. diss., University of Washington.

Andelman, S. J.; Else, J.; Hearn, J. P.; and Hodges, J. K. 1984. The non-invasive monitoring of reproductive events in wild vervet monkeys (*Cercopithecus aethiops*) using urinary pregnanediol-3-glucuronide and its correlation with behavioral observations. *J. Zool. Lond.* 205:467–77.

Anderson, S. S.; Burton, R. W.; and Summers, C. F. 1975. Behaviour of grey seals (*Halichoerus grypus*) during a breeding season at North Rona. *J. Zool. Lond.* 177:179–95.

Anderson, W. W. 1973. Genetic divergence in body size among experimental populations of *Drosophila pseudoobscura* kept at different temperatures. *Evolution* 27:278–84.

Anderson, W. W.; Levine, L.; Olvera, O.; Powell, J. R.; de la Rosa, M. E.; Salceda, V. M.; Gaso, M. I.; and Guzman, J. 1979. Evidence for selection by male mating success in natural populations of *Drosophila pseudoobscura*. *Proc. Nat. Acad. Sci. USA* 76:1519–23.

Andersson, M. 1982. Female choice selects for extreme tail length in a widowbird. *Nature* 299:818–20.

Ankney, D. C., and MacInnes, C. D. 1978. Nutrient reserves and reproductive performance of female lesser snow geese. *Auk* 95:459–71.

Arcese, P., and Smith, J. N. M. 1985. Phenotypic correlates and ecological consequences of dominance in song sparrows. *J. Anim. Ecol.* 54:817–30.

Arnold, S. J. 1983a. Morphology, performance and fitness. *Am. Zool.* 23:347–61.

———. 1983b. Sexual selection: The interface of theory and empiricism. In *Mate choice*, ed. P. Bateson, 67–107. Cambridge: Cambridge University Press.

———. 1986. Limits on stabilizing, disruptive, and correlational selection set by the opportunity for selection. *Amer. Nat.* 128:143–46.

Arnold, S. J., and Houck, L. 1982. Courtship pheromones: Evolution by natural and sexual selection. In *Biochemical aspects of evolutionary biology*, ed. M. Nitecki, 173–211. Chicago: University of Chicago Press.

Arnold, S. J., and Wade, M. J. 1984a. On the measurement of natural and sexual selection: Theory. *Evolution* 38:709–19.

———. 1984b. On the measurement of natural and sexual selection: Applications. *Evolution* 38:720–34.

Askenmo, C. 1979. Reproductive effort and return rate of male pied flycatchers. *Am. Nat.* 114:748–53.

Atkinson, W. D. 1979. A field investigation of larval competition in domestic *Drosophila*. *J. Anim. Ecol.* 48:91–102.

Atwood, J. L. 1980. Social interactions in the Santa Cruz Island scrub jay. *Condor* 82:440–48.

Baker, R. R. 1968. Evolution of the migratory habit in butterflies. *Phil. Trans. Roy. Soc. Lond.*, ser. B., 253:309–41.

Balen, J. H. van. 1973. A comparative study of the breeding ecology of the great tit *Parus major* in different habitats. *Ardea* 61:1–93.

———. 1980. Population fluctuations of the great tit and feeding conditions in winter. *Ardea* 68:144–64.

Banks, M. J., and Thompson, D. J. 1985. Lifetime mating success in the damselfly *Coenagrion puella*. *Anim. Behav.* 33:1175–83.

Bartholomew, G. A., Jr., and Hoel, P. G. 1953. Reproductive behavior of the Alaska fur seal, *Callorhinus ursinus*. *J. Mammal.* 34:417–36.

Bateman, A. J. 1948. Intra-sexual selection in *Drosophila*. *Heredity* 2:349–68.

Batt, B. D. J., and Prince, H. H. 1979. Laying dates, clutch size and egg weight of captive mallards. *Condor* 81:35–41.

Baumann, H. 1974. Biological effects of paragonial substances PS-1 and PS-2 in females of *Drosophila funebris*. *J. Insect Physiol.* 20:2347–62.

Beatty, J. 1980. Optimal design models and the strategy of model building in evolutionary biology. *Phil. Sci.* 47:532–61.

Begon, M. 1984. Density and individual fitness: Asymmetric competition. In *Evolutionary ecology*, ed. B. Shorrocks, 175–94. Oxford: Blackwell.

Bell, D. J. 1983. Mate choice in the European rabbit. In *Mate choice*, ed. P. P. G. Bateson,211–27. Cambridge: Cambridge University Press.

Bell, G. 1980. The costs of reproduction and their consequences. *Am. Nat.* 116:45–76.

Bellinger, P. F. 1954. Attraction of zebra males by female pupae. *J. Lepid. Soc.* 8:102.

Bellis, E. D. 1961. Growth of the woodfrog, *Rana sylvatica*. *Copeia* 1:74–77.

Benedict, B. 1972. Social regulation of fertility. In *The structure of human populations*, ed. G. A. Harrison and A. J. Boyd, 73–89. Oxford: Clarendon Press.

Benke, A. C., and Benke, S. S. 1975. Comparative dynamics and life histories of coexisting dragonfly populations. *Ecology* 56:302–17.

Bercovitch, F. B. 1985. Reproductive tactics in adult female and adult male olive baboons. Ph.D. diss., University of California at Los Angeles.

Berenstain, L., and Wade, T. D. 1983. Intrasexual selection and male mating strategies in baboons and macaques. *Int. J. Primatol.* 4:201–35.

Bertram, B. C. R. 1973. Lion population regulation. *E. Afr. Wildl. J.* 11:215–25.

———. 1975. Social factors influencing reproduction in wild lions. *J. Zool.* 177:463–82.

———. 1976. Kin selection in lions and evolution. In *Growing points in ethology*, ed. P. P. G. Bateson and R. A. Hinde, 281–301. Cambridge: Cambridge University Press.

———. 1978. *Pride of lions*. New York: Scribners.

———. 1979. Ostriches recognize their own eggs and discard others. *Nature* 279:233–34.

Berven, K. A. 1981. Mate choice in the woodfrog, *Rana sylvatica*. *Evolution* 35:707–22.

Betts, M. M. 1954. Experiments with an artificial nestling. *Brit. Birds* 47:229–31.

Bilewicz, S. 1953. Experiments on the effects of reproductive functions on the length of life of *Drosophila melanogaster*. *Folia Biol.* 1:177–94.

Björkland, M., and Westman, B. 1983. Extra-pair copulations in the pied flycatcher *Ficedula hypoleuca*: A removal experiment. *Behav. Ecol. Sociobiol.* 13:271–75.

Boag, P. T. 1983. The heritability of external morphology in Darwin's ground finches (*Geospiza*) on Isla Daphne Major, Galapagos. *Evolution* 37:877–94.

Boag, P. T., and Grant, P. R. 1981. Intense natural selection in a population of Darwin's finches (*Geospizinae*) in the Galapagos. *Science* 214:82–85.

Boer-Hazewinkel, J. den. 1980. Effect of the addition of food during the breeding season on the production of second broods by the great tit, *Parus major*. *Verh. Kon. Med. Akad. Wetensch., Afd. Natuurk.*, 2d ser., 77:6–9.

Boerma, J. T., and Vianen, H. A. W. van. 1984. Birth intervals, mortality and growth of children in a rural area in Kenya. *J. Biosoc. Sci.* 16:475–86.

Boggs, C. L. 1986. Reproductive strategies of female butterflies: Variation in and constraints on fecundity. *Ecol. Entomol.* 11:7–15.

Boggs, C. L., and Gilbert, L. E. 1979. Male contribution to egg production in butterflies: Evidence for transfer of nutrients at mating. *Science* 206:83–84.

Boness, D. J., and James, H. 1979. Reproductive behaviour of the grey seal (*Halichoerus grypus*) on Sable Island, Nova Scotia. *J. Zool. Lond.* 188:477–500.

Bongaarts, J. 1980. Does malnutrition affect fecundity? A summary of the evidence. *Science* 208:564–69.

Bongaarts, J., and Potter, R. G. 1983. *Fertility, biology and behavior: An analysis of the proximate determinants.* New York: Academic Press.

Bonnell, M. L.; Le Boeuf, B. J.; Pierson, M. O.; Dettman, D. H.; and Farrens, G. D. 1978. Summary report, 1975–1978. *Marine mammal and seabird surveys of the Southern California Bight area.* Vol. 3. *Pinnepeds.* Bureau of Land Management, Department of the Interior Contract AA550-CT7-36. Washington, D.C.: Government Printing Office.

Bonnell, M. L., and Selander, R. K. 1974. Elephant seals: Genetic variation and near extinction. *Science* 184:908–9.

Bonner, W. N. 1968. The fur seal of South Georgia. *Brit. Ant. Survey Sci. Rep.* 56:1–81.

Boppre, M. 1984. Chemically mediated interactions between butterflies. In *The biology of butterflies*, ed. R. I. Vane-Wright and P. R. Ackery, 259–75. Eleventh Symposium of the Royal Entomological Society of London. London: Academic Press.

Borch, H., and Schmid, F. 1973. On *Ornithoptera priamus caelestis, demphanes* and *boisduvali. J. Lepid. Soc.* 27:196–205.

Borgerhoff Mulder, M. 1983. Social organisation and biology. *Man* 18:786–87.

———. 1985. Polygyny threshold: A Kipsigis case study. *Nat. Geogr. Res. Rep.* 21:33–39.

———. 1987a. Adaptation in evolutionary biological anthropology. *Man* 22:25–41.

———. 1987b. Low reproductive performance and women's age at the onset of reproduction. In *Anthropologiai kozlemenyek*, ed. O. G. Eiben. Budapest; Hungarian Academy of Sciences. Forthcoming.

———. 1987c. On cultural and reproductive success: Kipsigis evidence. *Am. Anthropol.* In press.

———. 1987d. Resource defense polygyny in humans. In *Mating patterns*, ed. C. G. N. Mascie-Taylor and A. J. Boyce. Cambridge: Cambridge University Press. Forthcoming.

———. 1987e. Resources and reproduction of women with an example from the Kipsigis. *J. Zool. Lond.*, ser. B. In press.

———. 1988a. Kipsigis bridewealth payments. In *Human reproductive behaviour: A Darwinian perspective*, ed. L. L. Betzig, M. Borgerhoff Mulder, and P. W. Turke. Cambridge: Cambridge University Press. Forthcoming.

———. 1988b. Reproductive consequences of sex-biased inheritance. In *Comparative socioecology of mammals* and man, ed. R. Foley and V. Standen. Oxford: Blackwell *Scientific Publications.* Forthcoming.

Borgerhoff Mulder, M., and Caro, T. M. 1983. Polygyny: Definition and application to human data. *Anim. Behav.* 31:609–10.

———. 1985. The use of quantitative observational techniques in anthropology. *Curr. Anthropol.* 26:323–35.

Borgerhoff Mulder, M., and Milton, M. 1985. Factors affecting infant care in the Kipsigis. *J. Anthropol. Res.* 41:231–62.

Borgia, G. 1982. Experimental changes in resource structure and male density: Size related differences in mating success among male *Scatophaga stercoraria. Evolution* 36:307–15.

Bos, M., and Scharloo, W. 1974. The effects of disruptive and stabilising selection on body size in *Drosophila melanogaster*. 3. Genetic analysis of two lines with different reactions to disruptive selection with mating of opposite extremes. *Genetica* 45:71–90.

Botkin, D. B., and Miller, R. S. 1974. Mortality rates and survival of birds. *Am. Nat.* 108:181–92.

Bouletreau, J. 1978. Ovarian activity and reproductive potential in a natural population of *Drosophila melanogaster*. *Oecologia* 35:319–42.

Boutin, S. A. 1984. The effects of conspecifics on juvenile survival and recruitment of snoeshoe hares. *J. Anim. Ecol.* 53:623–37.

Bownes, M. 1983. Interactions between germ cells and somatic cells during insect oogenesis. In *Current problems in germ cell differentiation*, ed. A. McLaren and C. C. Wylie, 259–84. British Society for Developmental Biology Symposium 7. Cambridge: Cambridge University Press.

Bownes, M., and Partridge, L. n.d. Transfer of molecules from ejaculate to females in *Drosophila melanogaster* and *D. pseudoobscura*. *J. Insect Physiol.* In press.

Boyce, M. S. 1984. Restitution of r- and K-selection as a model of density-dependent natural selection. *Ann. Rev. Ecol. Syst.* 15:427–47.

Boyd, M. G.; Smith, E. J.; and Cooch, F. G. 1982. The lesser snow geese of the eastern Canadian arctic. Occasional Paper no. 46. Ottawa: Canadian Wildlife Service.

Bozcuk, A. N. 1978. The effect of some genotypes on the longevity of adult *Drosophila*. *Exp. Gerontol.* 13:278–85.

Brabin, L. 1984. Polygyny: An indicator of nutritional stress in African agricultural societies. *Africa* 54:31–45.

Bradbury, J. W., and Gibson, R. M. 1983. Leks and mate choice. In *Mate choice*, ed. P. P. G. Bateson, 109–38. Cambridge: Cambridge University Press.

Bradford, G. E.; Weir, W. C.; and Torrell, D. T. 1961. The effect of environment for weaning to first breeding on lifetime production of ewes. *J. Anim. Sci.* 29:281–87.

Brakefield, P. M. 1984. The ecological genetics of quantitative characters of *Maniola jurtina* and other butterflies. In *The biology of butterflies*, ed. R. I. Vane-Wright and P. R. Ackery. London: Academic Press.

Brierson, K.; Ekern, A.; and Homb, T. 1960. Relation of nutrition of the young animal to subsequent fertility and lactation. *Fed. Proc.*, suppl. 7, 20:275–83.

Broekhuysen, G. J. 1953. A post mortem of the Hirundinidae which perished at Somerset West in April 1953. *Ostrich*, December, 148–52.

Brooke, M. de L. 1978. Some factors affecting the laying date, incubation and breeding success of the Manx shearwater *Puffinus puffinus*. *J. Anim. Ecol.* 47:477–95.

Brown, J. L. 1983. Cooperation—a biologist's dilemma. *Adv. Study Behav.* 13:1–37.

Brown, J. L.; Brown, E. R.; Brown, S. D.; and Dow, D. D. 1982. Helpers: Effects of experimental removal on reproductive success. *Science* 215:421–22.

Brown, K. S. 1981. The biology of *Heliconius* and related genera. *Ann. Rev. Entomol.* 26:427–56.

Bryant, D. M. 1975. Breeding biology of house martins in relation to aerial insect abundance. *Ibis* 117:180–216.

———. 1979. Reproductive costs in the house martin *Delichon urbica*. *J. Anim. Ecol.* 48:655–75.

Bryant, D. M., and Turner, A. K. 1982. Central place foraging by swallows (Hirundinidae); The question of load size. *Anim. Behav.* 30:845–56.

Bryant, D. M., and Westerterp, K. R. 1980. The energy budget of the house martin *Delichon urbica*. *Ardea* 68:91–102.

———. 1982. Evidence for individual differences in foraging efficiency amongst breeding birds: A study of house martins *Delichon urbica* using the doubly labelled water technique. *Ibis* 124:187–92.

———. 1983a. Short-term variability in energy turnover by breeding house martins *Delichon urbica*: A study using doubly labelled water. *J. Anim. Ecol.* 52:525–43.

———. 1983b. Time and energy limits to brood-size in house martins. *J. Anim. Ecol.* 52:905–25.

Buckland, S. T. 1982. A mark-recapture survival analysis. *J. Anim. Ecol.* 51:833–47.

Bulmer, M. G. 1972. The genetic variability of polygenic characters under optimizing selection, mutation and drift. *Genet. Res.* 19:17–25.

———. 1980. *The mathematical theory of quantitative genetics.* Oxford: Clarendon Press.

Bulmer, M. G., and Perrins, C. M. 1973. Mortality in the great tit, *Parus major. Ibis* 115:277–81.

Bumpass, L.; Rundfuss, R.; and Janosik, R. 1978. Age and marital status at first birth and the pace of subsequent fertility. Demography 15:75–86.

Bumpus, H. C. 1899. The elimination of the unfit as illustrated by introduced sparrow, *Passer domesticus. Biol. Lect., Woods Hole Marine Biol. Station* 6:209–26.

Burnet, B.; Sewell, D.; and Bos, M. 1977. Genetic analysis of larval feeding behaviour in *Drosophila melanogaster*. 2. Growth relations and competition between selected lines. *Genet. Res. Camb.* 30:149–61.

Burrell, R. J. W.; Healey, M. J. R.; and Tanner, J. M. 1961. Age at menarche in South African Bantu school girls living in the Transkei Reserve. *Human Biol.* 33:250–61.

Bury, K. V. 1975. *Statistical methods in applied science.* New York: John Wiley.

Busse, C. 1982. Social dominance and offspring mortality among female chacma baboons. *Int. J. Primatol.* 3:267.

Bygott, J. D.; Bertram, B. C. R.; and Hanby, J. P. 1979. Male lions in large coalitions gain reproductive advantages. *Nature* 282:839–41.

Campbell, B. 1968. The Dean nestbox study, 1942–1964. *Forestry* 41:27–46.

Cantrelle, P. 1971. Is there a standard pattern of tropical mortality? In *Population in African development*, ed. P. Cantrelle, 33–42. Liège: International Union for the Scientific Study of Population.

Caraco, T., and Wolf, L. L. 1975. Ecological determinants of group sizes in foraging lions. *Am. Nat.* 109:343–52.

Carrick, R. 1972. Population ecology of the Australian black-backed magpie, royal penguin and silver gull. *U.S. Dept. Interior Wildl. Res. Rep.* 2:41–99.

Carrick, R.; Csordas, S. E.; and Ingham, S. E. 1962. Studies on the southern elephant seal, *Mirounga leonina* (L.). 4. Breeding and development. *CSIRO Wildl. Res.* 7:161–97.

Carrick, R., and Dunnet, G.M. 1954. Breeding of the fulmar, *Fulmarus glacialis. Ibis* 96:356–70.

Carrick, R., and Ingham, S. E. 1962. Studies on the southern elephant seal, *Mirounga leonina* (L.). 5. Population dynamics and utilization. *CSIRO Wildl. Res.* 7:198–206.

Case, T. J. 1978. On the evolution and adaptive significance of postnatal growth rates in the terrestrial vertebrates. *Quart. Rev. Biol.* 53:243–82.

Cervo, R., and Turillazzi, S. 1985. Associative foundation and nesting sites in *Polistes nimpha. Naturwissenschaften* 72:48–49.

Chagnon, N. 1974. *Studying the Yanomamo.* Studies in anthropological method. New York: Holt, Rinehart and Winston.

———. 1979. Mate competition, favoring close kin, and village fissioning among the Yanomamo Indians. In *Evolutionary biology and human social behavior: An anthropological perspective*, ed. N. A. Chagnon and W. Irons, 374–401. North Scituate, Mass.: Duxbury Press.

Chapais, R. 1983. Male dominance and reproductive activity in rhesus monkeys. In *Primate social relationships*, ed. R. A. Hinde, 267–71. Oxford: Blackwell.

Chapais, R., and Schulman, S. R. 1983. Fitness and female dominance relationships. In *Primate social relationships*, ed. R. A. Hinde, 271–78. Oxford: Blackwell.

Charlesworth, B. 1980. *Evolution in age-structured populations.* Cambridge: Cambridge University Press.

Chen, P. S. 1984. The functional morphology and biochemistry of insect male accessory glands and their secretions. *Ann. Rev. Entomol.* 29:233–55.

Cheney, D. L. 1981. Inter-group encounters in free-ranging vervet monkeys. *Folia Primatol.* 35:125–46.

———. 1983. Extra-familial alliances among vervet monkeys. In *Primate social relationships*, ed. R. A. Hinde, 278–86. Oxford: Blackwell.

Cheney, D. L.; Lee, P. C.; and Seyfarth, R. M. 1981. Behavioral correlations of nonrandom mortality among free-ranging female vervet monkeys. *Behav. Ecol. Sociobiol.* 9:153–61.

Cheney, D. L., and Seyfarth, R. M. 1981. Selective forces affecting the predator alarm calls of vervet monkeys. *Behaviour* 76:25–61.

———. 1983. Nonrandom dispersal in free-ranging vervet monkeys: Social and genetic consequences. *Am. Nat.* 122:392–412.

———. 1985. Vervet monkey alarm calls: Manipulation through shared information? *Behaviour* 94:150–66.

Cheverud, M. J. 1984. Evolution by kin selection: A quantitative genetic model illustrated by maternal performance in mice. *Evolution* 38:766–77.

Chitty, D. 1967. The natural selection of self-regulatory behaviour in natural populations. *Proc. Ecol. Soc. Austr.* 2:51–78.

Christensen, T. E., and Le Boeuf, B. J. 1978. Aggressive and maternal behavior of the northern elephant seal. *Behaviour* 64:158–72.

Christiansen, F. B., and Frydenberg, O. 1973. Selection component analysis of natural polymorphisms using population samples including mother-child combinations. *Theor. Pop. Biol.* 4:425–45.

———. 1976. Selection component analysis of natural polymorphisms using mother-offspring samples of successive cohorts. In *Population genetics and evolution*, ed. S. Karlin and E. Nevo, 277–301. New York: Academic Press.

Clapp, R.B.; Klimkiewicz, M. K.; and Kennard, J. H. 1982. Longevity records of North American birds: Gaviidae through Alcidae. *J. Field Ornithol.* 53:81–124.

Clark, A. B. 1978. Sex ratio and local resource competition in a prosimian primate. *Science* 201:163–65.

Clayton, F. E. 1957. Absolute and relative frequences of spermatogenic stages at different pupal periods in *Drosophila virilis*. *J. Morphol.* 101:457–76.

———. 1962. Effects of X-ray irradiation in *Drosophila virilis* at different stages of spermatogenesis. *Univ. Texas Publ.* 6205:345–73.

Clutton-Brock, T. H. 1983. Selection in relation to sex. In *Evolution from molecules to men*, ed. D. S. Bendall, 457–81. Cambridge: Cambridge University Press.

———. 1984. Reproductive effort and terminal investment in iteroparous animals. *Am. Nat.* 123:212–29.

Clutton-Brock, T. H., and Albon, S. D. 1982. Parental investment in male and female offspring in mammals. In *Current problems in sociobiology*, ed. King's College Sociobiology Group, 223–47. Cambridge: Cambridge University Press.

———. 1985. Competition and population regulation in social mammals. In *Behavioural ecology: Ecological consequences of adaptive behaviour*, ed. R. M. Sibly and R. H. Smith. Oxford: Blackwell Scientific Publications.

Clutton-Brock, T. H.; Albon, S. D.; Gibson, R. M.; and Guinness, F. E. 1979. The logical stag: Adaptive aspects of fighting in red deer (*Cervus elaphus* L.). *Anim. Behav.* 27:211–25.

Clutton-Brock, T. H.; Albon, S. D.; and Guinness, F. E. 1981. Parental investment in male and female offspring in polygynous mammals. *Nature* 289:487–89.

———. 1982. Competition between female relatives in a matrilocal mammal. *Nature* 300:178–80.

———. 1984. Maternal dominance, breeding success and birth sex ratios in red deer. *Nature* 308:358–60.

———. 1985. Parental investment and sex differences in mortality in birds and mammals. *Nature* 313:131–33.

———. 1986. Great expectations: Dominance, breeding success and offspring sex ratios in red deer. *Anim. Behav.* 34:460–71.

Clutton-Brock, T. H.; Guinness, F. E.; and Albon, S. D. 1982. *Red deer: Behavior and ecology of two sexes.* Chicago: University of Chicago Press.

———. 1983. The costs of reproduction to red deer hinds. *J. Anim. Ecol.* 52:367–84.

Clutton-Brock, T. H., and Harvey, P. H. 1983. The functional significance of variation in body size among mammals. *Spec. Publ. Amer. Soc. Mammal.* 7:632–63.

———. 1984. Comparative approaches to investigating adaptation. In *Behavioural ecology: An evolutionary approach*, 2d ed., ed. J. R. Krebs and N. B. Davies, 7–29, Oxford: Blackwell Scientific Publications.

Clutton-Brock, T. H.; Harvey, P. H.; and Rudder, B. 1977. Sexual dimorphism, socionomic sex ratio and body weight in primates. *Nature* 269:797–800.

Clutton-Brock, T. H.; Major, M.; Albon, S. D.; and Guinness, F. E. 1987. Early development and population dynamics in red deer. 1. Demographic consequences of density dependent changes in birth weight and date. *J. Anim. Ecol.* 56:53–67.

Clutton-Brock, T. H.; Major, M.; and Guinness, F. E. 1985. Population regulation in male and female red deer. *J. Anim. Ecol.* 54:831–46.

Cohet, Y., and David, J. 1966. Deleterious effects of copulation in *Drosophila* females as a function of growth temperature of both sexes. *Experientia* 32:696–97.

Collins, J. P. 1975. A comparative study of the life history strategies in a community of frogs. Ph.D. diss., University of Michigan.

Common, I. F. B., and Waterhouse, D. F. 1981. *Butterflies of Australia.* 2d ed. Sydney: Angus and Robertson.

Conde, J., and Boute, J. 1971. Size, structure and socio-economic characteristics of the population fertility-mortality. In *Population in African development*, ed. P. Cantrelle, 3–17. Liège: International Union for the Scientific Study of Population.

Cooke, F.; Abraham, K. F.; Davies, J. C.; Findlay, C. S.; Healey, R. F.; Sadura, A.; and Seguin, R. J. 1982. The La Pérouse Bay snow goose project: A thirteen-year report. On file, Canadian Wildlife Service, Ottawa.

Cooke, F.; Bousfield, M. A.; and Sadura, A. 1981. Mate change and reproductive success in the lesser snow goose. *Condor* 83:322–27.

Cooke, F., and Cooch, F. G. 1968. The genetics of polymorphism in the goose *Anser caerulescens. Evolution* 22:289–300.

Cooke, F.; Findlay, C. S.; and Rockwell, R. F. 1984. Recruitment and the timing of reproduction in lesser snow geese (*Chen caerulescens caerulescens*). *Auk* 101: 451–58.

Cooke, F.; MacInnes, C. D.; and Prevett, J. P. 1975. Gene flow between breeding populations of lesser snow geese. *Auk* 92:493–510.

Cooke, F., and McNally, C. M. 1975. Mate selection and colour preferences in lesser snow geese. *Behaviour* 53:151–70.

Cooper, C. F., and Stewart, B. S. 1983. Demography of northern elephant seals, 1911–1982. *Science* 219:969–71.

Coulson, J. C. 1959. The plumage and leg colour of the kittiwake and comments on the non-breeding population. *Brit. Birds* 52:189–96.

———. 1966. The influence of the pair bond and age on the breeding biology of the kittiwake gull *Rissa tridactyla. J. Anim. Ecol.* 35:269–79.

———. 1968. Differences in the quality of birds nesting in the centre and on the edges of a colony. *Nature* 217:478–79.

———. 1972. The significance of the pair-bond in the kittiwake. *Proc. XV Int. Ornithol. Congr.*, 424–33.

———. 1984. The population dynamics of the eider duck *Somateria mollissima* and evidence of extensive non-breeding by adult ducks. *Ibis* 126:525–43.

Coulson, J. C., and Horobin, J. M. 1972. The annual reoccupation of breeding sites by the fulmar. *Ibis* 114:30–42.

———. 1976. The influence of age on the breeding biology and survival of the arctic tern *sterna paradisaea*. *J. Zool. Lond.* 178:247–60.

Coulson, J. C., and Thomas, C. S. 1980. A study of the factors influencing the duration of the pair-bond in the kittiwake gull, *Rissa tridactyla*. *Proc. XVII Int. Ornithol. Congr.*, 822–33.

———. 1985a. Changes in the biology of the kittiwake *Rissa tridactyla*: A thirty-one-year study of a breeding colony. *J. Anim. Ecol.* 54:9–26.

———. 1985b. Differences in the breeding performance of individual kittiwake gulls, *Rissa tridactyla* (L.). In *Behavioral ecology*: Ecological consequences of adaptive behaviour, ed. R. M. Sibly and R. H. Smith, 489–503. Oxford: Blackwell Scientific Publications.

Coulson, J. C.; Thomas, C. S.; Butterfield, J. E. L.; Duncan, N.; Monaghan, P.; and Shedden, C. 1983. The use of head and bill length to sex live gulls *Laridae*. *Ibis* 125:549–57.

Coulson, J. C., and White, E. 1959. The post-fledging mortality of the kittiwake. *Bird Study* 6:97–102.

Coulson, J. C., and Wooller, R. D. 1976. Differential survival rates among breeding kittiwake gulls, *Rissa tridactyla* (L.). *J. Anim. Ecol.* 45:205–13.

Cox, C. R., and Le Boeuf, B. J. 1977. Female incitation of male competition: A mechanism in sexual selection. *Am. Nat.* 111:317–35.

Craig, R. 1983. Subfertility and the evolution of eusociality by kin selection. *J. Theor. Biol.* 100:379–97.

Cramp, S.; Bourne, W. R. P.; and Saunders, D. 1974. *The seabirds of Britain and Ireland*. London: Collins.

Critescu, M. 1975. Differential fertility depending on the age of puberty. *J. Human Evol.* 4:521–24.

Crow, J. F. 1958. Some possibilities for measuring selection intensities in man. *Human Biol.* 30:1–13.

Croxall, J. P. 1981. Aspects of the population demography of antarctic and subantarctic seabirds. *Colloque sur les Ecosystèmes Subantarctique 1981, Paimpont, CNFRA* 51:479–88.

Cundiff, L. 1972. The role of maternal effects in animal breeding. 8. Comparative aspects of maternal effects. *J. Anim. Sci.* 35:1335–37.

Curio, E. 1983. Why do young birds reproduce less well? *Ibis* 125:400–404.

Currie, N. L. 1961. Studies of the biology of *Erythemis simplicicollis* (Say) (Odonata: Libellulidae). Ph.D. diss., Ohio State University.

Cutler, S. J., and Ederer, F. 1958. Maximum utilization of the life-table method in analyzing survival. *J. Chron. Dis.* 8:699–713.

Daly, M., and Wilson, M. 1978. *Sex, evolution and behavior*. North Scituate, Mass.: Duxbury.

———. 1980. Discriminative parental solicitude: A biological perspective. *J. Marr. Fam.* 42:277–88.

Darling, F. F. 1937. *A herd of red deer*. London: Oxford University Press.

Darwin, C. 1859. *The origin of species*. London: Murray.

———. 1871. *The descent of man and selection in relation to sex*. London: John Murray.

David, J. 1963. Influence de la fécondation de la femelle sur le nombre et taille des oeufs pondus. *J. Insect Physiol.* 9:13–24.

David, W. A. L., and Gardiner, B. O. C. 1961. The mating behaviour of *Pieris brassicae* (L.) in a laboratory culture. *Bull. Entomol. Res.* 52:263–80.

Davies, J. C., and Cooke, F. 1983. Annual nesting productivity in snow geese: Prairie droughts and arctic springs. *J. Wildl. Mgmt.* 47:291–96.

Davies, N. B. 1978. Territorial defense in the speckled wood butterfly (*Parage aegeria*): The resident always wins. *Anim. Behav.* 26:138–47.

———. 1982. Behaviour and competition for scarce resources. In *Current problems in sociobiology*, ed. King's College Sociobiology Group, 363–80. Cambridge: Cambridge University Press.

Davis, D. E. 1940. Social nesting habits of smooth-billed ani. *Auk* 57:178–28.

———. 1941. Social nesting habits of *Crotophaga major*. *Auk* 58:179–83.

———. 1942. The phylogeny of social nesting in the Crotophaginae. *Quart. Rev. Biol.* 17:115–34.

Davis, J. W. F. 1975. Age, egg size and breeding success in the herring gull *Larus argentatus*. *Ibis* 117:460–73.

———. 1976. Breeding success and experience in the arctic skua, *Stercorarius parasiticus* (L.). *J. Anim. Ecol.* 45:531–36.

Dawkins, R. 1980. Good strategy or evolutionarily stable strategy? In *Sociobiology: Beyond nature/nurture*? ed. G. W. Barlow and S. Silverberg, 331–67. Boulder, Colo.: Westview Press.

DeVore, I. 1963. Mother-infant relations in free-ranging baboons. In *Maternal behavior in mammals*, ed. H. Rheingold. New York: John Wiley.

———. 1965. Male dominance and mating behavior in baboons. In *Sex and behavior*, ed. F. A. Beach, 266–89. New York: John Wiley.

Dhondt, A. A. 1971. The regulation of numbers in Belgian populations of great tits. *Proc. Adv. Study Inst. Dynamics Numbers Pop. (Osterbeek 1970)*, 532–47.

———. 1981. Post nuptial moult of the great tit in southern Sweden. *Ornis Scand.* 12:127–33.

———. 1985. Do old great tits forego breeding? *Auk* 102:870–72.

Dhondt, A. A., and Hublé, J. 1968. Fledging date and sex in relation to dispersal in young great tits. *Bird Study* 15:127–34.

Dhondt, A. A., and Schillemans, J. 1983. Reproductive success of the great tit in relation to its territorial status. *Anim. Behav.* 31:902–12.

Dittus, W. P. J. 1975. Population dynamics of the toque monkey (*Macaca sinica*). In *Socioecology and psychology of primates*, ed. R. H. Tuttle, 125–52. The Hague: Mouton.

———. 1977. The social regulation of population density and age-sex distribution in the toque monkey. *Behaviour* 63:281–322.

———. 1979. The evolution of behavior regulating density and age-specific sex ratios in a primate population. *Behaviour* 69:265–302.

———. 1980. The social regulation of primate populations: A synthesis. In *The macaques*, ed. D. G. Lindburg, 263–86. New York: Van Nostrand.

Dixon, W. S., ed. 1981. *BMDP statistical software*. 1981 ed. Berkeley: University of California Press.

Dobzhansky, Th. 1970. *Genetics of the evolutionary process*. New York: Columbia University Press.

Dominey, W. J. 1984. Alternative mating tactics and evolutionarily stable strategies. *Am. Zool.* 24:385–96.

Donegan, J. 1984. The courtship behaviour and songs of the *Drosophila virilis* species group. Ph.D. diss., University of Edinburgh.

Douwes, P. 1975. Territorial behaviour in *Heodes virgaureae* L. (Lep., Lyceanidae) with particular reference to visual stimuli. *Norwegian J. Entomol.* 22:143–54.

Dow, M. A., and von Schilcher, F. 1975. Aggression and mating success in *Drosophila melanogaster*. *Nature* 254:511–12.

Downhower, J. F., and Armitage, K. B. 1971. The yellow bellied marmot and the evolution of polygamy. *Am. Nat.* 105:355–70.

Drent, P. J. 1983. The functional ethology of territoriality in the great tit (*Parus major* L.). Dissertation, University of Groningen.

———. 1984. Mortality and dispersal in summer and its consequences for the density of great tits (*Parus major*) at the onset of autumn. *Ardea* 72:127–62.

Drickamer, L. C. 1974. A ten-year summary of reproductive data for free-ranging *Macaca mulatta*. *Folia Primatol.* 21:61–80.

Driver, E. D. 1981. Social class and fertility in metropolitan Madras. *Soc. Biol.* 28:30–40.

Dropkin, J. A., and Gamboa, G. J. 1981. Physical comparisons of foundresses of the paper wasp, *Polistes metricus* (Hymenoptera: Vespidae). *Can. Entomol.* 113:457–61.

Dunbar, R. I. M. 1980. Demographic and life history variables of a population of gelada baboons (*Theropithecus gelada*). *J. Anim. Ecol.* 49:485–506.

———. 1984. *Reproductive decisions: An economic analysis of gelada baboon social strategies*. Princeton: Princeton University Press.

———. 1985. Population consequences of social structure. In *Behavioural ecology*, ed. R. M. Sibly and R. H. Smith, 507–19. Oxford: Blackwell.

Dunbar, R. I. M., and Dunbar, E. P. 1977. Dominance and reproductive success among female gelada baboons. *Nature* 266:351–52.

Dunn, E. K. 1972. Effect of age on the fishing ability of sandwich terns *Sterna sandvicensis*. *Ibis* 114: 360–66.

———. 1977. Predation by weasels (*Mustela nivalis*) on breeding tits (*Parus* spp.) in relation to density of tits and rodents. *J. Anim. Ecol.* 46:633–52.

Dunne, T., and Dietrich, W. E. 1980. Experimental study of Horton overland flow on tropical hillslopes. *Z. Geomorph.* N.S., 35:40–59.

Dunnett, G. M. 1982. Ecology and Everyman. Presidential address to British Ecological Society, January 1981. *J. Anim. Ecol.* 51:1–14.

Dunnet, G. M., and Anderson, A. 1961. A method of sexing living fulmars in the hand. *Bird Study* 8:119–26.

Dunnet, G. M.; Anderson, A.; and Cormack, R. M. 1963. A study of survival of adult fulmars with observations on the pre-laying exodus. *Brit. Birds* 56:2–18.

Dunnet, G. M., and Ollason, J. C. 1978. The estimation of survival rate in the fulmar, *Fulmarus glacialis*. *J. Anim. Ecol.* 47:507–20.

———. 1979. The fulmar. Famous animals and plants series. *Biologist* 26:117–22.

———. 1982. The feeding dispersal of fulmars (*Fulmarus glacialis*) in the breeding season. *Ibis* 124:359–61.

Dunnet, G. M.; Ollason, J. C.; and Anderson, A. 1979. A twenty-eight-year study of breeding fulmars *Fulmarus glacialis* (L.) in Orkney. *Ibis* 121:293–300.

Dunnett, G. M.; Ollason, J. C.; and Anderson, A. 1983. Orkney fulmars. In *Orkney bird report, 1982*, ed. C. Booth, M. Cuthbert, and P. Reynolds. Stromners, Orkney: W. R. Rendall.

Efron, B., and Gong, G. 1983. A leisurely look at the bootstrap, the jackknife, and cross-validation. *Am. Statist.* 37:36–48.

Eickwort, K. 1969. Separation of the castes of *Polistes exclamans* and notes on its biology (Hym.: Vespidae). *Insectes Sociaux* 16:67–72.

Elliott, J. P.; Cowan, I. McT.; and Holling, C. S. 1977. Prey capture by the African lion. *Can. J. Zool.* 55:1811–28.

Emlen, S. T. 1968. A technique for marking anuran amphibians for behavioral studies. *Herpetologica* 24:172–73.

———. 1976. Lek organisation and mating strategies in the bullfrog. *Behav. Ecol. Sociobiol.* 1:283–313.

———. 1982. The evolution of helping. 1. An ecological constraints model. *Am. Nat.* 119:29–39.

———. 1984. Cooperative breeding in birds and mammals. In *Behavioural ecology: An evolutionary approach*, 2d ed., ed. J. R. Krebs and N. B. Davies, 305–39. Sunderland, Mass.: Sinauer.

Emlen, S. T., and Oring, L. W. 1977. Ecology, sexual selection, and the evolution of mating systems. *Science* 197:215–23.

Emlen, S. T., and Vehrencamp, S. L. 1983. Cooperative breeding strategies among birds.

In *Perspectives in ornithology*, ed. A. H. Brush and G. A. Clark, 93–120. Cambridge: Cambridge University Press.

Endler, J. A. 1986. *Natural selection in the wild*. Princeton: Princeton University Press.

Eshel, I. 1982. Evolutionarily stable strategies and viability selection in Mendelian populations. *Theor. Pop. Biol.* 22:204–17.

Estes, R. D., and Estes, R. K. 1979. The birth and survival of wildebeest calves. *Z. Tierpsychol.* 50:45–95.

Evans, M. E. 1978. Some factors influencing the use of a wintering site by Bewick's swans, studied through individual identification. M.Sc. diss., University of Wales.

———. 1979. Aspects of the life cycle of the Bewick's swan, based on recognition of individuals at a wintering site. *Bird Study* 26:149–62.

Evans, M. E., and Kear, J. 1978. Weights and measurements of Bewick's swans during winter. *Wildfowl* 29:118–22.

Everitt, B. S. 1977. *The analysis of contingency tables*. New York: Halsted Press.

Eyckerman, R. 1974. Some observations on the behaviour of intruding great tits, *Parus major*, and on the success of their breeding attempts in a high density breeding season. *Gerfaut* 64:29–40.

Fairbanks, L. A., and McGuire, M. T. 1984. Determinants of fecundity and reproductive success in captive vervet monkeys. *Am. J. Primatol.* 7:27–38.

Falconer, D. S. 1981. *Introduction to quantitative genetics*. 2d ed. London: Longman.

Fedigan, L. M. 1983. Dominance and reproductive success in primates. *Ybk. Phys. Anthropol.* 26:91–129.

Ferguson, G. W., and Fox, S. F. 1984. Annual variation of survival advantage of large juvenile side-blotched lizards *Uta stansburiana*: Its causes and evolutionary significance. *Evolution* 38:342–49.

Fincke, O. M. 1982. Lifetime mating success in a natural population of the damselfly, *Enallagma hageni* (Walsh) (Odonata: Coenagrionidae). *Behav. Ecol. Sociobiol.* 10:293–302.

———. 1984a. Giant damselflies in a tropical forest: Reproductive behavior of *Megaloprepus coerulatus* with notes on *Mecistogaster*. *Adv. Odonatol.* 2:13–27.

———. 1984b. Sperm competition in the damselfly *Enallagma hageni*: Benefits of multiple mating for males and females. *Behav. Ecol. Sociobiol.* 14:235–40.

———. 1985. Alternative mate-finding tactics in a nonterritorial damselfly (Odonata: Coenagrionidae). *Anim. Behav.* 33:1124–37.

———. 1986a. Lifetime reproductive success and the opportunity for selection in a short-lived damselfly (Odonata: Coenagrionidae). *Evolution* 40:791–803.

———. 1986b. Underwater oviposition, male vigilance, and female multiple mating in a damselfly (Odonata: Coenagrionidae). *Behav. Ecol. Sociobiol.* 18:405–12.

———. 1987. Female monogamy in the damselfly *Ischnura verticalis* Say (Zygoptera: Coenagrionidae). *Odonatologica* 16. In press.

Findlay, C. S., and Cooke, F. 1982a. Breeding synchrony in the lesser snow goose (*Anser caerulescens caerulescens*). 1. Genetic and environmental components of hatch date variability and their effects on hatch synchrony. *Evolution* 36:342–51.

———. 1982b. Synchrony in the lesser snow goose (*Anser caerulescens caerulescens*). 2. The adaptive value of reproductive synchrony. *Evolution* 36:786–99.

———. 1983. Genetic and environmental components of clutch size variance in a wild population of lesser snow geese (*Anser caerulescens caerulescens*). *Evolution* 37:724–34.

———. 1987. Repeatability and heritability of clutch size in lesser snow geese. *Evolution* 41:453.

Findlay, C. S.; Rockwell, R. F.; and Cooke, F. 1985. Does clutch size vary with cohort in lesser snow geese? *J. Wildl. Mgmt.* 49:417–20.

Finney, G., and Cooke, F. 1978. Reproductive habits in the snow goose: The influence of female age. *Condor* 80:147–58.

Fisher, H. I. 1969. Eggs and egg-laying in the Laysan albatross *Diomedea immutabilis*. *Condor* 71:101–12.

———. 1975. The relationship between deferred breeding and mortality in the Laysan albatross. *Auk* 92:433–41.

Fisher, J. 1952. *The fulmar*. London: Collins.

———. 1966. The fulmar population of Britain and Ireland, 1959. *Bird Study* 13:5–76.

Fisher, R. A. 1930. *The genetical theory of natural selection*. London: Oxford University Press.

———. 1958. *The genetical theory of natural selection*. 2d ed. New York: Dover.

Fitch, M. A., and Shugart, G. W. 1984. Requirement for a mixed reproductive strategy in avian species. *Am. Nat.* 124:116–26.

Fitzpatrick, J. W., and Woolfenden, G. E. 1986. Demographic routes to cooperative breeding in some New World jays. In *Evolution of animal behavior*, ed. M. H. Nitecki and J. A. Kitchell, 137–60. New York: Plenum.

Fitzpatrick, J. W.; Woolfenden, G. E.; and McGowan, K. J. n.d. Sources of variance in lifetime fitness of Florida scrub jays. *Proc. XIX Int. Ornithol. Cong.*, Ottawa. In press.

Ford, E. B. 1975. *Ecological genetics*. 4th ed. London: Chapman and Hall.

Fosbrooke, H. 1963. The *Stomoxys* plague in Ngorongoro. *E. Afr. Wildl. J.* 1:124–26.

Freedman, D. H. 1979. *Human sociobiology: A holistic approach*. New York: Free Press.

Fretwell, S. D. 1972. *Populations in a seasonal environment*. Princeton: Princeton University Press.

Friedel, T., and Gillot, C. 1977. Contributions of male-produced proteins to vitellogenesis in *Melanopus sanguinipes*. *J. Insect Physiol.* 23:145–51.

Frisch, R. E. 1978. Population, food intake and fertility. *Science* 199:22–30.

Frisch, R. E., and McArthur, J. W. 1974. Menstrual cycles: Fatness as a determinant of minimum weight for height necessary for the maintenance and onset. *Science* 185:949–51.

Gadgil, M. 1972. Male dimorphism as a consequence of sexual selection. *Am. Nat.* 106:-574–80.

Gadgil, M., and Bossert, W. H. 1970. Life historical consequences of natural selection. *Am. Nat.* 104:1–24.

Gamboa, O. J. 1978. Intraspecific defense: Advantage of social cooperation among paper wasp foundresses. *Science* 199:1463–65.

Ganetsky, B., and Flanagan, J. R. 1978. On the relationship between senescence and age-related changes in two wild-type strains of *Drosophila melanogaster*. *Exp. Gerontol.* 13:189–96.

Garnett, M. C. 1981. Body size: Its heritability and influence on juvenile survival among great tits *Parus major*. *Ibis* 123:31–41.

Gentry, R. L. 1970. Social behavior of the Steller sea lion. Ph.D. diss., University of California, Santa Cruz.

Geramita, J. M., and Cooke, F. 1982. Evidence that fidelity to natal breeding colony is not absolute in female snow geese. *Can. J. Zool.* 60:2051–56.

Ghiselin, M. T. 1974. *The economy of nature and the evolution of sex*. Berkeley: University of California Press.

Gibb, J. 1956. Food, feeding habits and territory of the rock pipit *Anthus spinoletta*. *Ibis* 98:506–30.

Gibo, D. L. 1978. The selective advantage of foundress associations in *Polistes fuscatus* (Hymenoptera: Vespidae) A field study of the effects of predation on productivity. *Can. Entomol.* 110:519–40.

Gibson, R. M. 1987. Bivariate versus multivariate analyses of sexual selection in red deer. *Anim. Behav.* 35:292–93.

Gibson, R. M., and Guinness, F. E. 1980. Behavioural factors affecting male reproductive success in red deer (*Cervus elaphus* L.). *Anim. Behav.* 28:1163–74.

Gilbert, C., and Gillman, J. 1960. Puberty in the baboon (*P. ursinis*) in relation to age and body weight. *S. African J. Med. Sci.* 25 (2/3):99–103.

Gillman, J. 1935. The cyclical changes in the external genital organs of the baboon (*P. porcarinus*). *S. Afr. J. Clin. Sci.* 32:342–55.

Gillman, J., and Gilbert, C. 1946. The reproductive cycle of the chacma baboon (*Papio ursinus*) with special reference to the problems of menstrual irregularities as assessed by the behaviour of the sex skin. *S. Afr. J. Med. Sci.* 11:1–54.

Goodman, D. 1984. Risk spreading as an adaptive strategy in the iteroparous life histories. *Theor. Pop. Biol.* 25:1–20.

Goodman, L. A. 1960. On the exact variance of products. *JASA* 55:708–13.

———. 1962. The variance of the product of K random variables. *JASA* 57:54–60.

Goss-Custard, J. D.; Dunbar, R. M.; and Aldrich-Blake, F. P. G. 1972. Survival, mating and rearing strategies in the evolution of primate social structure. *Folia Primatol.* 17:1–19.

Gouzoules, H.; Gouzoules, S.; and Fedigan, L. 1982. Behavioural dominance and reproductive success in female Japanese monkeys (*Macaca fuscata*). *Anim. Behav.* 30:1138–50.

Gowen, J. W., and Johnson, L. E. 1946. On the mechanism of heterosis. 1. Metabolic capacity of different races of *Drosophila melanogaster* for egg production. *Am. Nat.* 80:149–79.

Grafen, A. 1982. How not to measure inclusive fitness. *Nature* 298:425–26.

———. 1984. Natural selection, kin selection and group selection. In *Behavioural ecology: An evolutionary approach*, 2d ed., ed. J. R. Krebs and N. B. Davies, 62–89. Sunderland, Mass.: Sinauer.

———. 1985. A geometric view of relatedness. In *Oxford surveys in evolutionary biology*, ed. Richard Dawkins, vol. 2. London and New York: Oxford University Press.

Graham, C. E.; Kling, O. R.; and Steiner, R. A. 1979. Reproductive senescence in female nonhuman primates. In *Aging in nonhuman primates*, ed. D. M. Bowden, 183–202. New York: Van Nostrand Reinhold.

Gray, R. H. 1979. Biological and social interactions in the determination of late fertiliy. *J. Biosoc. Sci.*, suppl. 6:97–115.

Greig, S. A.; Coulson, J. C.; and Monaghan, P. 1983. Age-related differences in foraging success in the herring gull (*Larus argentatus*). *Anim. Behav.* 31:1237–43.

Grounds, J. G. 1964. Mortality of children under six years old in Kenya, with special reference to contributory causes, especially malnutrition. *J. Trop. Med. Hyg.* 67:257–59.

Gubhaju, B. 1983. Fertility differentials in Nepal. *J. Biosoc. Sci.* 15:325–31.

Guinness, F. E.; Albon, S. D.; and Clutton-Brock, T. H. 1978. Factors affecting reproduction in red deer (*Cervus elaphus* L.). *J. Reprod. Fert.* 54:325–34.

Guinness, F. E.; Clutton-Brock, T. H.; and Albon, S. D. 1978. Factors affecting calf mortality in red deer. *J. Anim. Ecol.* 47:817–32.

Gunn, R. G. 1968. Levels of first winter feeding in relation to performance of Cheviot hill ewes. 6. Lifetime production from the hill. *J. Agric. Sci. Camb.* 71:161–66.

Gustafsson, L. 1986. Lifetime reproductive success and heritability: Empirical support for Fisher's fundamental theorem. *Am. Nat.* 128:761–64.

Haartmann, L. von. 1949. Der Trauerfliegenschnäpper. 1. Orstreue und Rassenbildung. *Acta Zool. Fenn.* 56:1–104.

———. 1951. Successive polygamy. *Behaviour* 3:256–74.

———. 1956. Territory in the pied flycatcher, *Muscicapa hypoleuca*. *Ibis* 98:460–75.

———. 1957. Adaptation in hole-nesting birds. *Evolution* 11:339–47.

Hafernik, J. E., and Garrison, R. W. 1986. Mating success and survival rate in a population of damselflies: Results at variance with theory? *Am. Nat.* 128:353–65.

Haggard, C. M., and Gamboa, G. J. 1980. Seasonal variation in body size and reproductive condition of a paper wasp, *Polistes metricus* (Hymenoptera: Vespidae). *Can. Entomol.* 112:239–48.

Hails, C. J. 1982. A comparison of tropical and temperate aerial insect abundance. *Biotropica* 14:310–13.

Haldane, J. B. S. 1954. The measurement of natural selection. *Proc. IX Int. Congr. Genet.* 1:480–87.

Halliday, T. R. 1978. Sexual selection and mate choice. In *Behavioural ecology: An evolutionary approach*, ed. J. R. Krebs and N. B. Davies, 180–213. Oxford: Blackwell.

———. 1983. The study of mate choice. In *Mate choice*, ed. P. P. G. Bateson, 3–32. Cambridge: Cambridge University Press.

Hamilton, W. D. 1963. The evolution of altruistic behavior. *Am. Nat.* 97:354–56.

———. 1964. The genetical evolution of social behaviour, 1 and 2. *J. Theor. Biol.* 7:1–52.

———. 1966. The moulding of senescence by natural selection. *J. Theor. Biol.* 12:12–45.

———. 1970. Selfish and spiteful behaviour in an evolutionary model. *Nature* 228:1218–20.

Hamilton, W. D., and Zuk, M. 1982. Heritable true fitness and bright birds: A role for parasites? *Science* 218:384–87.

Hanby, J. P., and Bygott, J. D. 1979. Population changes in lions and other predators. In *Serengeti: Dynamics of an ecosystem*, ed. A. R. E. Sinclair and M. Norton-Griffiths, 249–62. Chicago: University of Chicago Press.

———. 1987. Emigration of subadult lions. *Anim. Behav.* 35:161–69.

Hannon, S. J., and Smith, J. N. M. 1984. Factors influencing age-related breeding success in the willow ptarmigan. *Auk* 101:848–54.

Hanwell, A., and Peaker, M. 1977. Physiological effects of lactation on the mother. *Symp. Zool. Soc. Lond.* 41:297–312.

Harper, J. L 1983. A Darwinian plant ecology. In *Evolution from molecules to man*, ed. D. S. Bendall, 323–45. Cambridge: Cambridge University Press.

Harris, M. P. 1980. Breeding performance of puffins *Fratercula arctica* in relation to nest density, laying date and year. *Ibis* 122:193–209.

Harrison, G. A.; Boyce, A. J.; and Platt, C. M. 1977. Body composition changes during lactation in a New Guinea population. *Ann. Human Biol.* 2:395–98.

Harvey, P. H.; Greenwood, P. J.; and Campbell, B. 1984. Timing of laying of the pied flycatcher in relation to age of male and female parent. *Bird Study* 31:57–60.

Harvey, P. H.; Greenwood, P. J.; Campbell, B.; and Stenning, M. J. 1984. Breeding dispersal of the pied flycatcher (*Ficedula hypoleuca*). *J. Anim. Ecol.* 53:727–36.

Harvey, P. H.; Greenwood, P. J.; and Perrins, C. M. 1979. Breeding area fidelity of great tits, *Parus major*. *J. Anim. Ecol.* 48:305–13.

Harvey, P. H.; Greenwood, P. J.; Perrins, C. M.; and Martin, A. R. 1979. Breeding success of great tits *Parus major* in relation to age of male and female parent. *Ibis* 121:216–19.

Harvey, P. H.; Stenning, M. J.; and Campbell, B. 1985. Individual variation in seasonal breeding success of pied flycatchers. *J. Anim. Ecol.* 54:391–98.

Hausfater, G. 1975. Dominance and reproduction in baboons: A quantitative analysis. In *Contributions to primatology*, vol. 7. Basel: S. Karger.

Hausfater, G.; Altmann, J.; and Altmann, S. A. 1982. Long-term consistency of dominance relations among female baboons (*Papio cynocephalus*). *Science* 217:752–55.

Hausfater, F., and Bearce, W. H. 1976. Acacia tree exudates: Their composition and use as a food source by baboons. *E. Afr. Wildl. J.* 14:241–43.

Hawthorn, G. 1970. *The sociology of fertility*. London: Macmillan.

Hayes, J. L. 1981. The population ecology of a natural population of the pierid butterfly *Colias alexandra*. *Oecologia* (Berlin) 49:188–200.

Haymes, G. T., and Blokpoel, H. 1980. The influence of age on the breeding biology of ring-billed gulls. *Wilson Bull.* 92:221–28.

Hazama, N. 1964. Weighing wild Japanese monkeys in Arashiyama. *Primates* 5:81–104.

Hegmann, J. P., and Dingle, H. 1982. Phenotypic and genetic covariance structure in milkweed bug life history traits. In *Evolution and genetics of life histories*, ed. H. Dingle and J. P. Hegmann, 177–85. New York: Springer-Verlag.

Heinzel, H.; Fitter, R.; and Parslow, J. 1979. *The birds of Britain and Europe*. 4th ed. London: Collins.

Hihara, F. 1981. Effects of the male accessory gland secretion on oviposition and remating in females of *Drosophila melanogaster*. *Zool. Mag*. 90:307–16.

Hjorth, I. 1967. Fortplantningsbeteenden inom hönsfagel-familjen Tetraonidae. *Var Fagelvärld* 26:193–243.

———. 1970. Reproductive behaviour in Tetraonidae. *Viltrevy* 7:181–596.

Hogstedt, G. 1980. Evolution of clutch size in animals: Adaptive variation in relation to territory quality. *Science* 210:1148–50.

Hoogland, J. L., and Foltz, D. W. 1982. Variance in male and female reproductive success in a harem-polygynous mammal, the black-tailed prairie dog (Scuridae: *Cynomys ludovicianus*). *Behav. Ecol. Sociobiol*. 11:155–63.

Horn, H. S., and Rubenstein, D. I. 1984. Behavioural adaptations and life history. In *Behavioural ecology: An evolutionary approach*, 2d ed., ed. J. R. Krebs and N. B. Davies. Sunderland, Mass.: Sinauer.

Houck, L. D.; Arnold, S. J.; and Thisted, R. A. 1985. A statistical study of mate choice: Sexual selection in a plethodontid salamander (*Desmognathus ochrophaeus*). *Evolution* 39:370–86.

Howard, R. D. 1978a. The evolution of mating strategies in bullfrogs, *Rana catesbeiana*. *Evolution* 32:850–71.

———. 1978b. The influence of male-defended oviposition sites on early embryo mortality in bullfrogs. *Ecology* 59:789–98.

———. 1979. Estimating reproductive success in natural populations. *Am. Nat.* 114:221–31.

———. 1980. Mating behaviour and mating success in woodfrogs, *Rana sylvatica*. *Anim. Behav*. 28:705–16.

———. 1981. Sexual dimorphism in bullfrogs. *Ecology* 62:303–10.

———. 1983. Sexual selection and variation in reproductive success in a long-lived organism. *Am. Nat*. 122:301–25.

———. 1984. Alternative mating behaviors of young male bullfrogs. *Am. Zool*. 24:397–406.

Howard, R. D., and Kluge, A. G. 1985. Proximate mechanisms of sexual selection in woodfrogs. *Evolution* 39:260–77.

Hrdy, S. B., and Williams, G. C. 1983. Behavioural biology and the double standard. In *Social behavior of female vertebrates*, ed. S. K. Wasser, 3–17. New York: Academic Press.

Hubbell, S. P., and Johnson, L. K. 1987. Environmental variance in lifetime mating success, mate choice, and sexual selection. *Amer. Nat*. 130:91–112.

Huxley, J. 1983. *Evolution: The modern synthesis*. London: Allen and Unwin.

Iason, G.; Duck, C.; and Clutton-Brock, T. H. 1986. The effect of gull colonies on red deer grazing and reproductive success. *J. Anim. Ecol*. 55:507–15.

Irons, W. 1979. Cultural and biological success. In *Evolutionary biology and human social behavior: An anthropological perspective*, ed. N. A. Chagnon and W. Irons, 257–72. North Scituate, Mass.: Duxbury Press.

Jacobs, M. E. 1960. Influence of light on mating of *Drosophila melanogaster*. *Ecology* 41:182–88.

Jacquard, A. 1974. *The genetic structure of populations*. Trans. from French by D. and B. Charlesworth. Biomathematics Series, vol. 5. New York: Springer-Verlag.

Jain, A. K. 1969. Fecundability and its relationship to age in a sample of Taiwanese women. *Pop. Studies* 32:169–85.

Jallon, J.-M., and Hotta, Y. 1979. Genetic and behavioral studies of female sex appeal in *Drosophila*. *Behav. Genet*. 9:257–75.

Janis, C. 1982. Evolution of horns in ungulates: Ecology and paleoecology. *Biol. Rev.* 57:261--318.

Jelliffe, D. B., and Jelliffe, E. F. P. 1978. *Human milk in the modern world: Psychosocial, nutritional and economic significance*. Oxford: Oxford University Press.

Johnson, L. K. 1982. Sexual selection in a tropical brontid weevil. *Evolution* 36:251–62.

Jones, P. J. 1973. Some aspects of the feeding ecology of the great tit *Parus major*. D. Phil. diss., University of Oxford.

Kaplan, E. L., and Meier, P. 1958. Non-parametric estimations from incomplete observations. *JASA* 53:457–81.

Kearsey, M. J. 1965. The interaction of competition and food supply in two lines of *Drosophila melanogaster*. *Heredity* 20:169–81.

Keith, L. B. 1963. *Wildlife's ten year cycle*. Madison: University of Wisconsin Press.

Kempthorne, O. 1957. *An introduction to genetic statistics*. New York: John Wiley.

Kenya Fertility Survey, 1977–78. 1980. Nairobi: Central Bureau of Statistics, Ministry of Economic Planning and Development.

Kerlinger, F. N., and Pedhazur, E. J. 1973. *Multiple regression in behavioral research*. New York: Holt, Rinehart and Winston.

Kidwell, J. F., and Malick, L. E. 1967. The effect of genotype, mating status, weight and egg production on longevity. *J. Hered.* 58:169.

Kimura, M. 1983. *The neutral theory of molecular evolution*. Cambridge: Cambridge University Press.

Kitching, R. L. 1983. Myrmecophilous organs of the larvae and pupae of the lycaenid butterfly *Jalmenus evagoras* (Donovan). *J. Nat. Hist.* 17:471–81.

Klahn, J. E. 1981. Alternative reproductive tactics of single foundresses of a social wasp, *Polistes fuscatus*. Ph.D. diss., University of Iowa.

Klein, D. F. 1978. The diet and reproductive cycle of a population of vervet monkeys (*Cercopithecus aethiops*). Ph.D. diss., New York University.

Klomp, H. 1970. The determination of clutch-size in birds: A review. *Ardea* 58:1–124.

———. 1980. Fluctuations and stability in great tit populations. *Ardea* 68:205–24.

Kluijver, H. N. 1951. The population ecology of the great tit, *Parus m. major*. *Ardea* 39:1–135.

———. 1971. Regulation of numbers of populations of great tits (*Parus m. major*). In *Dynamics of populations*, ed. P. J. den Boer and G. R. Gradwell, 507–23. Wageningen: Pudoc.

Koenig, W. D., and Albano, S. S. 1986. On the measurement of sexual selection. *Am. Nat.* 127:403–9.

Koford, C. B. 1965. Population dynamics of rhesus monkeys on Cayo Santiago. In *Primate behavior*, ed. I. DeVore, 160–74. New York: Holt, Rinehart and Winston.

Koford, R. R.; Bowen, B. S.; and Vehrencamp, S. L. 1986. Habitat saturation in groove-billed anis. *Am. Nat.* 127:317–37.

Konner, M., and Worthman, C. 1980. Nursing frequency, gonadal functions and birth spacing among! Kung hunter-gatherers. *Science* 207:788–91.

Koskimies, J. 1957. Polymorphic variability in clutch size and laying date in the velvet scoter, *Melanitta fusca* (L.). *Ornis Fennica* 34:118–28.

Köster, F. L. 1971. Zum Nistverhalten des Ani. *Bonn. Zool. Beitr.* 22:4–27.

Kraemer, H. C. 1975. On estimation and hypothesis testing problems for correlation coefficients. *Psychometrika* 40:473–85.

Krebs, C. J. 1978. A review of the Chitty hypothesis of population regulation. *Can. J. Zool.* 56:2463–80.

Krebs, J. R.; Erichsen, J. T.; Webber, M. I.; and Charnov, E. L. 1977. Optimal prey selection in the great tit. *Anim. Behav.* 25:30–38.

Krebs, J. R., and McCleery, R. H. 1984. Optimization in behavioural ecology. In *Behavioural ecology: An evolutionary approach*, 2d ed., ed. J. R. Krebs and N. B. Davies, 91–121. Sunderland, Mass.: Sinauer.

Kreulen, D. A. 1975. Wildebeest habitat selection on the Serengeti plains, Tanzania, in relation to calcium and lactation: A preliminary report. *E. Afr. Wildl. J.* 13:297–304.

Kriewaldt, F. H., and Hendrickx, A. G. 1968. Reproductive parameters of the baboon. *Lab. Anim. Care* 18:361–70.

Kruijt, J. P.; de Vos, G. J.; and Bossema, I. 1972. The arena system of black grouse. *Proc. XV Int. Ornithol. Congr.*, 399–423.

Kruijt, J. P., and Hogan, J. A. 1967. Social behaviour on the lek in black grouse, *Lyrurus t. tetrix* (L.). *Ardea* 55:203–40.

Kruuk, H. 1972. *The spotted hyena*. Chicago: University of Chicago Press.

Kvelland, I. 1965. Some observations on the mating activity and fertility of *Drosophila melanogaster*. *Hereditas* 53:281–306.

Labine, P. A. 1968. The population biology of the butterfly *Euphydrias editha*. 8. Oviposition and its relation to patterns of oviposition in other butterflies. *Evolution* 22:799–805.

Lack, D. 1954. *The natural regulation of animal numbers*. Oxford: Clarendon Press.

———. 1966. *Population studies of birds*. Oxford: Clarendon Press.

———. 1968. *Ecological adaptations for breeding in birds*. London: Methuen.

Lakhani, K. H. 1985. Inherent difficulties in estimating age specific bird survival rates from ring recoveries. In *Statistics in ornithology*, ed. B. J. T. Morgan and P. M. North, 311–21. Notes in Statistics 29. Berlin: Springer-Verlag.

Lamb, M. J. 1964. The effects of radiation on the longevity of female *Drosophila subobscura*. *J. Insect Physiol.* 10:487–97.

Lande, R. 1976. The maintenance of genetic variability by mutation in a polygenic character with linked loci. *Genet. Res.* 26:221–35.

———. 1979. Quantitative genetic analysis of multivariate evolution, applied to brain:body size allometry. *Evolution* 33:402–16.

———. 1980. Sexual dimorphism, sexual selection, and adaptation in polygenic characters. *Evolution* 34:294–305.

———. 1982. A quantitative genetic theory of life history evolution. *Ecology* 63:607–15.

Lande, R., and Arnold, S. J. 1983. The measurement of selection on correlated characters. *Evolution* 37:1210–26.

Lapin, B. A.; Krilova, R. I.; Cherkovich, G. M.; and Asanov, N. S. 1979. Observations from Sukhumi. In *Aging in non-human primates*, ed. D. M. Bowden, 14–37. New York: Van Nostrand-Reinhold.

Latter, B. D. H. 1960. Natural selection for an intermediate optimum. *Austr. J. Biol. Sci.* 13:30–35.

Lawn, M. R. 1982. Pairing systems and site tenacity in the willow warbler *Phylloscopus trochilus* in southern England. *Ornis Scand.* 13:193–99.

Laws, R. M. 1956. The elephant seal (*Mirounga leonina*, Linn.). 2. General, social and reproductive behaviour. *FIDS Sci. Rep.*, no. 13.

Le Boeuf, B. J. 1972. Sexual behavior in the northern elephant seal, *Mirounga angustirostris*. *Behaviour* 41:1–26.

———. 1974. Male-male competition and reproductive success in elephant seals. *Am. Zool.* 14:163–76.

———. 1981. The elephant seal. In *Problems in management of locally abundant wild mammals*, ed. P. Jewell, 291–301. New York: Academic Press.

Le Boeuf, B. J.; Ainley, D. G.; and Lewis, T. J. 1974. Elephant seals on the Farallons: Population dynamics of an incipient colony. *J. Mammal.* 55:370–85.

Le Boeuf, B. J., and Bonnell, M. J. 1980. Pinnepeds of the California Channel Islands: Abundance and distribution. In *The California islands*, ed. D. Power, 475–93. Santa Barbara: Museum of Natural History.

Le Boeuf, B. J., and Kaza, S., eds. 1981. *The natural history of Año Neuvo*. Pacific Grove, Calif.: Boxwood Press.

Le Boeuf, B. J., and Panken, K. J. 1977. Elephant seals breeding on the California mainland. *Proc. Calif. Acad. Sci.* 41:267–80.

Le Boeuf, B. J., and Peterson, R. S. 1969. Social status and mating activity in elephant seals. *Science* 163:91–93.

LeBoeuf, B. J.; Riedman, M. L.; and Keyes, P. 1982. White shark predation on pinnipeds in California coastal waters. *Fish. Bull.* 80:891–95.

Le Boeuf, B. J.; Whiting, R. J.; and Gantt, R. F. 1972. Perinatal behavior of northern elephant seal females and their young. *Behaviour* 43:121–56.

Lederhouse, R. C. 1981. The effect of female mating frequency on egg fertility in the black swallowtail *Papilio polyxenes asterius* (Papilionidae). *J. Lepid. Res.* 35:266–77.

———. 1982. Territorial defense and lek behaviour of the black swallowtail butterfly, *Papilio polyxenes*. *Behav. Ecol. Sociobiol.* 10:109–18.

Lederhouse, R. C.; Finke, M. D.; and Scriber, J. M. 1981. The contributions of larval growth and pupal duration to protandry in the black swallowtail butterfly, *Papilio polyxenes*. *Oecologia* (Berlin) 53:296–300.

Lee, P. C. 1981. Ecological and social influences on development of vervet monkeys. Ph.D. diss., Cambridge University.

———. 1983a. Context-specific unpredictability in dominance interactions. In *Primate social relationships: An integrated approach*, ed. R. A. Hinde, 35–44. Oxford: Blackwell.

———. 1983b. Ecological influences on relationships and social structure. In *Primate social relationships: An integrated approach*, ed. R. A. Hinde, 225–30. Oxford: Blackwell.

———. 1984. Ecological constraints on the social development of vervet monkeys. *Behaviour* 91:245–62.

Lefevre, G., and Jonsson, V. B. 1962. Sperm transfer, storage, displacement and utilization in *Drosophila melanogaster*. *Genetics* 47:1719–36.

Leridon, H. 1977. *Human fertility: The basic components*. Chicago: University of Chicago Press.

LeVine, R. A. 1962. Witchcraft and co-wife proximity in southwestern Kenya. *Ethnology* 1:39–45.

Ligon, J. D. 1981. Demographic patterns and communal breeding in the green woodhoopoe, *Phoeniculus purpureus*. In *Natural selection and social behavior*, ed. R. D. Alexander and D. W. Tinkle, 231–43. New York: Chiron Press.

Lind, E. A. 1960. Zur Ethologie und Ökologie der Mehlschwalbe *Delichon urbica* (L.). *Ann. Zool. Soc.* 21:1–127.

Lloyd, C. S. 1979. Factors affecting breeding of razorbills *Alca torda* on Skokholm. *Ibis* 121:165–76.

Lomnicki, A. 1978. Individual differences between animals and the natural regulation of their number. *J. Anim. Ecol.* 47:461–75.

———. 1980. Regulation of population density due to individual differences and patchy environment. *Oikos* 35:185–93.

Long, C. E.; Markow, T. A.; and Yaeger, P. 1980. Relative male age, fertility and competitive mating success in *Drosophila melanogaster*. *Behav. Genet.* 10:163–70.

Luckinbill, L. S.; Arking, R.; Clare, M. J.; Cirocco, W. C.; and Buck, S. A. 1984. Selection for delayed senescence in *Drosophila melanogaster*. *Evolution* 38:996–1003.

Lundburg, A.; Alatalo, R. V.; Carlson, A.; and Ulfstrand, S. 1981. Biometry, habitat distribution and breeding success in the pied flycatcher *Ficedula hypoleuca*. *Ornis Scand.* 12:68–79.

McCann, T. S. 1980. Territoriality and breeding behaviour of adult male antarctic fur seal, *Arctocephalus gazella*. *J. Zool. Lond.* 192:295–310.

McCauley, D. E. 1983. An estimate of the relative opportunities for natural and sexual selection in a population of milkweed beetles. *Evolution* 37:701–7.

McCleery, R. H., and Perrins, C. M. 1985. Territory size, reproductive success and population dynamics in the great tit (*Parus major*). In *Behavioural ecology: Ecological consequences of adaptive behaviour*, ed. R. M. Sibly and R. H. Smith. Oxford: Blackwell Scientific Publications.

McCracken, G. F., and Bradbury, J. W. 1977. Paternity and genetic heterogeneity in the polygynous bat, *Phyllostomus hastatus*. *Science* 198:303–6.

Macdonald, M. A. 1977. An analysis of the recoveries of British-ringed fulmars. *Bird Study* 24:208–14.

———. 1977b. The pre-laying exodus of the fulmar *Fulmarus glacialis* (L.). *Ornis Scand.* 8:33–37.

———. 1980. The winter attendance of fulmars at land in N. E. Scotland. *Ornis Scand.* 11: 23–29.

McGregor, P. K.; Krebs, J. R.; and Perrins, C. M. 1981. Song repertoires and lifetime reproductive success in the great tit (*Parus major*). *Am. Nat.* 118:149–59.

McNaughton, S. J. 1979. Grassland-herbivore dynamics. In *Serengeti: Dynamics of an ecosystem*, ed. A. R. E. Sinclair and M. Norton-Griffiths. 46–81. Chicago: University of Chicago Press.

McVey, M. E. 1981. Lifetime reproductive tactics in a territorial dragonfly, *Erythemis simplicicollis*. Ph.D. diss., Rockefeller University.

———. 1984. Egg release rates with temperature and body size in libellulid dragonflies. *Odonatologica* 13:377–86.

———. 1985. Rates of color maturation in relation to age, diet and temperature in male *Erythemis simplicicollis* (Say) (Odonata: Libellulidae). *Odonatologica* 14:101–14.

———. n.d.a. Despotic male settlement patterns and the resident's advantage in a territorial dragonfly. Unpublished MS.

———. n.d.b. Sperm competition, territory size, and the threshold for switching male mating tactics in the dragonfly *Erythemis simplicicollis*. Unpublished MS.

———. n.d.c. Temperature, oviposition rate, and optimal female reproductive tactics in libellulid dragonflies. Unpublished MS.

McVey, M. E., and Smittle, B. J. 1984. Sperm precedence in the dragonfly *Erythemis simplicicollis*. *J. Insect Physiol.* 30:619–28.

Maddock, L. 1979. The "migration" and grazing succession. In *Serengeti: Dynamics of an ecosystem*, ed. A. R. E. Sinclair and M. Norton-Griffiths, 104–29. Chicago: University of Chicago Press.

Malick, L. K., and Kidwell, J. F. 1966. The effect of mating status, sex and genotype on longevity in *Drosophila melanogaster*. *Genetics* 54:203–9.

Manly, B. F. J. 1985. *The statistics of natural selection*. London: Chapman and Hall.

Manners, R. A. 1967. The Kipsigis of Kenya: Culture change in a "model" East African tribe. In *Contemporary change in traditional societies. Vol. 1. Introduction and African tribes*, ed. J. Steward, 207–359. Urbana: University of Illinois Press.

Manning, A. 1967. The control of sexual receptivity in female *Drosophila*. *Anim. Behav.* 15:239–50.

Markow, T. A., and Ankney, P. F. 1984. *Drosophila* males contribute to oogenesis in a multiple mating species. *Science* 244:302–3.

Marks, R. J. 1976. Mating behaviour and fecundity of the red bollworm *Diparopsis castanes* Hmps. (Lepidoptera, Noctuidae). *Bull. Ent. Res.* 66:145–58.

Marlow, B. J. 1975. The comparative behaviour of the Australian sea lions *Neophoca cinerea* and *Phocarctos hookeri*. *Mammalia* 39:159–230.

Maxwell, A. E. 1961. *Analysing quantitative data*. London: Methuen.

Maynard Smith, J. 1956. Fertility, mating behaviour and sexual selection in *Drosophila subobscura*. *J. Genet.* 54:261–79.

———. 1958. The effect of temperature and of egg laying on the longevity of *Drosophila subobscura*. *J. Exp. Biol.* 35:832–42.

———. 1974. The theory of games and the evolution of animal conflicts. *J. Theor. Biol.* 47:209–21.

———. 1976. Sexual selection and the handicap principle. *J. Theor. Biol.* 57:239–42.

Maynard Smith, J., and Price, G. R. 1973. The logic of animal conflict. *Nature* 246:15–18.

Medawar, P. B. 1952. *An unsolved problem of biology*. London: H. K. Lewis.

Meikle, D. B.; Tilford, B. L.; and Vessey, S. H. 1984. Dominance rank, secondary sex ratio and reproduction of offspring in polygynous primates. *Am. Nat.* 124:173–88.

Metcalf, R. A., and Whitt, G. S. 1977a. Intra-nest relatedness in the social wasp *Polistes metricus*: A genetic analysis. *Behav. Ecol. Sociobiol.* 2:339–51.

———. 1977b. Relative inclusive fitness in the social wasp *Polistes metricus*. *Behav. Ecol. Sociobiol.* 2:353–60.

Michod, R. E. 1982. The theory of kin selection. *Ann. Rev. Ecol. System.* 13:23–55.

Miller, M. K., and Stokes, C. S. 1985. Teenage fertility, socioeconomic status and infant mortality. *J. Biosoc. Sci.* 17:147–55.

Mills, J. A. 1973. The influence of age and pair-bond on the breeding biology of the red-billed gull (*Larus novaehollandiae scopulinus*). *J. Anim. Ecol.* 42:147–62.

Mills, J. A., and Shaw, P. W. 1980. The influence of age on laying date, clutch size and egg size of the white-fronted tern, *Sterna striata*. *N.Z. J. Zool.* 7:147–53.

Montevecchi, W. A.; Blundon, E.; Coombes, G.; Porter, J.; and Rice, P. 1978. Northern fulmar breeding range extended to Baccalieu Island, Newfoundland. *Can. Field Nat.* 92:80–82.

Montgomery, G. G., and Sunquist, M. E. 1978. Habitat selection and use by two-toed and three-toed sloths. In *The ecology of arboreal folivores*, ed. G. G. Montgomery, 329–59. Washington, D.C.: Smithsonian Institution.

Mori, A. 1979. Analysis of population changes by measurement of body weight in the Koshima troop of Japanese monkeys. *Primates* 20:371–97.

Morris, F. O, 1891. *A history of British birds*. London: Nimmo.

Mosley, W. H. 1977. The effects of nutrition on natural fertility. Paper presented for the seminar on natural fertility, International Union for the Scientific Study of Population, Paris.

Mougin, J.-L. 1967. Etude écologique des deux espèces de fulmars—Le fulmar atlantique (*Fulmarus glacialis*) et le fulmar antarctique (*Fulmarus glacialoides*). *Ois R.F.O.* 37:57–103.

Moyles, D. L. J., and Boag, D. A. 1981.Where, when and how male sharp-tailed grouse establish territories on arenas. *Can. J. Zool.* 59:1576–81.

Mueller, W. H. 1979. Fertility and physique in a malnourished population. *Human Biol.* 51:153–66.

Mundot Bernard, J. M. 1975. *Relationships between fertility, child mortality and nutrition in Africa*. Paris: OECD Development Centre.

Murphy, G. I. 1968. Pattern in life-history and the environment. *Am. Nat.* 102:390–404.

Murray, M. G. 1985. Estimation of kinship parameters. The island model with separate sexes. *Behav. Ecol. Sociobiol.* 16:151–59.

Nag, M. 1980. How modernization can increase fertility. *Curr. Anthropol.* 21:571–88.

Nelson, J. B. 1966. The breeding biology of the gannet *Sula bassana* on the Bass Rock, Scotland. *Ibis* 108:584–626.

Nettleship, D. N. 1974 Northern fulmar colonies on the south coast of Devon Island, N.W.T., Canada. *Auk* 91:412.

Newton, I. 1980. The role of food in limiting bird numbers. *Ardea* 68:11–30.

———. 1985. Lifetime reproductive output of female sparrowhawks. *J. Anim. Ecol.* 54:241–53.

Newton, I., and Marquiss, M. 1979. Sex ratio among nestlings of the European sparrowhawk. *Am. Nat.* 113:309–15.

———. 1982. Fidelity to breeding area and mate in sparrowhawks *Accipiter nisus*. *J. Anim. Ecol.* 51:327–41.

———. 1983. Dispersal of sparrowhawks between birthplace and breeding place. *J. Anim. Ecol.* 52:463–77.

Newton, I.; Marquiss, M.; and Moss, D. 1979. Habitat, female age, organo-chlorine compounds and breeding of European sparrowhawks. *J. Appl. Ecol.* 16:777–93.

———. 1981. Age and breeding in sparrowhawks. *J. Anim. Ecol.* 50:839–53.

Newton, I; Marquiss, M.; and Rothery, P. 1983. Age structure and survival in a sparrowhawk population. *J. Anim. Ecol.* 52:591–602.

Newton, I; Marquiss, M.; and Village, A. 1983. Weights, breeding and survival in European sparrowhawks. *Auk* 100:344–54.

Newton, R. G., and Spurrell, D. J. 1967a. A development of multiple regression for the analysis of routine data. *Appl. Statist.* 16:51–64.

———. 1967b. Examples of the use of elements for clarifying regression analysis. *Appl. Statist.* 16:165–72.

Nicolson, N. 1982. Weaning and the development of independence in olive baboons. Ph.D. diss., Harvard University.

Nie, N. H.; Hull, D. H.; Jenkins, J. G.; Steinbrenner, K.; and Bent, D. H. 1981. *Statistical package for the social sciences.* 2d ed. New York: McGraw-Hill.

Nol, E., and Smith, J. N. M. 1987. Effects of age and breeding experience on seasonal reproductive success in the song sparrow. *J. Anim. Ecol.* 56:301–14.

Noonan, K. N. 1981. Individual strategies of inclusive fitness maximizing in *Polistes fuscatus* foundresses. In *Natural selection and social behavior: Recent research and new theory,* ed. R. D. Alexander and D. W. Tinkle, 18–44. New York: Chiron Press.

Noordwijk, A. J. van. 1984a. Problems in the analysis of dispersal and a critique on its "heritability" in the great tit. *J. Anim. Ecol.* 35:533–44.

———. 1984b. Quantitative genetics in natural populations of birds illustrated with examples from the great tit, *Parus major*. In *Population biology and evolution,* ed. K. Wöhman and V. Loeschke, 67–79. Heidelberg: Springer-Verlag.

Noordwijk, A. J. van; Balen, J. H. van; and Scharloo, W. 1980. Heritability of ecologically important traits in the great tit. *Ardea* 68:193–203.

———. 1981a. Genetic and environmental variation in clutch size of the great tit. *Neth J. Zool.* 31:342–72.

———. 1981b. Genetic variation in the timing of reproduction in the great tit. *Oecologia* (Berlin) 49:158–66.

Noordwijuk A. J. van, and de Jong, G. 1986. Acquisition and allocation of resources: Their influence on variation in life history tactics. *Am. Nat.* 128:137–42.

Noordwijk, A. J. van; Keizer, L. C. P.; Balen, J. H. van; and Scharloo, W. 1981. Genetic variation in egg dimensions in natural populations of the great tit. *Genetica* 55:221–32.

Noordwijk,, A.J. van, and Scharloo, W. 1981. Inbreeding in an island population of the great tit. *Evolution 35:674-688.*

Noordwijk, A. J. van Tienderen, P. J. van; Jong, G. de; and Balen, J. H. van. 1985. Evidence for random mating in the great tit *Parus major*, L. In *Behavioural ecology,* ed. R. Sibley and R. Smith, 381–85. Oxford: Blackwell.

Nunney, L. 1985. Female-biased sex ratios: Individual or group selection? *Evolution* 39:349–61.

Nur, N. 1984a. The consequences of brood size for breeding blue tits. 1. Adult survival, weight change and the cost of reproduction. *J. Anim Ecol.* 53:479–96.

———. 1984b. The consequences of brood size for breeding blue tits. 2. Nestling weight, offspring survival and optimal brood size. *J. Anim. Ecol.* 53:497–517.

O'Connor, R. J. 1984. *The growth and development of birds.* New York: John Wiley.

O'Donald, P. 1962. The theory of sexual selection. *Heredity* 17:541–52.

Ogilvie, M. A. 1972. Large numbered leg bands for individual identification of swans. *J. Wildl. Mgmt.* 36:1261–65.

Ojanen, M.; Orell, M.; and Väisänen, R. A. 1979. Role of heredity in egg size variation in the great tit *Parus major* and the pied flycatcher *Ficedula hypoleuca*. *Ornis Scand.* 10:22–28.

Ollason, J. C., and Dunnet, G. M. 1978. Age, experience and other factors affecting the breeding success of the Fulmar, *Fulmarus glacialis* in Orkney. *J. Anim. Ecol.* 47: 961–76.

———. 1980. Nest failures in the fulmar: The effect of observers. *J. Field Ornithol.* 51:39–54.

———. 1983. Modelling annual changes in numbers of breeding fulmars *Fulmarus glacialis* at a colony of Orkney. *J. Anim. Ecol.* 52:185–98.

Orchardson, I. Q. 1961. *The Kipsigis.* Nairobi: Kenya Literature Bureau.

Orians, G. H. 1969a. Age and hunting success in the brown pelican (*Pelecanus occidentalis). Anim. Behav.* 17:316–19.

———. 1969b. On the evolution of mating systems in birds and mammals. *Am. Nat.* 103:589–603.

Orsdol, K. G. van. 1981. Lion predation in Rwenzori National Park, Uganda. Ph.D. diss., University of Cambridge.

Otis, J. S.; Froehlich, J. W.; and Thorington, R. W. 1981. Seasonal and age-related differential mortality by sex in the mantled howler monkey, *Alouatta palliata. Int. J. Primatol.* 2:197–205.

Owen, M.; Black, J. M.; and Liber, H. n.d. *Pair bond duration and the timing of its formation in barnacle geese.* Waterfowl in Winter Symposium, Galveston, Texas, 1985. In press.

Owen-Smith, N. 1977. On territoriality in ungulates and an evolutionary model. *Quart. Rev. Biol.* 52:1–38.

Owens, M., and Owens, D. 1984. Kalahari lions break the rules. *Inte. Wildl.* 14:4–13.

Packer, C. 1979a. Inter-troop transfer and inbreeding avoidance in *Papio anubis. Anim. Behav.* 27:1–37.

———. 1979b. Male dominance and reproductive activity in *Papio anubis. Anim. Behav.* 27:37–46.

———. 1986. The ecology of felid sociality. In *Ecological aspects of social evolution*, ed. D. I. Rubenstein and R. W. Wrangham. Princeton: Princeton University Press.

Packer, C., and Pusey, A. E. 1982. Cooperation and competition within coalitions of male lions: Kin selection or game theory? *Nature* 296:740–42.

———. 1983a. Adaptations of female lions to infanticide by incoming males. *Am. Nat.* 121:716–28.

———. 1983b. Male takeovers and female reproductive parameters: A simulation of oestrus synchrony in lions (*Panthera leo*). *Anim. Behav.* 31:334–40.

———. 1983c. Cooperation and competition in lions. *Nature* 302:356.

———. 1984. Infanticide in carnivores. In *Infanticide: Comparative and evolutionary perspectives,* ed. G. Hausfater and S. B. Hardy, 31–42. Hawthorne, N. Y.: De Gruyter Aldine.

———. 1985. Asymmetric contests in social mammals: respect, manipulation and age-specific aspects. In *Evolution: Essays in Honor of John Maynard Smith,* ed. P. J. Greenwood and M. Slatkin, 173–86. Cambridge: Cambridge University Press.

———. 1987. Intrasexual cooperation and the sex ratio in African lions. *Am. Nat.* 130. In press.

Pamilo, P., and Crozier, R. H. 1982. Measuring genetic relatedness in natural populations: Methodology. *Theor. Pop. Biol.* 21:171–93.

Pardi, L. 1942. Richerche sui Polisstini. *Boll. Ist. Entomol. Univ. Bologna* 14:1–104.

Partridge, L. 1986. Sexual activity and lifespan. In *Insect aging: Strategies and mechanisms,* ed. K.-G. Collatz. New York: Springer-Verlag.

Partridge, L., and Andrews, R. 1985. The effect of reproductive activity on the longevity of male *Drosophila melanogaster* is not caused by an acceleration of senescence. *J. Insect Physiol.* 31: 393–95.

Partridge, L,; Ewing, A.; and Chandler, A. 1987. Male size and mating success in

Drosophila melanogaster: The roles of male and female behaviour. *Anim. Behav.* 35:555–62.

Partridge, L., and Farquhar, M. 1981. Sexual activity reduces lifespan of male fruitflies. *Nature* 294:580–82.

———. 1983. Lifetime mating success of male fruitflies (*Drosophila melanogaster*) is related to their size. *Anim. Behav.* 31:871–77.

Partridge, L.; Fowler, K.; Trevitt, S.; and Sharp, W. 1986. An examination of the effects of males on the survival and egg-production rates of female *Drosophila melanogaster. J. Insect Physiol.* 32:925–29.

Partridge, L., and Halliday, T. R. 1984. Mating patterns and mate choice. In *Behavioural ecology: An evolutionary approach*, 2d ed., ed. J. R. Krebs and N. B. Davies, 222–50. Sunderland, Mass.: Sinauer.

Partridge, L., and Harvey, P. H. 1985. Costs of reproduction. *Nature* 316:20–21.

Partridge, L.; Hoffmann, A.; and Jones, J. S. 1987. Male size and mating success in *Drosophila melanogaster* and *D. pseudoobscura* under field conditions. *Anim. Behav.* 35:468–76.

Paul, A., and Thommen, D. 1984. Timing of birth, female reproductive success and infant sex ratio in semi free-ranging Barbary macaques (*Macaca sylvanus*). *Folia Primatol.* 42:2–16.

Payne, R. B. 1979. Sexual selection and intersexual differences in variance of breeding success. *Am. Nat.* 114:447–52.

———. 1984. Sexual selection, lek and arena behavior and sexual size dimorphism in birds. *Ornithol. Monogr.* 33:1–52.

Payne, R. B., and Payne, K. 1977. Social organisation and mating success in local song populations of village indigo birds, *Vidua chalybeata. Z. Tierpsychol.* 45:113–73.

Pemberton, J.; Albon, S. D.; Guinness, F. E.; and Clutton-Brock, T. H. n.d. Selection revealed by an enzyme polymorphism in red deer. In preparation.

Pennycuick, C., and Rudnai, J. A. 1970. A method of identifying individual lions, *Panthera leo,* with an analysis of the reliability of the identification. *J. Zool.* 160:497–508.

Peristiany, J. G. 1939. *The social institutions of the Kipsigis.* London: Routledge and Kegan Paul.

Perrins, C. M. 1965. Population fluctuations and clutch size in the great tit, *Parus major. J. Anim. Ecol.* 34:601–47.

———. 1966. The effect of beech crops on great tit populations and movements. *Brit. Birds* 59:419–32.

———. 1970. The timing of birds' breeding seasons. *Ibis* 112:242–55.

———. 1979. *British tits.* London: Collins.

Perrins, C. M., and Jones, P. J. 1974. The inheritance of clutch size in the great tit (*Parus major* L.). *Condor* 76:225–29.

Perrins, C. M., and McCleery, R. 1985. The effects of age and pair bond on the breeding success of great tits, *Parus major. Ibis* 127:306–15.

Perrins, C. M., and Moss, D. 1974. Survival of young great tits in relation to age of female parent. *Ibis* 116:220–24.

———. 1975. Reproductive rates in the great tit. *J. Anim. Ecol.* 44:695–706.

Peterson, R. S., and Bartholomew, G. A., Jr. 1967. *The natural history and behavior of the California sea lion.* Special Publication 1: Stillwater, Okla: American Society of Mammalogists.

Petrinovich, L., and Patterson, T. L. 1983. The white crowned sparrow: Reproductive success (1975–1980). *Auk* 100:811–25.

Pfennig, D. W., and Klahn, J. E. 1985. Dominance as a predictor of cofoundress disappearance order in social wasps (*Polistes fuscatus*). *Z. Tierpsychol.* 67:198–203.

Pianka, E. R. 1978. *Evolutionary ecology.* 2d ed. New York: Harper and Row.

———. 1985. Lycaenid butterflies and ants: Selection for nitrogen-fixing and other protein rich food plants. *Am. Nat.* 125:888–95.

———. n.d. The influence of ants on the developmental rates of the larvae and pupae of *Jalmenus evagoras,* a myrmecophilous lycaenid butterfly. Unpublished MS.

Pierce, N. E. 1983. The ecology and evolution of symbioses between lycaenid butterflies and ants. Ph.D. diss., Harvard University.

———. 1984. Amplified species diversity: A case study of an Australian lycaenid butterfly and its attendant ants. In *The biology of butterflies,* ed. R. I. Vane-Wright and P. R. Ackery, 197–200. Eleventh Symposium of the Royal Entomological Society of London. London: Academic Press.

Pierce, N. E. 1987. The evolution and biogeography of associations between lycaenid butterflies and ants. In *Oxford surveys in evolutionary biology,* vol. 4, ed. Paul Harvey and Linda Partridge. Oxford: Oxford University Press.

Pierce, N. E., and Elgar, M. A. 1985. The influence of ants on host plant selection by *Jalmenus evagoras,* a myrmecophilous lycaenid butterfly. *Behav. Ecol. Sociobiol.* 16:209–22.

Pierce, N. E.; Kitching, R. L.; Buckley, R. C.; Taylor, M. F. J.; and Benbow, K. 1987. The costs and benefits of cooperation for the Australian lycaenid butterfly, *Jalmenus evagoras* and its attendant ants. *Behav. Ecol. Sociobiol.* In press.

Pierce, N. E., and Young, W. R. 1986. Lycaenid butterflies and ants: Two-species stable equilibria in mutualistic, commensal, and parasitic interactions. *Am. Nat.* 128: 216–27.

Pierson, M. O. 1978. A study of the population dynamics and breeding behavior of the Guadalupe fur seal, *Arctocephalus townsendi.* Ph.D. diss., University of California, Santa Cruz.

Pleszczynska, W., and Hansell, R. 1980. Polygyny and decision theory: Testing of a model in lark buntings (*Calamospiza melanocorys*). *Am. Nat.* 116:821–30.

Potts, G. R. 1966. Studies on a marked population of the shag (*Phalacrocorax aristotelis*), with special reference to the breeding biology of birds of known age. D. Phil. diss., University of Durham.

———. 1969. The influence of eruptive movements, age, population size and other factors on the survival of the shag (*Phalacrocorax aristotelis* L.). *J. Anim. Ecol.* 38:53–102.

Poulter, T. C., and Jennings, R. 1966. Annual reports of operations of Stanford Research Institute on Año Nuevo Island. Menlo Park, California.

Powell, J. A. 1968. A study of area occupation and mating behaviour in *Incisalia iroides* (Lepidoptera: Lycaenidae). *J. N.Y. Entomol. Soc.* 76:47–57.

Preston, S. H. 1975. Health programs and population growth. *Pop. Dev. Rev.* 1:189–99.

Price, G. R. 1970. Selection and covariance. *Nature* 227:520–21.

Price, T. D. 1984. Sexual selection on body size, territory and plumage variables in a population of Darwin's finches. *Evolution* 38:327–41.

Price, T. D., and Grant, P. R. 1984. Life history traits and natural selection for small body size in a population of Darwin's finches. *Evolution* 38:483–94.

Price, T. D.; Grant, P. R.; Gibbs, H. L.; and Boag, P. T. 1984. Recurrent patterns of natural selection in a population of Darwin's finches. *Nature* 309:787–89.

Prins, A. H. 1953. *East African age-class systems: An inquiry into the social order of the Galla, Kipsigis and Kikuyu.* Groningen: Free Press.

Prout, T. 1965. The estimation of fitnesses from genotypic frequencies. *Evolution* 19:546–51.

———. 1971a. The relation between fitness components and population production in *Drosophila.* 1. The estimation of fitness components. *Genetics* 68:127–49.

———. 1971b. The relation between fitness components and population prediction in *Drosophila.* 2. Population prediction. *Genetics* 68:151–67.

Pugesek, B. H. 1981. Increased reproductive effort with age in the California gull, *Larus californicus. Science* 212:822–23.

Pulliam, A. R., and Caraco, T. 1984. Living in groups: Is there an optimal group size? In

Behavioural ecology: An evolutionary approach, 2d ed., ed. J. R. Krebs and N. B. Davies, pp. 122–47. Oxford: Blackwell Scientific Publications.

Purser, A. F., and Roberts, R. C. 1964. The effect of treatment during first winter on the subsequent performance of Scottish blackfaced ewes. *Anim. Prod.* 6:273–84.

Pusey, A. E., and Packer, C. 1987. Dispersal and philopatry in lions. *Behaviour*. In press.

Pyke, G. H. 1978. Optimal size in bumblebees. *Oecologia* (Berlin) 34:255–66.

Pyle, D. W., and Gromko, M. H. 1978. Repeated mating by female *Drosophila melanogaster*. The adaptive importance. *Experientia* 34:449–50.

———. 1981. Genetic basis for repeated mating in *Drosophila melanogaster*. *Am. Nat.* 117:133–46.

Queller, D. C. 1984. Kin selection and frequency-dependence: A game-theoretic approach. *Biol. J. Linn. Soc.* 23:133–43.

Radford, K. W.; Orr, R. T.; and Hubbs, C. L. 1965. Reestablishment of the northern elephant seal (*Mirounga angustirostris*) off central California. *Proc. Calif. Acad. Sci.* 31:601–12.

Ralls, K. 1977. Sexual dimorphism in mammals: Avian models and unanswered questions. *Am. Nat.* 111:917–38.

Ransom, T. W., and Rowell, T. E. 1972. Early social development of feral baboons. In *Primate socialization*, ed. F. E. Poirier. New York: Random House.

Rasmussen, K. L. 1980. Consort behavior and mate selection in yellow baboons (*Papio cynocephalus*). Ph.D. diss., Cambridge University.

———. 1985. Changes in the activity budgets of yellow baboons (*Papio cynocephalus*) during sexual consortships. *Behav. Ecol. Sociobiol.* 17:161–70.

Rau, P. 1929. The habitat and dissemination of four species of *Polistes* wasps. *Ecology* 10:191–200.

———. 1931. The nests and nesting sites of four species of *Polistes* wasps. *Bull. Brooklyn Entomol. Soc.* 26:111–19.

Rees, E. 1981. The recording and retrieval of bill pattern variations in *Cygnus columbianus bewickii*. In *Proceedings of the IWRB Symposium* 105–19. Sapporo, Japan: IWRB.

Reiter, J. 1984. Studies of female competition and reproductive success in the northern elephant seal. Ph.D. diss., University of California, Santa Cruz.

Reiter, J.; Panken, K. J.; and Le Boeuf, B. J. 1981. Female competition and reproductive success in northern elephant seals. *Anim. Behav.* 29:670–87.

Reiter, J.; Stinson, N. L.; and Le Boeuf, B. J. 1978. Northern elephant seal development: The transition from weaning to nutritional independence. *Behav. Ecol. Sociobiol.* 3:337–67.

Reznick, D. 1985. Costs of reproduction: An evaluation of the empirical evidence. *Oikos* 44:257–67.

Rheinwald, G. 1975. The pattern of settling distances in a population of house martins *Delichon urbica*. *Ardea* 63:136–45.

———. 1979. Brutbiologie der Mehlschwalbe *Delichon urbica* in Bereich der Voreifel. *Vogelwelt* 100:85–107.

Rheinwald, G.; Gutscher, H.; and Hormeyer, K. 1976. Einfluss des Alters der Mehlschwalbe (*Delichon urbica*) auf ihre Brut. *Vogelwarte* 28:190–206.

Richdale, L. E. 1957. *A population study of penguins*. Oxford: Clarendon Press.

———. 1965. Biology of the birds of Whero Island, New Zealand, with special reference to the diving petrel and the white-faced storm petrel. *Trans. Zool. Soc. Lond.* 31:1–86.

Richdale, L. E., and Wareham, J. 1973. Survival, pair bond retention and nest-site in Buller's mollymawk. *Ibis* 115:257–63.

Richmond, R. C.; Gilbert, D. G.; Sheehan, K. B.; Gromko, M. H.; and Butterworth, F. M. 1980. Esterase 6 and reproduction in *Drosophila melanogaster*. *Science* 207:1483–85.

Ridley, M., and Thompson, D. J. 1979. Size and mating in *Asellus aquaticus* (Crustacea: Isopoda). *Z. Tierpsychol.* 51:380–97.

Robel, R. J. 1966. Booming territory size and mating success of the greater prairie chicken (*Tympanuchus cupido pinnatus*). *Anim. Behav.* 14:328–31.

Robel, R. J., and Ballard, W. B. 1974. Lek social organisations and reproductive success in the greater prairie chicken. *Am. Zool.* 14:121–28.

Robertson, A. 1966. A mathematical model of the culling process in dairy cattle. *Anim. Prod.* 8:93–108.

———. 1968. The spectrum of genetic variation. In *Population biology and evolution,* ed. R. C. Lewontin, 5–16. Syracuse, N.Y.: Syracuse University Press.

Robertson, F. W. 1954. Studies in quantitative inheritance. 5. Chromosome analyses of crosses between selected and unselected lines of different body size in *Drosophila melanogaster. J. Genet.* 52:494–520.

———. 1957. Studies in quantitative inheritance. 11. Genetic and environmental correlation between body size and egg production in *Drosophila melanogaster. J. Genet.* 55:428–43.

———. 1960. The ecological genetics of growth in *Drosophila*. 1. Body size and developmental time on different diets. *Genet. Res. Camb.* 1:288–304.

Robertson, H. M. 1985. Female dimorphism and mating behaviour in a damselfly *Ischnura ramburi*: Females mimicking males. *Anim. Behav.* 33:805–9.

Robertson, J. D. 1934. *Notes from a bird sanctuary*. Kirkwall: Orcadian Office.

Robinson, J. G. 1982. Intrasexual competition and mate choice in primates. *Am. J. Primatol.*, suppl. 1:131–44.

Rockwell, R. F., and Cooke, F. 1977. Gene flow and local adaptation in a colonially nesting dimorphic bird: The lesser snow goose. *Am. Nat.* 111:91–97.

Rockwell, R. F.; Findlay, C. S.; and Cooke, F. 1983. Life history studies of the lesser snow goose (*Anser caerulescens caerulescens*). 1. Influence of age and time on fecundity. *Oecologia* 56:318–22.

———. 1987. Is there an optimal clutch size in snow geese? *Am. Nat.* In press.

Rockwell, R. F.; Findlay, C. S.; Cooke, F.; and Smith, J. A. 1985. Life history studies of the lesser snow goose (*Anser caerulescens caerulescens*). 4. The selective value of plumage polymorphism: Net viability; the timing of maturation and breeding propensity. *Evolution* 39:178–89.

Roff, D. 1977. Dispersal in dipterans: Its costs and consequences. *J. Anim. Ecol.* 46: 443–56.

Romanoff, A. L., and Romanoff, A. J. 1949. *The avian egg*. New York: John Wiley.

Rood, J. P. 1975. Population dynamics and food habits of the banded mongoose. *E. Afr. Wildl. J.* 13:89–111.

Rose, M. R. 1984. Laboratory evolution of postponed senescence in *Drosophila melanogaster*. *Evolution* 38:1004–10.

Rose, M. R., and Charlesworth, B. 1981a. Genetics of life history in *Drosophila melanogaster*. 1. Sib analysis of adult females. *Genetics* 97:173–86.

———. 1981b. Genetics of life history in *Drosophila melanogaster*. 2. Exploratory selection experiments. *Genetics* 97:187–96.

Röseler, P.-F.; Röseler, I.; Strambi, A.; and Augier, R. 1984. Influence of insect hormones on the establishment of dominance hierarchies among foundresses of the paper wasp, *Polistes gallicus*. *Behav. Ecol. Sociobiol.* 15:133–42.

Ross, H. A., and McLaren, I. A. 1981. Lack of differential survival among young Ipswich sparrows. *Auk* 98:495–502.

Rowley, I. 1981. The communal way of life in the splendid wren *Malurus splendens*. *Z. Tierpsychol.* 55:228–67.

Rubenstein, D. I. 1981. Population density, resource patterning and territoriality in the Everglades pygmy sunfish. *Anim. Behav.* 29:155–72.

Rudnai, J. 1973. Reproductive biology of lions (*Panthera leo massaica* Neumann) in Nairobi National Park. *E. Afr. Wildl. J.* 11:241–53.

Rutowski, R. L. 1982. Mate choice and lepidopteran mating behaviour. *Fla. Entomol.* 65:72–82.

———. 1984. Sexual selection and the evolution of butterfly mating behaviour. *J. Res. Lep.* 23:125–42.

Ryder, J. P. 1975. Egg-laying, egg size, and success in relation to immature-mature plumage of ring-billed gulls. *Wilson Bull.* 87:534–42.

Sade, D. S.; Cushing, K.; Cushing, P.; Dunaif, J.; Figueroa, A.; Kaplan, J. R.; Laver, C.; Rhodes, D.; and Schneider, J. 1976. Population dynamics in relation to social structure on Cayo Santiago. *Ybk. Phys. Anthropol.* 20:253–62.

Sadleir, R. M. F. S. 1969. *The ecology of reproduction in wild and domestic mammals.* London: Methuen.

Salomonsen, F. 1965. The geographical variation of the fulmar (*Fulmarus glacialis*) and the zones of marine environment in the North Atlantic. *Auk* 82:327–55.

Sang, J. H. 1949. The ecological determinants of population growth in a *Drosophila* culture. 3. Larval and pupal survival. *Physiol. Zool.* 22:183–202.

———. 1956. The quantitative nutritional requirements of *Drosophila melanogaster. Ann. N.Y. Acad. Sci.* 7:352–65.

Saunders, C., and Hausfater, G. 1978. Sexual selection in baboons (*Papio cynocephalus*): A computer simulation of differential reproduction with respect to dominance rank in males. In *Recent advances in primatology*, vol. 1, *Behaviour*, ed. D. J. Chivers and J. Herbert, 567–71. New York: Academic Press.

Schaik, C. P. van, and Noordwijk, M. A. van. 1983. Social stress and the sex ratio of neonates and infants among non-human primates. *Netherlands J. Zool.* 33:249–65.

Schaller, G. B. 1972. *The Serengeti lion.* Chicago: University of Chicago Press.

Schifferli, L. 1973. The effect of egg weight on the subsequent growth of nestling great tits, *Parus major. Ibis* 115:549–58.

Schluter, D., and Smith, J. N. M. 1986a. Genetic and phenotypic correlations in a natural population of song sparrows. *Biol. J. Linn. Soc.* 29:23–36.

———. 1986b. Natural selection on beak and body size in the song sparrow. *Evolution* 40:221–31.

Schüz, E. 1957. Das Verschlingen eigener Junger ("Kronismus") bei Völgeln und seine Bedeutung. *Vogelwarte* 19:1–15.

Scott, D. K. 1978. Social behaviour of wintering Bewick's swans. Ph.D. diss., University of Cambridge.

———. 1980. Functional aspects of prolonged parental care in Bewick's swans. *Anim. Behav.* 28:938–52.

Scott, J. A. 1972. Mating of butterflies. *J. Res. Lepid.* 11:99–127.

———. 1974. Mate-locating behaviour of butterflies. *Am. Midl. Nat.* 91:103–17.

———. 1975. Mate-locating behaviour of western North American butterflies. *J. Res. Lepid.* 14:1–40.

Scott, J. W. 1942. Mating behaviour of the sage grouse. *Auk* 49:477–98.

Scott, L. M. 1984. Reproductive behaviour of adolescent female baboons (*Papio anubis*) in Kenya. In *Female primates: Studies by women primatologists,* 77–100. New York: Alan R. Liss.

Scott, P. 1966. The Bewick's swans at Slimbridge. *Wildfowl Trust Ann. Rep.* 17:20–26.

Scrimshaw, S. C. M. 1978. Infant mortality and behaviour in the regulation of family size. *Pop. Dev. Rev.* 4:383–404.

Seber, G. A. 1982. *Estimation of animal abundance and related parameters.* London: Griffin.

Seger, J. 1977, A numerical method of estimating coefficients of relationship in a langur troup. In *The langurs of Abu,* ed. S. B. Hrdy, 317–26. Cambridge: Harvard University Press.

———. 1981. Kinship and covariance. *J. Theor. Biol.* 91:191–213.

Selander, R. K. 1966. Sexual dimorphism and differential niche utilization in birds. *Condor* 2:113–49.

Serventy, D. L. 1967. Aspects of the population ecology of the short-tailed shearwater *Puffinus tenuirostris*. *Proc. XIV Int. Ornithol. Congr.* 165–90.

Severinghaus, L. L.; Kurtak, B. H.; and Eickwort, G. C. 1981. The reproductive behaviour of *Anthidium manicatum* (Hymenoptera: Megachilidae) and the significance of size for territorial males. *Behav. Ecol. Sociobiol.* 9:51–58.

Seyfarth, R. M. 1980. The distribution of grooming and related behaviours among adult female vervet monkeys. *Anim. Behav.* 28:798–813.

Shaikh, K., and Becker, S. 1985. Socioeconomic status and fertility in rural Bangladesh. *J. Biosoc. Sci.* 17:81–89.

Sharp, P. M. 1984. The effect of inbreeding on male-mating ability in *Drosophila melanogaster*. *Genetics* 106:601–12.

Short, R. V. 1976. Lactation—the central control of reproduction. In *Breast-feeding and the mother*. Amsterdam: Elsevier.

Silberglied, R. E. 1984. Visual communication and sexual selection among butterflies. In *The biology of butterflies,* ed. R. I. Vane-Wright and P. R. Ackery, 259–75. Eleventh Symposium of the Royal Entomological Society of London. London: Academic Press.

Silk, J. B. 1983. Local resource competition and facultative adjustment of sex ratios in relation to competitive abilities. *Am. Nat.* 121:56–66.

———. 1985. The evolution and adaptive consequences of altruism, competition, and aggression in primate groups. In *Primate societies,* ed. D. L. Cheney, R. M. Seyfarth, B. Smuts, R. W. Wrangham, and T. T. Struhsaker. Chicago: University of Chicago Press.

Silk, J. B., and Boyd, R. 1983. Cooperation, competition and mate choice in matrilineal macaque groups. In *Female vertebrates,* ed. S. K. Wasser, 316–49. New York: Academic Press.

Silk, J. B.; Clark-Wheatley, C. B.; Rodman, P. S.; and Samuels, A. 1981. Differential reproductive success and facultative adjustment of sex ratios among captive female bonnet macaques. *Anim. Behav.* 29:1106–20.

Silverin, B. 1980. Effects of long-acting testosterone treatment on free-living pied flycatchers, *Ficedula hypoleuca*. *Anim. Behav.* 28:906–12.

Simpson, M. J. A., and Simpson, A. E. 1982. Birth/sex ratios and social rank in rhesus monkey mothers. *Nature* 300:440–41.

Simpson, M. J. A.; Simpson, A. E.; Hooley, J.; Zunz, M. 1981. Infant-related influences on birth intervals in rhesus monkeys. *Nature* 290:49–51.

Sinclair, A. R. E. 1979. The Serengeti environment. In *Serengeti: Dynamics of an ecosystem,* ed. A. R. E. Sinclair and M. Norton-Griffiths, 31–45. Chicago: University of Chicago Press.

Singer, M. C. 1982. Sexual selection for small size in male butterflies. *Am. Nat.* 119: 440–43.

Skead, D. M., and Skead, C. J. 1970. Hirundinid mortality during adverse weather, November 1968. *Ostrich* 41:247–51.

Skutch, A. F. 1959. Life history of the grove-billed ani. *Auk* 76:281–317.

Sladen, W. J. L. 1975. The whistling swan. *Nat. Geogr.* 148:134–47.

Slagsvold, T. 1975. Annual and geographical variation in the time of breeding of the great tit *Parus major* and the pied flycatcher *Ficedula hypoleuca* in relation to environmental phenology and spring temperature. *Ornis Scand.* 7:127–45.

Smith, J. N. M. 1981a. Cowbird parasitism, host fitness, and age of the host female in an island song sparrow population. *Condor* 83:152–61.

———. 1981b. Does high fecundity reduce survival in song sparrows? *Evolution* 35:1142–48.

Smith, J. N. M.; Arcese, P.; and Schluter, D. 1986. Song sparrows grow and shrink with age. *Auk* 103:210–12.

Smith, J. N. M., and Dhondt, A. A. 1980. Experimental confirmation of heritable morphological variation in a natural population of song sparrows. *Evolution* 34:1155–58.

Smith, J. N. M.; Montgomerie, R. D.; Taitt, M. J.; and Yom-Tov, Y. 1980. A winter feeding experiment on an island population of song sparrows. *Oecologia* (Berlin) 47:164–70.

Smith, J. N. M., and Roff, D. A. 1980. Temporal spacing of broods, brood size, and parental care in song sparrows. *Can. J. Zool.* 58:1007–15.

Smith, J. N. M.; Yom-Tov, Y.; and Moses, R. 1982. Polygyny, male parental care and sex ratio in song sparrows: An experimental study. *Auk* 99:555–64.

Smith, J. N. M., and Zach, R. 1979. Heritability of some morphological characters in a song sparrow population. *Evolution* 33:460–67.

Smuts, B. 1982. Special relationships between adult male and female olive baboons (*Papio anubis*). Ph.D. diss., Stanford University.

———. 1985. *Sex and friendship in baboons*. New York: Aldine.

Snow, C. C. 1967. The physical growth and development of the open-land baboon, *Papio doguera*. Ph.D. diss., University of Arizona.

Sokoloff, A. 1966. Morphological variation in natural and experimental populations of *Drosophila pseudoobscura* and *Drosophila persimilis*. *Evolution* 20:49–71.

Southwood, T. R. E. 1978. *Ecological methods*. 2d ed. London: Chapman and Hall.

Spurr, E. B. 1975. Breeding of the Adélie penguin *Pygoscalis adeliae* at Cape Bird. *Ibis* 117:324–38.

Srivastava, I. N., and Saksena, D. N. 1981. Infant mortality differentials in an Indian setting: A follow-up of hospital deliveries. *J. Biosoc. Sci.* 13:467–78.

Stallcup, J. A., and Woolfenden, G. E. 1978. Family status and contributions to breeding by Florida scrub jays. *Anim. Behav.* 25:1144–56.

Stearns, S. C. 1976. Life history tactics: A review of the ideas. *Quart. Rev. Biol.* 51:3–47.

Steele, R. H. 1984. An investigation of male mating success in *Drosophila subobscura* Collin. Ph.D. diss., University of Edinburgh.

Steele, R. H. 1986a. Courtship feeding in *Drosophila subobscura*. 1. The nutritional significance of courtship feeding. *Anim. Behav.* 34:1087–98.

———. 1986b. Courtship feeding in *Drosophila subobscura*. 2. Courtship feeding by males influences female mate choice. *Anim. Behav.* 34:1099–1108.

Stein, D. M. 1984. Ontogeny of infant-adult male relationships during the first year of life for yellow baboons (*Papio cynocephalus*). In *Primate paternalism,* ed. D. M. Taub, 213–43. New York: Van Nostrand Reinhold.

Stenning, M. J. 1984. Population regulation in the pied flycatcher *Ficedula hypoleuca* (Pallas). M. Phil. diss., University of Sussex.

Stephens, D. W., and Krebs, J. R. 1987. *Foraging theory*. Princeton: Princeton University Press.

Stern, B. R., and Smith, D. G. 1984. Sexual behaviour and paternity in three captive groups of rhesus monkeys (*Macaca mulatta*). *Anim. Behav.* 32:23–32.

Strassmann, J. E. 1979. Honey caches help female paper wasps (*Polistes annularis*) survive Texas winters. *Science* 204:207–9.

———. 1981a. Parasitoids, predators and group size in the paper wasp, *Polistes exclamans*. *Ecology* 62:1225–33.

———. 1981b. Wasp reproduction and kin selection: Reproductive competition and dominance hierarchies among *Polistes annularis* foundresses. *Fla. Entomol.* 64:74–88.

———. 1983. Nest fidelity and group size among *Polistes annularis* foundresses. *J. Kans. Entomol. Soc.* 56:621–34.

Strassmann, J. E.; Lee, R. E., Jr.; Rojas, R. R.; and Baust, J. G. 1984. Caste and sex

differences in cold-hardiness in the social wasps, *Polistes annularis* and *P. exclamans* (Hymenoptera: Vespidae). *Insectes Sociaux* 31:291–301.
Strassmann, J. E., and Orgren, C. F. 1983. Nest architecture and brood development times in the paper wasp *Polistes exclamans* (Hymenoptera: Vespidae). *Psyche* 90:237–48.
Struhsaker, T. T. 1967. Social structure among vervet monkeys, *Cercopithecus aethiops*. *Behaviour* 29:83–121.
———. 1973. A recensus of vervet monkeys in the Masai-Amboseli Game Reserve, Kenya. *Ecology* 54:930–32.
———. 1976. A further decline in numbers of Amboseli vervet monkeys. *Biotropica* 8:211–14.
Strum, S. C. 1982. Agonistic dominance in male baboons: An alternative view. *Int. J. Primatol.* 3:175–202.
Strum, S. C., and Western, J. D. 1982. Variations in fecundity with age and environment in olive baboons (*Papio anubis*). *Am. J. Primatol.* 3:61–76.
Sugiyama, Y., and Ohsawa, H. 1982. Population dynamics of Japanese monkeys with special reference to the effect of artificial feeding. *Folia Primatol.* 39:238–63.
Sullivan, J. D., and Strassmann, J. E. 1984. Physical variability among nest foundresses in the polygynous social wasp *Polistes annularis*. *Behav. Ecol. Sociobiol.* 15:249–56.
Sutherland, W. J. 1985a. Chance can produce a sex difference in variance in mating success and explain Bateman's data. *Anim. Behav.* 33:1349–52.
———. 1985b. Measures of sexual selection. In *Oxford Surveys in evolutionary biology*, ed. Richard Dawkins, 1:90–101. London and New York: Oxford University Press.
Suzuki, Y. 1976. So-called territorial behaviour of the small copper, *Lycaena phlaeas daimio* Seitz (Lepidoptera, Lycaenidae). *Kontyu* 44:193–204.
———. 1978. Adult longevity and reproductive potential of the small cabbage white *Pieris rapae crucivora* Boisduval (Lepidoptera: Pieridae). *App. Entomol. Zool.* 13:312–13.
Takacs, D. 1982. An ethological analysis of hindquarter presentation in yellow baboons (*Papio cynocephalus*). B.A. honors thesis, Cornell University.
Tamm, C. O. 1948. Observations on reproduction and survival of some perennial herbs. *Botaniska Notiser* 3:306–21.
———. 1972. Survival and flowering of some perennial herbs. 2. The behaviour of some orchids on permanent plots. *Oikos* 23:23–28.
Tanner, J. T. 1978. *Guide to the study of animal populations*. Knoxville: University of Tennessee Press.
Tantawy, A. O. 1964. Studies on natural populations of *Drosophila*. 3. Morphological and genetic differences of wing length in *Drosophila melanogaster* and *D. simulans* in relation to season. *Evolution* 18:560–70.
Tantawy, A. O., and Rakha, F. A. 1964. 4. Genetic variances of and correlations between four characters in *D. melanogaster* and *D. simulans*. *Genetics* 50:1349–55.
Tantawy, A. O., and Vetukhiv, M. O. 1960. Effects of size on fecundity, longevity and fertility in populations of *Drosophila pseudoobscura*. *Am. Nat.* 94:395–403.
Taylor, C. E., and Condra, C. 1980. R- and K-selection in *Drosophila pseudoobscura*. *Evolution* 34:1183–93.
Temrin, H.; Mallner, Y.; and Winden, M. 1984. Observations on polyterritoriality and singing behavior in the wood warbler *Phylloscopus sibilatrix*. *Ornis Scand.* 15:67–72.
Thomas, C. S. 1980. Certain aspects of the breeding biology of the kittiwake (*Rissa tridactyla*). D. Phil. diss., University of Durham.
———. 1983. The relationship between breeding experience, egg volume and reproductive success of the kittiwake *Rissa tridactyla*. *Ibis* 125:567–74.
Thorndike, R. M. 1978. *Correlational procedures for research*. New York: Gardner Press.
Thornhill, R., and Alcock, J. 1983. *The evolution of insect mating systems*. Cambridge: Harvard University Press.
Thouless, C. 1986. Feeding competition in red deer hinds. Ph.D. thesis, University of Cambridge.

Tinbergen, J. M.; Balen, J. H. van; and Eck, H. M. van. 1985. Density dependent survival in an isolated great tit population (*Parus major*): Kluyver's data reanalysed. *Ardea* 73:38–48.

Tinkle, D. W. 1965. Population structure and effective size of a lizard population. *Evolution* 19:569–73.

———. 1967. *The life and demography of the side-blotched lizard,* Uta stansburiana. Miscellaneous Publications of the Museum of Zoology 132. Ann Arbor: University of Michigan.

Tompa, F. S. 1964. Factors determining numbers of song sparrows *Melospiza melodia* (Wilson) on Mandarte Island, B.C., Canada. *Acta Zool. Fenn.* 109:3–73.

Trail, P. 1985. The intensity of selection: Intersexual and interspecific comparisons require consistent measures. *Amer. Nat.* 126:434–39.

Trivers, R. L. 1972. Parental investment and sexual selection. In *Sexual selection and the descent of man, 1871–1971,* ed. B. Campbell, 136–79. Chicago: Aldine-Atherton.

———. 1976. Sexual selection and resource-accruing abilities in *Anolis garmani. Evolution* 30:253–69.

Trivers, R. L., and Willard, D. 1973. Natural selection of parental ability to vary the sex ratio of offspring. *Science* 179:90–92.

Turelli, M. 1985. Heritable genetic variation via mutation-selection balance: Lerch's zeta function meets the abdominal bristle. *Theor. Pop. Biol.* 25:138–93.

Turillazzi, S., and Pardi, L. 1977. Body size and hierarchy in polygynic nests of *Polistes gallicus* (L.) (Hymenoptera: Vespidae). *Monit. Zool. Ital.* 11:101–12.

Turke, P. W., and Betzig, L. L. 1985. Those who can do: Wealth, status, and reproductive success on Ifaluk. *Ethol. Sociobiol.* 6:79–87.

Turner, A. K. 1982. Timing of laying by swallows *Hirundo rustica* and sand martins *Riparia riparia. J. Anim. Ecol.* 51:29–46.

Turner, M. E., and Anderson, W. W. 1983. Multiple mating and female fitness in *Drosophila pseudoobscura. Evolution* 37:714–23.

Valen, L. van, and Mellin, G. W. 1967. Selection in natural populations. 7. New York babies (Fetal Life Study). *Ann. Human Genet. Lond.* 31:109–27.

Van den Berghe, E., and Gross, M. n.d. Mating strategies and reproductive success in Coho salmon. In press.

Varley, G. C., and Gradwell, G. R. 1960. Key factors in population studies. *J. Anim. Ecol.* 29:399–401.

———. 1962. The effects of partial defoliation by caterpillars on the timber production of oak trees in England. *Proc. XI Int. Congr. Entomol.* (Vienna) 2: 211–14.

Varley, G. C.; Gradwell, G. R.; and Hassell, M. P. 1973. *Insect population ecology.* Oxford: Blackwell

Vaz-Ferreira, R. 1975. Behaviour of the southern sea lion, *Otaria flavescens,* in the Uruguayan Islands. *Rapp. P.-v. Reun. Cons. Int. Explor. Mer.* 169:219–27.

Vehrencamp, S. L. 1977. Relative fecundity and parental effort in communally nesting anis, *Crotophaga sulcirostris. Science* 197:403–5.

———. 1978. The adaptive significance of communal nesting in groove-billed anis (*Crotophaga sulcirostris*). *Behav. Ecol. Sociobiol.* 4:1–33.

———. 1983. A model for the evolution of despotic versus egalitarian societies. *Anim. Behav.* 31:667–82.

Vehrencamp, S. L.; Bowen, B. S.; and Koford, R. R. 1986. Breeding roles and pairing patterns within communal groups of groove-billed anis. *Anim. Behav.* 34:347–67.

Vehrencamp, S. L., and Bradbury, J. W. 1984. Mating systems and ecology. In *Behavioural ecology: An evolutionary approach,* 2d ed., ed. J. R. Krebs and N. B. Davies, 251–78. Sunderland, Mass: Sinauer.

Verner, J. 1964. Evolution of polygamy in the long-billed marsh wren. *Evolution* 18:252–61.

Vos, G. J. de. 1979. Adaptedness of arena behaviour in black grouse (*Tetrao tetrix*) and other grouse species (Tetraoninae) *Behaviour* 68:277–314.

———. 1983. Social behaviour of black grouse: An observational and experimental field study. *Ardea* 71:1–103.

Wade, M. J. 1979. Sexual selection and variance in reproductive success. *Am. Nat.* 114:742–47.

———. 1985. Soft selection, hard selection, kin selection and group selection. *Am. Nat.* 125:61–73.

Wade, M. J., and Arnold, S. J. 1980. The intensity of sexual selection in relation to male behaviour, female choice and sperm precedence. *Anim. Behav.* 28:446–61.

Wall, R., and Begon, M. n.d. Individual variation and population density in the grasshopper *Chorthippus brunneus.*

Waser, P. M., and Jones, W. T. 1983. Natal philopatry among solitary mammals. *Quart. Rev. Biol.* 58:355–90.

Wattiaux, J. M. 1968a. Cumulative parental age effects in *Drosophila subobscura. Evolution* 22:406–21.

———. 1968b. Parental age effects in *Drosophila pseudoobscura. Exp. Gerontol.* 3:55–61.

Webber, M. I. 1975. Some aspects of the non-breeding population dynamics of the great tit. (*Parus major*). D. Phil. diss., University of Oxford.

Wegenen, G. van, and Simpson, M. E. 1954. Testicular development in the rhesus monkey. *Anat. Rec.* 118:231–43.

Wells, K. D. 1977. The social behaviour of anuran amphibians. *Anim. Behav.* 25:666–93.

West-Eberhard, M. J. 1969. *The social biology of polistine wasps.* Miscellaneous Publications of the Museum of Zoology 140. Ann Arbor: University of Michigan.

Western D. 1983. *A wildlife guide and natural history of Amboseli.* Nairobi: General Printers.

Western, D., and Praet, C. van. 1973. Cyclical changes in the habitat and climate of an East African ecosystem. *Nature* 241:104–6.

Whittaker, J. 1984. Model interpretation from the additive elements of the likelihood function. *Appl. Statist.* 33:52–64.

Whitten, P. L. 1982. Female reproductive strategies among vervet monkeys. Ph.D. diss., Harvard University.

———. 1983. Diet and dominance among female vervet monkeys (*Cercopithecus aethiops*). *Am. J. Primatol.* 5:139–59.

Wickman, P.-O. 1985. Territorial defense and mating success in males of the small heath butterfly, *Coenonympha pamphilus* (L.) (Lepiodoptera: Satyridae). *Anim. Behav.* 33:1162–68.

Wiklund, C., and Fagerström, T. 1977. Why do males emerge before females? A hypothesis to explain the incidence of protandry in butterflies. *Oecologia* (Berlin) 31:153–58.

Wiklund, C., and Solbreck, C. 1982. Adaptive versus incidental explanations for the occurrence of protandry in a butterfly, *Leptidea synapsis* L. *Evolution* 36:56–62.

Wiley, R. H. 1973. Territoriality and non-random mating in sage grouse. *Anim. Behav. Monogr.* 6:85–169.

———. 1974a. Effects of delayed reproduction on survival, fecundity, and the rate of population increase. *Am. Nat.* 108:705–9.

———. 1974b. Evolution of social organisation and life history patterns among grouse. *Quart. Rev. Biol.* 49:201–27.

Williams, G. C. 1957. Pleiotropy, natural selection and the evolution of senescence. *Evolution* 11:398–411.

———. 1966a. *Adaptation and natural selection: A critique of some current evolutionary thought.* Princeton: Princeton University Press.

———. 1966b. Natural selection, the costs of reproduction, and a refinement of Lack's principle. *Am. Nat.* 100:687–90.

Wilson, D. S. 1980. *The natural selection of populations and communities.* Menlo Park, Calif.: Benjamin/Cummings.

Wilson, E. O. 1971. *The insect societies*. Cambridge: Harvard University Press.

———. 1975. *Sociobiology: The new synthesis*. Cambridge: Harvard University Press.

———. 1980. Caste and division of labor in leaf-cutter ants (Hymenoptera: Formicidae: *Atta*). 2. The ergonomic organisation of leaf cutting. *Behav. Ecol. Sociobiol.* 7:157–65.

Wilson, M. E.; Gordon, T. P.; and Bernstein, I. S. 1978. Timing of births and reproductive success in rhesus monkey social groups. *J. Med. Primatol.* 7:202–12.

Wilson, M. E.; Walker, M. L.; and Gordon, T. P. 1983. Consequences of first pregnancy in rhesus monkeys. *Am. J. Phys. Anthropol.* 61:103–10.

Wittenberger, J. F. 1978. The evolution of mating systems in grouse. *Condor* 80:126–37.

Wolf, L. D. 1984. Female rank and reproductive success among Arashiyama B Japanese macaques (*Macaca fuscata*). *Int. J. Primatol.* 5:133–43.

Wood, J. L.; Johnson, P. L.; and Campbell, K. L. 1985. Demographic and endocrinological aspects of low natural fertility in highland New Guinea (1985). *J. Biosoc. Sci.* 17:57–79.

Woolfenden, G. E., and Fitzpatrick, J. W. 1978. The inheritance of territory in group-breeding birds. *BioScience* 28:104–8.

———. 1984. *The Florida scrub jay: Demography of a cooperative-breeding bird.* Princeton: Princeton University Press.

Woolfenden, G. E., and Fitzpatrick, J. W. 1986. Sexual asymmetries in the life history of the Florida scrub jay. In *Ecological aspects of social evolution,* ed. D. I. Rubenstein and R. W. Wrangham, 87–107. Princeton: Princeton University Press.

Wooller, R. D., and Coulson, J. C. 1977. Factors affecting the age of first breeding of the kittiwake *Rissa tridactyla*. *Ibis* 119:339–49.

Wrangham, R. W. 1981. Drinking competition in vervet monkeys. *Anim. Behav.* 29:904–10.

Wrangham, R. W., and Waterman, P. G. 1981. Feeding behaviour of vervet monkeys on *Acacia tortilis* and *Acacia xanthophloea*: With special reference to reproductive strategies and tannin production. *J. Anim. Ecol.* 50:715–31.

Wright, A. H., and Wright, A. A. 1949. *Handbook of frogs and toads of the United States and Canada*. Ithaca, N.Y.: Comstock.

Wright, S. 1921. Correlation and causation. *J. Agric. Res.* 20:557–85.

———. 1934. The method of path coefficients. *Ann. Math. Statist.* 5:161–215.

———. 1968–78. *Evolution and the genetics of populations*. Vols. 1–4. Chicago: University of Chicago Press.

Wrigley, E. A. 1966. *Population and history*. New York: McGraw-Hill.

Wynne-Edwards, V. C. 1952. Geographical variation in the bill of the fulmar (*Fulmarus glacialis*). *Scot. Nat.* 64:84–101.

———. 1962. *Animal dispersion in relation to social behaviour*. Edinburgh: Oliver and Boyd.

Yoshikawa, K. 1954. Ecological studies of *Polistes* wasps. 1. On the nest evacuation. *J. Inst. Polytechnics* (Osaka City University) 5:9–17.

Zahavi, A. 1975. Mate selection—a selection for a handicap. *J. Theor. Biol.* 53:205–14.

Zar, J. H. 1980. *Biostatistical analysis*. Englewood Cliffs, N.J.: Prentice-Hall.

———. 1984. *Biostatistical analysis*. 2d ed. Englewood Cliffs, N.J.: Prentice-Hall.

Contributors

S. D. Albon
Large Animal Research Group
Department of Zoology
Cambridge University
34A Storeys Way
Cambridge CB3 0DT
United Kingdom

Jeanne Altmann
Department of Biology
University of Chicago
940 East 57th Street
Chicago, Illinois 60637
USA

Stuart A. Altmann
Department of Biology
University of Chicago
940 East 57th Street
Chicago, Illinois 60637
USA

Sandy J. Andelman
Department of Ecology and Behavioral Biology
University of Minnesota
Minneapolis, Minnesota 55455
USA

J. H. van Balen
Institute for Ecological Research
Boterhoeksestraat 22
6666 GA Heteren
The Netherlands

Bonnie S. Bowen
Department of Biology C-016
University of California at San Diego
La Jolla, California 92093
USA

David Brown
AFRC Institute of Animal Physiology and Genetics Research
Cambridge Research Station
Babraham Hall
Cambridge CB2 4AT
United Kingdom

David M. Bryant
Department of Biological Sciences
University of Stirling
Stirling FK9 4LA
United Kingdom

J. David Bygott
Box 1501
Karatu
Tanzania

Sara J. Cairns
Section of Neurobiology and Behavior
Cornell University
Ithaca, New York 14853
USA

Bruce Campbell
West End Barn
Wootton
Oxford OX7 1DL
United Kingdom

Dorothy L. Cheney
Department of Anthropology
University of Pennsylvania
Philadelphia, Pennsylvania 19104
USA

T. H. Clutton-Brock
Large Animal Research Group
Department of Zoology
Cambridge University
34A Storeys Way
Cambridge CB3 0DT
United Kingdom

F. Cooke
Department of Biology
Queen's University
Kingston K7L 3N6
Canada

J. C. Coulson
Department of Zoology
University of Durham
Durham
United Kingdom

G. M. Dunnet
Department of Zoology
University of Aberdeen
Tillydrone Avenue
Aberdeen, AB9 2TN
United Kingdom

Mark A. Elgar
Department of Zoology
University of Cambridge
Downing Street
Cambridge, CB2 3EJ
United Kingdom

Ola M. Fincke
Department of Biology
University of Missouri
8001 Natural Bridge Road
St. Louis, Missouri 63121
USA

John W. Fitzpatrick
Division of Birds
Field Museum of Natural History
Roosevelt Road at Lake Shore Drive
Chicago, Illinois 60605
USA

Alan Grafen
Animal Behaviour Research Group
Department of Zoology
University of Oxford
South Parks Road
Oxford, OX1 3PS
United Kingdom

Fiona E. Guinness
Red Deer Research Project
Kilmory
Isle of Rhum
Inner Hebrides
Scotland PH43 4RR
United Kingdom

Jeannette P. Hanby
Box 1501
Karatu
Tanzania

Paul H. Harvey
Department of Zoology
University of Oxford
South Parks Road
Oxford, OX1 3PS
United Kingdom

Glenn Hausfater
Division of Biological Sciences
University of Missouri
Columbia, Missouri 65211
USA

Lawrence Herbst
Department of Ecology and Behavioral Biology
University of Minnesota
318 Church Street, S.E.
Minneapolis, Minnesota 55455
USA

Richard D. Howard
Department of Biological Sciences
Purdue University
West Lafayette, Indiana 47907
USA

Rolf R. Koford
Department of Biology C-016
University of California at San Diego
La Jolla, California 92093
USA

J. P. Kruijt
Zoological Laboratory
University of Gronigen
Postbox 14
9750 AA Haren (Gr.)
The Netherlands

Burney J. Le Boeuf
Department of Biology and Institute of Marine Sciences
University of California
Santa Cruz, California 95064
USA

Phyllis C. Lee
Subdepartment of Animal Behavior
University of Cambridge
Madingley
Cambridge CB3 8AA
United Kingdom

R. H. McCleery
Edward Grey Institute of Field Ornithology
Department of Zoology
University of Oxford
South Parks Road
Oxford OX1 3PS
United Kingdom

Margaret E. McVey
4012 South 18th Street
Arlington, Virginia 22204
USA

Monique Borgerhoff Mulder
Program in Evolution and Human Behavior
University of Michigan
524 Rackham Hall
Ann Arbor, Michigan 48109
USA

Ian Newton
Institute of Terrestrial Ecology
Monks Wood Experimental Station
Abbots Ripton
Huntingdon PE17 2LS
United Kingdom

A. J. van Noordwijk
Zoologisches Institut der Universität Basel
Rheinsprung 7-9
CH-4051 Basel
Switzerland

Janet C. Ollason
Department of Zoology
University of Aberdeen
Tillydrone Avenue
Aberdeen, AB9 2TN
United Kingdom

Craig Packer
Department of Ecology and Behavioral Biology
University of Minnesota
318 Church Street, S.E.
Minneapolis, Minnesota 55455
USA

Linda Partridge
Department of Zoology
Edinburgh University
West Mains Road
Edinburgh EH9 3JT
United Kingdom

C. M. Perrins
Edward Grey Institute of Field Ornithology
Department of Zoology
University of Oxford
South Parks Road
Oxford OX1 3PS
United Kingdom

Naomi E. Pierce
Department of Biology
Princeton University
Princeton, New Jersey 80544
USA

Anne E. Pusey
Department of Ecology and Behavioral Biology
University of Minnesota
318 Church Street, S.E.
Minneapolis, Minnesota 55455
USA

David C. Queller
Department of Biology
Rice University
Post Office Box 1892
Houston, Texas 77251
USA

Joanne Reiter
Department of Biology and Institute of Marine Sciences
University of California
Santa Cruz, California 95064
USA

R. F. Rockwell
Department of Ornithology
American Museum of Natural History
Central Park West at 79th Street
New York, New York 10024
USA

D. K. Scott
The Wildfowl Trust
Welney, Cambs. PE4 9TN
United Kingdom

Robert M. Seyfarth
Department of Psychology
University of Pennsylvania
Philadelphia, Pennsylvania 19104
USA

James N. M. Smith
Department of Zoology
University of British Columbia
6270 University Boulevard
Vancouver, British Columbia V6T 2A9
Canada

Martyn J. Stenning
School of Biological Sciences
University of Sussex
Biology Building
Falmer
Brighton, Sussex BN1 9QG
United Kingdom

Joan E. Strassmann
Department of Biology
Rice University
P.O. Box 1892
Houston, Texas 77251
USA

Callum S. Thomas
Bird Control Officer
Manchester Airport PLC
Manchester M22 5PA
United Kingdom

Sandra L. Vehrencamp
Department of Biology C-016
University of California at San Diego
La Jolla, California 92093
USA

G. J. de Vos
Zoological Laboratory
University of Gronigen
Postbox 14
9750 AA Haren (Gr.)
The Netherlands

Glenn E. Woolfenden
Department of Biology
University of South Florida
Tampa, Florida 33620
USA

Author Index

Subject Index